Scientific Information Transfer: The Editor's Role

Scientific Information Transfer: The Editor's Role

Proceedings of the First International Conference of Scientific Editors, April 24-29, 1977, Jerusalem

edited by

Miriam Balaban

National Council for Research and Development, Jerusalem
Center for Philosophy and History of Science, Boston University
Editor of Desalination

D. Reidel Publishing Company

Dordrecht : Holland / Boston : U.S.A. / London : England

Library of Congress Cataloging in Publication Data

CIP

International Conference of Scientific Editors, 1st, Jerusalem, 1977.
Scientific information transfer.

Includes bibliographical references and index.
1. Information storage and retrieval systems–Science–Congresses. 2. Communication in science–Congresses. 3. Editing–Congresses. 4. Technical writing–Congresses.
I. Balaban, Miriam. II. Title. III. Title: Information transfer.
Z699.5.S3I56 1977 029:9'5 78–16120
ISBN 90–277–0917–3

Published by D. Reidel Publishing Company,
P.O. Box 17, Dordrecht, Holland

Sold and distributed in the U.S.A., Canada and Mexico
by D. Reidel Publishing Company, Inc.
Lincoln Building, 160 Old Derby Street, Hingham, Mass. 02043, U.S.A.

Printed in The Netherlands

UNDER THE PATRONAGE OF

Ephraim Katzir, President of the State of Israel
The Weizmann Institute of Science, Rehovot

SPONSORS AND PARTICIPATING ORGANIZATIONS

National Council for Research and Development, Jerusalem
Institut für Dokumentationswesen, Frankfurt/Main
Bureau National de l'Information Scientifique et Technique, Paris
Commission of the European Communities
UNESCO
Israel National Academy of Sciences and Humanities
The Hebrew University of Jerusalem
Technion-Israel Institute of Technology, Haifa
Jerusalem International Book Fair
Center for the Philosophy and History of Science, Boston University
The Weizmann Institute of Science, Rehovot
Biological Abstracts - BIOSIS
Nordic Publishing Board in Science, Oslo
The Ciba Foundation, London
European Association of Editors of Biological Periodicals (ELSE)
Council of Biology Editors (CBE)
Association of Scientific Journals (ASJ)
Editerra
Association of Earth Science Editors (AESE)
Editeast
Aliterra
IUB Committee of Biochemical Journals (CEBJ)
Association of Philosophy Journal Editors
Ad hoc Council of Social Science Editors
Committee on International Publication in Anthropology (CIPA)
Commission of Editors of Journals Concerned with Clinical Chemistry (CEJCCC)

ORGANIZING COMMITTEE

Miriam Balaban, *Chairman*
National Council for Research and Development, Jerusalem;
Center for the Philosophy and History of Science, Boston University

Martin Cremer, Institut für Dokumentationswesen, Frankfurt/Main

Knut Faegri, International Union of Biological Sciences, President;
Botanical Museum, Bergen University

Karl Heumann, Federation of American Societies for Experimental Biology,
Rockville, Md.

Ernest Mann, Commonwealth Bureau of Dairy Science and Technology,
Shinfield, Reading, Berks.

Jacques Michel, Bureau National de l'Information Scientifique et Technique,
Paris

Gillian Page, Cambridge University Press, London

Udo Schützsack, Institut für Dokumentationswesen, Frankfurt/Main

ADVISORY COMMITTEE

* U.S. Program Committee

PREFACE

It was Faraday who in 1821 said that there are three necessary stages of useful research. The first to begin it, the second to end it, and the third to publish it.[1]

There has since indeed been so much research and publication that we have become increasingly alarmed by the galloping proliferation of scientific information produced in relation to the user's ability to retrieve and consume it effectively, conveniently and creatively.

In 1948, to deal with this concern, the Royal Society Scientific Information Conference held in London[1] spanned the whole realm of scientific information. Sir Robert Robinson, President of the Royal Society, in his opening address noted that "the study of scientific information services in all its ramifications has enormous scope", and the London conference dealt with scientific publication, format, editorial policy, subject grouping, organization, abstracting, reviews, classification, indexing and training of information officers.

It was about this time that information science began to develop more on the retrieval end, so it seems logical that the first editors' group founded in 1949 was ICSU AB, the International Council of Scientific Unions Abstracting Board.

In 1958 the National Academy of Sciences International Conference of Scientific Information in Washington[2] limited its interests and expanded on the later phases of the life cycle of information - storage and retrieval. As Sir Lindor Brown, Senior Secretary of the Royal Society said in his opening address there, whereas in London they had dealt with the subject from cradle to the grave, here (he dare not say), they were dealing with burial and disinterring. He further said that "the three parties in the cycle of information are the maker, the storer and the supplier, and the user who is often the maker also."

In the intervening years we have come to elaborate on these parties (links, estates or stations as they are alternatively called) as author, editor, publisher, librarian, documentalist (archivist, information scientist), and user

Although at the Royal Society editorial quality as well as technicalities were discussed in depth and detail, no mention is made of the editor's role. Somehow the editorial function is incorporated as an integral part of the process. The editor is swallowed whole. That scientists should strive to improve and maintain quality of writing and presentation is stressed but the role of the editor, methodology and training are not considered. And of course, in Washington the meeting, as declared, was devoted not to input but to output.

It has been just since 1958, however, that we have seen editors' associations spring up in various disciplines and regions. It was about that time that the National Science Foundation's Office of Science Information Services and the American Institute of Biological Sciences founded CBE, the Council of Biology Editors. Out of my attendance at the 1962 CBE annual meeting, the late Fred Cagle's encouragement, with CBE's blessings and

UNESCO's aid sprouted ELSE in 1964. UNESCO initiated the creation of editors' associations and cooperation of editors in Europe in various disciplines: life sciences, earth sciences, chemistry, physics, engineering and mathematics, and with the ultimate aim of a world information system, UNISIST was established in 1972.

In the earth sciences we have Editerra in Europe, the American Earth Sciences Editors (AESE), Alegeo in Latin America and Editeast in the Far East, strongly biased towards the earth sciences although dealing with science in general. There is the IUB International Committee of Editors of Biochemical Journals (CEBJ), the Commission of Editors of Journals Concerned with Clinical Chemistry (CEJCCC) and an embryo EdEuChem in Europe. The American Chemical Society and the Chemical Society in the UK, are ever experimenting and pioneering and are concerned with editors' affairs, and IUPAC is considering the formation of an international group for chemistry editors. Although there are no organizational structures for mathematics and physics editors, the American Mathematical Society, American Institute of Physics and UK Institute of Physics are actively involved in styles, standards and practices of editing and publishing, and the European Physical Society has a well-founded scheme for standards of Europhysics journals. In engineering the IEEE - Institute for Electrical and Electronic Engineers has sponsored important conferences in the US and the AEEP in Europe has been suggested but not activated. The Association of Philosophy Journal Editors has an active program, and anthropology and sociology editors' associations are enthusiastically forming as the Committee on International Publication in Anthropology (CIPA) and the ad hoc Council of Social Science Editors. Psychology editors have a strong reference point in the American Psychological Association.

Important meetings have been held by the Association of Scientific Journals (ASJ) in the US, and there is a proposal circulating to convert this into a US Association for Scholarly Publishing. Its meetings in 1973, 1975 and 1977, as its chairman James Lufkin introduced[3] were to "unite people who share a common interest in scientific journals but who are ordinarily separated by differences in their professional or business points of view."

There are regional and national groupings such as the Nordic Publishing Board in Science (NOP), the Japanese Association of Scientific Editors, and others interested in editors' affairs such as the German GID - Gesellschaft für Information und Dokumentation (formerly IDW), the French Bureau National de l'Information Scientifique et Technique (BNIST), the Commission of the European Communities (EEC), National Science Foundation (NSF), European Science Foundation (ESF) and the Pacific Science Association. Various ICSU groups such as IUGS, IUB, IUBS, IU Crystallography have been involved, and private companies such as the Institute for Scientific Information (ISI) and foundations such as the Ciba Foundation.

It is also interesting that the publishing counterpart at our input end of the cycle, STM - the International Group of Scientific, Technical and Medical Publishers, was founded in 1968, during this same later period of the development of infrastructures for parties involved in scientific information.

We see a continuing process of integration and subdivision in our field as in the disciplines of the journals we edit.

Lately there has been more and more cooperation between some of the groups in the same region or discipline such as ELSE's disciplinary relations with CBE or regional liaison with Editerra. The spark for this conference was indeed lit at Ustaoset, Norway at the ELSE General Assembly when I witnessed American and European biology editors trying to collaborate on style, a seemingly innocent unemotional subject. It occurred to me then and there that for this and for similar subjects, we had to have a flag which would join regions and disciplines and not confine our discussions to our inner sanctums nor to this particular detail, though basic and important, but to the myriad of matters where common ground was essential for loosening the clogs in the flow of information and reducing the cost and effort resulting from confusion, redundant effort and fragmentation.

ELSE's 1974 council meeting in Israel sealed the mandate to charge ahead, the organizing committee was formed, and we were off.

While each group would continue to deal with its own specific needs, and retain loyalty to its own sub-club, we aimed to find the common ground which was indicated for interdisciplinary and international cooperation to ease the system from the outset, set standards, practices and philosophies to introduce to authors and secondary systems for quicker and better service to users.

There was much self criticism and question of who is an editor and what is his role. There are many kinds of editors and many functions. He may be a career scientist devoting some of his time for some period of his life to editing. It has been expressed here that once he becomes full time editor, he is no longer a scientist nor an editor since editing is not a profession - not yet at least! This,despite his devoting full time to editing where his functions range from encouraging, inspiring scientists, identifying fields, creating and developing publications, moderating, monitoring, exercising fair discerning judgement, scientific and technical preparation of language and presentation of the author's creative efforts for publication, suitable, timely and economic production and dissemination of literature, and other essential functions. An editor can inspire, mediate, kill, encourage, discourage, transform, delay, advance, promote, hamper or help.

Only recently has there grown up a literature of communications on editors' problems by editors. Courses are given and degrees awarded. We are not yet subjected to the publish-or-perish syndrome although a few journals have already been founded. Thus at first we were reluctant to plan for publication even of these proceedings, but the interest expressed made it clear that we, whose primary function it is to communicate, should not fail to communicate. And perhaps now that we have the trappings of courses, degrees and publications of our own, ours too will be considered a profession. We hope always to maintain strong involvement of those career scientists among us as authors and editors and users.

In this conference we perhaps have had an overabundance of papers, but we wanted to give an opportunity to participants to express views and open up the various concerns as far as possible to lay the groundwork for continued dialogue towards cooperation. And we had no clear cut criteria for these as yet. At the next conference we can concentrate more on specifics and group interests and hold workshops to hammer out practical recommendations, criteria, and agendas for the future.

And indeed the conference did not cease its activity in Jerusalem. These Proceedings also include in the sequel suggestions resulting from a workshop sponsored by the Ciba Foundation in London on the consolidation and adoption of a rational system for references where we got down to the brass tacks appealed for at the conference. In June, August and November, the organizing committee met to plan founding of IFSEA, the International Federation of Scientific Editors' Associations at UNESCO in Paris in June 1978. Proposed statutes, still undergoing change, are presented here, as well as the current version of suggestions for reference practices. An ISI-hosted workshop for scientist-editors on primary journal-secondary service interface is envisioned, and plans for the Second International Conference in Amsterdam are underway for 1980 fostered by Elsevier and colleagues in Holland.

This volume is thus dedicated to the diverse aspects of editing towards advancing scientific information transfer - technical and sociological norms and practices, quality, refereeing and judgement, impact of new mechanical and organizational techniques, copyright, standards and style, economics, society and commercial publishing, primary-secondary-tertiary literature and their interfaces, and the philosophy and sociology of science. Scientists, editors, publishers and others concerned with information transfer have laid bare some of their attitudes, policies, practices, prejudices and problems.

Besides being of interest to full or part-time scientist-editors and technical scientific editors of all sorts, these Proceedings, we hope, will open insights into this developing discipline for research scientists as authors and users, for librarians and documentalists, students, publishers, printers, sociologists, philosophers, communication and management scientists, scientific administrators, forecasters and planners in universities, societies, government and industry.

The very structure of this book illustrates some of our problems. Since we were reluctant to publish, we didn't send out standard forms ahead of time. Once sent, styles spontaneously submitted were original from paper to paper. Economics dictated that we try to keep down the number of pages and cost, so we have a compact volume. Scientists are used to writing about original research and not about innovation or description of editorial practice, so much of what has been said had to be deciphered and interpreted from tapes. We do hope, however, that we have achieved the aim of making all the information available, and that this book can serve as a reference to the state-of-the-art of various facets of editing and scientific information transfer and provide a stimulus to change what should be changed and promote what should be promoted.

There are exciting prospects ahead!

February 1978 Miriam Balaban
Jerusalem

---

1. The Royal Society Scientific Information Conference, June 21 - July 2, 1948. Reports and Papers Submitted, London, 1948.

2. Proceedings of the International Conference on Scientific Information, November 16-21, 1958, National Academy of Sciences - National Research Council, Washington, D.C., 1959.

3. IEEE Transactions on Professional Communication, PC-18 (3), September 1975. (See also PC-20 (2), September 1977).

ACKNOWLEDGEMENTS

In addition to the special appreciation expressed in my introduction which follows to those who made this conference possible, I wish to thank all those who came for their active involvement and session chairmen for guiding their ships.

Thanks to Robert S. Cohen, Chairman of the Center for Philosophy and History of Science, Boston University for his kind support and keen advice, to Joseph Agassi of Boston and Tel-Aviv Universities for his inspiring confidence and for perspectives, to Ephraim Katzir, President of Israel and of the Weizmann Institute of Science and Shneior Lifson of the Institute for faithful guidance, and to Wayne A. McRae, Vice President of Ionics, my very personal thanks.

Viewing in this acknowledgement a link to the future, and in appreciation of foresight and implementation, I wish to thank R.E.M. van den Brink, Chairman of Elsevier Publishers for readily ensuring continuity in setting sites for our second conference, to Maeve O'Connor, senior editor of the Ciba Foundation for her initiative in organizing an international interdisciplinary workshop on references lending practical expression to the momentum spun in Jerusalem and ELSE, and to Eugene Garfield, president of ISI for a prospective workshop for scientist-editors on primary-secondary interfaces, to Wolfgang Löhner of UNESCO for hosting IFSEA.

My appreciation to Adina Basson for intelligent typing, to Eileen Ginsburg, Anne Biderman, Leora Amira and Gwenn Cohen for devoted assistance, and to Larry Lester and Paul Greenberg for proofreading some of the discussions. Special thanks to Janine Pellerin and Paula for help and warm cheer, and to my daughter Naomi for encouragement, criticism and comfort.

Again thanks to all who extended their moral, intellectual and financial support, and again and again my unlimited gratitude to ELSE and the organizing committee for past present and future.

Pelham, NH and Jerusalem MB

TABLE OF CONTENTS

SCIENCE, SOCIETY AND THE EDITOR
Shneior Lifson, *Chairman*

CREATORS AND GATEKEEPERS - AUTHORS, EDITORS, REFEREES
JOSEPH GILLIS, *Chairman*

REFEREEING AND JUDGMENT

STANDARDS AND STYLE
KNUT FAEGRI, *Chairman*

ECONOMICS AND ORGANIZATION OF SCIENTIFIC PUBLISHING
PAUL NIJHOFF ASSER, *Chairman*

INTERNATIONAL, NATIONAL AND SPECIAL PROBLEMS
WOLFGANG LÖHNER, *Chairman*

SOCIAL SCIENCES
JOSEPH BEN-DAVID, *Chairman*

AGRICULTURAL AND DEVELOPMENT PUBLICATIONS
JOSEPH PALTI, *Chairman*

EDITOR-SECONDARY SERVICES INTERFACE
Developments in Secondary Information Services and Information Transfer
UDO SCHÜTZSACK, *Chairman*

CLOSING SESSION:
Summary, Recommendations, Resolutions and Future
MIRIAM BALABAN, *Chairman*

SEQUEL AND PREVIEW

OPENING SESSION

Introductions and Greetings

Gillian Page, Chairman

INTRODUCTION TO THE FIRST INTERNATIONAL CONFERENCE OF SCIENTIFIC EDITORS

Gillian Page, *Chairman of Opening Session*
Director of Journals, Cambridge University Press,
London, United Kingdom

President Katzir, Ladies and Gentlemen. Welcome to the First International Conference of Scientific Editors. We are in the land of miracles. One miracle is that this conference is taking place. Miriam Balaban proposed it first almost exactly three years ago very close to where we are now. It seemed like a marvelous idea, something to dream about, but not something that it would be possible to bring about. How could it be organized and planned? Who would come? How would the money be raised? Who could organize it without an official body. Miriam saw the obstacles but she had the determination to overcome them. She has masterminded the whole thing, sometimes working from a distance in New Hampshire.

A dream has become a reality and we are here. This is the first international conference of scientific editors. In the world of scientific conferences it is very much a newcomer. Why should that be? Is it that scientific editors have no problems or that their problems are of specific subject or of regional difficulty. The program suggests that is not so. Are our problems of recent origin? When the Cambridge Philosophical Society set up a review journal in biology in 1923, the editors instructions to authors read: "Owing to the greatly increased cost of publication, authors are requested to present their pages in as concise a form as possible and so avoid delay, trouble and expense." So that there are several problems which we have had for more than fifty years. In spite of the increased costs some work has been done which suggests that the cost of publication, libraries, and information services represent only one or two percent of the cost of doing research. And research without publication is useless. That's a nice figure because we can use it in two ways. On the one hand it shows how much we achieve with so little resources. Alternatively if we take even one percent of the research budgets of the countries of the world we can see how very important we are.

Perhaps the reason why we have not had such a conference before is that until now we have lacked anyone with the vision to see the need, the enthusiasm to persuade others to support it, and the energy to carry it out. Miriam herself will offer our thanks to the many institutions and people whom she has persuaded to devote and contribute time and money and services to make this conference possible. We are very grateful to them all.

Having assembled this conference, we hope something will come out of it. That depends on you. We hope everyone will gain something. We have also allotted time at the end to consider the future. Should there be further meetings of this sort? Are there resolutions which should be made and which we can do something to achieve? Should there be an international association of journal editors of all subjects or would it be more efficient to divide by region and by subject. Those are questions for the final session, but perhaps we should bear them in mind throughout the meeting. I will now ask Miriam herself to say something about the conference and to introduce our other speakers.

Miriam Balaban (ed.), Scientific Information Transfer: The Editor's Role, xxi. *All Rights Reserved.*

INTRODUCTION TO THE FIRST INTERNATIONAL CONFERENCE OF SCIENTIFIC EDITORS

Miriam Balaban, *Conference Chairman*
Center for Philosophy and History of Science, Boston University
National Council for Research and Development, Jerusalem
Editor of Desalination

His Excellency, the President of Israel, Professor Ephraim Katzir, Gillian, honored friends from home and abroad. Welcome to Jerusalem. Welcome to the First International Conference of Scientific Editors.

After months - indeed years, of dreaming and planning together, it is exciting to see you gathered here from all ends of the earth representing a spectrum of scientific disciplines, links in the information chain, and national, regional and international groups.

Our hopes are high for using this opportunity to exchange ideas and experience - to recommend, and decide on ways to cooperate on an international, interdisciplinary and interinformational basis.

From the vantage point of the primary and secondary editor, we hope to further information transfer by creating bridges between editors, authors, publishers, librarians and information experts.

We hope to share proposals and experience on innovative practices and concepts, in all their aspects - from organizational and mechanical software and hardware, to theoretical ingenuity in order to alleviate the self-imposed burden which the scientific community has created, through proliferation of publications. We will have to find ways to publish less or absorb more.

There are 20 countries represented here and international organizations such as UNESCO, ICSU and the EEC. There are 10 editors' societies, and we will bestow our blessings on the founding of another. In suggesting to establish an international organization, the organizing committee did not have in mind, the substitution of existing local and regional structures and editors' societies. We intend to integrate, and coordinate, these activities - to provide a roof or clearinghouse rather than overlap and conflict. These national and regional bases and local identities should be strengthened and preserved, so as to be able to link readily into an international cooperative network. We aspire to a continuing dialogue, for intellectual and technical training, planning and implementation.

It is up to editors of primary and secondary journals, to create the standards, guidelines and keywords, for feedback to authors, and feedforward to the retrieval system, easily accessible to the mind of the user. Perhaps one day we will hold an instrument in our hands the size of a pocket calculator to alert us to new bits of information and direct us to the place from which they can be recalled. Editors should plan courses to give students the tools they need. Journals should prepackage their journals rationally and conferences their proceedings, rather than leave it to the computer to repackage them. Overlapping or regional journals might unite or coordinate their services where mutually beneficial, and overgeneralized ones twig off into more specific ones. Perhaps a universally used handbook or a series of existing guides uniting common ground could emerge, and forms for the preparation of camera ready copy could be obtained at every stationary store.

Miriam Balaban (ed.), Scientific Information Transfer: The Editor's Role, xxiii-xxiv.

This is far from proposing straightjackets, but on the contrary, the aim is to simplify the process of information transfer releasing the author and editor for his creative effort and individuality, enhance accessibility and operate economically by rationalizing the process.

I wish to thank the institutions which extended financial support and particularly to their representatives who guarded and encouraged our cause. Thanks to Dr. Eliezer Tal, Director of the National Council for Research and Development, Jerusalem, for his continued support and understanding, Dr. Martin Cremer, Director of the Institute for Documentation, Frankfurt, for his foresight and independence in furthering this event. Without him, this conference would not take place. I also wish to thank M. Jacques Michel of the National Bureau for Science and Technology, Paris, and Mr. J. Michel Gibb of the Commission of the European Communities for ready collaboration. Our appreciation is due to Prof. Aryeh Dvoretzky, President of the Israel Academy of Sciences and Humanities, to Prof. Nathan Goldblum, Vice President of the Hebrew University of Jerusalem, to Dr. Wolfgang Löhner of UNESCO, and to Daphna Avnon of the Jerusalem Book Fair.

And particular thanks to the organizing committee - Dr. Martin Cremer, Prof. Knut Faegri, Dr. Karl Heumann, Ernest Mann, Jacques Michel, Gillian Page and Dr. Udo Schützsack - who ever in good spirits, joy and pleasure in planning together, backed and furthered the idea to embark on this venture. My special thanks to Gillian. You say I persuaded others to support my idea of the conference, but it was you who persuaded me to get actively reinvolved in editors' affairs by inviting me to become your heir as ELSE's treasurer. Thanks too for providing the stimulating setting and gracious hosting for our planning meetings in your garden and library.

Thanks to the members of the advisory committee, to the U.S. program committee, and to all the societies of editors of organizations represented here.

We are proud to celebrate the 50th anniversary of *Biological Abstracts* which has been of such great service and example during its half century of activity. We are happy to have Dr. Edward Kennedy, its director, here with us. Thanks to Dr. Philip Altman for organizing the session on Standards and Style.

Although he is with us in spirit, Prof. Derek de Solla Price deeply regrets that he could not be with us because of a serious illness, but we are glad to hear that he is making a speedy recovery. Our warmest wishes are with him.

Israel has served as training ground and a prototype for me in the international ground we are now traversing. I hope we will continue to cooperate with the international community to improve participation in local and foreign publications. We also strive to further our publishing, printing, translation and information industry which is eminently suitable for a small country with a high concentration of scientific manpower.

With high expectations that through this meeting we will begin to accomplish our goals, that we will consolidate into an international body with an ongoing program to be inaugurated here. Thanks to you all.

I am honored to present His Excellency, the President of Israel, Prof. Ephraim Katzir, eminent scientist and editor in his own right. It is exciting that this morning is his first appearance as a Foreign Member of the Royal Society. It is significant that the Royal Society held the First Scientific Information Conference in 1948. Our congratulations to him. Ephraim.

נשיא מדינת ישראל
PRESIDENT OF ISRAEL

OPENING ADDRESS

Ephraim Katzir, F.R.S., *President of the State of Israel*
The Weizmann Institute of Science, Rehovot
Chairman of the Editorial Board, Weizmann Science Press, Jerusalem

Distinguished guests, Miriam,

It is certainly a pleasure to greet you here in my capacity as scientist, as well as President of the State. So many distinguished editors coming from abroad to this first international conference of scientific editors are surely most welcome guests. I understand that you will be dealing with scientific information transfer and the editor's role in tomorrow's world. The fact that you come from so many countries is additional proof - if proof is still required - that in science, as well as in any major cultural activity, political and geographical boundaries have no meaning.

You have chosen appropriately in selecting Jerusalem as conference site, for Jerusalem is a place where we have great respect for the written word, great respect for scholarly activity. And as perhaps some of you know it is this year that we are celebrating the re-unification of Jerusalem ten years ago. Eastern and Western Jerusalem were united for the benefit, I believe, of everyone concerned. Many a miracle is said to have occurred in Jerusalem, and I know that Miriam Balaban, for one, has performed many miracles during her life! Perhaps your meeting somehow falls into that category.

Information - that is receipt of information, evaluation of information, storage of information, retrieval, spread of information - is certainly one of the most important aspects of post-industrial society. The way we live in the future, the way we will carry out many of our individual and social activities, will be determined by our ability to have and to retrieve information. As a biologist I have always thought that information is, in a sense, something like an artificial gene. By being able to store information and add to it, we can actually transfer information from generation to generation. That is a unique feature of Homo Sapiens.

Miriam Balaban (ed.), Scientific Information Transfer: The Editor's Role, xxv-xxvii. *All Rights Reserved.*

Transmission of information surely depends in many a case on the editor. The editor must function as interlocuter and intermediary between the author and the audience; he must review scientific material for the author. He sometimes has to clarify the article submitted, as some of you mentioned in the abstracts of your papers. Here and there plastic surgery must be carried out on an article to make it readable and understandable. These are major tasks indeed.

I have been thinking about the way scientific journals started in the middle of the seventeenth century, the time of the revolution in science, when the number of scientists increased, and information was often transmitted via personal correspondence. Such correspondence actually transmitted scientific news not only to the recipient but to groups of interested persons. In England during this period the most important of these "letter boxes" was that of Henry Oldenburg, the secretary of the Royal Society. He was required to read aloud at meetings of the Society, any letters dealing with scientific matter that he received. In return he had to inform his correspondents of new scientific developments in England and in particular at the Royal Society. Finding this cumbersome, he decided in 1665 to establish a scientific journal, *The Philosophical Transactions of the Royal Society*, which he himself funded.

With the passage of time, of course, more and more journals and more and more books appeared. As Mr. Mann has so rightly put it, the history of information transfer should be divided into three periods. At the beginning of this century when we had relatively few scientists and relatively few scientific journals, new information was transmitted from scientist to scientist directly, via these primary journals. They even read the journals in those years; it was a good time for scientists - a true paradise.

The second period covers the last twenty-five years when scientists began to find it increasingly difficult to read original papers and instead read abstracts, resorting to bibliographical services. Now I think we all agree that we are in the third period when even abstracts have become so abundant that it is difficult to follow them and one has to look for new ways to solve the problem. I always recall my professor of theoretical physics, the late Professor Racah, who used to say, that when he had a problem and wanted to find the solution, if he had plenty of time he went to the library to find original papers, but if he had no time, would solve the problem himself.

I am afraid that times have changed and we have no choice but to use modern techniques. There is no doubt that computers will play a greater and greater role in the field of storing and retrieving information. But we have to take into consideration that usually computers have no refereeing system, so that we should be careful to feed the computer good articles and not material of indifferent value.

So, too, an editor can promote good science, but unfortunately also mediocre science. And the increasing number of journals may not always be of great service to science.

You will discuss creators and gatekeepers, the editors and the referees, and I don't envy editors that have to cope with referees and afterwards with the authors who get the criticism of the referee. These are days when standardization of style and form is dominant. It is obviously economical to standardize, even though scientific material thus presented, is dry and somewhat dehydrated. It is a far cry from the fascinating letters of Leeuwenhoek to the Royal Society. There everything is so human. I remember Leeuwenhoek describing how the women who came to shop in his store disturbed him while he was peering at microbes through his microscope.

It is gratifying to see that many of the participants in this conference are ignoring narrow national interests and trying to think in terms of regional international cooperation in printing and storing scientific information. I have no doubt that this is a most important trend from the point of view of science and also for budgetary reasons. And the importance of budgetary reasons cannot be lost on a man who has been president of this country for the last four years!

In conclusion let me wish you all success in your meetings. I am certain that scientific work is the noblest activity mankind has devised. Any way, any means, to promote science, the understanding of nature, the understanding of man, is a most noble activity. The International Federation of Scientific Editors' Associations you propose IFSEA surely has its part to play in the great scientific endeavor of our time. May all go well with it and you.

G R E E T I N G S

Nathan Goldblum, *Vice President, The Hebrew University of Jerusalem*
Department of Virology, The Hebrew University -
Hadassah Medical School

President Katzir, ladies and gentlemen. Thank you for saying that I'll say a few words only, because as you have heard Professor Katzir, it is very easy to speak before him, but so much more difficult to speak after he has spoken, and I am afraid that whatever I'll say will be repetitive of what he's said already, and I think redundance is one of the greatest enemies of this conference.

So I should like to take this opportunity to welcome to you in Jerusalem. Mr. Harman, President of the Hebrew University has asked me to convey to you his very warm wishes for a successful meeting and conference. It is indeed a great privilege to have the First International Conference of Scientific Editors held in Jerusalem. We at the university had wished that the meetings would have taken place on one of the campuses of the university so as to enable more of our professors and hopefully graduate students participate in the meetings, but this has been impossible because we are at present exactly at the height of the term of teaching, and there is not enough space left for a conference at our halls. I hope however that most of you will at least have the opportunity to visit the two or three large campuses we have in Jerusalem, spread all over the city. They can be seen from almost every place since some of them are high up on the hills.

One word about the conference itself. We as professors and teachers at the University are aware of the great and growing importance of scientific editing and the problems of scientific communication. You will be discussing novel ways of information storage and dissemination of scientific knowledge. Since one of our jobs is to publish papers and books we know the importance of the editors and their role as intermediaries between the author, the public, and the readers. We are often at the mercy of the referees and the editors as mentioned already by President Katzir.

Just one little figure about what we do at the University. The scientific body of the Hebrew University amounts to 1500 scientists who publish about forty five hundred papers and books per year. I have calculated very quickly that almost every hour we are awake, a paper or a book by someone from the Hebrew University is being published, and this can of course be tripled or quadrupled for the country as such. This shows what great weight is attached to the published word for the scientist at the University.

Well I hope that you will have a very creative and successful conference, and I wish you lots of luck and beautiful weather in Jerusalem.

Miriam Balaban (ed.), Scientific Information Transfer: The Editor's Role, xxix-xxxii. *All Rights Reserved.*

Knut Faegri, *Past President, International Union of Biological Sciences*
Botanical Museum, Bergen University

Ladies and Gentlemen. To those of us who have grown up in the Judaeo-Christian tradition, and I think most of us have, coming to this country is always a fantastic experience. Simply the topography of what we have read about so many times is enough to make this a unique experience. But if you stayed for a few days and met some of the people of this country one of the things that may strike you as much as it has struck me is that even in the most serious of situations it is never far to go for a smile. There is always a smile behind what is being done here, however serious, however important. There will be a smile behind this meeting too. We have already seen that. But that should not prevent us from realizing that it is a serious situation. Whether it is desperate or serious we may refer to as an editorial matter.

Science is many things. Science is also communication. In science we have had, for the last couple of hundred years, a means of communication that has served us so well that we almost can't think of any other. We know that this means of communication is changing. The aspects are changing. We don't know exactly what these changes are going to be, we don't know to what extent we can influence those changes. We have come here together to discuss these future changes from the background of the little innovations each one of us has been concerned with, trying to find our ways out of the difficulties of today, the difficulties we anticipate tomorrow, and into the world of the day after. There are many aspects of this.

It is easy enough to say that we can store our information on tape or in data banks; we will have to do that certainly. But how are we going to get it out again? I'm not thinking now of the great universities in the great countries. There are smaller universities and there are smaller countries. They represent problems that are easily forgotten in our joy of discovering the technical possibilities which are available but are, let's confess it, expensive. They may even be more expensive than our outdated methods of today. We don't know about that. These are some of the problems we, hopefully, shall not solve (that won't be possible) but discuss the solution of.

And speaking for all the non-Israeli members of this conference I think that I can express our profound gratitude to Miriam Balaban, to the Israel Academy of Sciences and Humanities and the Hebrew University of Jerusalem for making this possible, and to your Excellency President Katzir for coming here and giving this meeting an official welcome.

Gillian Page, *Cambridge University Press*

We begin now with a topic which concerns us all, as authors, editors, publishers, information scientists and readers. The journal has served us well for three hundred years, yet it is difficult today to be unaware that there are people who question whether the journal as we know it is right for all the purposes it serves. The winds of change are blowing, changes range from relatively minor ones as from hot metal to cold setting, to the electronic journal. They affect all areas of the transfer of information from the author to the reader. Some have already been adopted by many journals, others have pilot projects and yet others have schemes for the future, perhaps themselves depending upon some technical breakthrough. We need to be aware of the possibilities and to try to evaluate honestly their advantages and disadvantages. This can be difficult to do from written reports, travellers' tales and anecdotes which reach us on wonders in other lands.

We hope that from today's discussions by people who have studied various innovations in depth we shall get a clearer picture of what is going on, what might happen, what looks doubtful and how we are affected by these possible changes in our different roles. We are very fortunate in having Martin Cremer from the Institute of Documentation at Frankfurt to introduce the whole of this subject.

Martin Cremer, *Director, Institut für Dokumentationswesen, Frankfurt (Main), Germany*

Madam Chairman, ladies and gentlemen. After the introduction of the President of the State and after some other introductions it is a little bit difficult to give another, the fourth or the fifth introduction. What I did in preparing my so-called opening remarks was to find out if there is a kind of thread through the papers of the first session.

On first view it seems more or less a kind of cocktail of very different problems and points, but I think there is a thread. It may be that a first session is the general introduction to the whole meeting, so it may be that the thread is the thread of the whole conference. I have only put together some points to introduce the papers of the first session. I don't know if my colleagues, the experts who will give the papers agree, but I'll try my best to give the essence in five short points.

The first point is very obvious - it is that in the past, now, and in the future, the main carriers for the dissemination of information are publications. Publications of any kind - monographs journals, secondary journals, published data collections, new kinds of secondary services, synopsis journals and so on, library catalogues and published means printed or not printed. I think that for the first point there is no difference if the publication is printed or available in microform or in computerized form. It may be in the full text storage.

The second point, pointed out very clearly and very impressively by Professor Katzir is that it seems indeed that in our society the need for information in general is growing more and more not only in the field of research and science but in other sectors of the society - in administration and participation of the public in public affairs. That information is indeed one of the important resources of the society. It seems that information becomes more and more a kind of product or commodity which enters the picture more and more. It is true for the printed primary publication and also for the secondary publication and for stored information.

The third point is that in the past ten years we have made very important technological developments which influence this kind of new market, make necessary new structures of the information market and offer many problems as do other main new resources,for instance energy, transport and so on which has a big influence on the structure of society, which makes some problems, which has some reactions of the society, positive reactions and negative reactions. You know that in many countries the action of the public against energy is very strong, and we sometimes have a feeling that the reaction of the public against the computerized and the fully informed society will be a real problem.

My fourth point is, what are the new technologies we will discuss in the first session? First is the new photosetting device to be considered for the editor and the publisher. Second is the rapidly developing copying technology - telocopying and others. I think these will have very great influence on publishing primary and secondary information. The third is the development of the computer. We see the storage capacity of computers becoming bigger and bigger and cheaper and cheaper. This will for instance influence the question of whether it is technically possible to store full texts in a computer.

The next technology is a micro-technology. There will be papers in the field of ultra-micro techniques, and on another technology - the combination between ultra-micro techniques and computer storage and printing by photosetting and so on. And the last one is the development of communication techniques in principle. Some countries are working on the development of technical networks with the application of modern PTT technology with a tremendous influence not only on the input of the information but on the output - that is on the use of information. In the European community and in some member states of the European community we are trying to work out the technical problems of networking. We see that the influence of technical networks on the information community is terribly important.

My fifth and my last point is the question - what is the influence of new technology on the editing and the production of information. We know that we have the problem of the new structure of printed journals. We have the problem of the so-called dual-journals - a combination between the printed version of the journal and the microfilm version as an alternative or as an addition. We have a second reaction to the new technology I mentioned before through the new organization of publishing. I think it is very interesting to be informed about the new ideas on editorial processing centers which will I think very much influence the whole technical and economic structure of publishers.

In the same way the library is influenced by new technologies, not only in automation of library activities, but the role of the library as a medium between the author and the user of the information are changing more and more. The libraries are forced to play a role in the transmission of computerized information resources to their users. They are having more and more to handle not only micro-materials but computerized materials. Then we must consider the competition between printed materials - the conventional way of publishing - and new techniques for the automated storage of information.

There are many discussions in many countries between publishers and the so-called information politicians on at least two aspects of the problem. One is the economic aspect, and the second one is the legal aspect. I think the legal aspects in connection with copyright - with the right to copy articles - and the question of what happens with the copyright and author's rights when full texts are stored presents an important challenge of the new technology and the structure of publishing and publications.

I may mention one experience in this context. Sometimes it is said that there is competition between a printed version and a computerized version of information. From our experience with secondary services, we find this not to be true. I have some examples. When we offer abstract versions in the printed form and then we offer the same on tapes for retrieval for preparing dialogues and on license and protection, against our expectations, we find that there is no competition. On the contrary, the possibility to get very narrow and profiled information from the computer very quickly has had some impact on encouraging users to subscribe to the printed version. I think, although I'm not sure it is true for every discipline, that there is no real competition between selling a printed secondary service and the tapes.

The last point and a very important point is the reaction of the users. What is the user's reaction to all this technology? Is he willing, is he prepared, is he trained and educated, or is he forced to accept the new technologies? Is he willing to accept microforms? Is he willing to accept synopsis journals? Is he willing to sit on the terminal and use a dialogue system? In my country we don't have much experience with the psychology of the users on the new technology. It may be, and it is our hope, that the new technology will encourage a growing number of users to make more use of information which is offered. But to repeat, I think that users are not willing to abandon their use of printed materials. I'm quite sure that we have to accept the new technologies, we have to accept microforms, ultra-microforms, and computerized information, but I think that it is not necessary to forgo the printed version and to be able to browse through the printed version or the secondary versions. What I say now applies more to the secondary services, and of course, it may be that some legal texts are stored in the full text or abstracts are stored in full texts. But there is no danger I think that the printed or the published primary publications are under the pressure of the new technology. But we have to be aware of developments, and I hope that the outcome of the papers from our first session will be to understand that we are in a changing society with an impact of new technology on editors, on information scientists, and not least, on the users themselves. And now I propose giving the floor to the real experts.

EDITORS AND THE IMPACT OF NEW TECHNIQUES:

Mechanization and Organization

Gillian Page, Chairman

SOME ASPECTS OF DATA RECORDING AND MICROFORM OUTPUT FOR INFORMATION DISTRIBUTION

Klaus-Dieter Lehmann
Universitätsbibliothek Studentbucherei, Frankfurt, FRG

ABSTRACT

The growing rate of information poses increasing problems for publishers and editors and also for libraries and documentation centers. Data processing techniques provide several possibilities for improvement.

This paper describes various techniques for data recording; the application of OCR-readers is demonstrated by a special project. Also included is a discussion of some character set extension techniques.

For data output a comparison is made between CRT photosetting and microforms. This based on actual experience in the Federal Republic of Germany.

1. BASIC PURPOSE

1.1 Scientific journals, reports, books and pamphlets represent a pool of information which grows from year to year. The increasing rate of knowledge has led to a production of about 5 million publications per year in science and applied science alone.

In the Federal Republic of Germany, about 700 million characters have been recorded, printed and distributed in the last year.

1.2 To this primary, you must add the secondary information of libraries and documentation centers. Their catalogues, bibliographies, information services and data bases also increase by billions of characters per year.

1.3 For these specialized applications - editing, printing and distributing of primary and secondary information - an efficient hardware configuration is developed. Of particular interest is the peripheral equipment for data recording and data distribution.

2. DATA RECORDING

2.1 For data to be communicated between people and the computer, means must be provided for its conversion from human readable form to machine language (input) and from machine language to printed form (display and output).

The results of data processing depend on the efficiency of hardware, software and - last but not least - on the quality of the data to be processed. Data are an important and sensitive link in a computerized system. The requirements of data recording and data output vary from application to application, but there are four characteristics of printed data we speak of:

Miriam Balaban (ed.), Scientific Information Transfer: The Editor's Role, 3-8.

- they are voluminous
- they consist of alphabetic letters, numbers and other symbols
- they are of different type fonts
- they are prepared in a variety of ways.

2.2 Data are recorded
- for direct typographic processing
- for computer-based information and retrieval systems.

There are several criteria for the selection of data recording machines in copy processing systems:
- suitable and enlargeable character set
- likely rates for typing and input
- powerful methods for error detection and correction
- flexibility of format.

The available equipment for input processing has changed in recent times. Key-to-tape and key-to-disk stations have replaced the punched card and punched tape equipment, either as stand-alone systems or as mini computer-controlled data collection systems.

2.3 In some fields of application on-line CRT-terminals are becoming more and more important. The significant advantages are:
- the direct response of the computer to the input data
- the function as an information system (data base)
- the processing to check the input data for consistency, accuracy and reliability.

On-line input equipment is not intended for high-volume input situations. But if the input operator needs a lot of centrally stored information in order to do his job, or if there is a high rate of data changes and modifications, an on-line system is advantageous.

There are two alternative possibilities for on-line data processing: to be a participant in a central computer or to use a stand-alone computer.

2.4 For some computer users, not able to afford their own computer and yet not requiring the interactive response of an on-line computer system, remote-batch terminals are offered.

2.5 A new situation is created by the development of package mini-computers to support specific publication functions.

The mass production of integrated circuits hastened the development of mini-computers. Today the efficiency of mini-computers is comparable in some fields of application to that of large computers.

In order to use the mini-computer for data recording purposes it has to be equipped with:
- disk or floppy disk units with sufficient capacity
- magnetic tape unit to transfer the stored data
- visual display unit with sufficient capacity for representing a scope of a document as large as possible
- printing and hard-copy facilities.

The mini-computer is able to work independent of or in connection with a large computer.

2.6 An interesting kind of application in the field of data recording is the combination of optical character recognition with the following correction by on-line CRT-terminals. For example, this mode of input is used at Brönner Umschau Verlag in Frankfurt (Fig. 1). The printing type OCR-B is applied. Just as before, the editor or author produces his texts according to a set of standardized work-related instructions using a typewriter at home, in a branch office, or in the firm itself. These texts, including the codes for typesetting instructions, are simultaneously machine readable on the OCR paper.

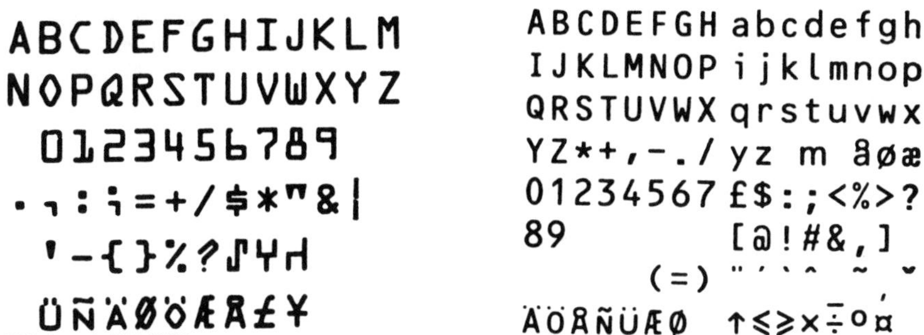

OCR-A según DIN 66 008. OCR-B según ECMA-estandard.

Figure 1

Optical character readers convert the alphnumeric characters, symbols and codes on the document into electronic signals; the recognized signals will be stored on magnetic tape or disk, and then processed in the central computer. Thus OCR is a by-product of regular operations.

To make a text available for further work, it is called to a CRT-display by way of the job number and if necessary an additional tag. Correction, editing and addition of setting instru-tions can be directly carried out. It is also possible to make corrections by way of print-out, producing a new OCR-script. The editor's revision includes text corrections as well as the electronic codeswhich give the computer program its directions. Thus prepared, the text is now ready for automatic computer processing. The computer program distributes to certain text components the correct type face.

OCR is a good method of input in high-volume input situations, combined with on-line keyboarding for low-volume situations (error detection and correction in two-way conversations). It is also a good method of input for copy processing systems with a large number of authors and correspondents outside the publishing house, for OCR makes it possible to use the typewriter as a keyboard. An OCR-scanner can handle the production of more than 400 typewriter personnel.

It is also a good method of input where an institution must deal with organizing, updating and maintaining material.

3. COMPREHENSIVE CHARACTER SETS

A special problem for data recording is the limited font of characters. The solution of this problem is very important for all editors of scientific journals with different alphabets and special mathematical or chemical characters and also for libraries and documentation centers with their catalogues and information services containing international literature.

Various codes exist today, standardized and non-standardized, built up by 6-bit, 7-bit or 8-bit strings. I do not want to analyze existing codes, but I want to tell about some suitable code extension techniques. Some of these have been proposed in the draft standard ISO/DIS 2022.

Extension by use of free codes. In this case, free positions of the basic set will be filled. This method is limited by the number of free code positions.

Extension by substitution. In this case characters of the basis set are replaced by substitute characters. This method does not lead to a code extension, but to a new standard.

Extension by shift-out/shift-in mechanism. In this case control characters allow the temporary substitution and replacement of complete character sets. By using multiple shift-out sets, any non-Roman alphabet can be represented. (ISO/DIS 2022)

Extension by escape sequences (Fig. 2). In this case, an escape sequence consists of two or more 7-bit combinations to announce the code extension facilities. It can be used for single characters or character sets. (ISO/DIS 2022)

The escape sequence is built up as follows:

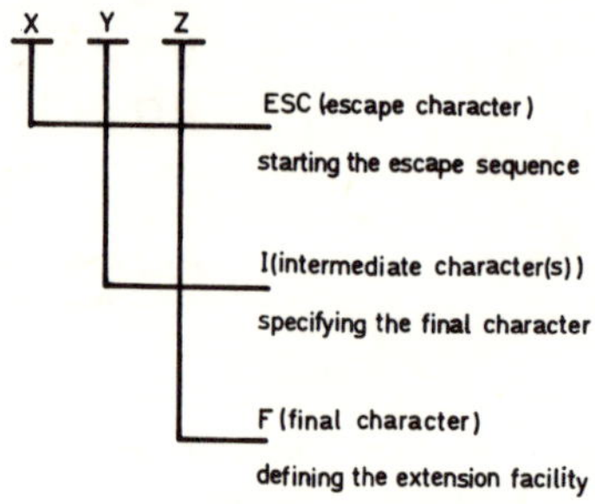

Figure 2

Extension of prototypes (Fig. 3). In this case, a protype stands in place of the original character. Protypes are composed in the following form: a basic character and a two-digit number; in the text the protypes are marked by a preceding control character. The protype method allows the representation of about 6000 different characters. It has been developed by the Zentralstelle für maschinelle Dokumentation, Frankfurt am Main. For example, the German Bibliography, Food Science and Technology Abstracts, the South African National Bibliography, the Hessian Union Catalogue for Serials, and others use this protype method for data recording. After processing the protypes are replaced by the original characters for printing.

These considerations are necessary, for there is no balance of character fonts between data recording equipment like photocomposing machines.

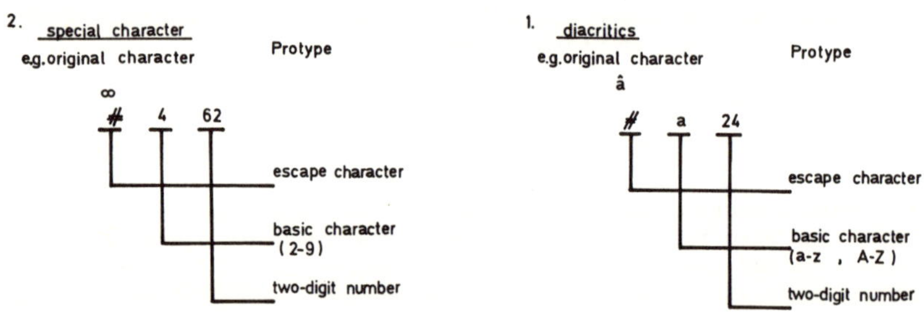

Figure 3

4. DATA OUTPUT

4.1. At the end of every data processing job is its result: the computer output. For the production of books, periodicals or library catalogues, the line printer is not sufficient. The conversion of the processed data in a typographic quality is usually done by CRT photocomposing machines.

Control data and text data - recorded on magnetic tape - make a photographic exposure of the pages possible. The peak capacity is about 2 million letters per hour. The digitized characters are taken from a disk storage unit into the core storage unit during processing. Then the data are taken from the core storage unit and the letters or garphics are recorded by luminous spots on the cathode ray tube. The available recording or output materials are photographic paper and film. Modern CRT photocompositors can set more than 1000 different characters in different type faces and different display types.

4.2 The other important mode of output is COM computer output by microform, especially microfiche. The use of COM has been primarily based on the need to escape from the avalanche of paper emanating and to minimize the escalating costs of printing, duplication and distribution. In many instances it is most significant to have great amounts of information easily accessible which change rapidly. For example this applies to comprehensive registers of books in print, catalogues of libraries, parts lists of industries, financial reports and others.

COM is basically a character generating system from magnetic tape, whose preparation is typically offline from computer. In practice there exist two standards: 42 and 48 reduction. (The alpha-numeric recorders are usually restricted to one character font).

The main properties of COM are:
- the speed of output (10× more than impact printer)
- the cost reduction
- the improvement of retrieval
- the ability to intermix text with drawings
- the easy way of duplication
- the wide range of ancillary equipment including high speed printout on paper
- the wide range of layouts available including page size.

4.3 The publishing group Brönner Umschau has found an interesting method this year which combines the advantages of photocomposition and COM. Contrary to the COM procedure, where the character font is usually restricted to one, the microfiche is here produced via a Digiset photocomposition machine (Fig.4)

This allows the insertion of all kinds of typographic fonts. For example, the user is able to combine boldface, italics and normal type and insert any number of additional characters. In addition, the lengths of the text can be influenced by "compressing" the print type. The microfiche has the reduction 125, making possible the placement of 3000 pages size DIN A 4 (210× 297 mm)

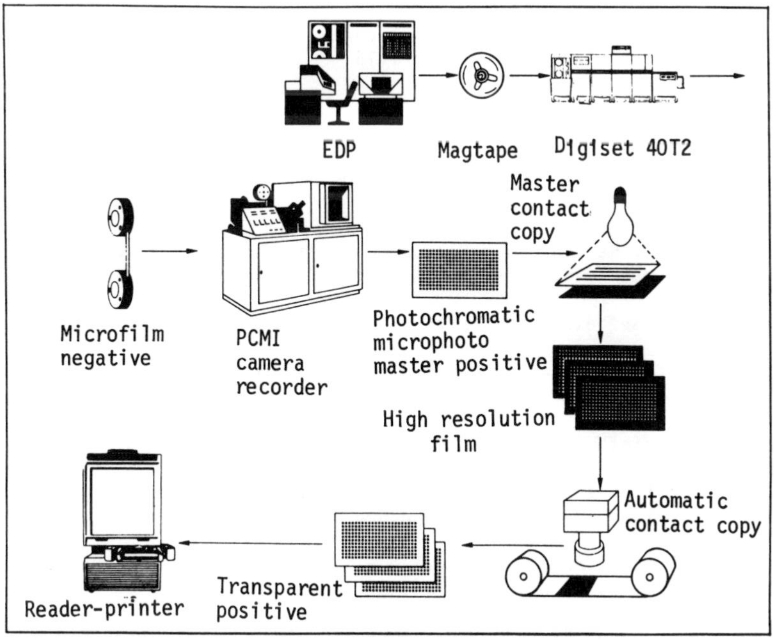

Figure 4 Production flow of a "Typoprint-Ultrafiche" from computer to reader.

on postcard-sized microfiche (DIN A 6 = 105 × 148 mm). The convenience of a single "Typoprint-Ultrafiche" as a data carrier in place of 10 conventional microfiches makes the handling much easier for the user. The development was carried out together with the book trade processing center. The readers were developed by NCR. The newest machines also carry an interchangeable lens for 48 reduction and 125 reduction.

The Typoprint-Ultrafiche has been put to use in the Verzeichnis lieferbarer Bücher (Index of Available Books), which is available in printed form (4 volumes)as well as in Ultrafiche.

CONCLUSION

The development of methods of information distribution is moving along at a rapid pace. With certainty the near future holds increasing possibilities for on-line retrieval; data telecommunication should play an important role too. But also the procedures I have described will surely continue to have their job-related justification.

REFERENCES

1. Program der Bundesregierung zur Förderung der Information und Dokumentation 1974-1977. Bonn 1975.

2. Das Stuttgarter Modell. In: Deutscher Drucker. Nr. 8/24-2-1977. p.10.

3. E. Fritz: Was leisten online-Computersysteme für Redaktion und Setzerei. In: Deutscher Drucker. Nr. 6/10-2-1977. p.10.

4. Arbeitsstelle für Bibliothekstechnik: Neue Verfahren für die Dateneingabe und Datenausgabe in Bibliotheken. München 1973.

5. F.K. Olf: Lesender Computer. Köln 1973.

6. R. Bernhardt: Zeichenvorräte und Zeichendarstellung für das EUDISED-Netzwerk. In: EUDISED. Straszburg 1976.

7. Draft ISO/DIS 2022. Genf 1972.
   Draft ISO/DIS 2375. Genf 1973.

8. ZmD: Typen und Protypen für den LINOTRON 505. 2 ed. Berlin 1975 (ZmD - A - 22).

9. K.-D. Lehmann: Bericht über Planungen fürden Einsatz der EDV im Bibliothekswesen des Bundeslandes Hessen. In: ABT-Informationen 20(1976). p.34.

10. 2500 Seiten auf einer Postkarte. In: Umschau 76(1976), 14, p.445.

11. Review of developments in COM and micrographic technology, present and future. AGARD Lecture Series. No. 85. 1976.

12. G. Thiele: Einführung in die Reprographie. Frankfurt 1976. (IDW).

13. K. Jordan: Lichtsatz. In: 4. Internationaler Kongresz für Reprographie und Information 1975, p. 243. München 1975.

14. R.J.D. Johnston: An experimental on-line retrieval system using ultrafiche. In: Program Vol. 9, no. 2, p. 56.

15. C. Deider u. R.H. Klar: COM-Verfahren in Wissenschaftlichen Bibliotheken. Berlin 1974 (ABT).

EDITORIAL PROCESSING CENTRES: A VIEW FROM THE UNITED KINGDOM*

A.M. Woodward
Aslib R & D Department, 3 Belgrave Square, London SW1X 8PL, UK

ABSTRACT

Faced with mounting economic pressures and changing market conditions, publishers of scholarly works, especially scientific societies, must look for different products, cheaper production methods, new technologies and different systems of information dissemination. In the short term, a move away from conventional printing to microforms, synopsis journals, etc. will provide greater flexibility while adoption of data-processing techniques during the editorial stages can help reduce costs of conventional printed products. In the main, individual learned societies are too small to take advantage of available new technologies because of the investments and the high volumes of throughput that are required.

The concept of Editorial Processing Centres (EPCs) was developed in the USA under the auspices of the NSF. It postulates a cooperative venture between societies to provide computer text handling capabilities, communications control and photo-typesetting facility. In a series of studies, Aspen Systems Corp. and Westat Inc. developed 4 progressive configurations of EPCs and an economic model which showed cost savings of up to 15% over conventional methods (depending on the configuration adopted and number of journals processed).

Aslib R&D Department has recently completed a project funded by the British Library R&D Department to review and evaluate the potential EPCs in the UK. Despite the difficulties of generalising costs and practices of learned society publishinga simple comparative exercise showed there are potential savings to be gained from the technical innovations suggested, even when one of the requirements of the American studies (for authors to submit manuscripts in OCR-type) is eliminated because of impracticality in the UK. The cost savings in smaller EPCs arise almost exclusively at the typesetting stage whereas editorial costs are slightly increased.

Rapidly developing technology and continuing falls in the real cost of computing make the prospects of using these technologies much more viable for smaller processing units than even a short time ago. Problems are more likely to occur in the social and organizational areas than in technical and economic ones. Developments will have to be coordinated with changes in social expectations and availability of suitable equipment elsewhere in the community. A particular problem relates to the position of honorary subject editors and referees; current systems to handle their requirements are clumsy and costly. Future systems might rely on widespread availability of dial-up facilities or on professionalization. In either case substantial modifications will be required in work patterns and may involve erosion of the inbuilt values of present systems, possibly counterbalanced by quality improvements elsewhere.

* The studies were funded by the British Library R & D Department.

Miriam Balaban (ed.), Scientific Information Transfer: The Editor's Role, 9-20.

Introduction

The concept of editorial processing centres to help contain learned society publishing costs through co-operative ventures was developed in the USA from the original ideas published by Bamford (1) The subsequent studies by Aspen Systems Corp. and Westat Inc. (2) showed that the concept was technically feasible and that the co-operation of the several societies could provide a viable, faster and cheaper method of publishing for the learned societies. These studies were based on American data and were tailored to the prevailing economics, technology and publishing structures in the USA. It was therefore necessary that this exciting concept should be evaluated in a United Kingdom context to see whether similar advantages were likely to accrue to UK publishers by the development of EPCs.

Aslib Research Department was commissioned by the British Library to review the Aspen/Westat studies and to evaluate the potential of EPCs as a means of assisting UK learned society publishers. The study had three components: an analysis of the technology and assumptions of the original reports; a study of UK learned societies, especially in contrast to their American counterparts; an outline design for a British version of an EPC, based on the results of the other two parts of the study. These studies have been reported in British Library R & D Report No. 5271 (3).

An initial assessment of the Aspen/Westat reports showed that the results and conclusions from the background studies in the main appeared to reflect practices in the UK with only minor differences. Discussions with learned society publishers generally confirmed this for the actual operations and manuscript flows. The one difference appeared to be the use of editorial boards (not mentioned at all by the Americans) to collectively vet and advise on acceptance of submitted papers. However, this is essentially the same as refereeing and either replaces it or becomes a second stage filter.

Learned society publishing in the UK

The publishing activities of UK learned societies were studied by discussion with the directors of publishing of several large societies and by mail survey of approximately 100 smaller societies. The extent to which scientific publishing is carried out by commercial publishing houses on behalf of learned societies was also examined for both British and American societies.

The learned society publishing in the UK encompasses a wide variety of sizes of undertaking, styles and methods but on the whole there is a marked break in the publishing process between editorial control and selection and the production process beginning with 'marking up' of manuscripts right through to distribution. On the editorial side, control in smaller journals is usually effected by a voluntary or honorary editor often with the assistance of an editorial board, usually chaired by the editor. Refereeing of some description is normally employed to help in the manuscript selection process, although the referees are often the members of the editorial board. Nevertheless there are many small journals where the editor has almost total control over the contents. In the larger societies, day-to-day control of the journal and, to a large extent, selection of the manuscripts is handled by a full-time paid editor. He too usually has assistance from an editorial (advisory) board which sets the general editorial tone of the journal and can act as an arbitration board where manuscript selection is difficult. The refereeing system is more formally organized with two referees as the norm. However, for both large and small societies or journals, informal methods are quite extensively employed, especially where manuscript selection involves controversy or questions of taste and expediency.

On the production side, hot metal processes still predominate but there is an increasing trend towards the use of offset-litho printing methods. The societies usually show considerable conservatism towards changing to new methods or even new printers, partly because of a lack of awareness of new methods and partly, in the case of small societies or journals, because of lack of time to get competitive tenders or to study new techniques.

From the results of the survey of the smaller societies, it seems that many societies are actively restricting the numbers and length of papers they will accept, and there is a noticeable reduction in the length of papers published over the past ten years. Circulations still seem to be rising steadily in contrast to the experience of larger (especially American) societies. Above a certain price (about £10 pa) circulation seems to be inversely related to price; circulations also appear to be similarly related to price per page. Overseas sales constitute a large part of the total circulation (40% in our survey) but this can be higher than 80% for the larger society or commercially published journals.

The dichotomy between editorial and production processes was reflected in the extreme difficulty of obtaining meaningful costs for learned society publications. Experience elsewhere had shown that it is almost impossible to get even rough costings that are comparable between societies or journals. Part of the difficulty lies in the inability of small societies to provide much information in this area. However, a far more important point is connected with the traditional role of the society in providing a number of services to members for a fixed membership fee. This means that editorial staff, if employed as paid staff, may be included in the general running costs of the society, whereas the profitability of the journals is often measured in terms of recouping the paid production costs (i.e. composition, printing, binding and distribution) from journal subscription income from non-member subscriptions. In many cases this can produce an artificially optimistic view of the economics of each journal. The production costs which are paid for outside of the society appear to be increasing faster than internal costs and the fairly massive general level of current subscription increases (37.5% between 1975/6 in our sample) is a reflection of this. In small societies, a much higher proportion of the total production costs is affected by these costs than in those societies where much of the editorial work is properly costed into the production costs.

Further difficulties arise in establishing any useful costs because of the widespread use of voluntary or honorary editorial staffs.

Even where honoraria are paid, these seldom cover the actual value of the editor's time involved in his editorial work. Secretarial services are usually provided by his parent institution as part of his normal working arrangements and the parent institution also carries other overheads such as space, heating, telephones, copying and postage (which may even extend to distribution of the journal itself). Some editors are entirely of amateur status, paying the costs of their editorial work entirely out of their honoraria. Much of this type of editorial work is of course carried out in spare time, or on the train. This is an important consideration if and when an EPC is contemplated since the increased communication between editor and the EPC, possibly involving on-line terminals for enquiry and editing, will make this type of arrangement much less practical than at present.

On the whole, although there is concern about cost inflation, societies appear to be relying on the apparent inelastic demand for journals and raising prices to compensate for increases in production costs. There is relatively little active effort in looking for different or innovative methods of publishing or disseminating the results of research, although cost-reduction exercises in relation to conventional publication are widespread. The difficulty is often a lack of staff and ignorance of potential advantageous innovations.

UK-USA differences in learned society publishing

The main difference in the societies themselves is that of size. American societies generally are much larger than their British counterparts. This gives them a substantially larger base of subscriptions for journals and in fact circulations are considerably higher. This should produce several effects.

Firstly, the larger circulation immediately gives a much larger income with the implications of employing a larger number of professionals to perform editorial functions within the society. However, in the current situation where printing, paper and distribution costs are increasing faster than other costs, larger circulations will accentuate the effects of these cost increases.

Secondly the number of manuscripts submitted and published should be higher leading to the necessity of employing more, paid, professional staff by the society. The 'small' hypothetical journal used by Aspen/Westat has 1200 pages per annum; only 13.7% of the journals in our survey reached this size in 1975. (NB This survey was designed to locate small societies.) This means that there should be a lower proportion of 'hidden' costs in US society publications than might be expected in UK society journals where there is a great deal of voluntary assistance in journal preparation. A third effect is that, given that there is a certain inelastic subscription base of journals by institutions worldwide, wherever the journal is published, a larger personal membership will produce a much larger segment of the total circulation which will be sensitive to price. In other words, there is a certain proportion of subscriptions for which the demand/price ratio is inelastic and this proportion is surmised to be lower in the USA than the UK. Fourthly, there will be a much larger proportion of discounted subscriptions for society members. Because of the differential pricing policies adopted, prices charged to members may not be sufficient to cover the run-on costs (ie paper, printing and distribution) especially when, as is the case now, these are inflating fast. This turns conventional publishing economics upside down so that large circulations become a liability rather than an asset. In summary therefore, economic pressures may be greater for American than British societies simply through the effects of size.

Conversely, ignoring the effects of 'donated' costs (voluntary editors etc), British societies should have a higher proportion of pre-run than run-on costs. EPCs affect almost exclusively the pre-run costs and would have a greater potential for smaller societies. However, the greater use of voluntary labour will make implementation more difficult, technically, socially and economically.

The other major difference in learned society publishing practices between the UK and the USA is the imposition of page charges in the latter country. This practice not only has direct effects on the feasibility of the EPC concept in the UK and USA, but also has subsidiary effects which themselves would have an influence on the viability of EPCs, particularly in the UK.

Although page charging had been practised by some societies (notably the American Institute of Physics) since before the Second World War, it was the ruling in 1961 by the Federal Council for Science and Technology allowing the payment of page charges for publication of work generated under Federal research and development contracts to be offset by the conditions of grant that opened the way to widespread use of page charging to underwrite the cost of producing society-published journals. In the UK, the Royal Society refused to allow page charging on the grounds that this would impose financial discrimination on the publication of research. One of the probable effects of this action was that American societies were able to remain economically viable with the subscription price tending to cover just the marginal costs.

With respect to the adoption of EPC-type operations, the page charging system means that there is a direct method of influencing authors to produce OCR copy by offering a reduction in page charges. However, since a large proportion of page charges levied on authors is directly offset by grant income, it is not clear how effective a lever this might prove to be. In the absence of page charging in the UK, this method of attempting to influence the kind of copy that is received from authors does not exist. Furthermore, even with financial inducements in the form of reduced page charges, not all authors would be able to produce acceptable OCR copy and, because of the non-standard nature of the input and the re-typing required, delays would be experienced in the processing of such material in an EPC, thus doubly penalising authors without access to good OCR copy production. A further direct effect of the page charging system is that the US Federal Government, through the grant-aiding of page charges to authors is effectively subsidising a large part of American learned society publishing and thus has a considerable financial stake in the process of research publication. This means that the US Government, through the National Science Foundation, is in a much more influential position with respect to journal publishing than is the case in Britain and it can justify a substantial investment in the development of EPC-type ventures if it can thereby save money on grant -aided payment of page charges.

There are two other differences between the UK and the USA which could be indirectly associated with the page charging system although other factors are undoubtedly at work. Firstly, there is a much greater use of commercial publishers by UK learned societies (47% vs 30% in the US in our two surveys). There are several reasons why this is so, one of them being with page charging outlawed in this country all revenue must be obtained from membership or journal subscriptions. There is inevitably a greater financial risk for the societies in this situation; the answer here is to use the services of the commercial publisher who is prepared to take the financial risk and can promote the journal to non-members, especially libraries and institutions. Another major reason relates to the structure of learned societies described above where the editorial services of the society are financed from within the society or outside voluntary editorial assistance is used. The production side of journal publishing is viewed as a revenue-raising process, balancing journal subscription income against production costs. In this situation, farming out the entire process to an outside contractor who takes the risks and provides a proportion of the proceeds is an attractive proposition. In some respects, the use of commercial publishers in this way pre-empts the need for EPCs as the commercial publisher is better able to take advantage of innovations in the printing industry (and elsewhere) and offer the advantages of large-scale processing, professional staff and organised marketing and distribution.

The other difference is the greater international authorship of journals published in the UK. The UK has traditionally always had a large international influence and prestige, particularly with the Commonwealth countries. This tradition is possibly reinforced by the page charging system in the USA as authors from developing countries and elswhere, who are unable to pay page charges and who wish to publish in prestigious journals, will turn to UK society journals or purely commercial journals. It would, of course, be much more difficult to obtain good OCR copy from foreign contributors and there would be much less scope for interactive communications with authors, and this would further restrict the scope of advanced or maximum EPC ventures. Moreover, honorary editors and referees for UK-produced journals also tend to be more internationally based, causing further complications in communications.

Viability of EPCs

Taking account of the attributes of UK scholarly publishing discussed above an outline systems design for an EPC-type operation was produced. In this design, authors are not required to submit manuscripts in OCR although OCR was used as the input technique within the co-operative centre. Broad costs for development, establishment and operation were computed and compared with production costs of a generalised, hypothetical journal produced as one of a stable by a large learned society. The analysis showed that the break-even point between conventional and EPC costs occurred at a throughput of around 11-12000 pages per annum. Furthermore, such a centre would be able to 'contract out' a complete publishing service to smaller societies at rates comparable to those charged by commercial publishers. In these studies, a deliberate decision was taken to ignore specific problems of system design detail, such as coding for extended character sets, full page make-up, display conventions and compatibility between different suppliers or operators, since these problems on the whole are in the process of solution and relatively susceptible to 'technological fix'. Each situation will of course have a different set of these problems depending on the type of system adopted. The problems that these details pose are considerable and the solutions adopted would have to be very carefully evaluated. Similarly the problems of participants in, organizational structure of and financing of EPCs were by-passed as it was felt that each situation would be determined to a great extent by the pressure of societies wishing to pursue such a course. Interested societies must, of course, determine for themselves the participants, organization and financing of such a venture together with the appropriate control and review mechanisms. What this design exercise served to show was that the problems of implementation of EPC-type operations are probably not those of technology or economics but rather those of human engineering, organization of work processes and social acceptability. In particular, the major savings that can be achieved using an EPC-type approach arise

from the automation of typesetting rather than in any other area.
Other aspects of the publishing process may be more expensive if EPC-type procedures are used. Computer typesetting viewed as just a different form of composition technology, means that it can be purchased from a printer offering this method of typesetting at an economical price (but competitive pricing may upset the cost advantages of different technologies). Computer typesetting as purely a means of producing printed versions of manuscripts must be viewed in comparison with typewriter composition or other cheap methods and the advantages carefully evaluated. For computer typesetting these include good graphic quality (almost letterpress quality) and possibilities of using the text in other ways as well as increasingly competitive costs.

As the Americans emphasize, the earlier the data is captured the greater the cost benefits which are likely to accrue in the long-term. This is a reflection of the greater efficiency of on-line editing and the use of data contained within the manuscripts themselves to control editorial and production functions. Furthermore, printing is increasingly becoming just another output mode from all types of data processing and it is possible to envisage 'publishing' houses or divisions which process text by computer as a prime objective with printing or reprography as one output or even a by-product. However, we must not lose sight of the fact that the major difficulties with EPCs will occur in the pre-production stages. There are several reasons for this but the most important are that this area largely involves intellectual assessment and discussion, that the work is often voluntarily undertaken and, on a detailed level, there is little logical correspondence at the moment between machine-readable inputs or printouts and the final printed version. The imposition of an EPC within the communications net of authors, editors, referees, copy editors and printers is likely to complicate these links. Furthermore, there would be at the present stage of development, an increased need for referees and editors, who are often voluntary, to have access to specialized equipment which would limit the role that these people could play in the transmission of scholarly information.

Nevertheless, data processing is becoming cheaper, easier to use by the individual and can be expected to form a part of personal office systems in the future. Totally electronic methods of communication may become both practical and cheaper than paper/postal systems. This suggests that publishing operations should be evolving systems to handle these developments in the future. More immediately, other economic effects and changing attitudes by purchasers and users mean that different forms of dissemination of scholarly materials may assume relatively greater importance than hitherto and result in a desire for greater flexibility in the way that publications are purchased and processed. This in turn will demand a greater flexibility on the part of publishers to cater for different markets. The availability of texts in machine-readable form at an early stage of the publishing process could mean that the same material is published in several different formats. For example, microforms may be economical for journals with an expected low use in an individual library whereas the paper version is necessary elsewhere; but the optimum format for microform is somewhat different from the optimum paper format, and the availability of text in machine-readable form would allow re-processing of the text to produce both versions. Similar arguments apply to the production of synopsis journals, repackaged journals, selective dissemination of documents according to profiles, and the production of machine-readable data files for sale.

Acknowledgements

The author is indebted to Miss G Yska and Mr J Martyn who colloborated on these studies and performed many of the analyses. In addition, members of the publishing departments of several learned societies contributed time and data to the study.

REFERENCES

1. Bamford, HE, Jr, 'A concept for applying computer technology to the publication of scientific journals'. J. Wash. Acad. Sci., 6 (4) pp 306-314 (1972) also 'The editorial processing centre'. IEEE Trans. Prof. Commun. PC-16 (3) pp 82-83 (1973)

2. Aspen Systems Corp, Editorial processing centres: feasibility and promise. Washington, DC: National Science Foundation, Office of Science Information Services. 70 pp (1975)

3. Woodward, AM, Editorial processing centres: Scope in the United Kingdom. London: British Library Research and Development Department. BL R & D Report No. 5271 HC, 31pp (1976) ISBN 0 85350 139 4.

EDITORIAL PROCESSING CENTRES AND INTERNATIONAL SCIENTIFIC JOURNALS

A.A. Manten, *Senior Editor, Elsevier Scientific Publishing Company, Amsterdam, The Netherlands*

ABSTRACT

International scientific journals differ in several respects from American journals: smaller average size, wider international distribution of editors-in-chief, referees and editors, and more linguistic complications. These differences make the setting up of editorial processing centres (EPC's) for international journals more complex. The concentration of many international journals with large publishing houses, the existence of centralised editorial secretariats, available experience in automated journal business management and a high percentage of pre-run costs, are factors which, on the other hand, may encourage the development of some sort of EPC configuration for international scientific journals.

INTRODUCTION

The work done in the U.S.A. on the development of Editorial Processing Centres (EPC's) (Bamford, 1973; Berul et al., 1974; Aspen, 1975) has stimulated people elsewhere to study the prospects which an EPC might offer in their publishing activities.

Woodward (1976) examined the scope of the EPC for Great Britain. He noticed a number of differences in journal operations between the U.S.A. and Great Britain and his design of an EPC which might be feasible under British conditions has only about the level of complexity and automation of the minimum EPC configuration designed by Westat/Aspen in the U.S.A.

DIFFERENCES BETWEEN AMERICAN AND INTERNATIONAL JOURNALS

An investigation to assess the feasibility and economic viability of EPC-type operations in the publishing of international research journals must start with an inventory of differences that exist as compared to journal publishing in the U.S.A. The next step is to establish what modifications these differences would necessitate in published EPC configurations. Only after this has been done, can the technical and economic practicability be realistically reviewed. The present article will concentrate on the first phase, and list aspects specific to the publishing of international journals.

Journal size

The American studies make a distinction between small and large journals (see, for instance, Aspen, 1975, p.35). They regard as a typical small journal one that publishes 156 articles per year in 1210 pages, and as a typical large journal one printing 540 articles in 4169 pages.

An examination of international journals reveals that, measured against American standards, the great majority of international journals must be characterised as being very small. Over 70% of the journals handled by the Elsevier/North-Holland group in Amsterdam are smaller than the typical small American journal, and only about 3% are of the same size as the typical large

Miriam Balaban (ed.), Scientific Information Transfer: The Editor's Role, 21-24.

U.S. journal. With the journals published by Springer Verlag the situation is not essentially different.

This means that an EPC for international journals could, with a few exceptions, only be considered for a fairly large group of journals, even in the case of a minimum EPC system.

Pooling of journals

The EPC concept is clearly based on the assurance that journals which combine their operations in an EPC need not submit to unwanted standardization, nor sacrifice any of the autonomy which editors and publishers otherwise enjoy. Nevertheless psychological obstacles that may be met in attempts to bring thusfar separately operating journals together should not be underestimated.

In the domain of international scientific journals there are a number of publishing houses which publish many such journals. This may prove to be an important organizational advantage.

Editors-in-chief

The editor-in-chief of an international journal need not be a resident of the country in which the journal is published, and in actual practice very often is not. A group of 16 Elsevier journals in the agricultural sciences published in Amsterdam shows the following geographical distribution of their editors-in-chief: 4 Great Britain, 4 The Netherlands, 1 Belgium, 1 Italy, 1 Canada, 2 U.S.A., 3 Australia.

In some cases the editorial secretariat is located with the editor-in-chief. In other cases most of the administrative functions are delegated to a centralised secretariat in the publisher's headquarters. This normally works very well, though it often implies that decision-making stages in the processing of submitted articles involve long-distance communication.

One of the study projects in which I am currently involved concerns the short-term possibilities for computer assistance in the operation of editorial secretariats for international journals.

Referees

The referees used by international journals are normally distributed all over the world. Table 1 gives a picture of the situation in 1976 with one large journal and two groups of smaller journals. In each of these three cases the total numbers of papers processed is large enough, according to published U.S. criteria, to warrant an EPC.

A wide geographical distribution of referees serves a number of purposes: (1) it underlines the truly international character of the journal; (2) it helps to adjust the contents of a journal to the demands of a world-wide readership (e.g.by proposals to make authors provide more background information, clearer definitions, more broadly based discussions, etc.); (3) it may provide the editor-in-chief with elements in the judgement of papers that require some knowledge of regional conditions; (4) it increases the chances of article submissions from particular parts of the world; (5) it ensures that the content of a journal does not become controlled by a scientific "elite".

Authors

The geographical distribution of authorship of papers published in international scientific journals shows a similar sort of picture to that shown for the referees.

In 1975 a group of 12 international agricultural-science journals received 16% of its published articles from Great Britain and Ireland, 29% from continental Europe, 25% from North America, 13% from Asia, 11% from Australia and New Zealand, 5% from Africa and 1% from Latin America.

The difference between authorship in international and national journals becomes clear when we compare these percentages with the following figures. A study of a sample of journals in the agricultural sciences showed that the national U.S. journals in the sample contained for 95% articles by U.S. authors; of the remaining 5% about two-thirds came from Canada. The

TABLE 1
Geographical distribution of referees employed by international primary journals (in percentage of total)

Geographical area	Biochimica et Biophysica Acta	23 geo-science journals	16 agric. science journals
Great Britain and Ireland	20.1	9.9	9.7
Nordic countries	4.6	5.9	4.2
Low countries	7.4	5.9	10.7
France	3.0	3.5	5.7
Fed. Rep. Germany	4.9	6.2	8.1
Switzerland and Austria	1.7	2.6	2.8
Southern Europe	1.0	2.7	3.8
Eastern Europe	1.3	5.4	8.3
Israel	1.3	1.6	1.6
India	0.1	0.8	1.8
Japan	1.3	4.0	2.8
remaining Asia	0.2	0.8	1.6
Africa	0.2	1.9	4.2
Canada	2.7	7.5	4.2
U.S.A.	47.0	32.5	18.6
Latin America	0.5	1.3	3.0
Australasia	2.7	7.5	8.9
Total of referees	2255	627	504

British journals contained for 70% articles by authors from Great Britain; 17% came from authors in Commonwealth or former Commonwealth countries; 13% was by authors from a great diversity of nationalities in Europe and elsewhere. The national French journals had for 93% articles by French authors. National journals in the Federal Republic of Germany had 78% domestic contributions, those in the German Democratic Republic 90%.

Authors and referees in a great many countries have more difficulty in producing typescripts in optical character recognition (OCR) type than U.S. authors. There are still far lesser possibilities for an interactive computerised communication between editor (or editorial secretariat), referees and authors. This restricts the perspectives for the introduction of advanced or maximum EPC ventures in international journal operations.

Distribution

It is likely that the differences between international and national journals will be smaller with regard to the geographical picture of journal circulation than they are in the distribution of authors and referees, although such differences do exist. However, these will have little effect on the feasibility of incorporating business management in an EPC operation. Larger publishing houses, moreover, already have quite some experience in computer--based journal business management.

Economic aspects

Most scientific journals in the U.S.A. are society journals. In addition to library subscriptions these normally have a substantial criculation to private scientists. International journals are very largely independent journals, and if they are sponsored by an international learned society and that society is not a federation but one with individual membership, the membership fee of the society only includes a subscription to the journal in exceptional cases. International journals, therefore usually find a very high percentage of their subscribers in institutional libraries. Library copies of scientific journals have a high average readership. In consequence, however, the quotient of the number of journal users and number of subscriptions placed is generally less favourable for international journals than for national society journals. In recent years this situation has been further aggravated by the widespread photocopying of journal articles.

International journals do not usually receive any substantial income through page charging. Most international journals do not practise page charging at all, and those which do, generally restrict this to asking some contribution to the costs of printing lengthy articles or to cover the costs of pages published in excess of the budgeted number of pages per volume.

Society journals often have a greater benefit of donated income (voluntary labour, grants, etc.) than independent international journals. In this respect, the latter category of journals is therefore likely to be the group which has to bear the highest budgetary burdens of all.

As a consequence of all this, a high percentage of the costs of international journals are pre-run costs, that is costs incurred between submission of articles for publication and the beginning of the actual printing of the journal issues. An EPC has a much greater effect on pre-run costs than on run-on costs. As such, an EPC might thus have a greater potential for international journals.

Linguistic matters

American journals are virtually monolingual throughout their entire contents. International journals are more complex from a linguistic point of view.

Quite a number of international journals accept papers for publication in any out of two, three or even more official languages, or provide abstracts in two or more languages.The author's address to be used in correspondence may in another language than the address printed in the article heading. Papers may contain quotations in another language. Bibliographies added to the papers may list titles in a great variety of languages.

Special problems originate from the fact that many contributors to international journals have to write in a language that is foreign to them, often causing the need for linguistic correction by the editor or publishers' staff.

CONCLUDING REMARKS

Much more work still needs to be done to explore the applicability of EPC's in the editing and publishing of international journals, and to study specific problems in greater detail.

On the basis of current insight, it seems likely however, that, if further experiences in the U.S.A. prove the viability of EPC operations for national American journals, the EPC concept could well also be applied to international journals.

The EPC configuration to be adopted is likely to differ from those described in American studies. The advanced and maximum configurations in the Westat/Aspen reports (e.g.Berul et al., 1974; Aspen, 1975) are difficult to apply to journals with a world-wide distribution of authors and referees. Computerisation offers the best short-term prospects for all those handlings performed in the offices of editorial secretariat, publisher and typesetter.

REFERENCES

Aspen Systems Corp., 1975. *Editorial processing centers: feasibility and promise.* Division of Science Information, National Science Foundation, Washington, D.C., 70 pp.

Bamford Jr., H.E., 1973. The editorial processing center. *IEEE Transactions on Professional Communication,* PC-16(3):82-83.

Berul,L.H.,King,D.W.and Yates,J.G.,1974.*Editorial processing centers: a study to determine economic and technical feasibility. Final report.*Westat In and Aspen Systems Corp.,Rockville,M.D.,Report PB 234 959, 111 pp.

Manten, A.A., 1976. *Editorial processing centres: a general review.* Elsevier, Amsterdam, internal report, 39 pp.

Woodward, A.M., 1976 . *Editorial processing centres: scope in the United Kingdom.* The British Library Research and Development Report 527 1 HC, British Library Lending Division, Boston Spa, Wetherby, 31 pp.

EDITING A PHOTOGRAPHICALLY REPRODUCED JOURNAL

Earl J. Taft
Editor, Communications in Algebra
Department of Mathematics, Rutgers University, New Brunswick, N.J.

This is not the formal paper that I have prepared. I will give it on Thursday as scheduled. These are a few extra informal remarks that I was asked to give at the last moment in this session on new techniques.

I am speaking here as a mathematician and as editor of *Communications in Algebra*, a journal produced by photographic reproduction directly from the manuscript of the author so that my remarks are going to be somewhat different from the others this morning.

Now I don't presume to be an expert on the various problems in publising and reproducing mathematics. There are of course very important and tricky problems connected with that, but I will be speaking mainly as the editor of a photographically reproduced journal. In the title of this session are the words "new technique". This is hardly a new technique in scientific publishing. In fact it is an old technique, but I think I should point out that from the point of view of mathematics, at least mathematics journal publication, it is not such an old technique. There are comparatively few mathematics journals that are published in this manner. I don't know the exact number, but it is certainly very small compared to the total number of mathematics publications. So it seems to be reasonable in today's session to perhaps try and explain in a somewhat informal way why this might be, why mathematicians have perhaps resisted to a certain extent, this type of publication and perhaps at the end make some guesses or comments about what might happen in the future.

Now I'm going to be discussing the differences between mathematics and other types of scientific publication and some of these remarks will probably be misunderstood but it is the best I can do in a short time. President Katzir in his introductory remarks, remarked that prior to formal scientific journals, the traditional method of communication was for scientists to send letters to each other describing what they had done. This is in a certain sense not so changed in mathematics today. There are some mathematicians - fortunately not many - but some quite good ones actually, to whom this is still a standard method of communicating. They really don't believe in formal publication, they don't want to be bothered with it, and when they do find something, they prefer to write it up perhaps even in letter form and send it to colleagues whom they consider interested in the subject. A modern equivalent would perhaps be to xerox some informal scribblings and send it off to people. That would suffice for some people, certainly not for very many. But it gives you some idea of the attitude of mathematicians.

Now another feature of course you might say is that there are fewer mathematicians than there are scientists. That is perhaps true, but there is still quite a large number of them, and mathematics publications are growing. I believe this same type of technical problem is faced by everyone. And I con't think that that is a particular explanation of the resistance of mathematicians to new techniques. Basically I think that the difference can be

Miriam Balaban (ed.), Scientific Information Transfer: The Editor's Role. 25-27. *All Rights Reserved.*

explained on an unfortunately philosophical basis, and these type of remarks are hard to present in five minutes. It should probably take five years to prepare such remarks, but basically mathematics is an abstract science, and in perhaps distinction with experimental science, the nature of mathematical reasoning and the nature of the final product is regarded as a more permanent affair. That is, mathematics is somehow out there and one is going to do it, and once you have done it, it is then in a form that is going to stay that way.

Now this is, in some simplified way, in contrast to results in the physical sciences where you do an experiment, you observe something, you make an explanation - it may be even a mathematical explanation - and then you predict and so on. You see whether you think it is right or wrong. You think it is right for a while, then in the future it is going to change, or it will be considered wrong and there will be a better explanation. So in a very very simplified way, this is some explanation of the mathematician's view of his subject as a permanent one, as opposed to an experimental scientist's view of the subject as a somewhat temporary one.

I'm not trying to say that one is more important than the other. It is a sort of difference in kind or nature. Anyway, the result of this is that a mathematician will insist that they are going to write down what was at the final moment the theorem, and that it should be published in what they consider a permanent form.

Now as the editor of a photographically reproduced journal you see one meets resistance to the idea, and it is perhaps somewhat irrational. I believe that the idea that photographic reproduction from typing is not as permanent as traditional typesetting, will somehow not last as long. Do not misunderstand me. I'm not talking about questions of taste for example. Certainly in mathematics as in other fields, some things are considered more important than others. There are fads at the moment. Some subjects will lie and will not be studied for centuries and then be taken up again, but I'm not denying this. It is also not a question of whether the material is right or wrong. Naturally when an author has done something, he believes it is right, and it of course happens frequently that it is proven wrong later. I am assuming that at the moment everyone thinks that what he has done is right and therefore deserves a permanent form.

Besides editing *Communications in Algebra*, I am mathematics advisor for the mathematics series of the publisher Marcel Dekker. Now there are basically two other types of operations. One is lecture notes in mathematics and the other, of course, is a normal series of books and monographs. Now a good contrast I think would be the attitude towards lecture notes. Now the term "lecture notes in mathematics" to most people implies something temporary in contrast to books or journals. A mathematician regards these as some sort of permanent record of what they've done. The lecture notes are quite popular among mathematicians. Springer Verlag has for many years an enormously successful series of lecture notes and is considered by mathematicians to be a service to them, the idea being that they will produce very quickly results of conferences, seminars, lecture notes in the usual sense, things that are not considered in a final, permanent form but that one would like distributed quickly to one's colleagues. Everyone knows that by the time the paper actually appears, several years have gone by, most people have known about it anyway and are on to other things, so the idea about lecture notes is considered a very valuable service to the mathematical community. The lecture notes series of course are photographed directly to accomplish the purposes that I just mentioned.

I've told you some reasons as to why perhaps some mathematicians would resist photographic offset. I'll talk more on Thursday about my general experience with editing the journal from other points of view, but let me say here that of course the good features in practice I feel have weighed more heavily in the minds of most authors. Most authors feel that because it comes out more quickly, it is cheaper. For mathematicians it is mainly a matter of time, getting the results out more quickly. Most people overcome their resistance to the photographic method. I don't know how many authors didn't send me papers because they didn't like the process but I can say in looking at other journals that are traditionally typeset, that we certainly get the same type of papers as they do and it has generally been a successful operation. Mathematics was the last field of the sciences that my company attempted. They had had many years of publishing such communications - photographed journals in other sciences - chemistry, biology, medicine, more recently in physics. Mathematics was the last field that they tried, perhaps a bit skeptically. There has since been another one, doing very well on partial and differential equations. A new one will start next year in numerical analysis. So they apparently regard it as a successful procedure.

So as far as the future goes I would think that photographing mathematics journals would become a more and more widespread popular thing. Naturally I cannot predict the general trends whether people are going to revolt against ecology or whether it is inevitable, I have my own feelings that are irrelevant about this and I think mathematicians will go along with them in general. They do want their things to come out very very early and they do want them to appear in a nice and what they regard as a permanent form. In general I will welcome more technical advice about the particular problems in publishing mathematical manuscripts.

THE DUAL JOURNAL CONCEPT

D.H. Michael Bowen, *Director, Books and Journals Division, American Chemical Society, 1155 Sixteenth Street, Washington, D.C.*

About 18 months ago, the American Chemical Society received a grant from the National Science Foundation to undertake an experiment concerning the so-called dual journal.

The background to this is rather simple. Like everybody else, we were subject to economic pressures, and we knew that composition was one of the single most expensive steps we go through, so we wanted to prove in practice that anything that diminishes the amount of composition in a journal was likely to save money. We also knew, as everybody else does, that the traditional journal of four rather long articles - all typeset - apart from being expensive, was a rather inefficient way to transfer information to many sorts of readers though very efficient in transferring information to libraries and to large numbers of readers.

We wanted to draw on previous experience we had in the ACS with miniprint. When we started the experiment we had had quite a few years of experience with microfiche and microfilm, and like other societies and publishers, discovered that readers do not care for microform. Therefore, in formulating the concept and designing the experiment, we decided that microforms would not be part of it.

What we did was to publish two concurrent versions of the *Journal of the American Chemical Society*, our biggest primary journal with a circulation of some 15,000 world-wide. One was a summary version with typeset summaries of about 2 pages for each paper. The second version is the archival version which contains the typeset summaries and the entire manuscript in author-furnished form reduced to a print size half way between miniprint and full size reproduction. These two versions embody the dual journal concept.

We published three consecutive issues of the *JACS* in its regular form together with the summary and archival versions. To get as much factual and quantitative information as possible about the reception of the new format we sent out questionnaires, and our staff interviewed four groups of people. The first and second were authors and readers, both subscribers and non-subscribers to the journal. The third was referees chosen from the reviewer files of the editors of the *JACS*. A fourth group was librarians because they would be expected to be impacted by any change of this sort both by the relative complexity bibliographically of having two versions of the same journal and of course because they would be expected in large part to buy them.

Now to summarize the results. We substantiated the reading patterns of basic journals that we've seen for years and confirmed that nobody reads a big journal like the *JACS*. In fact very few people read all the way through more than one or two articles. Now there are a number of questions posed to the different groups above with respect to converting the existing journal to the new format. Our journal has a large number of subscribers very very zealously supporting it for a long time. We thus expected strong opinions

Miriam Balaban (ed.), Scientific Information Transfer: The Editor's Role, 29-30.

and we got strong opinions. We learned as much psychologically as we did technically in the study, and the staff was subjected to a certain bombardment of rather hostile questions such as "why are you wasting money doing this experiment?", but we learned quite a bit.

The group that most liked this format was non-subscribers! They opted 78% in favor. (The way the questions were phrased is of course very important. The questionnaires were prepared by a professional firm experienced in these things, and I hope the questions were not slanted in such a way as to bias the respondents' response.) The group that was most against conversion of the journal to this format was the reviewer group. Only 32% of them liked the idea and 66% did not. This is largely due to their feeling that to read both a summary of a paper and the paper introduces an added element of complexity to the voluntary and unpaid process of review. As all you editors know, it is difficult enough at the moment to get people to referee papers. With the new format, they would have to gauge not only the scientific merit of the paper but also whether the summary was a good summary of the entire paper. They didn't like that.

Recategorizing the types of persons who liked the new format, we found that non-United States faculty members graduate students all over liked it. An age analysis of the respondents showed that the younger he was, the more he liked the idea, and the older he was, the more he disliked it. And this was a very strong trend.

A very important aspect with authors and reviewers of course is, were the journal converted, would they continue to submit papers. 44% of the coterie of *JACS* authors said that they would be less likely to submit manuscripts and 46% of the reviewers said they would be less likely to referee papers. Although it's difficult to say what this really means, clearly people were trying to tell us , "do not convert this journal into this format."

As I mentioned before, the librarians felt that essentially they didn't like the bibliographic aspects, and they didn't like the fact that the storage costs for the archival version would probably be more since it would contain more pages. Although it costs less because there is less typesetting, it uses more paper. Also librarians, like chemists, are resistant to change.

In summary. We did prove (through spending a lot of money in 3 consecutive issues!) that you can save 25-30% production costs by going to this concept, but we did encounter enough opposition to conversion to rule it out for ACS. We did, however, determine a very considerable interest in summary journals. People liked the summary journal though not as a part of the entire package. Therefore one of the avenues we are going to explore is summary journals.Of course a summary journal by itself, outside the dual journal concept has a different set of economics because these are new and typeset summaries. In our original concept, the cost of typesetting is shared between both versions, but we are looking into this.

The *Journal of Chemical Research* published by the British Chemical Society on behalf of the British, French and German chemical societies is a little different than our experiment in several respects. For one thing, it is a new journal begun this year and not a conversion of an existing one. I believe that was considered, but probably, although I don't know the inside stories, the same forces came into play. The supporters of the existing primary chemical journals in the three countries said"I don't want my -ournal to be converted", and the new one was started. The archival version there is either miniprint or microfiche.

INNOVATION IN THE PUBLICATION OF LENGTHY RESEARCH ARTICLES

A.A. Manten, *Senior Editor, Elsevier Scientific Publishing Company, Amsterdam, The Netherlands*

ABSTRACT

It is possible to save much space in the publication of lengthy research articles by conventionally printing only a synopsis of the article and reproducing all further information in miniprint or on microfiche. An early enquiry provided indications about what parts of the article need not be traditionally reproduced. Authors' attitudes play a major role in deciding what can be put into practice. An experimental publication, containing a research report in the synopsis plus miniprint mode is being used as a basis for exploring what is both practically feasible and acceptable.

INTRODUCTION

The concept of synopsis publishing implies a separation of a scientific paper into two items: (1) a synopsis, that is a brief version of the article, containing the most essential information, and able to stand alone; (2) the full paper, reproduced in miniprint or microfiche form. The synopsis is therefore an additional element. The fact that publication of the complete package requires less space is due to the technically reduced reproduction of the full paper.

It is easy to understand that the achieved saving in space is greatest in the case of long papers. Even if the miniprint alternative is chosen, the total space required by the full package may be reduced by about 40% for a 10.000 words article, to about 60% for a 30.000 words contribution.

My studies have, therefore, concentrated on possible innovations in the publication of lengthy articles.

AN EARLY ENQUIRY

In 1970, that is before the present concept of synopsis publishing was born, I made an enquiry among an international group of scientific authors to find out whether they would accept a system in which longer scientific articles were published in two parts: (1) a conventionally printed part, containing the most essential elements of a lengthy contribution, (2) a microfiche, containing the remaining parts of the article. Both parts would be included in the same journal issue and be distributed to all subscribers.

The enquiry involved scientists who were members of the editorial advisory board or active contributors to three international journals which used to receive many long papers (over 6000 words, illustrative material included), viz. *Geoderma,* an international journal of soil science, *Sedimentary Geology* and *Review of Palaeobotany and Palynology.*

Motives

The idea of publishing long articles in this fashion was based on a number of considerations, which were:

(1) There was increasing pressure from various sides to give preference to

Miriam Balaban (ed.), Scientific Information Transfer: The Editor's Role, 31-33. *All Rights Reserved.*
Copyright © 1978 by D. Reidel Publishing Company, Dordrecht, Holland.

the printing of shorter papers. This might cause authors to divide their research results over a greater number of papers, appearing in different media, making access to the total information more difficult. Authors who kept their information together in one long article were sometimes confronted with longer publication delays than was the case for shorter papers.
(2) Readers more frequently complained about the continuously increasing flow of primary publications and demanded that greater attempts be made to control this flow.
(3) It had become clear that readers of primary journals generally read only parts of the published articles, notably the abstract, a definition of the studied problem, essentials about the methods used, and the general conclusions reached. It would help readers if these portions were more clearly separated from the remaining parts of scientific articles, which were mainly published because such information requires public storage.

Content of microfiche

The enquiry showed that readers in a great majority (85%) would favour receiving long articles as a combination of a shortened article and a microfiche.

Information commonly mentioned as suitable for removal to a microfiche supplement included:
(1) A review of earlier work done on the problem field in question.
(2) Text and tables about material studied, giving quantitative and qualitative detail, particularly if these parts are elaborate.
(3) Detail about techniques and methods, insofar as these are not essentially new, are not the object of scientific discussion, or contribute no new arguments.
(4) All circumstantial background information which is only important to those few colleagues of the author who concern themselves with a similar geographical area, the same soil types, directly related groups of fossils, or the like.
(5) Results obtained, particularly all that is assembled in tables, and the discussion thereof.
(6) Literature resources other than the most important ones, particularly if the bibliography comprised many titles.

Furthermore, it was suggested that certain information might be added to the microfiche supplement, which is commonly eliminated from conventionally published articles, such as:
(7) Tables listing the original individual observations, in addition to tables which are usually published and show totals, averages, highest and lowest values, etc.
(8) Large maps.

Objections

There were also several objections against the proposed procedure. These were to a large extent the usual complaints about microfiche: need of a reading apparatus, risk of fiches becoming lost, poor reproduction of halftone illustrations, etc. A warning was made that some authors might try to abuse the system by taking a lesser effort to eliminate non-essential information and redundancies from the part of the paper to be reproduced on microfiche.

Authors attitude

The 1970 study revealed a notable difference in attitude between readers and authors, although these were often embodied in the same individuals. When asked whether they would accept the construction of a brief printed article plus a microfiche supplement for their own longer papers, or would first explore the possibilities of getting their full paper conventionally published elsewhere, a significant number of authors (approx. 25%) replied they would take the latter course. Among them were many authoritive authors. Approx. 7% gave no opinion.

It was this authors' attitude that led at that time to my company deciding to shelve the project. In a world-wide scientific community in which many learned journals co-exist in competition, there is a risk that those journals who prematurely embark on an experiment lose quality of content and thereby part of their acquired international prestige and circulation.

SUBSEQUENT DEVELOPMENTS

In the nineteen-seventies a variety of reasons, including not in the last place major economic problems in scientific publishing, gave rise to the concept of synopsis publishing. Several learned societies in both North America and Europe did pioneering work in this respect. Their activities are likely to influence readers' and authors' attitudes in primary publishing.

My own work in these years was extended from publishing mainly for an academic market to also providing publishing services for certain of the learned professions.

Primary literature production in the professions

Investigations made in the learned professions most usually result in reports, which receive only a very limited circulation. Most authors are so busy that they rarely find the time to subsequently work up the information in their reports in the form of journal articles or conference contributions. A certain proportion of the report literature is of far more than incidental importance. The problem is of how to get this information to the people who need it.

One possible solution is to produce a synopsis of an important report, backed-up by the relevant parts of the full report in miniprint or microfiche. An experiment in this context has recently been made. It involves a report in synopsis and miniprint form (Jones et al., 1976), preceded by an explanatory article and enquiry form (Manten, 1976). The experimental publication has also been used as a basis on which to carry out a new enquiry among scientists working in fields in which lengthy articles are commonly produced, notably the earth, life and agricultural sciences. Reactions are presently coming in. A survey of the outcome of the experiment and the additional enquiry will be published in the journals *Landscape Planning* and *Journal of Research Communication Studies*, respectively.

CONCLUDING REMARKS

It is hoped that the combined results of the early enquiry and the more recent studies mentioned will now lead to devices which are applicable in practice to the publication of lengthy research papers.

Probably these devices could also be of use in finding alternatives to conventional monograph publishing, which is presently under severe economic restraints.

REFERENCES

Jones, G.R., Ady, J. and Gray, B.A., 1976. Scenic and recreational highway study for the state of Washington. *Landscape Planning,* 3:151-302.

Manten, A.A., 1976. Experiment in synopsis publishing. *Landscape Planning,* 3:147-150.

ON-LINE PREPARATION OF TYPESCRIPT

U. Bloch and K. Shmueli
P.O.B. 962, Haifa, Israel

ABSTRACT

A small NOVA computer is used to prepare reports. The typed "pages" are stored on a floppy disk and printed on a Versatec 100 points/in. electrostatic printer plotter. The system serves 2 typists but can be expanded to many more stations. Among its special features are intermixed English and Hebrew letters, choice of different type fonts, fixed page background, an option to overlay on one page curves and graphs which the computer has prepared in a separate Fortran computation, merging of 2 separate files, etc. The stand-alone cost is about $40,000 + $2,000-4,000 for every typing station.

Further development envisaged includes: line and page spillover, expanded type set of 256 different symbols (128 now), possibility to print pictures overlayed with printed data using an optical scanner as the input device with 4 to 8 shades of grey.

Actual operation has proved remarkably failure-free (the printer is believed to have an MTBF of 5000 hrs. and typists have easily taken to the new station.

Since keyboarding is a very labour-intensive operation, it is one of the more expensive operations in publishing. This is especially so where highly professional workers are employed, but even with simple typists it is not negligible. Roughly a third of the average typist's time is expended on error correction (which consists not only of the actual overtyping but also of entering the paper/matrix and positioning it, erasing the error and counting letters to obtain an eye-pleasing output).

It is natural to see whether that indefatigable worker - the computer - can help with this problem by saving us some of the effort. We use a terminal, connected on-line to a NOVA minicomputer, as an effective typewriter. The typist types at the terminal (the computer need not be near but in any case it takes up little space) and sees what she is typing on the screen. Errors are corrected simply by overtyping. The finished page is stored on a (floppy) disk. After the whole manuscript has thus been keyboarded it is printed onto paper (at 6 pages/min) and given to the author to proofread. His corrections are entered very simply (instead of erasing a code is punched for deletion, entering, changing, etc.) thus cutting the error correction time to a fraction of taht needed with regular typewriters. When the final draft has been approved, small runs are run directly off the printer thus saving also the cost of collating the pages. Alternatively a single copy is produced which is then photographed onto a paper/aluminium plate and sent to press.

The process decribed above is not new and commercial systems have been available for many years (although earlier models were limited by the printer - a fast electric typewriter). Our system is characterized by a more sophisticated approach.

Our main departure from the well-trodden path is in using an electrostatic printer-plotter as the output device. These use a comb of closely

Miriam Balaban (ed.), Scientific Information Transfer: The Editor's Role, 35-36.

spaced needles (100/inch), which charge electrostatic paper which is then developed just like in any electrostatic copier. The letters or indeed any pattern are made by charging the needles in the correct sequence as the paper passes beneath the needles, that is they are software created. Therefore by simply changing the software, one can change the typefont or even use other alphabets such as Cyrillic, Greek, Hebrew or Arabic.

In the present version one can use at one time - up to 128 different characters (this is easily extended to 256) but, as will be shown, the number of different characters which can be used in one document is virtually infinite owing to the overlay feature. Most computerized typesetters can merge two files, that is insert part of one page into another. This is done either with pointers or, more often, by merging the files at the character level with the system making up the pages anew. But in principle it is possible to remember the matrix position of each dot the printer-plotter will make on the page. This information about the page make-up can then be overlayed from a different file, e.g. that created when using letters outside the set of 128 characters. Thus on the final printed page, letters from both sets will appear. Actually there is no reason why the overlay should be limited to only two files, thus giving in fact an unlimited set of characters.

There remains still the inconvenience of the limited characterset on the visual display unit. This can be overcome, if so desired, by using a RAM instead of a ROM. There, each time the operator calls a different set of characters, the VDU memory is fed the new set as well and the typist sees at all times the types of characters in actual use.

The overlay feature can be extended further. A curve computed by the computer from any data whatsoever and properly formatted can be overlayed just as easily. In this case the final product would be a page with the graphic art already positioned properly. We believe that this facility is unique to our system. It is limited at present to curves created inside the computer, but it is envisaged to extend this feature to any pictorial material in the following way: the picture is scanned by a vidicon camera and the signals sent through a digitizer to the computer where they are processed and output to the printer-plotter. Incidentally, if one is prepared to forgo some resolution one can use 4-9 shades of grey by considering a 2×2 or 3×3 matrix of needles as one dot in the make-up of the picture (a pixel).

The final product has not, of course, the appearance of a high class typeset journal; but for the small circulation journal, often typewritten and run off a cut offset press, this we believe a viable alternative. The appearance of the characters can be improved by using a 200 points/inch printer-plotter or simply by reducing the original output during the plate-making process.

The system has proved remarkably failure-free. The printer-plotter is believed to have an MTBF of 5000 hrs. and the only other moving parts are the keyboard and the - continuously rotating - diskdrive.

The system is also convenient for writing form letters, and in fact our typist likes to type all letters on this gadget. She has taken to it like a duck to water and cannot stop enthusing, though to what extent this is a matter of the Hawthorne effect we do not know.

Finally, it must be stressed that a system like this is flexible and does not consist of one single version. For a small operation a simple system with only 1-2 typing stations might suffice whereas others may want a larger unit with 6-12 typists (or even more) each storing on her own floppy disk or alternatively (as we do) using one central disk. In the latter case, it is easier to use several typefonts. In either case the system is practically the same, but with more typists a more extensive supervisory program is necessary. Otherwise the difference is mainly one of storage size and number of terminals. Either way, throughput does improve dramatically with such a machine.

DISCUSSION on "Some Aspects of Data-recording and Microform Output for Information Distribution" by K.-D. Lehmann

"Editorial Processing Centres: A View from the United Kingdom" by A. M. Woodward

"Editing a Photographically Reproduced Journal" by Earl J. Taft

"The Dual Journal Concept" by D.H. Michael Bowen

"On-Line Preparation of Typescript" by U. Bloch & K. Shmueli

LARRY LESTER - To Dr. Bowen. Now you have your summaries. Is there any reason why you must publish a full paper. Can you not have it on a request basis only?

MICHAEL BOWEN - No. There is no reason at all. As a matter of fact, one of the disadvantages of doing an experiment and having it funded by an external agency means you have to define all the variables exactly what you are going to do before you start. Now frankly before we even started this thing, we were entirely sick of the whole idea. And there were lots of other things we would like to do however we were obliged, obligated to go through and do the things that we wanted to do. Now there is on-demand a lot of reprints. There is a lot of things that you can do in fact.

LARRY LESTER - The whole idea is that you're trying to save money and you are trying to reduce the amount of paper in your libraries. And you're not doing it. You are expanding the amount of papers so what. I see it as a rather unsuccessful experiment. You are providing summaries which is a wonderful idea, and then you're clogging up the libraries again aren't you?

MICHAEL BOWEN - Well actually it wasn't a case of that. The libraries I think always feel that they have to have the entire amount of whatever it is you're giving. Libraries want everything. Librarians always want everything and never have the money for it but they want it. Our idea, of course, in its simplest form was, summaries to individuals and the archival version to the libraries. It got complicated because the libraries wanted the archival version; sometimes they also wanted the summary version even though they had it within the archives. And individuals wanted the archival version.

LARRY LESTER - I have another question for Prof. Taft. If you are working with camera ready copy provided by the authors and if comes in terrible condition-let's say it comes from a European whose English is very poor and it must be reworked or it was poorly typewritten, what do you do? Do you throw it away or do you retype or do you have the author retype and if it's in poor English and the author is incapable of putting it into correct English what do you do?

EARL TAFT - There are several facets to that question. The point of view of the company is that they will photograph anything as long as it is in any sense legible to the reader. As editor, if I believe that the manuscript really will look very bad when photographed I will of course return it to the author together with a list of instructions for preparation of manuscripts (which he hopefully should have had before but perhaps he didn't, or didn't know about) and requesting him to retype it. Now these instructions appear on the cover of the journal but not everybody sees it. They look rather formidable at first glance, but I usually tell the authors that its really not necessary to follow every little detail in the instructions.

The main thing is that the paper be of good quality and that the typing is clear enough and dark enough so that it will photograph reasonably well.

Most of the instructions concern style which isn't really necessary but is just put in to encourage a certain uniformity in the appearance in the published article. It is not necessary from a technical point of view for photographing the article. The company itself of course will not retype the paper of a particular author and I certainly don't have the facilities to do that. So I believe it's the author's responsibility.

As far as poor English goes that is a trickier question. You see one of the advantages, or you might in this case consider it a disadvantage, of photographic reproduction is that whatever the author writes from the mathematical point of view, or from the style point of view or language point of view, is what is going to appear. So it is basically the author's responsibility. I probably will not object if there are occasional mispellings, misuses of English as long as I feel that the paper is mathematically comprehensible to the reader. By the way it doesn't have to be in English of course since it is photographed it can be essentially in any language. We even have papers in Cyrillic letters. Whichever language it is, if I feel that the misuse of language interferes with the comprehension of the paper, then naturally I will return it to the author with perhaps specific suggestions as to how the English could be cleaned up and so on. I hope that answers most of your points.

IRVING AMRON - I wanted to confirm what Anthony Woodward spoke about concerning the use of OCR. We use the OCR on a very small scale of course not in the same way as an Editorial Processing Center. The quality control, the line alignment of the OCR ribbons cause such a high error rate that we discarded it and transferred to tape cassette units which I'll be talking about. These tape cassette units greatly reduce the error rate and the cassettes were used with CRTs transmissioned to our main frame. The error rate was reduced three or four times in comparison with the OCR and I think this is very important in the EPC.

A second item concerning the editorial processing center: At one time a united information center was conceived in the United States. Lots of money was spent by National Science Foundation, and it was found that it is very difficult to get organizations working in various fields to get together. I myself tried to get civil engineering abstract journals set up with only American civil engineering societies, and it is impossible to get them to work together in producing something. It is a very good idea theoretically but your conclusions, Dr. Woodward, concerning the future editorial processing will probably be correct and each one of these organizations, each one of these professional societies, will work individualy setting up their own systems and using the available software and hardware for setting up their own processing of composition.

TONY WOODWARD - I have another comment to make on that. In the United Kingdom we haven't even bothered to try to talk to these societies about getting together. But what we have noticed is that there are a number of entrepreneurial firms who are looking at ways of extending the publisher-printer interface, not necessarily with learned societies in mind. I think this is possibly a new trend for getting new sevices and it follows in the UK situation I think very much on from the fact that the pressure to have something like an EPC is much less because UK learned societies traditionally have always used commercial publishers much more than their American counterparts, due to various changes in the economics of publishing I think. So in fact, in some way, the society is passing the financial problem to the commercial publisher and the problem is the small society who still has not been able to find a commercial publisher to take them under their wing.

JOHN ZIMAN - I'm a little concerned that the very beginning of this unique conference which is on a very important general theme is starting off immediately plunging into this technological push and pull game. So I'd like to make just a little general remark which applies to what I think we've been saying this morning and I think it will apply to some of the other sessions later on. If you will excuse me pushing forward that way and I'll put it from a professional scientist's point of view rather than a professional editor's, and I'll make it provocative. We all know that there is enormous range of quality in what passes for a scientific paper. A tremendous range. And we also know, those of us with some experience in the scientific game rather than just publishing the stuff, that it's not so very difficult to judge in advance whether the paper is going to be wanted widely or not. I don't say that it is perfect judgement but it is not too difficult to know which are the ones which are really high quality even at the earliest stages. Now the thing we know is that the best scientific papers turn out to be priceless. They are the things that really make the whole thing go. They are beyond their value rate in pearls or rubies or whatever it is that a good woman is supposed to be valuable in. (It's rubies. Thank you). We also know that as a community we want to have these as quickly, as widely and as well produced as possible. These are the needs of the scientific community in these things.

From that stage, in most circumstances the economics are irrelevant. the actual cost of conventional techniques of publishing are way below the real value of that type of material if we can get it. And so I'm going to pose the thing this way. What are we after here in the scientific world, in scientific publishing? Are we producing Rolls Royces or Fords? And I'm listening to this morning's session and I feel that this technological push here is all the way towards vast quantities of tin lizzies and not nearly enough thinking of whether the need mightn't be for Rolls Royces even though none of us are oil sheikhs. So I'm putting it in that sort of way but I do believe that we should get this sort of thing in focus all the way through this meeting and I think it follows with the technical issues that start from the very beginning. The big thing we are told is that we need all this stuff. We have got to produce all these papers, and papers and papers and papers. Therefore we need all this machinery with which to do it and we put it all through an optical character reader and it produces two million letters a second or something, three thousand pages or something like that. I'm being very provocative and rude about it, but let's take it in that term.

URI BLOCH - You will permit me to be provocative as well. I couldn't agree with you more that if you scientists could agree on what is a good paper and could agree to refrain from publishing all the rest of the tripe all this would be unnecessary, but the fact is that you insist of publishing because you don't want to perish. And so we are trying to give you the best deal economically and I don't think you are talking to the right people here.

KLAUS HELBIG - I want to speak as the editor of a multi-national, multilingual journal. We have a multi-national authorship and a multi-national readership and I want to address myself to a few simple questions Prof. Taft has raised. I cannot for instance send back a paper written in bad English to an Italian author and ask him to write better English. I cannot send back a paper from Eastern Europe where typewriters do not have parenthesis signs and ask them to put parenthesis signs in. They always put slashes instead of parentheses and there is nothing I can do about it. I have to do it myself. Many of these things we have been suggesting today have been suggested on a rather narrow basis if you look at it from an international standpoint. There have been suggestions how to deal in the framework of the U.S.A. or the framework of the U.K. but as soon you come to multi-national or multicontinental publishing, it doesn't have to be a big journal but as soon as it draws its authors and has its readers in several continents, you run into difficulties if you start this kind of thing.

EARL TAFT - I'll try to respond briefly on two points. Yes of course if one is to return a paper in bad English to an Italian author as I said I will make suggestions you know. I'm not just going to return it cold to the author and get it back in equivalently bad form. Many authors in that situation of course will seek help. They will get advice from English speaking colleagues or from friends who are better at it than they are and I haven't had any experience where the paper comes back in just as bad a form as when I sent it on. Now as far as lack of technical facilities, of course it is a technical question. Yes the best I can answer is this. If the lack of a parenthesis is not too widespread, or if it is what I consider a minor thing, I will have it done myself. I will ask my own secretary to do it or a secretary in the company - if it is not too time consuming and fairly minor they will do it. If it is really a major one then there is a problem and the only thing I can say is that ultimately it is the author's responsibility to take care of this and perhaps the larger operations could handle it themselves but my particular operation would not be able to undertake that sort of revision and that basically it would be the author's responsibility.

MIRIAM BALABAN - I edit a multi-national monthly journal, *Desalination*, published by Elsevier. Here I try to adapt flexibly to serve the community with the best quality and providing maximum information swiftly at a manageable price within the framework of the journal. It is a properly edited, carefully refereed, finely typeset journal. In order to capture and instantly disseminate progress in the rapidly expanding desalination community, we are including some of the central conferences in the field. To do this we are publishing these from camera-ready copy as is at very short notice - two volumes, 1000 pages in about 6 weeks - delivered to participants of the meeting wherever it may be - Mexico City, Tokyo, Jerusalem. Since there is very little time to do cosmetics on the author-supplied copy, and in one case not even time to send out forms and instructions, we have some slashes, no brackets and some smile evoking English, but they served the purpose and the users were pleased by the prompt and full information (the camera-ready copy enabled the publishers to maintain the price and yet include the significantly greater number of pages required).

And from meeting to meeting repeat authors become trained to provide better copy earlier so that we could request improved text or illustrations. So the multi-national aspect might present problems but challenges with solutions as well. Readers have been happy to receive good coverage of information in the field - most volumes carefully refereed and beautifully typeset and the prompt information on meetings in one package even if it is camera ready.

By confirming that the organizers of the meeting are screening the material (they are usually members of the editorial board), we have reasonable assurance we are not sacrificing quality of content - but rather some lingual and technical aesthetics.

The other thing I mentioned before which applies here is training. One of our main roles here is to agree on standards and guidelines which can help editors unify basic requirements to get better copy from authors whether it be for typesetting or instant use as camera ready copy.

NORMAN CHIGIER - I think in considering all these computer processes and camera ready processes we should try to take into account what the users are doing with this. People who are teaching at the universities and scientists are being faced with a deluge of material which is coming out and the main problem now is one of selectivity. And I think that before we consider the use of computers we should recognize that the camera ready process is already leading to a huge increase of material which isn't being properly edited. We are very concerned about the fact that because it comes so easy it is now possible for conference organizers to simply provide material which is unedited. And the net result of this is a reduction in the quality. I know that from my own student's point of view, they are looking very much more towards some guidance on how they can find material which is worthwhile reading. The fact is if they go to a library and they try to find how many papers have been published in the area of their interest they find it impossible to read all the abstracts, never mind read the papers themselves, so the main emphasis I think should be to try and select materials and not to try and make it so easy to produce a large quantity of material which is not worth publishing.

PRIMARY PUBLICATIONS AND MODERN INFORMATION SYSTEMS
COPYRIGHT PROBLEMS FROM THE VIEWPOINT OF EDITORS OF SCIENTIFIC JOURNALS

Paul Katzenberger
Max-Planck-Institut für Ausländisches und Internationales Patent-Urheber- und Wettbewerbsrecht, Munich, West Germany

ABSTRACT

In the age of flood of information, secondary information systems seem to offer the needed control. Those systems, however, are supplemented by new means of supplying sceintists with the full text of the documented literature which, on the one hand, have become indispensable for the scientists' work, but, on the other hand, are jeopardizing the economic basis of primary publishing. Insofar, particularly dangerous is the mailing of reprographic reproductions of journal articles on demand by libraries and documentation centers. Copyright law has to guarantee the economic basis of the likewise indispensable primary scientific literature. As to this, endeavours for copyright law reform in some countries are reported in this paper.

I. Flood of Information and Modern Information Systems

In the age of flood of scientific and technical information, the primary objective should not be a limitation of primary publications but rather the control of the flood of information by modern documentation and secondary information systems. Modern technologies and methods in this field have already reached a high level. Particularly the technics of reprography, of electronic data processing and data transmission have revolutionized the traditional documentation system. Today, we find in addition to customary reference books, bibliographies, card index systems, and reference services very fast and highly current title services such as Current Contents, novel documentation services such as citation services, and particularly the storage of such services in electronic data banks, which allows automatic literature searches and continuous information by "selective dissemination of information". Therefore, many of the traditional printed documentation services now have computer-readable siblings. Large data banks of documentation centers contain a number of such services thus making possible, by means of data telecommunication and use of terminals, automatic long-distance literature searches.

The modern technics and methods of documentation are supplemented by new means of supplying scientists with the full text of the documented literature. The traditional approach was to go to one's own library to find the respective work, or to lend the book or journal from a public scientific library, or even the purchase of the books, journals, or off-prints, in cases

Miriam Balaban (ed.), Scientific Information Transfer: The Editor's Role, 43-48.

of constant need of articles in a specific journal possibly the subscription to such a journal. Today, there exist other possibilities, and they are utilized to an increasing extent.

The electronic full text storage and thus the retrieval of whole books or articles from a central data bank appear to remain a dream. However, central libraries and documentation centers offer good services: they supply the scientists quickly, inexpensively and reliably with copies, especially of articles in journals, on demand, thus making unnecessary the walk to the nearest local or regional library. The best known of these activities are those of the British Library Lending Division which fills annually hundreds of thousands of such copying requests, and similar services exist in other countries. However, the future seems to belong to central EDP-supported microfilm banks, which will make possible in their final stage the automatic retrieval of full texts of articles, research reports, etc., even over long distances. Telecopy and viewdata are even now a technical reality.

II. Economic Basis of Primary Publishing and Its Protection by Copyright

The publication of scientific primary information requires substantial effort in work and capital. It is not only the labor of the scientist which is expensive, with his wages being paid primarily by public or private employers. Rather, the publication as such is also expensive, the printing, organization of publication and the quality control. The source of the necessary capital is primarily the sale of copies of the work, in cases of scientific journals especially the price paid for subscriptions. Other income is only of a secondary nature, such as income from advertisements, grants by scientific organizations, government subsidies or page charges paid by the authors or their employers.

The sale of copies at a price corresponding to the cost, and with it the essential economic basis of primary publication, are legally secured by copyright. Copyright grants the authors, and in the same way the publisher as their successor in title, the exclusive right to exploit the work, especially the right to make and distribute copies of the work, in this case of books and journals. Authors and publishers may proceed on this basis above all against commercial reprinters or copyers. This is the historical basis for the copyright. The legal monopoly of exploitation held by the owners of copyrights is, additionally, effected also in those cases where, by means of modern reprographic technologies, their works are reproduced in a manner which does not primarily pursue the purpose of commercial dissemination of the work. This applies in principle for all means of reprographic reproduction of copyrighted works for one's own internal use, such as in schools, universities, enterprises, and public authorities. In addition, this applies especially in those cases where the copies are made not on the basis of one's own original copies of the work but rather - similar to commercial reprints and even more similar to "on-demand publishing" - they are offered to the public for sale on demand or where other means of an immaterial access to centrally stored works are made available, which have a similar economic effect.

Thus, it is not surprising to see that in many countries the problem of reprographic reproduction of copyrighted works belongs to the most-discussed copyright issue of the last years. Especially disputed is the mailing of copies of articles in scientific journals on demand, such as it is practised within the framework of modern secondary information services by central libraries and documentation centers. This way of distributing journal articles which constitutes a genuine competition to the sale of original issues of the journal, is very dangerous to the sale of the journals particularly for the reason that it is usually offered at very low, subsidised prices and may as a result in many cases make unnecessary even the subscription to the original by local or regional libraries. Similarly disputed is the storage and retrieval of copyrighted literature by electronic data banks and microfilm archives. On the other hand, in principle there are no similar copyright problems with regard to the documentation as such, to the extent that it merely refers to the literature but does not constitute a substitute for it.

III. The Affected Interests and Rights. The Copyright of Editors of Scientific Journals

In searching for a suitable copyright solution of the problem, the affected interests and legal positions must be balanced. On the one hand, there are the interests of the public in a free flow of scientific information, an interest which is a great one in view of the flood of information mentioned earlier. This interest demands, among others, that all scientists have quick and reasonably easy access to scientific journals, also to those which are needed only occasionally. On the other hand, there are the interests of the authors and publishers in maintaining the economic basis for their important primary publications.

The publishers are highly aware of this interest and, therefore, they are the main group of interests claiming to be damaged or jeopardized. On the other hand, the authors usually assert the position of free flow of information. This is understandable for several reasons: authors generally do not publish because of the expected remuneration for publication, rather they publish primarily to be read, an interest served as well by distribution of reprographic copies of their articles. The authors also are very well aware of the advantages of secondary information services and unrestricted copying, since they all use scientific literature of others in their work. It is only after closer inspection and consideration that the authors will recognize the problem and the dangers resulting from unlimited secondary information services for the economic basis of many primary publications, as a result of which these publications may disappear and the authors may lose the important possibility of publishing their works.

These interrelations are quite apparent to the editors of scientific journals, since they always consider in the same way both the scientific objective of a dissemination as large as possible of the scientific work for which they are responsible as well as the economic basis for the existence of their journals. As to their legal position, the editors of scientific journals are, however, substantially dependent on the rights of authors and publishers. Indeed, editors of so-called collective works, to which the journals belong, may acquire primary copyrights relating to their contributions in selecting and arranging the articles in the collective work (see Art. 2(5) Berne Convention in its Paris version, Sec. 4 of the German Copyright Act). But the modern secondary services in practice until now, such as especially mailing of copies on demand, usually relate to individual contributions only, such as individual articles in journals, and not to whole issues or annual volumes of a journal. The editor does not have an independent copyright in these individual contributions. On the other hand, separate rights of the editor may be involved when, in the future, whole issues of journals or annual volumes are stored in the electronic databanks or microfilm archives. Nevertheless, even today the editors of scientific journals are called upon, in their intermediary position between authors and publishers and, on the basis of their specific expertise and responsibility, to take a position on the copyright questions, they are called upon also for the reason that they, different from the publishers, are without suspicion to be the representatives of partisan economic interests.

IV. Solutions in Individual Countries. International Protection

There are no uniform international solutions in view. Nevertheless, the two universal international copyright conventions, the Berne Convention and the Universal Copyright Convention, contain important guidelines in their most recent versions. According to Art. 9(2) Berne Convention, the member countries may allow the copying of protected works without authorization by the proprietor of the right only "in certain special cases". Furthermore, such reproduction must not conflict with a normal exploitation of the work and not unreasonably prejudice the legitimate interests of the author. According to Art. IV bis (2) of the Universal Copyright Convention, legal exceptions to the exclusive right of reproduction of the author must "not conflict with the spirit and provisions of this Convention".

Turning to domestic legislation, the copyright problems of secondary information services have been approached in a few countries and have found a solution only in a very limited manner. Thus, the new Copyright Act of the United States of 1976 contains a special provision regarding copying by libraries and archives (Sec. 108). This section provides, inter alia, that the systematic reproduction or distribution of single or multiple copies of, inter alia, journal articles, is in principle unlawful. An exception is made for interlibrary arrangements, which, however, must not have as their purpose or effect that the libraries or archives receiving such copies do so in such aggregate quantities as to substitute for a subscription to or purchase of such work (Sec. 108(g)(2)). The American Commission on New Technological Uses of Copyrighted Works (CONTU) has attempted in a guideline to circumscribe this limitation more closely. This guideline states, for example, that a library may, within the framework of interlibrary arrangements, receive not more than five copies of articles anually out of the last five annual volumes of a given journal. Objections against such a solution are raised primarily in view of the administrative expense of this program.

In the United Kingdom, the long expected report of the Whitford-Committee proposes a system of blanket licenses in the framework of which a general remuneration for reprographic copies may be established. If the copyright proprietors do not, within reasonable time, offer such blanket licenses, the copying is to be possible without any remuneration.

A different solution is presently being discussed in the Federal Republic of Germany. For one, it is suggested that all lawful reprographic copies of copyrighted works are subject to a legally established remuneration. Presently, the law provides for such a remuneration only for copies serving a commercial purpose (Sec. 54(2) of the German Copyright Act). On the other hand, it is discussed whether the publishers themselves might not establish a cooperative copying center and thus close an essential gap in the present system of the private enterprise literature market. Such a publisher-owned copying center might offer copies of articles from their journals at a reasonable price based on cost. The sale of copies at such a price would, on the one hand, reinforce the economic basis of scientific journals, and, on the other hand, not jeopardize the subscription as the general way of distributing or marketing scientific journals. To the extent that this publisher-owned copying center is offering its services, the protection of copyright law would have to be reinstituted and the distribution of copies of journals on demand by libraries and documentation centers would have to be considered as unlawful and enjoinable. As an alternative, it would also be possible to provide that libraries may sell copies of articles on behalf of the publishers at prices established by the publishers, but there does not appear to be any willingness on the side of libraries to accept this approach.

In general, it has to be expected that different solutions will be adopted by the individual countries. Thus, non-copyright solutions, which would favor exclusively the national publishers, are possible. This is demonstrated by the tax-like fee for reprographic machinery introduced in France by the Finance Act of 1976. This will bring with it the problem of reciprocal protection in the relationship between countries having a different level in their copyright protection. The international copyright conventions do not offer an adequate solution, since they are all based on the principle of national treatment which does not allow to discriminate against foreigners even if the own national cannot expect similar protection in the respective foreign country. Thus, countries with a high level of protection are faced with the problem whether they should not attempt, alongside to or in place of a copyright solution of the problem of reprographic copies of protected works, a solution based on a new neigbouring right in printings held by publishers. New international agreements may be made regarding such a new neighbouring right, similar to the Rome Agreement on the Protection of Performing Artists, Producers of Sound Recordings and Broadcasting Organizations of 1961, which may bring about reciprocity of protection. We already have a model for such a new neighbouring right in the English and Australian Copyright in Published Editions of Works.

V. New Forms of Primary Publications?

If it should not be possible to find an adequate and reasonable solution to the problem of the relationship between primary publication and secondary information services, one will not be able to avoid having to find new forms of primary publications for at least a part of the sciences. One possibility, which is partially even practised today, is the deposit of manuscripts and the sale of copies of such manuscripts on demand, one version of "on-demand-publishing". It is apparent that this will bring with it significant disadvantages for the authors. Furthermore, if such literature deposits are established and administered by public authorities, the problem of freedom of the press and censorship is raised. In order to secure the necessary quality of scientific publications, in any event those independent persons will be indispensible who today guarantee the high standard of scientific journals as their editors. However, it is another question whether they might be willing to serve such new forms of primary publications.

DISCUSSION on "Primary Publications and Modern Information Systems. Copyright Problems from the Viewpoint of Editors of Scientific Journals" by Paul Katzenberger

CYRIL BELSHAW - I'd like to raise a question about international copyright, particularly regarding the adherence of the Soviet Union the International Convention. I've gained the impression - because we have quite a number of Soviet contributors in our journal - that there is a great deal of confusion in the Soviet Union as to just really what they ought to do in relation to scientific material in this respect. I think that the scholar doesn't care very much at all but the scientific administrator does.

And we've had an experience recently in which permission was granted by a scientific administrator to publish in terms that were quite unrealistic and indicated that he really didn't understand what the whole copyright thing was about.

I wonder whether you have any experience of this. Do you know whether the administration of copyright in the Soviet Union is really coming close to conformity with our own expectations or whether there are problems in it that you may happen to know about.

PAUL KATZENBERGER - As I know, the Soviets intend to correspond with their international obligations following from the membership to the Universal Copyright Convention. As I understood your question, you want to know what is the policy of the Soviets when Soviet scientists publish in western channels.

The legal situation would be that those articles are protected by copyright laws of the western states and also are protected by the Universal Copyright Convention. Then from this it follows that the Soviets themselves must in principle be in accordance with the Universal Copyright Convention and they may not reproduce such articles if the Soviet scientist has transferred his rights to the western publisher. But here, I think are problems because VAAP, the copyright agency of the USSR, presumably does not allow the transfer of exclusive rights to western publishers. And then western publishers cannot make claims in the Soviet Union in a copyright case.

NORMAN CHIGIER - I think the question of copyright in scientific journals is very different to all forms of publication and perhaps we here should address this problem. As far as the scientist concerned, his interest is that as many people as possible should read what he has written. And therefore he has no interest whatsoever in copyright. He doesn't get any payment for what he has written and therefore the question of copyright seems to be taken over by the publishers for they are the only people who are getting any remuneration for this. I have for instance a case now where an article has been submitted to me by an author who I would like to publish in my journal. He has also submitted this to another journal which I believe has an entirely different readership because in science we are very widely split. As far as I'm concerned, I would like to cater for my readers. The editor of the other journal would also like to cater for his readers. The author is delighted that both of us wish to publish his article twice. Yet the publishers are arguing among each other about the question of copyright. And I think possibly here we should reconsider this whole question. Seeing that the publishers are already making a profit from authors who get no remuneration, I do not see why this question of copyright should be applied to authors of scientific articles.

PAUL KATZENBERGER - I see the problem from another point of view. As you think, the author doesn't need copyright for economic reasons. He doesn't earn any money by publishing. In Germany the situation is partially another one. But in principle he doesn't earn money and therefore he has no interest in copyright protection. But the publisher has an interest in copyright protection, and his right is acquired from the author. He needs this right.

MARTIN CREMER - Publishers say that if you have a very free flow of copying, and if you have no protection against everything being copied by students, research workers, and institutes, then the number of subscriptions decreases. The reason for copyright is not for getting additional money but to avoid the decreased distribution of the printed journal.

With new microdevices and copying devices, there is a great deal of copying, which might be dangerous for the sale of publications. Therefore publishers are very concerned about this problem because they see not only the problem of copying and the restriction of the journal, but they see the problem of the full text in storage which is another limitation to the sale of subscriptions. Publishers tell the research community that it must be interested in having the printed journals. So I think it is not the problem of getting additional revenue, but the problem of the very existence of the printed scientific journal.

URI BLOCH - May I say one more word to Dr. Katzenberger. I would like to take issue with the whole concept of copyright in libraries from the following, purely economic point of view. Libraries are not given a budget consistent with their needs because the amount of money allocated to libraries does usually not come from a zero based budget but relates to what they spent last year and the year before and to what the university actually has available. Therefore I conject that libraries can spend a fixed sum on journals if you insist that some of it will be spent on copyright it simply means that less will be spent on the acquisition of journals. Therefore the publishers defeat their own purpose when they insist that the copying should be paid as well.

GILLIAN PAGE - Thank you. I think I ought to put the record right. I hope it does not abuse my position as chairman; that is not my position on copyright. I believe in copyright. I believe there are rights which, if they are infringed, there should be payment.

LIBRARIANS' REACTIONS TO NON-CONVENTIONAL PUBLISHING METHODS

A.M. Woodward
Aslib R & D Department, 3 Belgrave Square, London SW 1X 8PL, UK

ABSTRACT

A survey, by structured interviewing, was made of 46 librarians in the academic, public, research and industrial sectors to ascertain both the degree of difficulty induced by current economic trends and their responses to the use of non-conventional publishing methods which appear probable due to economic, social and technical pressures

The responses showed that library bookfunds in general are not keeping pace with inflation, but they are not falling behind as badly as feared.

Alternative publishing methods are not as acceptable as conventional paper publications largely because of user problems and increased processing requirements in libraries.

Nevertheless if sufficient cost advantages can be introduced for non-conventional methods, these could become preferable for certain types of material. Experiences with interlibrary lending and microfiche repositories show how library resources can be extended and made more responsive to user needs by the selective adoption of new methods of information dissemination.

Introduction

Scholarly publishing seems to be entering a period of change and uncertainty as the financial basis on which it operates is threatened by rapidly increasing costs and limited library funds, especially in Britain and the USA. There are several underlying reasons why these pressures are suddenly manifest. Paper is becoming a scarce commodity, both publishing and library operations are labour-intensive with little scope for increases in productivity, and the rapid expansion in higher education in Western countries is coming to an end. The more general economic crisis means that there is simply less money available for purchasing library materials at the moment. However, overt general crises merely mask the

Miriam Balaban (ed.), Scientific Information Transfer: The Editor's Role, 51-58. *All Rights Reserved.*

underlying trend of scholarly publications steadily becoming relatively more expensive compared with other products.

In addition to these purely economic stresses, there are also changes in the ways that information is used, its ever increasing volume, its growing diversity, coupled with narrowing user needs, growing requirements by decision makers and a growing awareness of the real use made of a publication. There are also developments in the technology of publication or dissemination of information such as repography, microforms, cheaper data processing, increasing telecommunications facilities and facsimile transmission. The combination and interplay of all these factors is producing a gradual move away from conventional publishing by multiple print runs on paper towards non-conventional publishing methods. These include microfilm and microfiche versions of existing publications, synopses with paper or microform full-text back-up, various types of on-demand systems, repackaged journals for small groups of clients, and electronic dissemination of information. (Information is used in a wide sense to include all types of scientific and social science publishing). Various types of non-conventional publishing already in use or contemplated have been described by Line & Williams (1) and by Lea (2).

In order to explore the kinds of problems or advantages that non-conventional media may have in the library, 46 librarians from academic, research and industrial organisations in the UK were interviewed during the winter of 1975-76. To ensure that all interviewees were aware of the various types of non-conventional publishing a preprint of the paper by Line & Williams (1) was circulated in advance. The study was not designed to study users' reactions, and concentrated on the alternatives as applying to conventional journals.

Overall reactions

In general terms, most of the librarians interviewed accepted that there may be a need for these various types of non-conventional media but felt there would be practical difficulties in libraries if they were introduced on a large scale. The main arguments against the general acceptance of non-conventional methods were:

a) the numbers of different media, sytems and procedures are likely to proliferate which will cause problems in location of sources and in physical handling,

b) there is an inherent resistance to change by both librarians and users irrespective of relative advantages and disadvantages,

c) the overall costs to libraries were considered on the whole to be higher than for conventional publications because of increased secondary costs (handling, equipment, postage, etc) unless the primary cost advantages are much better than presently contemplated; several librarians felt that a microfiche price at 80% of the paper price, as currently offered by Pergamon, was too high,

d) most non-conventional methods will introduce delays in getting material into the hands of the users or require specific equipment; these will impose addition barriers between the user and the information.

When asked to say which of six alternatives would be most preferable (or least objectionable), 12 librarians opted for journals on microfiche, 10 for synopses plus microfiche, 6 for abstracts plus microfiche, 5 for repackaged journals, 4 for repositories, 1 for separates and 8 had no preference. This lack of consistency was reflected in scattered positive reactions but fairly consistent negative reactions to the alternative methods, reflecting the lack of long-term practical experience of these forms. Nevertheless in some locations specific adaptations had been made to capitalize on the attributes of systems already being used. For example, in one library, microfiche copies of reports are distributed to users for retention with little in the way of paperwork being required and no cataloguing; the occasional second or third acquisition of the same item can be tolerated as this is offset by low cost and an improved service to users.

Media effects

So far the only medium used in any quantity other than paper are the various types of microform (audio-visual materials are considered here to be additional, not alternative, forms of presentation). As is well known, microforms are not well received by the user, partly because of the need to use specific equipment and therefore not readily portable for the user and partly because the user cannot easily move from page to page or annotate it,etc. In many cases users will only tolerate microforms if they are able to get paper copies; in this way microforms are regarded more as an interim storage device than as a communications medium. Very little microform material is bought where a paper version exists except for back runs or bulky material, such as newspapers.

Paper as a communications medium has been a key element in our society for a very long time and its relative cheapness, portability and permanence means that we have adapted it for many purposes. It can be used for all types of communication including letters and for sketching out thoughts and permits filing with other types of communication.

Above all, it requires no special equipment for its use, it can be carried anywhere and sent to anyone who can read. Inevitably other media will be unfavourably compared with paper and there is a strong undercurrent that even if print quality has to be reduced by the use of typewriter setting, camera-ready-copy or miniprint, this will still be preferable because of the versatility and ease of use of paper.

Though it was not specifically studied there was a general awareness of the future possibilities of electronic dissemination. This seems, at the moment, to offer far more to the user than microforms or possibly even paper, not the least of which is a wider use of electronic systems outside purely scientific communication including normal office communications. Electronic media offer versatility with a capacity for annotation, filing, mailing etc. Although it is a system dependent on equipment the wide range of possible applications inherent in the electronic system means that the equipment will be widely available and familiar to the ordinary user. Furthermore, the possibilities of direct communication and interaction between 'author' and 'user' would seem to provide a totally new, socially acceptable, means of scientific communication whilst retaining desirable features of the present journal system such as quality control and organised, coded information (3).

System effects

The choice of medium affects the user whereas the system employed to get the information to the user has implications for the library. This may explain the preference by librarians for full-text parallels of conventional serials on microfilm or fiche although their users may not agree. Scepticism was expressed as to whether synopses could really replace the full text of a paper and there would be difficulties in keeping track of two different parts of the same item. Nevertheless synopsis journals are sufficiently similar to the present journal system to be viewed with reasonable tolerance by the librarian.

All the other systems considered involve more work by the library, firstly in keeping track of what is being published and where (think of the bibliographic problems of a 'stable' of repackaged journals where the contents of each journal are overlapping subsets of a common pool of articles) and secondly because often a two-way traffic is involved to get a single article with the attendent problems of postal costs, delays and unreliable service.

If individual items have to be ordered via the library staff because they are not on site or are in closed access (as is often the case with microfiche) or specific equipment has to be used, the user is disadvantaged in terms of delay etc., whereas the library must provide a higher level of staffing and/or more equipment. Users cannot often assess the relevance of a particular item until it has been seen, so that where material is not carried on-site there is likely to be a higher level of requests for material which will not actually be used, thus further increasing staffing levels and operating costs.

Nevertheless decreased storage requirements and lower total materials costs (if these can be related to actual use) make some of the alternative methods highly attractive for low use material. However, what is marginal, low use material for one library may well be high use core material for another. It was continually stressed that core materials and, especially, current material cannot readily be replaced by any of the alternatives since they do not allow for full-text browsing and circulation amongst an interested clientele. This would indicate that markets may develop for the same material in different formats or packages. Libraries may be prepared to pay highly for really important material in conventional format and accept inconvenient low-cost methods for their marginal material. Unfortunately, the economics of publishing the same material in two different formats are not attractive so that costs might be pushed even higher in an attempt to provide a limited proportion in a cheaper version.

In the scientific publishing sphere the needs of users are only part of the total requirements of the system. Publishers and learned societies have a role which, like that of libraries, is to serve the needs of both the author and the end user; they must operate under the prevailing economics and either make profits or balance costs and income in order to serve the system. Users and, more particularly, authors are not bound by these constraints and consequently demand more from the system. It may serve the library adequately and keep the publisher in business to use camera-ready-copy but the user will not be so happy and the author will certainly feel that a cheaper production method reflects on the quality of his own work. These considerations will apply with more force to the alternative systems considered here.

Conclusions

Current economic trends are such that the increasing costs of producing scholarly publications are likely to continue to outstrip the increases in the means to buy them by libraries. This means that non-conventional methods are likely to grow in importance. Certainly on the results of this study increasing this proportion will pose problems for libraries and by extension, for readers. There is the problem that in trying to minimise his own costs and therefore prices, the publisher in fact transfers part of his burden to other sectors of the total system of communication. By using camera-ready-copy the burden of composition is passed back to the author, by using microfiche the burden of providing equipment and probably reproduction is passed onto the library or even the user. There is a danger that in trying to fight the adverse economic trends in scholarly communication from the publishing aspect alone, the costs of the whole system become greater than before.

Why should this be so? When discussing the various alternative systems, it becomes apparent that they lack an immediate attractive logic, are clumsy and are too obviously compromises enabling existing organizations and patterns of use to be maintained without too much disruption, whilst reducing or containing only the costs of producing the primary publication.

Some of the alternatives are quite acceptable for some types of material, eg. repositories for very low use material such as reports, or microfiche for out of-print works, bulky material or quick reference materials. Thus the use of some alternatives is likely to grow in some limited areas. The main system of primary publishing is likely to remain unchanged unless and until a totally new dissemination system using a single medium evolves. This at the moment seems to be the concept of distribution of scholarly materials via electronic networks providing the tremendous problems of economics and human acceptability can be overcome.

Acknowledgements

I am indebted to my former colleague, Mrs D Jardine, who carried out most of the interviews and performed some of the analyses and to the British Library Research & Development Department who commissioned the study in connection with their 'Trends in Scholarly Publishing' programme.

References

1. Line, MB & Williams BJS 'Alternatives to conventional publication and their implications for libraries'. Aslib Proceedings vol. 28 (3) 109-115 (March 1976)

2. Lea, PW 'Trends in scientific and technical primary journal publishing in the USA'. British Library Research & Development Department. BLR & D Report 5272 HC (March 1976)

3. Woodward, AM 'Methods of information dissemination from the academic viewpoint' Paper presented at Blackwells Second Periodicals Conference, Oxford, March 1977.

P A N E L : USERS' REACTIONS TO NEW TECHNIQUES IN SCIENTIFIC PUBLISHING

Authors: Shneior Lifson, *Department of Chemical Physics, The Weizmann Institute of Science, Rehovot, Israel*

Librarians: Anthony M. Woodward, *Aslib R & D Department, London, UK*
Uri Bloch, *Scientific Department, Ministry of Defense, Haifa, Israel*

SHNEIOR LIFSON - What we heard this morning was extremely interesting. As scientists we should be interested in all these wonderful promising technologies. But you realize that a big part of it is interesting but not of interest to the scientist - how it is published, is it published with this technique or that technique. What we are interested in is really the product, its availability, speed of publication, and first and foremost - the quality.

I think the biggest problem is that you editors are too lenient, maybe too numerous and publish too much. The first lecturers this morning reported that within thirty years the number of publications has grown by an order of magnitude of ten. Now the population of the world is only doubling every thirty years. If publications grow ten-fold every thirty years, obviously there will be more publications than human beings born every year, and this is certainly an indication that this is a transient state rather than a trend which will go on. Maybe it would be appropriate to point out that the publication process is the end product of a process which starts with the student studying at the university and trying to climb up the ladder by publishing or perishing. He produces papers, some of which are rejected, and some of which are published. And I would predict that there are several reasons why in the coming decades the number of publications per annum will not increase but must eventually decrease.

The main reason for this assumption is that everybody knows that universities exploded in the sixties and stopped growing in the seventies. In those days, there were so many young people who tried to climb up the ladder by publishing so the rate of publication and the pressures on journals were enormous. Just look at the volumes of say *Chemical Physics* which I know very closely. How many pages per annum did it publish ten years ago, twenty years ago and today? You see an enormous growth because of the pressure to publish.

I don't think the interest in science is going to dwindle, but I do think that the growth of universities has leveled off and the availability of jobs has gone down, and therefore the publication rate will go even farther down. There is another reason which is even more important. In the old system, science was based on two elements which were distinguishable - teaching and research. They always went together. Now we have come to a period where teaching is a mass phenomenon and we have mass production of students. It is no longer possible that every academic teacher be an academic producer of new knowledge of relevance. Therefore the universities are moving in various ways to separate between the duty of teaching and the chance or duty of producing new research results.

Just look at the Open University in Great Britain which has become the biggest single producer of education on an academic level and which claims to sponsor research but does not essentially do it. You can see that in many other places. Academic education is becoming less and less intertwined with research. Therefore I think there will be a period where the production of Cadillacs will increase at the expense of the production of Fords or Volkswagens, or small Fiats. I do believe it is in the biggest interest of us as scientists to create a system in which publications are really worth reading. We just can't follow the literature - not only because it is enormously huge, but because it is so diluted in content. I could show the number of wrong, irrelevant, or repetitive publications which could well be done without, but the question is how to do it. This is a subject for the next session when we talk about refereeing.

Miriam Balaban (ed.), Scientific Information Transfer: The Editor's Role, 59-70.

This is a terribly crucial problem for us as readers. We just can't cope with the literature. Before you read a paper you don't know if it's good or not. You have to read twenty papers to find one good one. You sometimes don't get to the good one unless you find out through other ways that somebody published a very good paper and then you go and read it.

This brings me to one comment which was not introduced in the technology aspect. I think that scientific journals cannot be only a business. They must be very heavily subsidized enterprises which spend a lot of money on much better editing. Usually the editor looks at the paper, decides on the referees, sends it to them; two referees reject it, two referees say take it. It's taken. He has no time or sometimes no interest in this particular field or subject to read and judge for himself, and sometimes he's just incapable. Just take for instance the editor of *Chemical Physics*. He's just incapable of covering all these fields which are covered there. Now the point is that many papers, even if they are good, are written in a poor repetitive way, in a vague jargon which forces you to look up initials and terms which are not used. This cannot be done by the referees. They are too busy, they are not paid, they are not interested.

A referee is usually a forced reader who is purely hostile to the subject unless he is caught by the interest. Because the paper is impossible, he doesn't want to read it, but he has to read it to reject it. Why should he spend time improving the style? Or maybe he says it's good because he doesn't want to spend time on it. Let it be published. Why bother? Usually if it is rejected, it comes back to you - rebound!

So all this implies that if you talk about the technology of editing, it is not only the electronics and the machinery. The much bigger part of it is how the editorial work is done so that you can give the reader very good papers, not only in innovation of discovery and content. But you must help the scientist if he is an excellent physicist but poor as a writer. Somebody must help him or force him to produce really good papers in journals.

TONY WOODWARD - Technology is not the only problem facing publishers or authors or readers. There is a tremendous problem of economics and this already is beginning to have some effect because of the increase in the volume of publishing. Unlike my colleague here, I don't believe it will decrease in the near future - for exactly the same reasons. But I think if there is a shortage of jobs, the publication rate will go up. I suppose we can argue that one later.

Anyway what is happening of course is that in order to get material published, whether it is good or not, publishers and their authors or producers are turning to different media. And some of these are quite well known and in fact have had them around for some time without accepting that they are alternative media - microfiche, the dual-journal or the synopsis journal. There are systems or repositories, on-demand systems.

One system which operates in Britain is the repackaged journal. This means that there is a stable of journals which draws articles from a common pool. But the subset of this pool that goes into any one journal may well overlap with one, two, three or four journals. The direct distribution of separates is the other conventional alternative.

As I mentioned, in our study on what librarians think about all these developments, we came up with four points. Firstly, that the number of different types of things they will have to deal with will increase. They'll be different media which means different types of equipment. You'll get them from different suppliers, not necessarily the major publishers, but the new reprint houses. Some of the new methods are effectively do-it-yourself publishing—printing where there is no obvious source of supply at all. And of course there are bibliographic problems. Those of dual-journals were mentioned this morning, and think of the bibliographic problems of a repackaged journal where you've got the same paper appearing in more than one journal, possibly at random. And they will require different methods of physical handling. Microfiche require one method of storing, and separates another set of storage.

Secondly, there is an inborn resistance to any kind of change. People want to use paper or whatever they're used to. It doesn't matter whether something else comes along that's advantageous or cheaper. They can't be bothered to relearn.

Thirdly, and this is related to the first point, is the cost of actually getting information. I'm talking here of alternatives to conventional journals, not new types of material, but a method of containing the cost of publishing a journal, making it more efficient or just keeping the cost right down.

And in fact the total cost of actually processing stuff may actually rise in the library because of the necessity to provide equipment, provide extra staffing levels, especially where you've got material that requires a two-way transaction to get at it, as might happen with an on-demand system or a repository system. Where the user has to go to a member of the staff and say "look I think I'm interested in this, will you get it for me?" The librarian has to then send away for it, wait for it and get it back. When it arrives, in many cases of course the user looks at it and says, "no I wasn't interested in it anyway, which compounds the problem.

Certainly the microfiche prices put out by the current publishers as an alternative to their paper journals are just not conducive for switching over from one medium to the other by library subscription. Pergamon offer eighty percent of the paper price and it is considered by many libraries that this extra inconvenience in costs is just not warranted to save twenty percent on the price.

The other final factor is that a lot of these methods introduce delays. They inconvenience the user, they need equipment, they need to go to the staff, they may have a delay before they can get hold of the material. They can't scan it, they can't get it on circulation into their offices. Nearly all of these particular systems are clumsy and they don't help the user to use the information.

Now why does this situation prevail? I think the answer is because there are tremendous problems, but the only people who have really been trying to solve them so far, even fairly sporadically are the publishers who comprise only one sector of the total communication system. And they are trying to solve their problems at the expense of the scientists, I think quite unwittingly. But take the simple case of camer-ready-copy. This keeps the publishers prices down, but it imposes on the author. Perhaps it is necessary discipline, but it certainly imposes something that wasn't there before--to produce material in a somewhat better way, while on the other hand, the user gets a slightly inferior product. He gets a typescript which may vary from page to page in a journal, not as easy to read as a properly composed page. So the publisher is solving his economic problem at the expense of other sectors. The same goes for microfiche--it's cheaper to produce and to sell, but at the other end the user may well want to blow the microfiche back onto paper. In effect all he's using microfiche for is a substitute for the paper, and in fact the cost of producing the paper copy falls on the library or the user. What we need to do is to look at all the sectors of communication and not just try and solve some of these distributions problems from the publisher's point of view.

URI BLOCH - I agree with most of the things which have been said so far. I couldn't agree more with Prof. Lifson. I don't know whether the number of publications will increase or decrease, but I have no doubt that the onus of making publications a viable proposition lies with the editors and the publishers. The rpoblem is that publishers use the credibility of the scientific community to create a large number of unnecessary papers which are not even worth a Vespa. I do not propose to list all the reasons for that. There is the publish or perish syndrome; there is the publisher who thinks that, if he can make good profit, why not do so even though society does pay for it. Incidentally, this is one of the questions we should address later in this conference: is the major duty of the publisher towards his shareholders or has he some duty also to society which is, after all, footing the bill in the end?

I am going to stick out my neck and be unpopular by being a bit pro-microfiche. I am fully aware of the unpleasantness of microfiche to the reader and I will admit that I also try to wriggle out of using it, if I can. Still, what are the major points against microfiche? There is, first of all, the problem of annotation, because at present it is impossible to add to a microfiche which is a serious problem. I do not know the answer but there are new ways of creating microfiche and I think that the vescicular technique does enable one to update an existing microfiche and I can imagine that in the future it will be possible to get even annotations onto a microfiche. This is definitely star gazing, but I wouldn't offhand decide that it is impossible. The next problem that is always raised is: "I've got a bibliography at the end of my paper and each time I want to look up one of the references I've got to go back and forth which is extremely annoying on a microfiche reader and much more simple when I have a hard copy in my hands." If I remember correctly, bibliographies are sometimes called footnotes and were originally at the foot of the page. Now, if we were to go back and have the footnote also at the foot of the page that whole problem would dissolve itself. I know some who still do it, sometimes with a comprehensive bibliography at the end, but most have the bibliography only at the end. I am quite sure that the publishers and editors do know why, however, they might remember that they may have to go microfiche, so would it not be wise to bring back the good old footnote the way it used to be?

The next problem is what I call B&B, namely, you can't read the microfiche neither in the bus nor in the bath, which is apparently where most people read, or so I am told. Assuming this to be true, I suppose that if technology would really address itself to solve that problem it could be done. At the end of the sixties Harold Wooster was very much infavour of the cuddly reader, something you could cuddle on your knee and read in the bus or, conceivably, even in the bath if you didn't mind being electrocuted. It did not come off, because whoever got the contract from the U.S. government made a shoddy job of it and the cuddly reader was never convenient totuse. This brings me to the last point. People claim that a reader is not convenient to use. It's noisy, it's hot, it's not light enough, the focus is not sharp enough. In a way this is tough luck, because there exist today some readers that are quite reasonable. Their resolution is fairly good; they are very quiet, you'd have to be in a very quiet library to notice a fan; they are sufficiently cool, and I have no doubt that with a little bit of pressure on the manufacturers a decent product can be made, and at a reasonable price, and without the inconvenience that I remember from my own time. I remember the first microfilm reader we built. It was an awful thing, we were using something like a one kilowatt lamp and you had to be in a very small cubicle. You can imagine how we felt in a heavy Hamsin in Israel; we drew the curtains mainly so that you shouldn't see in what state of undress you were reading, if you were reading.

So far the negative aspects which are, I repeat, annotations, footnotes, bus & bath and convenience. But what are the advantages? Well, the major advantage is a tremendous saving in space. You save about ninety eight percent. If you are a librarian, that is very meaningful, but even in your own private library this is worthwhile. Just imagine that instead of having that whole wall covered with scientific publications you have something like a shoebox in one of the drawers and that is all you need! I admit it won't work for books, but why not keep reports and such things that way? In fact, our engineers do so. It is also far cheaper. To copy a microfiche costs literally a few pennies, whereas, if you want to order that report, you have either to xerox a hundred pages, or you can reorder it. So it is far cheaper, both in space and cost, to have your private library on microfiche.

It should also be cheaper in a library, bit if a journal on microfiche is priced at eighty percent of the hardcopy price, I can very well understand why a library will not be prepared to go to the expense of buying a new reader, in order to save just twenty percent of the cost of a journal. But if we talk about reports, that is another matter. NTIS sells you a report on paper for, let's say, $7.50, but if you buy it as a microfiche it costs you just 83¢, that is just about 10% of the HC price. Thus, if you buy a few reports you have already saved yourself the money for one reader and very soon it gets worthwhile to order most of your stuff that way. I do not know why NTIS can sell you the fiche at 10% of the price and the publisher at 80%. It would be interesting to find out.

There is one problem when it is senseless to use microfiche and that is when the reader comes and says: "Look, I have got this microfiche I want to read, please blow it back. This simply won't do. Tony mentioned, quite rightly, that if the onus of producing a paper copy is thrown back onto the library, the whole process becomes much more expensive than were you to buy a paper copy right from the beginning. The solution is simply to resist the demand to blow it back. Educate the reader to read a microfiche, to see whether he really and truly needs the whole paper blown back or whether perhaps some other solution can be found. This is really a problem for librarians, whereas this is a meeting of editors, so I won't go into it.

The last advantage of microfiche is its speed. You can receive a microfiche from NTIS within a week because they send it airmail, but if you want a paper copy it will be sent by fourth class mail and in Israel that means anything from two to six months! I wonder how many scientists prefer to wait six months rather than read a microfiche? The funny thing is that even the publishers themselves are not quite aware how important speed is. We obtain our microfiche often from BLLD, the lending library in Boston Spa, but Tony told me this morning that an English library which should be able to receive the microfiche within a day or two may not order from BLLD, but has to go to some NTIS agency in Britain which orders from the States in bulk, with the result that they, too, wait eight weeks. In other words, because the publishers are not aware of the importance of speed they don't mind making you wait eight weeks, whereas I, since I do not live in Britain but in Israel, can get it within ten days. I wonder where the logic of all this is and why publishers should have the right to demand actions which are detrimental to society, just because it will increase their own profits. During the conference I intend to come back again and again to the fact that publishers and editors have also a duty to society and not only to their shareholders.

DISCUSSION on Panel: Users' Reactions to new Techniques in Scientific Publishing
Shneior Lifson, Anthony Woodward & Uri Bloch

BASIL WALBY - I'd like to tell you a sad tale. A certain American publisher who shall remain nameless sends about three times a week some of his famous cheaply produced books which unfortunately still cost about sixty dollars each. They are exceedingly shoddily produced and we've heard they produce them shoddily because they want to keep the price down. Well they don't succeed. This is one of the main complaints that users have, but of course what the publisher doesn't realize is where these review copies end up. He doesn't bother to find out that we don't review books in most of our journals. That's common to most publishers. We put these unrequested copies into the CSIRO library which has been the *de facto* national science and technology library for many years. So in fact he sent me this book for a review I can't give him. It now goes into the Australian library system and is perhaps the only copy that will be available in the Country. I hope you publishers will take this as a slight warning on overpricing the product.

KNUT FAEGRI - I'd like to support Bloch in his lively defense of the microfiche, but he forgot two very important aspects. Point one is when the microfiche is the only thing you can get. This sometimes refers to old publications and often to publications which are not particularly old. And I got into it in a completely different way. Namely the microfiches of famous herbaria, like the Linnean Herbarium. It is on microfiche, and we can use it that way instead of having to go to London (the Linnean society is understandably rather reluctant to send those sheets out of their rooms). The other thing. I don't know if any others of you have entered at any time in historical research, but anybody who has the choice between trying to turn the pages of an eighty year old newspaper and using the corresponding microfiche knows what to choose. These two things have nothing to do with the main theme of what we are discussing today, but in an overall picture of what librarians should be able to do with their microfiches. I think this is important. If we want the microfiche we should of course use them where they are also at advantage.

PETER LENGYEL - Even as an editor, I could feel I entirely agree with Professor Lifson, and also as a reader of what I edit concerning the quality of the material one offers and the need for radical birth control. I must say however that blaming editors is putting the shoe on the wrong foot. Editors are merely reacting, in some cases perhaps even revolting against the standards set by the scientific profession itself. If these standards had not been lowered in the academies, if there had not been so such emphasis on "publish or perish", if there were not, as the lady who spoke before me said, three articles instead of one, if one were not so impressed by long lists of publications in the profession, then we editors would be in a better situation. Personally, when I see a long bibliography of articles by an author in the social sciences I become excessively suspicious. I think a couple of good books and a few major articles is what anybody respectable should show, even if he has published much more. If he presents a long list it is likely to be mediocre in quality.

ANDY JESSE - We have heard a couple of times the argument of a "good" publication. I claim that it is very difficult to give a sound definition of a "good" publication. Let me describe it in a different way. An author's paper is like a mother's baby. Now, you tell a mother she can't produce good babies! I agree with the former speaker that it is suspicious if somebody has a long list of publications; but if you say one of his books is a really good book - well, this also depends on the "receiver", that is on the question "good for whom?" I think if one writes an unscientific article, it still may serve its purpose for somebody else who is not working in the field. What I should like to suggest is that authors state in their list of publications the contents and purpose for each article and book, perhaps also where their paper has been cited. And a final point: a very important purpose for an author writing a paper is for himself. So why not give him a fair chance to write something, in order to learn?

MYRA KAYE - I think those of us who are grassroot editors must be very familiar with the paper which begins, "Some preliminary results have been obtained" followed three months later by a paper, "in our continuing study", followed three months later "we have studied". In other words, three papers for the price of one. One way in which it might be possible to try to cut down on this excess of inadequate and often misleading publication would be to encourage the trend, which I think has started, of asking scientists who are presenting their work for promotion assessment not to present a list of all their publications, but their three of four principal publications. In this way we might be able to get them to put these three papers together to make one *good* paper. I'd also like to make another comment which I think is a nasty one for us editors. In my experience, when a scientist wants to know something, what he does is, he goes and talks to a man. If he wants to know something about physical chemistry, at the Weizmann Institute he goes and talks to Professor Lifson. If he wants to know about something else he goes to some other Department Head and so on. And he goes to conferences. Scientists still prefer to obtain their information by *talking* to other scientists. What they want from journal articles is not really information but stimulation toward the creative thinking which will help them meet the challenges in their area of research. And here I think we have made a case for the review article as the prime target of the journal. It is only in such comprehensive articles that the scientist can find a range of ideas sufficiently wide that he can select what seems stimulating to him, from related research areas as well as from his own narrow one.

DIMITRI VIZA - Mrs. Kaye forgot to say that the same paper can be published more than three times. It can be published in several languages. I think the comments made so far clearly show that we all want to get rid of the rubbish, which is appearing more and more often in scientific journals. As yet, I have not heard any proposition of how to do it, but I don't worry about this because if the need is there, the propositions will come, I hope, sooner or later. What I'd like really to draw attention to is that if we desperately want to get rid of the rubbish, we might eventually impose a kind of control resembling censorship, and this we don't wish to see in our countries.

The next thing I would like to take up is the statement that the publisher should disregard the interest of his shareholders, and look more to the interest of science. This, I think, is an unrealistic proposition. I believe that in our countries, with our political system, we should accept that the first duty of publishers is to their shareholders, and it is therefore clear that it is up to the scientists and to the editors to control the inflation of publications. We don't want to impose any kind of censorship, and I insist on that, or any sort of control over publications, but we could, as editors, try to influence scientific policies, and I will say something about that tomorrow when I shall expose my views on the influence of editors upon scientific policy making. What I believe is necessary and vital for a healthier science is to get rid of the idea "publish or perish" which is the credo of today's scientist.

Finally, to answer the question posed by one of us previously on how to use microfiche in the bath, I suggest you have a ceiling projector and command the buttons with your toes.

KLAUS HELBIG - I have been a little bit challenged by Dr. Viza's statement that "no proposal has ever been made to cut down on multiple publications". We have made one and have tried to operate it among a group of journals which are closely related - they have a common set of authors and a common set of readers: we exchange the titles of all 'manuscripts received' together with authors names and affiliation. There have been at least two cases in the last two years where people had submitted the same article to different journals which had exactly the same readership. In those cases we could at least tell them:"Make up your mind in which journal you want to submit this article; we are not going on before that decision." Of course it doesn't help very much, because several times I have found in the same set of journals closely related articles: the same author, slightly different titles, about the same subject, some pictures were identical, others different (but could be found in still another article). So I don't say the system works, but at least it is an *attempt by editors* to avoid republishing the same material again and again.

CARL KEREN - I'd like to speak from the point of view of an information broker in this area. I would like to disagree with Professor Lifson's prognosis that at least in the foreseeable future, say the next ten to fifteen years, the rate of increase in the number of publications will slow down considerably. A discussion of the facts on which I base my statement would lead too far and should not be made here in plenum. But an even more important point has not emerged as yet from this discussion concerning the techniques which, from the point of view of the user, should in fact influence the process through which the user selects what is relevant to him. Whether a million or three million or five million items of information are produced annually is immaterial for the individual user. The typical user of scientific and technological information reads as a maximum something like two to two and a half papers per working day or about five hundred to seven hundred papers per year. This is his capacity at best never mind how much is being published. The question is how to select these five to seven hundred articles which are the most pertinent to an area of interest. This selection process is continuous because users change fields of interest or have to draw on very different types of information such as catalogues of manufacturers etc. This problem of selection has also some relationship to the question of copyright mentioned by Dr. Katzenberger. I believe that editors at present do not use all possible means which are at their disposal in order to ease the selection process. Whether the material which is of interest to the user is supplied on microfiches or on paper or as a photocopy etc. is not of importance to him once he has established an interest in a certain piece of information. Mostly he will take it in any form in which it is available. In this connection I would like to recite a story which I read a few days ago which I believe that some editors might want to ponder. I read that the ten commandments were written in less than one hundred words, the regulations for the transport of farm eggs between the common market countries amount to more than eighty two thousands words, thank you.

CYRIL BELSHAW - I just want to make a couple of points by way of correcting the record a little bit. It seems to me that in my own fields the publishers are not making a lot of money out of journals. Indeed if the publishers were to be governed by profit in the sense that apparently applies in other fields, we wouldn't have the Cadillacs in my field at all. That's one point.

Now the second point that I'd like to make is this - that I think that part of the explosion in publication, at least again in my field of anthropology, is due to the probability that the Cadillacs are not covering the special interest of many groups of scholars in the kind of way that those scholars want their interests covered. The Cadillacs of the industry and the Rolls Royces of the industry cannot deal with the ethnology of Peru in the detail that the Peruvians want it. Nor can they deal with some of the other specialized interests such as urban studies in the way that urban studies and scholars want it. In other words it is a reflection of specialization and specialization of interests. And it seems to me that this ties in with the precise points that I think have been brought up in the technology discussion. It seems to me that this ties in with the precise points that I think have been brought up in the technology discussion. It seems to me that the technology, the new forms of technology might well be oriented towards the solution of the questions of the specialized journals with rather limited numbers of scholars involved rather than the problems of the Rolls Royces. It seems to me that there is a real challenge for the technology element. Because in many cases the very people who would be the consumers of this are the people who can't afford the technology.

JOHN ZIMAN - Talk about new technology reminds me of the story of the old lady who was asked whether she would like to travel by car and she said she would go by train as God intended her to do. In other words we are of course always in danger of being appraising the methods in which we were brought up and thinking that's the only natural way things can go. However when we are looking at a proposal for a rather radical change of technique you might ask what the need, what the reasons for them are.

I am still a little puzzled in listening to this discussion as to whether they belong to anyone of the following three causes that I can find. One of the suggestions is that there is the need to solve what you might call a practical crisis. People call this the information explosion, I think I call it the information constipation which is rather a different way of thinking about it. Or perhaps in this context we call it the traffic jam. Is there a traffic jam in our whole system? People say there is, but I would have liked to have had more evidence on that and on the absolute need therefore for a completely new way of dealing with these problems. We haven't quite been given that. I mean when one looks at the numbers, one knows that. But is there really actually such a difficulty that it is impossible to proceed the way we're doing. That would be the first reason - practical.

Another reason would be economic. Is our whole system being ruined, I mean in some sort of way going to ruination economically. Well there again some people say so and some people say not. There are the problems of publishers of small societies, there is a whole range of information. Nobody seems to know whether microfiches are expensive or cheap. Some people are producing them very cheaply and some people are producing them expensively in relation to the old style of things. It would really be helpful to know the overall economics of various typical publication systems just to see whether there was a very big gain in them or whether it was a relatively small one which would perhaps not justify the many other changes in the system. I'll come to what these changes will be in a moment.

The third reason is simply convenience. Yes I can see the point in having a sort of cuddly microfiche reader. Wouldn't you start on the telephone directory or the daily newspaper actually as being the clumsiest, most often type of document which everybody wants to look at. Must we in this scientific information system be the leaders of new technologies of this kind. Are we in the worst position from this point of view most needful of this, and therefore should we be putting the most effort into it? This again is something which I haven't heard. Certainly if these things existed widely in society in the way that the pocket calculator does now, we would take them over. But should we be the innovators in that because we have the greatest needs. That's something we haven't heard. What worries me over all is this. When we make a substantial technical change, we do change the balances in the system. Now there are a number of balances. There is the author to reader balance. The author who wants every word he writes undiluted to be distributed throughout the world to everybody, and the reader who is very selective knowing just what he wants and nothing else. Well that balance is changed by the new system. We haven't quite heard how that balance will work then.

Again there is a relationship between what I call the original and the critical. That's something of the author to reader one, but it's a question of how far a system which you use actually criticizes, redistributes the knowledge and so on before it gets to the reader. And finally and most importantly it is a balance between the individual and the bureaucracy - the individual and the organization. Many of the things that have been proposed I feel surely give greater strength to various large organizations and are not necessarily invaluable to the individual's initiatives which may be more important in science than almost any other human institution.

So it seems to me that there are a great many of these issues which we must keep in our minds before we start saying, three papers for the price of one and all the other things that we know about.

SHNEIOR LIFSON - One of the speakers felt that I did not evenly blame the three participants in the journal - the editor, the referee and the author. Well I wanted to even it out but still this is a conference of editors and not of referees. If they would be here I would talk to them, but still I feel that I wanted to say that the weakest point in the system is the refereeing. If I could make a suggestion which is perhaps not practical but very necessary, it is that the editor should be assisted by a panel of experienced professional referees on style, on form and organization of material etc. so that the advice to an author to revise his manuscript would not come from laymen in this field. I can quote my experience. I get a letter from a journal and there are ten questions. Is the style proper, is the number of pictures too little or too much, is this and that and most of them are technical. They are not asking me only to judge whether what is written there is right or wrong, but they ask me can it be improved in form and this is a professional question which should be answered by professionals in this field. And this is not a small journal. For example one of the journals who asks all these questions from the referee is the *Journal of Chemical Physics* one of the biggest. *Physical Review* does the same.

Here is a very weak element. It is a sacred cow essentially that an author has the right to publish the paper as he sees fit because nobody has the right to interfere in the style and form if it is an original publication. I suggest that it is a wrong practice. There might be journals which should adopt it but there should be also some journals who say we are an elite journal and we accept only papers which conform with our demand that there is no repetition, that there is enough information, that it is not too short, not too long. I remember a paper by Terrell Hill which was completely ignored in the profession although it was a very important innovative paper but it was only one page and people didn't understand what there was in it - it was too short. But I can mention papers which are really too long. So somebody has to take care of it and a good journal, an excellent journal, is a journal that will not publish until the journal is satisfied that the paper is really written well. And it is not the referee who is not paid for it and is too busy to put a lot of work into it and who is unknown to the editors. There are so many referees. Some of them go into a lot of detail and don't care about the subject and others care only about the idea and don't care anything about the rest and some of them are too lenient, some of them are too obnoxious, some of them are too pious, etc. The editors don't know them essentially. They choose them sometimes from the bibliography. So I think this is an element of professionalism which is lacking in many journals and the reason might be that it is too expensive, and that's why I felt that the journals should be subsidized. The authorities which subsidize research should subsidize the journals. It will cost them five percent more to add to the expense which they pay for the research itself and the produce will be maybe fifty percent on the average presented better to the reader.

MIRIAM BALABAN - It is precisely the role of the editor to see to it that the paper is adequately written and presented. He may do this by obtaining the proper advice from referees and then using it with discretion, by guiding authors and referees or by actual rewriting. Here too strengthening the role of our editor's associations in instituting training programs and cooperation on standards is just what we are here for.

TONY WOODWARD - In fact going back to your private library, I think I'd rather have a few bookshelves of nicely bound journals than an old shoebox full of microfiches.

URI BLOCH - Well that is a matter of opinion of course. If you had to pay for them out of your own pocket I wonder which you would really prefer, especially if you have to pay also for the bookcase.

TONY WOODWARD - Which may be cheaper than wallpaper. Another misconception of course is that we seem to be talking about publishers as being big wicked capitalistic ogres. Of course many of the biggest publishers are in fact learned societies. And if they have shareholders they are of course members of the scientific community. And they are not big capitalistic ogres. They are just trying to survive I think. They have got to keep their economics in balance in order to be able to do the job that the scientific communications system is asking of them. Sorry I've got back onto an economics thing, but I think it is very important. The economics are bad. The author and the user are not terribly interested in the economics. Publishers and libraries are. They have to balance their costs and their incomes, wherever these may come from and for commercial publishers they have to be in it to make a living as well. But on the whole this is where the problems are because of the need to balance costs and income. On the problem of differential microfiche prices I suspect that journal microfiche prices are much much higher than report ones, A) because on the whole the journals are much nicer produced, and B) because the microfiche is very much a subsidiary byproduct of the main product whereas NTIS reports are pretty poorly produced in the first hand and the microfiche on the whole is the main product.

On the question of birth control for articles, if we were able to institute an effective birth control I wonder what would happen before then. After all if we have effective birth control we might end up with an orgy of scientist to scientist informal communication and editors a bit like nannies might become obsolete. Just a passing thought. Finally I think we perhaps tend to take these things a bit too hard to heart. It is a social system, an organic system, and it grows and it decays and I think that really what we are experiencing and part of the problems that we have now are because it happens to be changing and we are not quite sure how it is going to change. And the problems are posed for people all the way along from authors, editors and publishers, libraries and users. If in the end we end up with a new system, probably most of the problems of communication would be half way solved.

URI BLOCH - I'd like to take issue with Mr. Jesse about the question how you are going to say whether a paper is good or not. I think that Professor Lifson has given some hints on how this can be evaluated. I assumed that Professor Ziman knew what he was talking about when he mentioned the Rolls Royces. After all, when I see a Rolls Royce I can see it from far away and I assume that a good editor is very capable of seeing the same thing, not by the length of the paper, but by its content. Editing is a professional job and I assume that the editor knows what it is all about. I'll try to go even further. I think Myra Kaye mentioned what makes a bad paper, one that has been produced in an itty bitty way in order to make three instead of one. I think of rehash papers. Now there is a price for a rehash. There are journals which say this is a journal which gives you rehash, which explains something to the layman or to somebody who needs to be initiated. But to publish a rehash paper in a primary journal is unfair. We have heard Prof. Helbig spotting them even with the same pictures! Now this is a system which, you say, is organic and works. We know how it came about, we all mention publish or perish, and the making of profits. But why don't we try to find a way to change? Why is there at the end of this congress, on Friday, a meeting on the code of working editors should have, but nothing on a code of responsibility or even how to prevent people from perishing because they don't publish. This is, basically, at the root of the problem and this is what I've been hearing in the couloirs during the whole day. Then why don't we take it to heart?

SHNEIOR LIFSON - Academic education need not necessarily be linked to research. Not everybody who wants to be a good teacher has to publish a novel in order to get tenure.

URI BLOCH - Today it is still so that you do have to.

SHNEIOR LIFSON - Then it is important to change it for the benefit of both education and research.

URI BLOCH - Would you suggest or come up with some concrete proposal which this congress will adopt and will then be published so that the Hebrew University should change its attitude?

SHNEIOR LIFSON - I wrote an article about it in *Trends in Biochemical Science*.

GILLIAN PAGE - Thank you all very much indeed. Thank you first our panelists. Thank you everyone who listened, who contributed to the discussion, once again I am sorry that many of you were not able to make your points in the full discussion. I am delighted though that we provoked such a lively discussion.

MULTIPLE CHOICE, A PUBLICATIONS OPTION FOR A DIVERSE SCIENCE *

Alfred P. Fishman, *Chairman of the Publications Board of the American Physiological Society*
Director, Cardiovascular-Pulmonary Division,
University of Pennsylvania Hospital,
3400Spruce Street, Philadelphia, Pa. 19104, U.S.A.

and

Stephen R. Geiger, *Publications Manager and Executive Editor*
The American Physiological Society,
9650 Rockville Pike, Bethesda MD20014,U.S.A.

After eighty years of evolution, the journals of the American Physiological Society have undergone a major reorganization to meet current needs of the discipline, its specialities, individual authors and readers, and libraries

Why should a society whose journals have been serving the scientific community with distinction for many years decide on a major reorganization of those journals? Why should a society make such a change, involving user acceptance of the new format and hence sizeable financial risk, when the journals have been operating at the desired non-profit break-even level? At several meetings and in the pages of the American Physiological Society publications, the reasons for such a reorganization, and the plan itself, have been described** but neither the implementation nor the initial encouraging results have yet been recorded. This article is intended to provide a progress report on what has been done so far. We hope the story is an interesting one, telling of a publications option for a very diverse science. The field concerned became increasingly specialized; the journals of a major publisher in that field had to evolve to meet current needs.

Throughout the period of reorganization, the primary concern was to satisfy the scientific needs of authors by creating a type of publication that would be attractive to individual subscribers and librarians. For each of these constituencies, requirements were different and habits concerning publications were ingrained. Once these needs were identified and analysed, a mechanism was sought for satisfying them in a way that would be financially self-supporting.

* Much of the first half of this article is based on material previously published and copyright © 1977 by The Institute of Electrical and Electronics Engineers, Inc. Reprinted, with permission, from IEEE Transactions on Professional Communication, September 1977, vol. PC-20, no. 2.

** See A.P. Fishman, 'The journals of the American Physiological Society,' *American Journal of Physiology* 232:C1-C3, 1977; A.P. Fishman, 'The changing journals of the American Physiological Society (abstract), 'First International Conference of Scientific Editors, the Israel Academy of Sciences and Humanities, Jerusalem, 24-9 April 1977, p. 20; S.R. Geiger, 'Specialization: times change, journals change,' IEEE Trans. Professional Comm. PC-20, no. 2: 87-9, 1977.

Miriam Balaban (ed.), Scientific Information Transfer: The Editor's Role, 71-76. *All Rights Reserved.*

This order of priority reflected the responsibility and strength of the non-profit society as a publisher. In order to appreciate the nature of the recent changes in the journals of the American Physiological Society, it is necessary to understand their previous state.

The American Physiological Society was founded in 1887. The first issue of the *American Journal of Physiology* was published in 1898. Initially it was financed by a member of the society, and it was not until 1914 that the society assumed the ownership and management of the journal. It is proper, however, to look on this journal as a publication of the American Physiological Society from its first issue, since the ties between the society and the journal were extremely strong even before the society became the publisher of the journal. Articles covering a broad range of subjects appeared in its first issue, and that broadness of scope has continued.

In 1921 the need for a review journal was met with the initiation of *Physiological Reviews*. It continues to serve that role and thus differs from the other journals, which are primary research journals. It is not part of the present reorganization.

After the second world war the need for an outlet to publish research on 'applied' physiology, that is, research emphasizing man in relation to his environment, became apparent. In response, the society began publishing the *Journal of Applied Physiology* in 1948. This journal has continued to grow since that time. It satisfies all the criteria normally used to determine that a journal is in an active growth phase of its life cycle. It has an actively involved set of authors and editors, a healthy rejection rate, and an increasing number of subscribers and advertisers. The number of manuscripts submitted to it has increased steadily; it has grown from a bimonthly to a monthly publication with about thirty articles per issue.

The growth and specialization of physiology continued. In 1962 this was met in two ways by further expansion of the publications program. The society purchased the *Journal of Neurophysiology*, which already had its own tradition; it has retained its individual nature to this day. It has tended to receive longer manuscripts than the other primary research journals of the society, and has published a different kind of neurophysiology article from that which has appeared in the *American Journal of Physiology* or the *Journal of Applied Physiology*. The *Journal of Neurophysiology* has become a highly respected specialty journal in a rapidly growing field. The Publications Committee of the society and the editorial board of this journal have for several years been trying to broaden its scope, but they are not anxious to see it become over generalized. They see a distinct need for a solid, respected journal in a field that is in flux.

The second event that occurred in 1962 to accommodate the growth and specialization of physiology was the formation of sections within the *American Journal of Physiology* and the *Journal of Applied Physiology*. From the combined editorial boards of these two journals, eight sections representing subfields of physiology were formed. Each section was headed by a section editor, able to select referees with the best competence in the areas covered by the manuscripts being considered for publication.

Additional sections were added and new section editors appointed through the 1960s and 1970s. The journals continued to grow; about one thousand articles per year have been published in recent years. Specialization among physiologists became even more marked.

Within the society, and especially within the Publications Committee, many plans were proposed to help alleviate the problem of publishing a general journal to satisfy specialized needs. Many believed that an article in the *American Journal of Physiology*, with about fifty diverse articles per issue, was lost and that the section editor system did not provide the rapport between author and editor necessary for a strong journal.

A simple questionnaire was sent to members of the society. Their response indicated that 50% wanted separate specialty journals, 25% wanted sections within the existing journal, and 25% did not want any change. On the basis of this response, efforts were made to formulate a plan that would best serve physiology and that would work. Among the plans seriously considered were two that were eventually rejected, for reasons that may interest the reader.

One plan was to combine the *American Journal of Physiology* and the *Journal of Applied Physiology* into a single *American Journal of Physiology* and to arrange it in sections within one cover or to publish the sections as separate journals. This plan was rejected for several reasons. A single journal containing sections would not emphasize the separateness of the units. On the other hand, by publishing only separate journals the society might lose the support of generalists and librarians who had a long history of subscribing to the *American Journal of Physiology*. Also it was considered unwise to eliminate the *Journal of Applied Physiology*, which had been experiencing in recent years increases in the number of manuscripts received, rejected, and published; increases in the number of subscribers; increases in the number of advertisers; and improvement in the quality of what was being published. Further, as this journal had been started in response to the needs of members of the society who are still active, the potential for unhappiness within the society was great if its publication were to be stopped.

A second plan was to leave the *Journal of Applied Physiology* unaltered, but to bud off, one by one, the sections of the *American Journal of Physiology* as separate journals and thus cautiously develop a series of specialty journals, while gradually reducing the scope of the parent journal. This plan was not tried since slow budding would not satisfy the present needs of the field for specialization and might seriously sap the strength of the parent *American Journal of Physiology* if the budding was spread over an extended period. The Publications Committee was confident that the larger, stronger sections of the journal - for example, renal physiology or circulatory physiology - could compete successfully with specialty journals in those areas, if published as separate journals. They were uncertain, however, of the vitality of what would be left - articles on many different subjects in a journal without a strong author-editor tie. This would be especially marked when the new journals received their own editors and editorial boards, and the attention that goes with the establishment of new journals. Clearly, the experiment would result in several areas of physiology being provided with specialty journals and most areas finding themselves with weaker publication outlets.

A different plan was selected as best serving physiology. It has the following essiantial elements: (1) The scope of the *Journal of Applied Physiology* is restricted to articles on respiratory, environmental, and exercise physiology, with the title modified to reflect this change. Most manuscripts in other areas formerly published in the *Journal of Applied Physiology* will find a home in the *American Journal of Physiology*. (2) The *American Journal of Physiology* is divided into five individual specialty journals. (3) Each of the journals has its own editor, associate editors, and editorial board. (4) Great freedom is given to each editor to develop a journal that is responsive to and best fits the needs of the subfield it serves. As part of this plan, each editor is allowed up to one hundred pages per year to use at his discretion - for example, for editorials, invited reviews, and abstracts. (Most editors are using these pages for articles intended for a broader readership than that of the specialized volunteered research reports that predominate in the journals. Particular care is taken in reviewing and editing these articles to ensure that effective communication is achieved with this broader audience.) (5) The *American Journal of Physiology* continues to appear as a consolidated journal that combines the five specialty journals.

Journals of the American Physiological Society

reorganized journals	1976	1977
American Journal of Physiology (AJP)	monthly	monthly
AJP: Cell Physiology		bimonthly
AJP: Endocrinology, Metabolism and Gastrointestinal Physiology		monthly
AJP: Heart and Circulatory Physiology		monthly
AJP: Regulatory, Integrative and Comparative Physiology		bimonthly
AJP: Renal, Fluid and Electrolyte Physiology		monthly
Journal of Applied Physiology: Respiratory, Environmental and Exercise Physiology	monthly	monthly
unchanged continuing journals	**1976**	**1977**
Journal of Neurophysiology	bimonthly	bimonthly
Physiological Reviews	quarterly	quarterly

By this plan, the four journals published by the American Physiological Society are replaced by nine journals (table). This plan has numerous advantages, among them: (1) An author's article appears as one of only ten to twenty-five papers rather than as one of fifty in each issue. (2) Separate editors, associate editors, and editorial boards replace the section editors and the large combined editorial board. (3) Specialty journals can be added or dropped as the needs of physiology change. Indeed, two new journals were added in the process of reorganization, one covering cell physiology and the other regulatory, integrative, and comparative physiology. (4) The consolidated journal continues to exist and to serve generalists and librarians who do not wish to break the existing series of the journal. (5) Each specialty journal can be purchased at a price more affordable by individuals than that of the large journal.

Thought had been given to the editorial arrangement, but the actual selection of editors could not begin until the reorganization had been approved by the Council of the American Physiological Society. This permission was granted in October 1975 at the society's meeting in San Francisco. The Publications Committee decided that it would select an editor for each journal and then give him great freedom in deciding the number of associate editors and editorial board members that he believed would best suit the needs of his individual journal. The selection varied from two to six associate editors and none to fifty members on the editorial board. Recommendations for appointments were made to the Publications Committee, which then issued the formal invitations. Two members of the Publications Committee agreed to become the editors of the two new bimonthly journals, and were succeeded by two new members on the committee.

An organizational meeting was held for each new journal at the end of 1975 or the beginning of 1976. Each meeting was attended by the editor, associate editors, publications manager, executive editor, and managerial assistant. Each editor was encouraged to operate in a manner that would best meet the needs of his journal, without contradicting the general policies of the society. The scope of each journal was discussed. Although a statement of its scope was eventually developed for each journal, it was considered desirable to have authors decide which journal would provide the best audience for reports on their research, and thus shape the contents of the reorganized journals.

Operating procedures for handling manuscripts were established at these organizational meetings, and manuscripts received in the Publications Office after 31 March 1976 were considered for the reorganized journals. (In the transition period, an author who did not designate a specific journal was telephoned and asked to choose the appropriate journal for which his manuscript should be considered.) After a manuscript is checked for completeness and logged in, it is sent to the editor or associate editor who is responsible for having it reviewed. With four of the journals, manuscripts are sent to the editor; however, two groups decided that the Publications Office should assign the manuscripts to the appropriate associate editor. A statement of expertise is provided by each editor and associate editor to assist in the proper routing of manuscripts. Review procedures have not been changed.

Once the decision had been made to reorganize the journals and the editors and editorial boards had been selected, a myriad of decisions regarding the printing of the journals had to be made and implemented. In each case, care was taken to reorganize in a manner that would provide the maximum advantage to physiologists, librarians, and information retrieval specialists.

The Publications Committee decided that the scientific information in the individual *American Journals of Physiology* should not vary from that in the consolidated journal. Yet some material should be in the individual journals that need not appear in the consolidated journal. For example, 'Information for Authors' should appear in each journal, but it is not desirable to repeat it five times in the consolidated journal. It was decided that the best way to produce the journals would be to have a pressrun sufficient for the individual and consolidated journals. Each individual journal is planned so that signatures can be collated to produce the consolidated journal. Material that is to be removed from the individual journals is printed on separate signatures, and care is taken so that excessive blank pages do not result when signatures are collated to make the consolidated journal. (Up to three blank pages could result between journals since the smallest signature used by the printer has four pages.) Advertising pages and front matter that are added to the consolidated journal are also printed on four-page signatures. Author indexes, subject indexes, and front matter for the end of volume issues generate their own problems, which are solved by applying the general rules just described for removing and adding signatures.

The high quality and legibility of the type and illustrations were not changed. However, a fresh look was given to the covers, front matter, and mastheads that identify categories of articles. The cover design is similar for each of the six individual journals and the consolidated journal, varying only in exact titles and colour. The colours were selected in a rainbow sequence that corresponds to the alphabetical order of the individual *American Journals of Physiology*. The front matter and overall table of contents, with tabs to help locate the component journals, emphasize the individual nature of the separate journals.

Restructuring the journals necessitated some reorganizing of the editorial staff. Individual or pairs of copy editors were made responsible for one or two journals. Because the consolidated journal cannot be assembled until all its components are ready, it was essential to introduce a more formal production system. The managerial assistant for publications took over the role of production manager of this multijournal program and designed a system of tracking the many deadlines for manuscripts, galleys, page proof, and the like. Potential delays can now be spotted readily and remedied.

Once the reorganization had been sanctioned, the next essential action was to inform subscribers of the impending changes. Only fifteen months were allotted between the approval by Council and the publication of the first issues under the new plan in January 1977. Articles were written for *The Physiologist*, the society newsletter, and four-page inserts were placed in the *American Journal of Physiology* and the *Journal of Applied Physiology*, with a banner on the cover of each drawing attention to the insert. Direct mail pieces were sent to libraries and individual subscribers. In concert with this informational program, a subcription campaign was begun. A firm that specializes in promotional work for scientific publishers was selected to handle this aspect of the reorganization. Librarians and generalists were encouraged to maintain their collections of the *American Journal of Physiology* and *Journal of Applied Physiology* as an unbroken series. Individuals and specialty libraries - for example, cardiovascular and nephrology units in hospitals - were encouraged to purchase one or more of the individual journals. This message was spread by direct mail, by advertising in physiology journals and in subscription agency catalogues, and through the display of the journals at scientific meetings. Records were kept of sales traceable to specific lists used in the direct mail campaign. Analyses of these records and those derived from the newly installed computer fulfilment service proved useful in designing a promotional plan for 1978.

In 1978 we will try to increase the number of subscribers to the individual journals. A theme has been developed emphasizing that in a science as diversified as physiology, each researcher's needs can best be met by providing flexible publications options - multiple choice. The new publishing program of the American Physiological Society offers such flexibility: for the specialist there is a journal or group of journals of choice, for the generalist or librarian there is the consolidated journal.

The most encouraging aspect of the new program has been the sizeable increase in receipt of new manuscripts. From 1971 to 1975 the number of manuscripts received for the *American Journal of Physiology* and the *Journal of Applied Physiology* remained at about 1650 per year. In 1976 the number rose to 1827 and at the rate of receipt as this was written promised to be over 2000 in 1977. Each of the specialty journals is developing in a way best suited to meet the needs of the field it serves. This is most visible in the hundred pages per year allotted to each editor. Some of the special features that have appeared are: selected abstracts of important articles in the field of specialty of the journal; a series of articles on the philosophy of science, as it relates to one of the journals; brief reviews that evaluate a technique or research area; a series of reviews on one topic; articles based on a symposium; and companion papers on opposite sides of a controversial subject.

The time between the receipt of manuscripts and publication has been shortened. New individual and library subscribers have been gained for the individual journals, which balance the anticipated loss in subscribers to the consolidated *American Journal of Physiology*. Those who gave up the consolidated journal often chose several of the individual journals. Now that the product is on the market, we anticipate a continuing increase in the number of subscribers.

Researchers and librarians have written suggesting improvements. Often these have been adopted or shortly will be. We welcome additional comments. Only through constant evaluation and modification can such a program succeed.

THE JOURNAL AS A CENTRE FOR INTERNATIONAL COMMUNICATIONS: THE EXPERIENCE OF *CURRENT ANTHROPOLOGY*

Cyril S. Belshaw, *Editor, Current Anthropology*
University of British Columbia, Vancouver, B.C., Canada V6T 1W5

ABSTRACT

Current Anthropology was conceived as an international interdisciplinary journal to which, subject to practical constraints, every professional anthropologist had the right of access. To maximize the possibility, special steps have been taken to overcome the financial and political barriers to distribution, to establish an agreed upon language policy, and to counterbalance the dominant flow of scientific information. As part of this process, *Current Anthropology* has an international refereeing and commentary-discussion process of unique dimensions, and professional subscribers constitute an organization of Associates. Each issue has a bound-in Reply Letter which is used to inform the Editor of editorial, subscription, professional and scientific matters. Developments now being studied include the creation of a computerized system to replace directories and bibliographies, the formation of an association of anthropology editors, and the establishment of an Institute for International Communication in Anthropology to resolve the many questions of international communication which cannot be handled now. The disappointing record of international organizations in this and perspectives of Third World countries is discussed.

From its inception in the late fifties *Current Anthropology* has been conceived of as a printed journal, a conception which is now being seen to be important, but to have limitations. It has had three primary objectives, and one upon which the others must be founded.

The first objective is that the journal must speak across the sub-disciplines of anthropology, a subject which is defined differently in different countries in different generations. Some idea of the variety of fields can be obtained from the following partial list: prehistory, physical anthropology, human biology, human evolution, social evolution, social change, cultural anthropology, economic anthropology, urban social organization, values, methodology. The journal does not compete with specialist periodicals in these fields, and there is an obligation on authors to present their material in an educative way, that is to speak to those scholars in other sub-disciplines who may not be familiar with the detailed subject matter of discussion.

The second objective is currentness, that is to present materials as the growing points of the various subjects, both by presenting new results of significance, and by reviewing critically the state-of-the-art, particularly in such a way as to raise new questions and hypotheses. There is inevitably a speculative quality in some of the presentations, and in some respects the journal is like the review journals of other disciplines. However, there is a considerable additional emphasis upon debate, and special techniques are

Miriam Balaban (ed.), Scientific Information Transfer: The Editor's Role, 77-80.

used to encourage discussion. The idea of currentness extends to professional as well as to scientific matters, professional information being communicated in a Newsletter section.

The third objective is that the journal be completely international. When it was founded, Dr. Sol Tax, its first Editor, made it clear that in principle the journal should be available to all anthropologists everywhere, *regardless of capacity to pay*. That principle has been maintained. It has not been unduly expensive, but would not have been possible but for the back up provided by the Wenner-Gren Foundation, and the willingness of scholars in the "have" countries to support scholars elsewhere. The results can be seen in the extraordinary reach of the distribution of the journal, much wider than that of any other journal in anthropology and indeed in other social sciences except those which go routinely to governments.

The Editor of the journal thus obviously has a vested interest in the world infrastructure of anthropology, its networks of communication, and the organizations which give visible manifestation to it in various countries. He has a mandate then to work to support the formation of scholarly and professional institutions, to make the pages of the journal available for their intercommunication, and to keep in physical touch through visits and discussions. A great part of my time is thus spent in travel to obtain first hand knowledge of conditions of work in various parts of the world, to see for myself the institutions and colleagues wherever they may be, and to encourage the publication of work and ideas which may not be internationally known or which may be operating at a disadvantage by comparison with scientific work in the richer countries. *Current Anthropology* has stood in support of the formation of national societies, for example, and is almost the sole regular communicator of information about the International Union of Anthropological and Ethnological Sciences and the anthropological work going on in the United Nations agencies.

On a very broad definition of anthropology, there are about some 3,000 journals which publish in the field. Concern with the communications infrastructure has led *Current Anthropology* with its sponsor the Wenner-Gren Foundation for Anthropological Research, to study the problem of relationships between journals and editors, to create a Committee for International Publication in Anthropology, and to envisage an International Association of Anthropology Editors which will become affiliated with this body. Colleagues in India are proposing to establish a computerized up-dating directory of journals, with the ultimate aim, among other things, of enabling editors to re-direct mis-directed manuscripts to more appropriate journals, and to provide a basis for later rationalizations.

I referred above to the question of world distribution, one of the major elements in the *outflow* of information. The problems of national boundaries and economics, to say nothing of language, are very serious, and we have only partially overcome them. Still, the journal goes to individual scholars in 110 countries, including the Peoples Republic of China, USSR, Mongolia, Afghanistan, Nepal, wherever in fact anthropologists are present. To do this, we must beat the financial restrictions which prevent scholars from obtaining hard currency subscriptions. Like the journal of the Society for International Development, our subscription rates are substantially reduced for countries outside of North America, Europe, Australasia and Japan; in effect the richer countries subsidize the poorer. This is a situation which must continuously be reviewed, and the rates for the OPEC countries where individual scholars have good income will shortly be raised to the richer level. In some countries (Poland, India) we encourage scholars to place their subscriptions in a local fund, since they cannot send hard money outside their country, to be used for the benefit of anthropology, although on our books this must be counted as free subscriptions. In others (USSR) senior scholars pay subscriptions, and for each subscription we send free copies to persons who do not have official access to foreign funds. In yet others (Brazil and Indonesia) we have taken scholars off the free list as they have been able to gain access to currency, on the understanding that we would use the released money to help in countries (e.g. Peru) where restrictions are suddenly applied. There is evidence to the effect that such measures are understood and greatly appreciated, adding to the goodwill with which the journal is received, and adding to the likelihood that scholars in those countries will contribute actively to the inflow of ideas and research. This goodwill has also meant that political questions have not so far stood in the way of exchange of ideas and circulation.

By decision of the readers who were polled, the lingua franca of the journal is English. We do however reprint from other languages; we print abstracts in French, Spanish and Russian as well, and we print short items in the language of submission as well as in English. We translate the titles of non-English publications received, and their contents into English. We have produced one selection of articles in Spanish and are studying the possibility of such selections in a variety of other languages. This is extremely important because although English is a lingua franca, it is by no means universally known or used, and indeed the importance of other languages - Russian, Spanish, Hindi, Chinese, Arabic, Hebrew, Italian - at the undergraduate level is actually increasing. It is also becoming evident that work in those languages is less and less entering the international mainstream unless it is translated into English. English then takes on the function of interface between different languages.

The outflow questions are essential to the journal's concept of active scholarly interaction, the other side of which is *inflow* of materials. Only if a world wide community of scholars receives the journal will they have any chance or expectation of contributing to it, attaining international recognition, and correcting the ethnocentric imbalances which are as rife in the world of science as in any other world. Professional anthropologists are in fact accorded the status of Associates in Current Anthropology, who in various ways influence the policies of the journal, which thus takes on a peer-cooperative character. Each issue of the journal contains a bound-in air letter, which Associates can fill out and mail easily. The Editor can put questions of policy and also fact-finding questions to them, using the Reply Letter to get answers. This can be used for market research, for testing policy ideas, for mobilizing support when colleagues are being persecuted for scientific activity, for obtaining professional data for inclusion in the world directory, and many similar purposes.

Another element in inflow is the peer review and commentary system, which has been adapted for use in the new Cambridge journal *Behaviour and Brain Science*. Each article submitted is sent to a panel of twenty referees in various parts of the world; U.S., USSR, Western Europe, Latin America, Asia, Africa, are always represented in the panel ensuring an international structure. After acceptance, the successful article is then sent to a further thirty scholars, making fifty in all, for inviting comment. Not all reply, for various reasons, but usually there are from twenty to forty responses, to which the author adds his final word. The effects are interesting. The international distribution ensures that points of view are expressed from the whole variety of scientific and philosophical traditions, and many scholars are drawn into discussion networks from which they would normally be excluded by politics and geography. While some of the comments are banal, many add new data and dimensions which would not have been read in conjunction with the review article under normal circumstances. It often happens that the article plus comments plus author's reply encapsulate all possible points of view on a developing scientific topic, a package of inestimable benefit for graduate discussions and teaching.

A major inhibitor in the process is language, and we do everything we can to encourage scholars who may have difficulty writing in English, who may not even have a typewriter, to contribute in their own handwriting, in their own language. We find ourselves translating from Turkish, Croatian, Slovak, Indonesian.

The modern technologies which are being discussed here have important possibilities and limitations. We cannot afford to use a technology which is unevenly distributed in the world if this means excluding scholars who do not have access to it; thus dependence on a world distribution of computer terminals is out. But this does not mean rejection of the computer. In fact we are developing a system which will enable us to use our world directory of anthropologists with the utmost flexibility not only to assist our direct journal processes, but also to assist further in world communication. We will hope that it will be possible for a scholar in Africa or Java to write to us if he wants information about scholars interested in pertinent topics in other parts of the world, and obtain a printout of relevant names. Thus a world satellite computer system can operate to the disadvantage of poorer countries, but a system centralized in one or a few places, coupled to the world distribution of the mails, can actually improve their comparative position.

This applies to other fields, and is one reason why *Current Anthropology*, although committed to print for the review articles, is not committed to print for further developments. The same computer system can be used for bibliographies, abstracts, and the like. We have our eye on this. At the moment our bibliographical services are in print, and so expensive that third world countries can't afford them. In a few years we hope to be able to use the same combination of computers and mail to reduce the inequities and make a much more flexible and efficient service available. Anthropologists are very wordy in their writing. We are trying to develop in the columns of *Current Anthropology* a demonstration of the utility of brevity in Research Conclusions. If successful, these too can be put on computer and coupled to data centres storing the longer ethnographic and similar papers which can be replicated and distributed on demand after identification through abstracts. I fully believe that the main handicap at the moment is conservatism in the profession plus the high costs of computer storage and search. The latter factor is likely to change dramatically in the next ten years, particularly with the perfection of silicon chips. If so, *Current Anthropology* wants to be ready with a system that can be applied inexpensively and with international equity.

The system is in fact basically in place, and waiting for the economies of technological development to turn what has been a journal into, in effect, a centre for communication, electronic as well as printed.

DISCUSSION on "Multiple Choice, a Publications Option for Diverse Science" by Alfred P. Fishman and Steven R. Geiger

The Journal as a Center for International Communications - the Experience of *Current Anthropology*" by Cyril S. Belshaw

ANDREW SHERRINGTON - To Prof. Fishman. Did I understand you to say that you could get each of the five sections for fifteen dollars a piece but you are charging the librarians two hundred?

ALFRED FISHMAN - The whole system would fail then, wouldn't it? The members can get each one of their specialty journals at half price. For example *Circulation Research*, a distinguished journal of the American Heart Association that attracts the same material as one of the new journals of the Society, *Heart and Circulatory Physiology*, costs fifty dollars per year. The American Physiological Society's journal costs twenty-seven dollars and fifty cents per year for members.

PETER LENGYEL - I would like to ask Prof. Belshaw how many anthropologists do you suppose there are in the world and do you think that professions with many more members than that could adopt a system similar to the one of *Current Anthropology*?

CYRIL BELSHAW - Well that is an interesting question Peter. First of all, I've revised my opinion about the number of anthropologists in the world. Before I took on the journal and got involved with it, I thought the number of anthropologists in the world was something like the order of seven or eight thousand which is not a very big community. I now realize that it is totally and completely wrong. Furthermore I had thought that the largest number and certainly the largest number in any one country still is, in the United States, and this is the kind of information that generally circulates. That I know is not true. There are nearly as many anthropologists in the Soviet Union as there are in the United States. There are nearly as many in Latin America as there are in the United States. Now what this adds up to is a total population of I would say somewhere between fifteen thousand and twenty thousand - pure guess. We are tapping seven thousand of those as subscribers with an additional number as readers. In other disciplines, certainly in other social science disciplines, I know there is thought being given to this for example in political science which has a smaller number of professional people in a smaller number of countries. They are thinking about it for example.

URI BLOCH - I have a suggestion about this difficulty you seem to have in getting your data base into the computer. If you could typeset with the aid of a computer, which is not necessarily any more expensive than typesetting any other way, you would get your data base for free and you could then use it very conveniently. You could give a tremendous service by pointing out to every one of the fifteen thousand anthropologists the actual papers which might interest him directly. The storage itself, I believe, is not very expensive. I think that a disc with a hundred megabytes, that is a hundred million characters, costs something like a hundred and fifty dollars. I cannot believe that this is too expensive for you.

CYRIL BELSHAW - I am not an authority on this but we've been through it with the computer centre, and it so happens that I agree with you. It is not the expense. It is the demand and the particular computer that I happen to be linked to or able to use which has reached saturation point. So that requires a much more difficult expansion level. This is basically the issue. But also you have raised another question which interested me very much this morning and that is the capacity to feed the material in. As a matter of fact I want to put this question to you, whether the people you were referring to this morning could also be used to put the data into computer storage in your system? That is, have we reached the point of machine readable scanning which could actually be used to store the material? If that could be done economically it would certainly save some of the problems, deal with some of the problems that I am facing, although we are managing.

ALFRED FISHMAN - I am intrigued by your approach. Unfortunately, it is not easily transferable to the American Physiological Society. But, I do have one concern: where do you get the assurance that your reviewers are first rate? The American Physiological Society has difficulty in finding two good reviewers for each scientific paper. Not infrequently, a third reviewer is necessary to resolve differences between the two. How do you mobilize 20 or 30 reviewers?

CYRIL BELSHAW - The first point you should understand I think is that the articles by definition are broad. They are very broad. For example here I have, *A Review of World Archaeology* between years X and Y, as seen from Leningrad. And another one dealing with cultural resource management. Well that is a broad article. It is not so difficult to really get people who can interpret what is going on and as you say they write extensive critiques usually.

YEHUDAH WERNER - Now I was fascinated by the economic aspect of *Current Anthropology*. We had such an experience in Israel where the Journals published locally were priced with one price for people who could pay in dollars and at a much lower rate for people of a somewhat underdeveloped country who could not afford the same prices but who were interested in the material - the local people. We reaped two kinds of results: first, we heard that people abroad resented the fact that they were subsidizing the local people. Second, people from abroad tried (and very often succeeded) to locate Israeli colleagues who would buy the journal for them at the lower rate. So gradually the publishing institution, the Weizmann Science Press, is raising the local rates in order to gradually equate them with the rates abroad and I wonder whether you are not going to experience something similar.

CYRIL BELSHAW - Well I think I can answer that fairly directly. The journal has now been in existance for some eighteen years with this kind of rate structure. And what you are afraid of hasn't happened to my knowledge. It may have happened in one or two instances, but I really don't believe that it has happened. In the first place the addresses are a giveaway and for someone in the United States or Britain to order *Current Anthropology*, to have it come to Israel and to have it go back to them, I don't think it would pay anyone. The resentment is certainly not there to my knowledge at all. I think if there were real resentment, it may be there among the non-subscribers, but these issues have been put to anthropologists in groups who have been able to discuss them, as I mentioned in the case of Brazil for example. No, I really believe that the anthropological community is seriously interested in getting the journal out and it has built up a kind of a mystique which representes the journal as an instrument of international dialogue.

FLAGGING AND TAGGING SCIENTIFIC DATA

EDGAR F. WESTRUM, JR.; SECRETARY GENERAL, CODATA
DEPARTMENT OF CHEMISTRY, UNIVERSITY OF MICHIGAN
ANN ARBOR, MICHIGAN 48109 USA

Abstract

Although indexing of the data content of the scientific literature has been scarcely attempted, its desirability has been acknowledged and experimental projects in nuclear science, in thermodynamics, and elsewhere have shown how concise statements of data content can be constructed and used. There are yet unanswered questions as to whether data indexing will be of enough utility to the scientific community to offset the added cost and--if so--whether data indexing should be introduced into the primary literature, or the secondary literature, and upon whom the responsibility for their generation will devolve. Should data flags--as an aid to retrieval--or more detailed data tags--as an aid to a fuller description of data content of a paper--be added? Should data flags or tags have both and/or other functions?

CODATA has, since its inception in 1966, promoted the work of data centres and data-conscious scientists. The efficiency of data centres, compilers, and evaluators could be significantly improved by appropriate indications of data content of articles in the scientific literature by the processes of flagging and tagging data. Together with ICSU AB, CODATA has sought to explore the basic philosophy, the fineness of grid, the economic and personnel requirements, the interdisciplinary and international compatibility, and the promotion of international coordination of flagging and tagging.

Miriam Balaban (ed.), Scientific Information Transfer: The Editor's Role, 83-94.

INTRODUCTION

Although indexing of the data content of the scientific literature has been little done, almost two decades ago(1) it was mentioned and nearly a decade ago in a study of scientific and technical activities in the United States(2) it was noted:

> "The operations of existing data systems and document systems should be conducted so that they complement and supplement one another. Existing document handling systems should augment current indexing of conceptual content of documents to include adequate indexing of the data content of documents."

Yet even in 1977 we find that the primary literature--the scientific journals--is not particularly receptive to data! In fact, it treats data as something alien to itself.

Most editors tend to suppress data--on an argument of shortage of space--and favor textual material, speculation, and conclusions as more relevant. As a matter of fact, many authors have a similar point of view.

In the secondary literature--the abstract literature--the abstracting itself is largely on the basis of concept and it is often not possible to tell whether an article does or does not contain any original, new data.

The tertiary literature consists of tabulations, compilations, and critically-evaluated data. It provides an important service in the compression and the compaction of the literature making it accessible to the scientific and technological data users. The Committee of ICSU (International Council of Scientific Unions) most involved in this compaction--or implosion--of the burgeoning scientific literature is CODATA. This acronym stands for the Committee on Data for Science and Technology. It has, since its inception in 1966, promoted the work of data-conscious scientists and of data centers. Several hundred such centers exist and--unlike A and I services--are poorly organized and have little contact with each other, especially since many of them deal only with a rather narrow field of science. It is, of course, noted that individual scientists and groups of scientists on occasion are involved in a non-recurrent data compilation endeavor and thus also are intimately interested in the work of CODATA.

Is data indexing of enough utility to the scientific community to offset the additional cost? Should data tags be introduced into the primary literature? the secondary literature, or elsewhere? Upon whom does the responsibility for their generation devolve? Should data flags--as an aid to retrieval--be used or should more detailed data tags--to provide a more fuller description of data content of a paper be produced? What other functions, if any, should data flags and tags have?

BUT WHAT ARE FLAGS AND TAGS?

A data flag--as designated by the ICSU AB/CODATA Joint Working Group on Data Flagging and Tagging(3)--is a code symbol which indicates the presence of data and may--but need not--identify the type of data reported in a primary publication.

Perhaps the most simple-minded scheme might consist of the three symbols: a plus sign, a negative sign, and a question mark (+,-,?). Such flags could be designated to indicate the presence, the absence, and uncertainty as to the presence of data in a given article. One might, however, envisage one or more appropriate symbol at the end of an abstract in a primary publication or in an abstract journal which had been established internationally and which would indicate the various data types in that paper.

A data tag is a more detailed identifier which is used to characterize scientific observations, measurements, or calculations on the substance or system under discussion. In some respects, it might be convenient to regard data tags as "key words" for data.

EXTANT AND EMERGING FLAGGING SCHEMES

A representative sampling of existing flagging schemes will be seen in the four publications noted below.

1. Bulletin of Thermodynamics and Thermochemistry[4]
2. Nuclear Science Abstracts (discontinued)
3. Current Physics Information Data Base[5]
4. Physical Review C ("keyword abstracts"--a tagging scheme)

This is by no means an exhaustive list, but it does indicate typical current schemes, some of which have been utilized already for a number of years. The developing and emergence schemes of the same sort are indicated in the subsequent list:

1. American Institute of Physics (OSIS-NSF)
2. Chemical Abstracts (OSIS-NSF)
3. Engineering Abstracts (OSIS-NSF)
4. IUPAC - Analytical Chemistry Division

Details are not yet available on all of these schemes since final project reports have not in all cases been issued.

PRINCIPLES IN FLAGGING AND TAGGING OF DATA

Although in many ways the rather parochial schemes developed--or underdeveloped--in the above lists, may be said to provide a useful service for the organizations (or periodicals) presently employing them, to what extent is there a case for the development of a more general, inter-disciplinary, international flagging and/or tagging scheme for the scientific literature? What are the fundamental questions which need resolution?

Although there are many peripheral questions of importance, the core questions involved would seem to be the following:

THE FUNDAMENTAL QUESTIONS

- Who are the potential users?
- How useful is the scheme?
- How complex a scheme should be devised?
- The fineness of grid problem?
- What about the economics of the operation?

Taken together, all of these questions represent a formidable array of interrelated concerns. It is questions such as these that the Joint Working Group has addressed, and I can only hint at the principles that are devised in CODATA Bulletin 19[(3)] but further details are available in the copies of this Bulletin made available to all of the auditors of this presentation.

The potential users of the scheme have already been enumerated and any author who has searched at great expense, delay in time, etc. a particular article in the obscure reaches of the scientific literature will appreciate the importance of knowing whether or not the article contains new data--and thereby justifies the search procedures for it--or whether it is simply a review or a discussion of already known data, etc. And, therefore, probably not worth the procurement procedures in some instances. As has been indicated--not only do data centers benefit by this knowledge --but occasionally individual scientists involved in a compilation endeavor likewise benefit by it.

Potential users of the systems are first and foremost the data centers and those who routinely or simply from time to time--or indeed only once in a lifetime--prepare a compilation or a critical evaluation of data in some particular area. The potential beneficiaries--i.e., the users--will, therefore, be seen to include not only the full-time compilation and evaluation professionals (--data centers--) but as well individual scientists. That secondary services appreciate the need is indicated by the response from half of those questioned that they were indeed considering the design of a flagging or a tagging scheme.

The question of the ultimate utility of the system has been suggested in part by the answers to questionnaires submitted to 319 entities presumed to be data centers by the Joint Working Group. The questionnaires indicated that approximately 48% of the sample represented data centers which were heavily dependent on the primary literature for their input data. About a third of these centers, in fact, relied totally on printed input from the primary and secondary literature. Most of them used the secondary literature primarily as a retrieval tool about; 40% of the centers did use computerized searching to a greater or lesser extent--typically for between 5 and 25% of their input. An actual small-scale test of the scheme may be necessary in order to really judge its ultimate utility.

The degree of complexity that can be tolerated in the scheme depends on whether it is envisaged that flags and/or tags will be attached by professional information specialists or by authors or editors of primary journals. The latter groups--whatever their degree of professionalism in their own fields--cannot be expected to cope with a scheme of great complexity. Certainly, a complex scheme would here meet with passive resistance and at best there would be frequent errors in applying it. Common sense and experience would suggest that five characters are about the limit for non-specialist uses. For example, telegraphic and cable codes are "words" of five letters; four-letter mnemonic codes are much more acceptable than current non-mnemonic six-letter varieties. If a telephone number is more than five digits, it is broken down in various ways into groups of at most five and usually in groups of three of four. On the other hand, information specialists can be expected to master schemes of any required degree of complexity.

But perhaps the most crucial question of all in the consideration of data indexing schemes is the question of the fineness of the grid under which the flagging/tagging is to be done. The answer must be sought in terms of the need of the prime users--in general, the data centers and/or compilers and other "users". However, it is obvious that a signficant relationship exists between the fineness of grid and the cost of tagging--as well as with the precision of retrieval. Consequently, the basic problem is to find the optimum in flagging/tagging detail at which the greatest benefit is obtained at the lowest cost. Although flagging is obviously the cheaper procedure, its cost benefit may well be much less than that of other more expensive detail tagging. On the other hand, diminishing returns may be expected to result from increasing the fineness of the tagging grid too much. Ascertaining the optimum will require experimentation with feedback from the user community.

I hold it to be obvious that the finer the grid the more useful the system will be to someone, the more it will cost, and the smaller will be the likelihood of that system ever coming into existence.(6,7)

The question of economics is a very complex one and the actual economics will depend very much upon the parameters involved in the choice between the adjuvant parameters of the section.

SOME ADJUVANT QUESTIONS

- Where should flags and tags be located?
  - the primary literature?
  - the secondary literature?
  - computer-searchable abstract tapes only?
- What are the personnel requirements of the different schemes?
  - Authors? Editors? Referees? Abstractors? Specialists?
  - Computer-aided schemes?
  - Coding?
  - Standardization?
  - Internationalization?
  - A truly inter-disciplinary scheme?

The question of where flags should be located is very much an economic one. Flags of reasonable complexity could, without significant increased cost, be added to the abstracts of articles in the primary literature and simply be carried along into the secondary literature without signficantly increasing the cost insofar as the abstract itself is concerned. If, however, the flag is to be carried into the abstract indexes, on each entry of a given substance, compound or property, the sheer printing costs become a significant factor. If, however, such entries are to be added only to computer - searchable abstract tapes, the cost can again be made more tractable.

Another rather important economic question is that of the personnel requirements for the different schemes. Who creates the flags and the tags? Can the authors be entrusted with this responsibility just as they are entrusted with the creation of a title and an abstract for their article? Possibly so if the flagging scheme is kept simple enough. Editors would probably not be willing to become very much involved in this. Could referees play a role? Or if it is not done in the primary literature, could regular abstractors provide this tagging and flagging as they prepare the abstracts for the abstract journal? On the other hand, if the scheme is such as to require information specialists to create the tags and flags the cost becomes a more significant factor. However, there is evidence that computer-aided schemes may indeed be feasible and as effective or more effective than some mentioned above.[8,9]

The question of the language of the flag or tag is also an important consideration. Natural language flags or tags are really rudimentary abstracts. The disadvantage of natural language tags are that they are uneconomical, that there are hazards of error in their translation and that they are difficult to computerize in search as compared to the use of an "idiot code". Moreover, the question of whether a "free vocabulary" or a "controlled vocabulary" would be utilized also needs consideration. Natural language tags are easy to apply, to check, to understand <u>in their native language</u>. They are unhampered in the sense that one can add new ones without prior agreement. It is to be noted that idiot codes do in a sense surmount the language barrier because in principle even a natural language is an idiot code in a foreign language. Utilization of mnemonic codes would, however, help prevent confusion arising between fields.

The important questions of standardization, internationalization, and the creation of a truly inter-disciplinary scheme as compared to a collection of parochial schemes obviously require careful consideration by internationally oriented groups. The importance of these principles has been stressed in the Joint Working Group's report. If these three criteria are to be involved it would be helpful if the important decisions could be made before the literature is cluttered with a welter of independently created parochial schemes.

It is to be hoped, therefore, that the Joint Working Group--or some successor operation--will be enabled to continue the discussion of the principles involved in the formulation of a tagging scheme, the ascertainment of its urgency, and the creation of a suitable scheme, taking into account the different requirements within the different area of science for data indexing. Is it too much to hope that a thermodynamicist, for example, would be able to use the same flags in seeking out a thermodynamic entity whether he was seeking data in chemical abstracts, physical abstracts, engineering abstracts, or biochemical abstracts?

In closing, I commend to your attention the considered conclusions of the Joint Working Group.

REFERENCES

1. "Proceedings of the International Conference on Scientific Information", Volume 1, National Academy of Sciences, 1959.
2. Cited by John Murdock in Chapter 4 for Unesco-GIP Guide.
3. "Flagging and Tagging Data", CODATA Bulletin No. 19 (1976), CODATA Paris, France.
4. Bulletin of Thermodynamics and Thermochemistry, No. 1 through No. 19 (1958-1976), E. F. Westrum, Jr., ed., University of Michigan Publications Distribution Service, Ann Arbor, Michigan 48106, U.S.A.
5. "Nuclear Data Sheets", (1974), 13, (1) iii-iv et seq., American Institute of Physics, New York, U.S.A.
6. "Energy Data Accessing and/or Retrieval", Report of the Working Panel on Data Tagging convened at the Energy Research and Development Data Workshop held May 6-7, 1974 at the National Bureau of Standards, Gaithersburg, Maryland, U.S.A.

7. "Energy Data Accessing and/or Retrieval", CODATA Bulletin No. 12 (1974), CODATA Paris, France.

8. "Automated Law Research", R. May, ed., American Bar Association, Chicago (1973).

9. J. Myers, "Computers and the Searching of Law Texts in England and North America: A Review of the State of the Art", J. Documentation (1973), 29, 212-28.

"By a fiction as remarkable as any to be found in law, what has once been published, even though it be in the Russian language, is spoken of as known, and it is too often forgotten that the rediscovery in the library may be a more difficult and uncertain process than the first discovery in the laboratory."

--Lord Rayleigh (1884)

DISCUSSION on "Flagging and Tagging Scientific Data"
by Edgar F. Westrum, Jr.

IRVING AMRON - One of our major problems now is standardization of nomenclature. The same subject handled on two sides of the water has different units and often different abbreviations. How can you retrieve some of these? How can you use index terms? Is there any mechanism by which one can standardize the international nomenclature?

EDGAR WESTRUM JR. - Through the committee of ISO the nomenclature problem I think is gradually being handled at least in some of the core of scientific domains, I'm not sure that it is true for all of the domains such as the medical, the bio and geo domains, but in the core sciences - at least in physics and chemistry - this is being brought to order. It is not at all certain though that one could hope to use the ISO nomenclature symbols as a basis for the actual flagging and tagging. It's under consideration, it's a possibility, but ISO symbols don't seem to be well suited to this. They are a bit complex and involve superscripts and subscripts and the like and are not necessarily well adapted to computer searching.

JOSEPH AGASSI - I think you are right in saying that much depends on the ability to persuade authors to flag and tag the way they now abstract. I suggest that the success or failure of such a venture depends on successful and intuitive and immediate and well-formulated criteria of what are data, what are criteria of novelty, what are criteria of significance. You didn't have of course in your schedule to speak of this, but I think sooner or later there must be a discussion. These are controversial questions. You know in every scientific sub-subliterature there is a question of what is the datum, or whether a specific bit is a datum or not too. So you know it can't be resolved either way; it's the same with novelty, and the same with significance. I think a discussion of these things ought to be taken up.

EDGAR WESTRUM JR. - I agree. You will find that for each sentence I have given you today, you'll find almost a page in Codata Bulletin 19 of which there are enough copies for everyone. It is true that authors are a source of cheap labor in this connection and if the scheme is not too complex it might be a feasible solution. I take considerable comfort in the fact that the American Institute of Physics in the United States has done a remarkable job in educating authors in the proper manner of writing titles and abstracts for articles and I think they have shown that the authors can be educated to do a task of reasonable complexity in this respect. However there are many ways of achieving a solution to this problem and one way that I didn't mention -- but is certainly on the horizon -- is the possibility of having it done largely by computer. Experiments have already been done to demonstrate when full typescripts are in computer-readable form, the computer is perhaps better at this game of tagging and flagging than are human flagging.

DIMITRI VIZA - I wonder what effect it would have if you asked the authors to 'flag' their own manuscripts. This might provide a sort of self-censorship. Indeed, every author has an idea of the importance of his ideas and results in his particular field or in his own discipline and even of his contribution to science as a whole, but nobody has ever asked him to evaluate and thus criticise his own work - except for appraising it when he's trying to scrounge a grant. So, no scientist has ever publicly evaluated his own work, say, as 'first', 'second' or 'third rate', or as 'repetition and confirmation of known data'. Now, do you think that many authors would be prepared to 'flag' the papers they submitted for publication as 'third rate', or 'repetition confirmation of trivial data'? Hopefully they would rather abstain from submitting their publications. This sort of self-censorship might be a solution to the current data inflation. Flights of imagination have ever had room to put the good housekeeping seal of approval on the data to attempt to classify it as to quality. I agree this is a difficult question and one shouldn't expect the author to be forced to do this, although we do expect the author to tell us - and this is another totally different matter - something about the precision and the accuracy of his data, but not in the classification scheme. Even if one uses a four and five character scheme it would be difficult enough to tell what properties were measured and what kind of substances over what range of temperature. There would not be any characters left over to indicate the quality of the data.

EDGAR WESTRUM JR. - That is a good point! As a matter of fact, no scheme that the committee has ever conceived even in its wildest flights of imagination have never considered putting the "Good Housekeeping seal of approval" on the data, i.e., to attempt to classify it as to quality. I agree this is a difficult question and one shouldn't expect the author to be forced to do this, although we do expect the author to tell us -- and this is another totally different matter - something about the precision and the accuracy of his data, but not in the classification scheme. Even if one uses four or five character schemes it would be difficult enough to tell just to take a simple example from physics or chemistry, what properties were measured on what kind of substances over what range of temperature. There would not be any characters left over to indicate the quality of the data.

A.J.C. WILSON - Professor Westrum has said about half of what I wanted to say in connection with the tagging of quality. I do not think that that is a job that either the author or the secondary service should be asked to undertake. What they should do is to tell the reader what has been measured over what range, or what field of geology has been worked on, or what dates in history are involved, as it is up to the reader or the compiler of the tables to make his own judgment, on the basis of what is said in the rest of the article, whether the data are good or bad. One other thing that the author might be asked to do in his flag is to say whether the data are his own or whether he is quoting somebody else. That is an objective fact, not a matter of judgment.

I was a member of the group that Professor Westrum has mentioned, but there is something that I should like to say as an editor, and not as having had anything to do with the group. The instructions for authors for the journals of the International Union of Crystallography tell them that as far as possible they should put the new numerical data in the abstract, and not merely say that there are data in the paper. We have generally managed to get them to do this as far as the primary crystallographic data are concerned, though obviously in any detailed structure determination there are a lot of things that will not go into the abstract. However, if anybody looks at an abstract in Acta Crystallographica he will see immediately whether there are numerical data or not - he doesn't need a flag. A flag of five-character complexity might of course tell him more about what he could expect to find if he could understand the flag and if he then bothered to read the article. If I could just put in a personal opinion, I think that five characters is about as much as one could expect the average author to provide and the average reader to understand. Anything beyond that would need a professional flagger or tagger to produce it and need a computer to do the sorting on it. It would not be a thing that the ordinary human reader or writer could be expected to use.

SCIENCE, SOCIETY AND THE EDITOR

Shneior Lifson, Chairman

TREATMENT OF 100 MAN-RELATED SCIENTIFIC PROBLEMS: SCIENCE AS AN INDICATOR OF PUBLICATIONS[1]

Jacques Richardson, *Editor, Impact of Science on Society*
UNESCO, 7 Place de Fontenoy, 75700 Paris, France

ABSTRACT

There are outlined a hundred problems facing mankind that have social significance -- problems which are all based on science or technology or both. The typology can serve as a useful guideline to scientific editors and publishers.

INTRODUCTION

Within the context of Unesco's current programmes on Research and Human Needs and the Human Implications of Scientific Advance, there follow six major groupings of science-based problems confronting mankind. These problems, whose nature is intricately complex and whose solutions are inevitably transdisciplinary, are in the process of being measured, analyzed or resolved through the intervention of Unesco as well as other organizations. The sum of the six groups of problems comes to roughly a hundred -- a figure an order of magnitude larger than the principal problem zones of all kinds that confront humanity at the international level.

The scientific communicator needs to remain astutely aware of the social directions and possible cultural implications of new research and continuing innovation of the types indicated on the list. The role of the scientific editor or other science communicator (who could be the research scientist or practising technologist himself) is briefly discussed for each classification of problem and for many of the specific problems.

THE PROBLEMATIC SERIES

Let us look, first, at the problem zones just mentioned, condensed slightly from Unesco's official Medium-Term Plan (1977-1982). These zones are:

---

1. For 'publications,' one can also read 'TV or radio programmes, or audiovisual presentations.'

Miriam Balaban (ed.), Scientific Transfer Information: The Editor's Role, 97-104.

Assurance of human rights
Reinforcement of the conditions for peace
Man as the centre of development
Application of science and technology to man and society
Education in response to individual and social requirements
Opportunity for new achievement by disadvantaged groups
Man and his environment
Population (in terms of demographic growth and distribution)
Communication between persons and peoples
Tools and systems for the exchange of information

Most communicators will readily recognize many, if not all, of the items on this list as the principal, generic pre-occupations of all of our species. One type is related to another, and all are interrelated. Detailed examination of these problem zones results in the collation of six, discrete but associated, assemblies of problems.

A HUNDRED PROBLEMS

(1) The expansion of science and technology

Most of these problems pertain, often acutely, to developing nations and their present level of development.

1. The rapid growth of scientific knowledge and the quick obsolescence of its resulting technology. This affects the industrialized as well as the less developed countries.
2. Science-technology for national development via integrated technology transfer. There is a critical need for endogenous science-and-technology education.[1]
3. Slowness in transfer of skills. An important sociocultural facet of the process of technology transfer.
4. The science-technology gap separating poor nations from the 'developed' countries. The gap is growing. (Combines with the two preceding questions.)
5. The pertinence to human needs of university-level teaching and research in science and technology. Can research be oriented towards man's needs more than is currently the case?
6. The imperatives of conservation and growth. Can these seemingly opposed compulsions be reconciled?
7. The brain drain. How serious is it?
8. Can the future be simulated? Science fiction aside, how systematically can we prepare for and provide for the future?
9. Integrating futures research in academic programmes. Are there social, cultural or ideological constraints?

---

1. This subject will be a principal pre-occupation of the United Nations Conference on Science and Technology for Development, 1979.

10. Improved systems of informatics and the social options available for their application.
11. Secrecy and privacy of the individual in the computer age. (Some nations do not interpret the invasion of privacy as a problem.)
12. Integrating econometrics in the formulation of public policy. Many legislators eschew 'the dismal science' and mathematical expression.
13. Transdisciplinary, cross-model, multipurpose information systems. Computer science is here to stay. Why not exploit it to the fullest as a systemic tool?
14. Forecasting future technology. If we do it wisely, we may be spared errors we can no longer afford.

(2) The physical setting of the human condition

15. Optimization of the patterns in human settlement. This is related directly to:
16. Rational design of our working and living areas which, in turn, relates to:
17. Suitable technology for alternative communities of the future. Where science and technology can be expected to have increasingly appropriate roles.
18. Demographic growth. Although the problem seems to be sociocultural, are there scientific-technical solutions and can they be applied?
19. Rational water management seems to be a crucial factor in all economic and social development, everywhere.
20. The criteria of water quality, whatever the uses of water might be.
21. Changes in the hydrological regime induced by human activities, and their environmental consequences.
22. Fluctuations in the hydrological regime relating to climatic variation (especially floods and droughts, and their environmental impact).
23. The development of water sources in the world's arid regions. The planetary supply is finite, so the question is how to redistribute it.
24. The global problem of fluvial and marine pollution.
25. Water supply in urban areas. It is still deficient in 1978.
26. Air pollution over urban areas. It can be controlled, e.g. the United Kingdom's Clean Air Act of 1956.
27. Science, technology and mass-transit systems.
28. Urban, interurban, and national transport grids 25 years hence. Based on electricity, they were well-planned a century ago. How can we correct the system breakdown in most areas of the world?
29. Attenuation of damage by natural cataclysm. We have slowly developed synoptic systems to foretell floods, tsunami, typhoons and some earthquakes. Are there limits to optimization of the processes?

(3) Research on the ecosphere

The following problems correspond roughly to those identified in Unesco's Man and the Biosphere programme and the work of Unesco's divisions concerned with the marine, hydrological and earth sciences.

30. How constraining on human endeavour are our natural resources? The term 'resources' includes energy.
31. The need to compensate for the uneven distribution of natural resources. Are there scientific or technical means to exploit?
32. The strain on the planet's fossil fuel supplies. Are they, in truth, destined to disappear?
33. How to assess nuclear fuel resources qualitatively and quantitatively? On the basis of foreseeable nuclear development...
34. Managing the extraction, transport and allocation of fuels.
35. Development of non-polluting sources of energy.
36. Substitute sources of non-fuel minerals. The problem applies to both ferrous and non-ferrous metals as well as non-metallic minerals.
37. Inseparability of the problems related to minerals, energy, and the environment.
38. Thermal loading of the planet. Are there limits?
39. Problems of fission power. Primary among these are (a) disposal of radioactive wastes and (b) the potential for nuclear terrorism.
40. Long-range supply of important minerals. This includes the recovery of metals.
41. Understanding how and where mineral ores are formed, including the minerals formed on the sea-bed.
42. Mining the sea-bed for minerals: developing rationalized procedures of investigation and extraction processes.
43. Social, economic and political aspects of ownership of the sea-bed, a problem expected to be as thorny as that of claims concerning the continental shelf.
44. Tracing man-made and volcanic particulates in the ocean-atmosphere system. Natural and artificial pollution combine to form a single threat.
45. The chemistry and physics of the ocean-atmosphere system.
46. Technology required to measure the chemistry-physics of, and the particulates in, the ocean-atmosphere system. There remains much instrumentation to be invented and perfected.
47. Developing better models of the chemistry-physics of the ocean-atmosphere systems. The permutations seem to be unlimited.
48. The irreversible entry into the environment of pollutants, such as asbestos, fission products, fly ash, heavy metals and talc. What happens to these, what damage do they do?
49. The response of ecosystems to large-scale, man-wrought change, which is directly related to:
50. The determination of tolerable levels of environmental contaminants, in its turn directly relevant to:
51. Deterioration of the biosphere which is attributable to technological evolution. Can this degradation be halted?
52. Rust and corrosion as natural technical pollutants. Can the economic constraints they impose be discounted?
53. Wastage of existing natural resources. These include sunlight, wind, tides, waves, and fluvial current.
54. Rational utilization of solar energy, in the form of direct sunlight, winds, tides, etc.

55. Finding new substitutes for energy resources. This complements problems 30, 32, 34 and 35.
56. Natural materials coming into short supply. (If appropriate substitutes are found in sufficient quantity, this becomes a non-problem.)
57. The new international economic order, a problem which may be defined in terms of available natural resources, relative national wealth, and the manageability of the indexing of the costs of raw materials.

58. New models of development. The search for paradigms other than those taken from the North Temperate Zone (excluding the People's Republic of China), identifiable as industrial production to satisfy personal consumption.

(4) Living organisms within the biosphere

These problems are commonly associated with basic research objectives in biology and chemistry.
59. Has the microcosm of life finite limits? Can we identify the lowest level at which 'life' commences?
60. Improvement of our basic knowledge of uni- and multicellular life. The process has momentum, but can its rate be increased?
61. Decision-making at cellular and sub-cellular levels, or determining how the neurophysiological options are taken at life's primitive stages.
62. Detecting environmental threats caused by chromosomal aberration in living organisms, whether animal or plant.
63. Molecular structures capable of triggering mutagenesis, one facet of the complex question of the causes of mutations.
64. The problem of genetic engineering, including (a) rules for the creation of life in the laboratory and (b) the socioeconomics of genetic manipulation as an ethical problem.
65. What are the biological bases of the immunological processes? Research has made small dents in the mystery of the immune processes.
66. A code of research ethics applicable outside the clinical setting. If we learn how to control degenerative diseases, how long will we live, how actively, even how productively?
67. Enzyme engineering. This, too, is fundamentally a socioeconomic problem.
68. Is eugenics a valid field of research in the intergovernmental framework? If not, to whom can be ascribed the authority therefor? Or is the question unanswerable?
69. How brain mechanisms work in memory and learning. This implies that man must 'reach inside' in order to understand his own brain.

70. The neurophysiology and psychobiology of human aggressiveness and violence: does sociobiology explain everything... or nothing?
71. The use of custom-made microbes to improve our management of natural resources and the environment: one form of controlled biological engineering.
72. The extinction of certain zoological elements: whales, seals, sites for bird-nesting. (Derivatives of DDT, to take another example, have so weakened the egg-shells of certain birds of prey -- eagles, falcons,

hawks, owls -- that most of their eggs do not hatch.)
73. Sexual differentiation in animals and crops, by which the evolutionary processes can be better controlled.
74. Photosynthesis: understanding biological energy transfer therein.
75. Man-made, new crop plants, to help combat famine.
76. Nitrogen fixation, a problem in chemical energy transfer (especially in the non-legumes).
77. Improvement of both natural and synthetic pest controls, with particular attention paid to secondary negative effects.
78. National self-sufficiency in food production. This is a requisite antecedent to industrial development and the growth of technology in the industrialized countries. The utility of the model for developing nations is clear.
79. Optimizing the use of chemical fertilizers. The challenge is well known, but meeting it remains a serious problem.
80. The substitution of vegetable for animal protein. The problem is sociocultural as well as technical.
81. Research and development of synthetic protein. The solution to the problem may have more sociocultural overtones than technical significance.

(5) Research policy, information, and international co-operation

The following category consists of the types usually referred to as problems in science policy and technology assessment.
82. A code of ethics for the scientific community. A perennial problem concerning the morality of research, its recent manifestations have included the renewed call (February 1977) by an industrial researcher, Arthur Kantrowitz, for a 'science court.' A recent civic implication of this ethical question has been municipal action by the city of Cambridge (Massachusetts) warning the biologists and geneticists of Harvard and MIT to observe certain safeguards in the development of further genetic research in their genetic institutions -- precautions even more severe than those already prescribed by the National Institutes of Health.
83. The role of science in modern society: bridging the 'two cultures': an everyday challenge to anyone involved in explaining science and technology to the layman, or even the intricacies of a given discipline to a non-specialist.
84. Scientific and technical information: is there an overload? The very question is capable of provoking an endless dialogue.
85. The efficiency of international organizations in promoting scientific and technical development. Whether we speak of the 'Common Market', the Club of Dakar, or the Organization of American States, a review of programme and activities on a regular basis seems justified.
86. The public's confidence in scientific research, one of the many social reasons for the creation of this list.
87. Science fiction and other visions of future worlds. Are we sure, in our own minds, of the difference between futurology and science fiction?
88. The obligations and limitations of scientific research.

Obviously related to problem 82, this is a constant obsession of many researchers in microbiology and high-energy physics.

89. Academic meritocracy and the publish-or-perish road to success in science. The issue is not parochial, as all of society must help provide the resources to train technical specialists.

90. Transdisciplinary stimulation of life sciences through physics, chemistry and mathematics. Integrated science teaching has only begun.

91. The effectiveness of communication within and between the 'invisible colleges.' Related to the question concerning an overload of scientific information (84), scientific communication is still relatively poor three centuries after the founding of the Royal Society.

92. Youth's expectations of scientific research and technical innovation. The gaps between generations and the 'two cultures' are frequently perversely great.

93. Discrimination against women in science and technology. This is a persistent social problem in most cultures, where often one-half of humanity has not the opportunity to participate directly in research or engineering.

(6) Threats posed by armaments and nuclear attack

The following category, while readily admissible in problem analysis, poses in turn another problem:

94. Who, or what, will reduce the threat of armed conflict or will attenuate the peril of nuclear war? Any variant of a reply must take into account the role of the United Nations, one of the reasons for the creation of which was explicitly to reduce or eliminate the resort to battle for (a) conflict resolution or (b) the settlement of disputes.

95. Conditioning the human mind so as to make warfare unacceptable. The obvious counter-problem is to neutralize politico-psychological systems tending to render warfare acceptable in the public mind.

96. The functions of science and technology in nuclear disarmament. How effectively can they be enrolled?

97. Reducing/eliminating the threat of nuclear war, by all means.

98. Attenuation of the potential of biological and chemical warfare, including, of course, radiological warfare.

99. The principles and systematic stages of technical disarmament.

100. Counteracting, in the world of science and technology, the influence of the military-industrial complex. The problem is primarily economic and social.

101. The role of military establishments in future society: alternative use of their research and development capacities.

102. Disarmament in relation to improved co-operation in science and technology, especially between nations having different ideological systems.

SIGNIFICANCE

Careful consideration of the problems described

above helps underline some of the reasons why scientists and engineers themselves have increasingly questioned their professional roles in relation to all the other walks of life. Public scepticism, anguish and even anger about what layman construes as misuses and abuses of research and innovation become clear as one reviews and re-examines the list.

Civic outcries to bring research to a stop are not likely to bring work in the laboratory or at the test bench to a dead halt, but both the scientist and the scientific communicator have the obligation to ponder the ethical and social foundations of each of the problems identified above. Only in this way will the science communicator be able to chart his way justly and usefully through tomorrow's world -- a world of endless research and exciting applications of every new discovery.

The author of this document, Jacques Richardson, is editor of the quarterly journal, impact of science on society, published by Unesco in five language editions. The ideas expressed are not necessarily those of Unesco or of impact of science on society.

INFORMATION TRANSFER AND MANAGEMENT IN THE EUROPEAN COMMUNITY

J.M. Gibb
Directorate-General for Scientific and Technical Information and Information Management, Commission of the European Communities, DG XIII, European Centre, Kirchberg, Luxembourg

ABSTRACT

An explanation is given of why the European Communities have, from the very start, developed a concern for the dissemination of information. It stems from the research programmes which were launched by the European Community for Coal and Steel, by Euratom and, more recently, by the European Economic Community. The Communities' interest in this subject has gradually broadened out into a recognition of the "Knowledge industries" as an important feature of the economy and a determination to create a true common market for information in the nine Member States. This led to the launching, in 1974 , of a "Plan of Action in the Field of Scientific and Technical Information", the major component of which is EURONET, the European On-Line Information Network. More recently, a programme was started covering the development of technical aids to the solution of the language barrier problem, such as machine-translation systems. In parallel to the Plan of Action, efforts have been devoted to the improvement of "primary" communication in Europe, i.e. of the actual generation of the materials to which information systems and networks give access. Here too the language barrier has been a source of concern. A seminar was held on the way it effects publishing in a multilingual society. Action has been taken on the basis of the recommendations which emerged.

In order to arrive at policy decisions in as rational a way as possible in the sector of primary communication, a special "issue-based information system" is being built up.

In the title of this paper are found the words "European Community". What is meant by this is a grouping of nine European countries (Belgium, Denmark, the Federal Republic of Germany, France, Ireland, Italy, Luxembourg, the Netherlands and the United Kingdom) which are in the process of moving towards economic, and possibly some form of political union.

There are in fact three European Communities. The first one, the European Community for Coal and Steel, was set up by a treaty signed in Paris in 1951. It is in Rome in 1957 that the other two were instituted, that is the European Economic Community, better known as the Common Market and the European Atomic Energy Community, better known as Euratom.

Miriam Balaban (ed.), Scientific Information Transfer: The Editor's Role, 105-109. *All Rights Reserved.*

Dissemination of information - a duty of the European Communities

How does scientific and technical information fit into the general framework of the European Communities? Its does so in two ways. Let us look first at the place taken by science and technology in the day-to-day work of the Communities. Research on coal and steel technology, as well as on certain health and safety problems affecting the coal and steel industries, is an important facet of the activity of the first-born of the European Communities. For over twenty years now, a substantial part of the research conducted within the Member States is funded by this Community. As for nuclear research, it is part of the raison d'être of Euratom, under which a joint programme is conducted. More recently, this concern for the development of joint research efforts has spread to other fields. Decisions were thus taken, under the E.E.C. treaty this time, to launch Community research programmes in the areas of environmental protection, agriculture and energy in general, to mention the most important sectors only.

The executive body of the European Communities, namely the Commission, has the duty to ensure the adequate dissemination of the results of all this research. Although some of the results are given confidential treatment, most of them are quite simply published.

This is done in various ways. The Commission has its own series of scientific and technical reports, divided into sub-series, one for each major subject area. It also publishes a series of conference proceedings.

As far as possible, however, the Commission prefers to rely on normal commercial channels for the publication of the information it has to disseminate. Thus research workers active within the joint programmes are encouraged to publish results in journals. The publication of longer reports and of conference proceedings are frequently entrusted to commercial publishers. This policy not only saves money, but - and that is certainly the major advantage - ensures that the information is satisfactorily disseminated, inasmuch as a publisher will only accept material which he is in a position to market satisfactorily.

I should like to come back for a moment to our own report and conference proceedings series. We feel somewhat uneasy about them, inasmuch as they tend to take care, in a way, of what publishers of books and journals have no use for. This does not mean that we consider these series to be rubbish-bins. In actual fact, some effort is put into excluding second-rate material. It is just that many research activities are so specialised that, however great the quality of the results which they produce, they can never be of interest to more than a few hundred or even a few dozen experts. Their distribution under normal commercial conditions is therefore impossible. Is this a reason for simply burying them ? I doubt whether this is an honest solution. The only course open to an organisation which wishes to make sure at the same time that valuable information is not lost and that too much of the taxpayers' money is not wasted over conventional reproduction of reports, is to look for alternative methods of dissemination. I hasten to add that we have not found an ideal solution. It consists in publishing the reports in microfiche form essentially, but I must admit that there has been no rush on them. Maybe this should not be interpreted as a condemnation of microfiche; after all, by definition, we are dealing here with highly specialised material about which only a limited number of people can get excited. The standard marketing techniques of publishers therefore tend to be uneconomic; I see no alternative to relying on the so-called "secondary" information services, namely abstracts journals and computerised bibliographic data-bases, to ensure that the availability of these reports is well-known.

A "common market" for information

The European Communities are sometimes looked upon as a mere European customs union. Let us for a moment dwell on the concept of a Common Market, with which the European Communities are frequently associated.

It will become apparent, upon reflection, that a Common Market cannot be set up simply by abolishing tariff barriers. What about other taxes, which can distort trade ? What about the social hardship which can be created as a consequence of changing the patterns of trade ? These, among other problems, are part of the official brief of the European Communities. Similarly, a Common Market cannot be confined to agricultural produce and industrial goods. It has to apply to services. I am glad to say that information services or, to use a more forceful term, the "knowledge industries" have received special attention. As a result, practical steps could be taken.

The measures I am referring to are contained in a "Plan of Action in the Field of Scientific and Technical Information", launched in 1974. The best known element of this plan is no doubt the creation of the European On-Line Information Network, which has been given the name of "EURONET". The basic component of EURONET will be an efficient international data-transmission network that will be the first such system to be set up jointly by the PTT administrations of the nine Member States. ts ultimate purpose is to provide users anywhere in the Community via screen or typewriter with direct access to scientific, technical and socio-economic data, to integrate existing and future on-line information services in the Community into a commonly shared network and to make available for the first time a large variety of services operated on widely different host computers in many different countries, via one single network.

The Plan of Action also covers the operation by the Commission, on its own host computer, of a number of European data bases, which are in existence or are being developed.

A third component of the plan covers what we call "horizontal" projects, serving as back-up to EURONET. They include the development of various standards, affecting software systems, for instance, the training of information specialists, particularly those who will be required in the special context of EURONET, and the provision of a number of multilingual aids, such as thesauri of indexing descriptors in several languages.

More recently, the multilingual facet of the Plan of Action has been considerably extended, in fact well beyond the immediate needs of EURONET. It has grown into a Plan of its own aimed at finding appropriate technical solutions to the whole problem of improving communication within a multilingual society. It includes fairly ambitious projects on machine-aided or completely mechanised translation, not only of texts where a standardised syntax is used but also of completely free-language texts.

"Primary" communication in Europe

After this brief sally into the EURONET and translation fields, let us turn to the subject of "primary" communication. However efficient information networks may become from a technical point of view, their intrinsic value ultimately depends on the quality of the primary material which they handle. They are to a great extent faced with a "garbage in/garbage out" problem.

Apart from the question of quality, the sheer quantity of published material is also continuing to give difficulties. Modern data bases and the ingenious search devices now available to users are helping them to live with the information explosion. Some users are however on the verge of complaining about the very efficiency of certain systems, which provide them in answer to their queries with extremely large numbers of references to documents, most of which are relevant : paradoxically, that is the trouble : the user simply cannot extract from the mass of unevaluated material an immediate and accurate answer to his problem.

If this is the kind of situation we are getting into, correctives are required. There is evidence that some are already being applied. Referral systems and inventories of research projects, many of them computerised, are being operated, which ultimately bring us back to the old-fashioned practice of getting information by talking to people.

The conclusion to be drawn from the point of view of editors and publishers is surely that there is a growing need today for review articles, state-of-the-art reports and the like ; also that conferences will remain for long a useful method of communication.

The Commission is endeavouring, in connexion with the primary communication process, to spot problems which it ought to tackle within its general duty to ensure that a true European Common Market for information is set up. The field for action is immense. Selectivity is therefore the order of the day. In any case not only are publishers and editors to a considerable extent quite capable of solving their own problems, but useful measures are being taken at a national level. This means that the Commission can concentrate on questions which its international vantage point enables it to tackle most readily.

It is therefore not surprising that one of the first topics it selected for consideration was that of scientific and technical publishing in a multilingual society. A seminar was devoted to it in November 1976. It produced recommendations, some of which are already giving rise to practical measures. The basic problem underlying the seminar was the growing predominance of the English language as a vehicle for the communication of scientific and technical information. I am glad to say that the subject, although potentially quite an awkward one, was discussed without unnecessary passion. No revolutionary conclusions were reached but it is probably true to say that a good deal of clarification was secured : more especially, it was realised that different solutions applied, depending on the type of publication considered and the subject matter it covers. For instance, whereas an article in English is probably the most satisfactory way of reporting on a piece of research in the field of physics, a good state-of-the-art report on a technological subject may only achieve the penetration it deserves if its is made available in several languages.

Another Commission venture I should like to mention in the field of primary communication is an experiment covering a new form of review journal. The journal was initially to be called "Growing points in Veterinary Science". The objective is to pin-point the most interesting recent breakthroughs in veterinary research which have occurred in the European Community and to map out the further developments and applications they could possibly lead to. The end-products are, as it were, review articles in reverse, or, if you like, prospective review articles.

Although originally conceived as an independent periodical, "Growing Points in Veterinary Science" has in the end become a section of a new journal recently launched by Elsevier Scientific of Amsterdam entitled "Veterinary Science Communications".

The Commission has not ignored the trickly problems raised by reprography and copyright. We are however only at the beginning of our efforts here. Ideally, the Commission would like, under its general brief to propose measures leading to the harmonisation of legislation between the Member States, to work up to a directive of the Commission or of the Council of Ministers in this area, which is so vital to the interests of the knowledge industries.

An information system for policy-making

We have, for the sake of efficiency, to be selective about the initiatives we take. The most popular way of settling such policy matters is, as everybody knows, to rely on hunches.

Feeling uneasy about this, we decided to see whether we could not arrive at decisions in a more rational manner. We ended up entrusting the Institut für Systemforschung of Heidelberg in the Federal Republic of Germany with the task of building up an Issue-Based Information System on the whole area of primary communication.

For the moment this system carries the name STIEC (Scientific and Technical Information in the European Community) and is on the way to being computerised. It differs from the more conventional data-bases in several ways. The scope, for instance, is defined in terms of issues, which can be either factual or deontic. An issue such as "How many articles have been published annually on astronomy in a reference period x?" is a factual issue. So is the issue "What different types of scientific conferences are there?", although opinion starts playing a role here. On the other hand "What is the relative usefulness of the different types of conference?" is a deontic issue inasmuch as the material which is gathered in its content will suggest to the user what type of conference he ought to be organising.

The material collected under each issue does not consist of bibliographic references but of facts and statements of opinion, directly relevant to the issue, which have been extracted from the literature or obtained through interviews.

It is impossible yet to make big claims for the efficiency of such a system. It still has to prove itself. We can only hope that it turns out to be sufficiently useful to warrant loading onto EURONET for general consumption.

It has been possible here to give a glimpse only of the European Communities' activities in the field of information management. I hope it has become apparent that one of the Commission's aims is to do everything which it can to improve the environment in which editors and publishers have to work.

DISCUSSION on "Treatment of 100 Man-Related Scientific Problems: Science as an Indicator of Publications" by Jacques G. Richardson

"Information Transfer and Management in the European Community" by J. Michel Gibb

JEFFREY WATSON - I think it was a real pity that Mr. Richardson ran out of time. I would like to know if you have any pointers to us on what does UNESCO think the editor's role is in tomorrow's world in relation to the problems that you raised.

CARL KEREN - I would like to ask Mr. Richardson (and perhaps also call on UNESCO and other international organizations) to heed a little bit of what has been said this morning about refereeing. We often find that once we go out of a strictly limited range of scientific journals, we are flooded with papers, reports, etc. which due to the absence of proper refereeing convey such as amount of gobbledygook, as it was called this morning, that it is nearly impossible for the information worker to cope with it or for the user to digest the material. Another topic about which I had some discussion with Mr. Gibb concerns the role of countries which are outside the rather limited number of countries which belong to the industrialized nations. The interfacing of this group of nations (North American hemisphere, Western Europe, Eastern Europe and Japan and a few others) with the rest of the world as far as communication in science is concerned, needs additional attention and maybe some investigation. At present, one gets the impression that the gap between the developed world and the developing world is increasing in spite of all the attempts to alleviate the situation.

JACQUES RICHARDSON - I think that the two questions are related.
To Jeffrey Watson - We have none. The organization that I represent does not have *pro forma* pronouncement on what the role of the editor should be because I think that those of us who are inside the organization, those of us who are editors and who work mainly with people outside the organization, share the same sympathies that all of our colleagues here have expressed in the past two days. On the one hand, we have the John Ziman approach which is "science as popular knowledge." There are countless arguments for this thesis. There is the other, second, approach that only excellence can justify publication. If you believe that science is part of the general culture, you immediately have to lower your standards and criteria and of maximization of excellence, of optimization of communication, in order that prople who would not normally have access to this level of information can share it in some way. This lays you open immediately to the accusation that you are diminishing the quality of your product, that you are prostituting your role as a communicator of science and technology.
To Carl Keren - I doubt that anyone can give an Olympian response to your observation. As one of the speakers this morning said (I think it was John Ziman), it is the editor who has to be the person who makes the final decision on what to use. My colleague, Peter Lengyel, who is in the social sciences, and I agree that it is indeed the editor's responsibility. How good the editor is, how much he is willing to bend his standards of ethics, how much he is willing to "play" with the rules that have been established for him or he has established himself -- this is a matter of individual, professional conscience. I don't know if many of you share this opinion, but I believe that one part of my mission is to help communicate science as broadly as possible, especially to those who do not belong to the informed elite. And this is a terribly, terribly vexing moral social problem that any editor faces. To understand certain aspects of certain kinds of human knowledge requires a certain kind of equipment that not every one has; this makes the person who has the equipment a little different from the one who does not. On the other hand, it is equally immoral to withhold from those that could benefit from what happens in science and technology simply because it is believed that he cannot understand.

HOW EDITORS CATALYZE THE PUBLICATION EXPLOSION

Yehudah L. Werner*, *Editor, Israel Journal of Zoology*
The Weizmann Science Press of Israel, P.O.B. 801, Jerusalem
and Department of Zoology, The Hebrew University of Jerusalem, Israel

ABSTRACT

As seen in zoology, science editors catalyze the publication explosion of primary literature in five main ways but could refrain from doing so. (1) Because of the manuscript pressure on journals, they press for papers being brief; authors respond by producing (a) halved articles, (b) series of papers representing monographs, and (c) split-off paperlets on techniques. Theoretically this vicious circle should be broken by universities, who should measure a scientist's achievement by quality of best publications. Practically, editors can (and some do) help, by accepting papers with extradisciplinary components and details of techniques. Authors must be encouraged to publish comprehensive papers or they will oppose the wanted revolution in appraisal methods lest they fail by its criteria. (2) Some publish shallow, redundant, book reviews, so as to earn the book. (3) Universal separation of preliminary notices from short notes would reduce the swarms of the former. (4) Prohibition to adequately cite unpublished documents sometimes begets dispensible publications. (5) Page charges paid by the author to the journal (by any formula) increase, through several mechanisms, the total number of pages and papers published; they should be replaced by direct support from funding agencies to journals. In conclusion, editors should delight in the quality rather than quantity of what they publish.

CONTENTS

*Presently Visiting Associate Professor, Department of Anatomy, University of Chicago, partly supported by the National Science Foundation, USA (grant no. NSF DEB 76 18987 to Dr. R.E. Lombard).

Miriam Balaban (ed.), Scientific Information Transfer: The Editor's Role, 113-121.

INTRODUCTION

Science editors appear to agree that there occurs a publication explosion of primary scientific literature and that somebody, i.e. somebody else, should do something about it. I refer only to the explosion in the number of titles because this is what affects the secondary services and the user in the worst way. The effects of the length of each publication depend largely on its structure, so that length *per se* is not objectionable to me. My examples are derived from my experience as both author and editor in zoology, but I believe that the principles apply to certain other natural sciences, at least, as well.

It has been suggested, that this publication explosion could and should be curbed by filtering out the mediocre papers, and publishing only the few important and good ones. This would put the burden of the culling to a considerable extent on the referees, the elite of the author population. However, the matter is not that simple, and the term 'mediocre' comprises contributions of at least two very different kinds. Those which are poor (because either the research, or the presentation, or both, are poor), and those which represent absolutely sound research, perhaps perfectly presented, but of little intellectual novelty.

The potential importance of the latter type is demonstrated by the books of Charles Darwin, in which he convinced the world of his case for evolution. These had been based to a large extent on a huge enterprise of private correspondence, through which he gathered the required data, the building stones for his monumental works. Nowadays correspondence that would be comprehensive to a comparable degree is almost unthinkable, and one would rely on the orderly retrieval of *published* data. Hence papers presenting nothing more than new data are important, certainly so in the fields of animal and plant taxonomy, and they better be good and accurate. Their appearance in print and distribution should not be prevented.

However, I shall point out several other ways in which, to my mind, editors are catalyzing the publication explosion and could refrain from doing so.

THE CRUSADE AGAINST LONG PAPERS

Editorial rules, guidance and pressure lead authors to accept, or at least to yield to, the belief that papers should be short in pages and perhaps short of details. This dogma represents part of a 'positive' (self-reinforcing) feedback mechanism of the publication explosion. It stems from the increasing pressure of offered manuscripts on journals. It leads to an increase in the number of submitted and published manuscripts, through the following mechanisms.

(a) The author of a long report often responds by subdividing it into two seemingly independent reports (example: Werner, 1972 and 1976), sometimes printed contiguously.

(b) The author of a thesis or monograph very commonly anticipates this pressure by publishing a series of independent papers. This may be relatively constructive when chapters presenting different aspects are best suited to different journals (example: Werner, 1965, 1967, 1968 and 1971a, all representing Werner, 1961); it may be relatively harmless when serial numbering clarifies the situation (example: Haacke, 1975, 1976a, 1976b); or it may be awkward for the user , when the journal prohibits such numbering (example: Ben-Eliahu, 1975a, 1975b, 1976a, 1976b, 1976c, 1977a, 1977b).

(c) The editorial crusade against 'minutiae' of technique means that often an author will seek to serve the public by publishing his technique in a separate paperlet (example: Werner, 1977). On other occasions an author may publish a technique because he feels pressured to provide a justifying precedent for its use in another paper in a conservative journal (example: Werner, 1971b).

Admittedly, these three phenomena often represent the direct psychological effect of the 'publish or perish' pressure on the author. However, in other cases (including most or all of the actual examples given), they represent the indirect effect of this slogan, through the increased general manuscript pressure it exerts on journals and the resulting pressure of journals on authors to be brief in each manuscript. Needless to say, each separate paper has its separate introduction, acknowledgements and references, at the very least, resulting in widespread redundancy, aggravating the publication explosion.

In theory, this vicious circle should be broken by the academic establishment (mainly universities), which ought to revise its personnel appraisal philosophy, and revolutionarily begin to measure a scientist's achievement and worth by the quality of his best publications. The publications to be evaluated would be selected by the candidate himself, in accordance with the standard number required by the establishment in each particular type of case.

In practice, however, editors can help, and in fact many, perhaps repentingly, do accept long papers with extradisciplinary ("irrelevant") components or with details ('minutiae') of technique (thus a 'zoological' paper by Lipkin, 1975, discusses also the distribution of seagrass species).

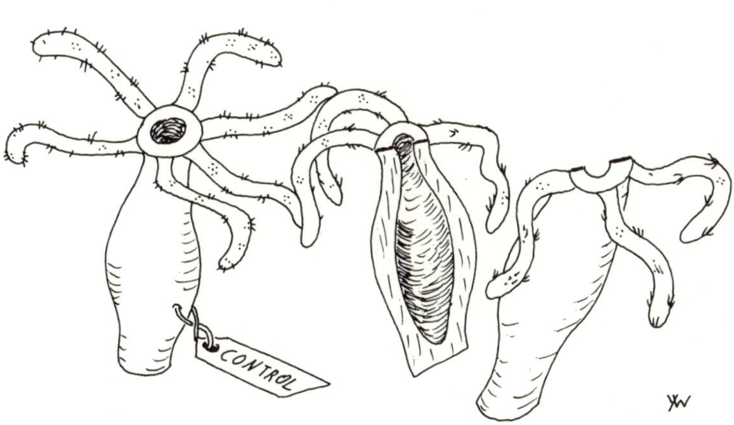

"He expects each of new-you to regenerate all missing parts, just like a bissected scientific paper."

It is extremely important that authors (especially young authors) be further encouranged in this direction, to publish comprehensive papers each of which is a whole, meaningful entity with its own merit. Otherwise the community of authors will oppose and prevent the so-desirable revolution in appraisal methods: Each author will be afraid to appear defficient by the new criterion of quality, becuase his long list of short, interdependent publications includes none that could be singled out as truly weighty.

THE ACQUISITIVE BOOK REVIEW

Perhaps the least weighty and most futile titles that infest the literature are certain book reviews. Some publishers of scientific books generously send copies of these to journal editors without prior arrangements, in the hope of generating a review that might promote sales. The receiving editor, or his specifically interested accomplice, then often composes and publishes a lip-service review that would entitle him to the book copy. (Thus the Israel Journal of Zoology used to publish such reviews until the 1973 revision in editorial policy, when book reviews were discontinued as yet another measure towards improving the journal.) The harm done by these reviews is not so much in their occupation of paper - some are so brief as to omit data on the book's format, pagination and price (example: Haas, 1972) - but in their littering the secondary information services. Usually the same book is additionally covered by serious reviews, which become more difficult to retrieve because of this littering.

THE REDUNDANT SHORT REPORT

Most of the short reports that abound so in the primary literature may be categorized as either (a) preliminary reports which are (sometimes justifiably, sometimes unjustifiably) by definition redundant; or (b) short notes, each a complete and unrepetitive presentation of a minor but

-"He says we multiply just like little scientific papers."
-"I am really famished!"
-"He is too busy arranging for new cages."
-"Soon he won't need <u>any</u> cages..."

interesting point (example: Safriel, 1975), or of a weightier point that can be expressed briefly enough (example: Bustard, 1968).

I believe that many redundant preliminary reports would not be written at all except for the hope that they would not be recognized as such but taken for short notes. Universal, clear-cut, separation of the two types, preferably by both name and publication-outlet, would greatly reduce the swarms of preliminary reports.

A special case of short report is the abstract of a paper (or talk) presented at a meeting. This abstract belongs to either category, or it may even reiterate a paper already published (or in press) by the time the collection of abstracts, making up the proceedings of the meeting, is published. Such a redundant abstract should be replaced, in the published proceedings, by a mere reference to the main paper (example: Zoological Society of Israel, 1976: p. 199). It is the responsibility of the editor of the scientific journal in which the proceedings appear, to prevent the publication of redundant material. Recent and future advances in information retrieval make redundancy both superfluous and increasingly harmful.

THE TABOOED 'UNPUBLISHED' DOCUMENT

A taboo widespread among the 'Instructions to Contributors' or practices of scientific journals, prevents any author from listing certain types of items in the References to his paper. The commonest victim of this taboo is a would-be reference to the author's own manuscript (or to one of a cooperative colleague) which has not attained official 'in press' status by the time his paper with its References goes to press. Instead of listing such a manuscript in the References by its full title, and marked 'in MS', the author is told to mention it in the text as an MS, without title. In such cases this taboo does not aggravate the publication explosion; it is a mere disservice to users who, in the case of a truly existing manuscript which probably will indeed be published, would prefer to know what title to look up.

However, where this taboo applies to various internal reports or other existing but unpublished documents, the effects may be worse. Occasionally an author may feel a strong need to broadcast the nature of his supporting material. Forbidden to publish the title, pagination and whereabouts of the unpublished document, he may feel obliged to see to its publication. Thus comes into being another paper that could have been done without.

THE CHARGED-FOR PAGE

Scientific journals, almost without exception, have long ceased to pay their authors; in recent years increasing numbers of journals have joined the new trend of making authors pay for the publication of their papers, to a varying extent and by various formulas (Van Valen, 1975). The main motive for this is to enable the sale of the journal, in the face of rising costs, at a price that would not reduce its circulation. The main justification is that the research grant which enabled the work itself, presumably could and should cover publication costs of the results as well. The catch is that adequate grants are not that universal. There are therefore many processes through which such page charges sensu lato increase the publication rate, i.e. cause or enable the publication of additional pages. As will be seen, most of these processes indeed arise because many authors feel that they do not have adequate funds; but one process is intensified in proportion to the money they do pay.

(a) Paradoxically, the most obvious process through which page charges generate additional papers involves the particular type of page charge which has the additional objective of reducing the length of individual contributions, so that more authors can be accommodated in the same journal issue. This type says, "We will publish for you X pages free of charge but each excessive page will cost you Y dollars." The author who feels he both has to and is able to do so will respond by cutting his too-long paper in two, again producing redundant introductions, references etc.

(b) The Wistar Press, so commendable in other respects, has a policy of underwriting "all tabular and illustration costs except for pages that exceed 50% of the total number of printed text pages." (Here 'illustration' includes line drawings, usually regarded as not more expensive than text.) Thus every page of text 'earns' the author half a page of illustrations or tables. This policy puts some authors under tremendous pressure to try to avoid excessive illustrative and tabular material by putting into words (and the more words the better) matters that the reader would prefer to see in a graph or a table.

(c) More subtle but also more widespread are the effects of the so-called 'honour system' page charges of many American journals issued by learned societies. The submitted manuscript is refereed and accepted, or rejected, regardless of the payment of page charges; the author is automatically billed for his pagination but automatically exempt from payment if he pleads lack of funds. Theoretically, the system is not only fairness incarnate, but would seem to reduce the publication of unnecessary pages or even articles. Practically, it also has two effects that lead to the publication of many more, rather than fewer, pages.

First, the income from page charges enables the learned society to publish more pages than it otherwise could; in this case, the more money authors can pay, the stronger the effect. The inflated pagination reduces the selectivity of the journal; a higher percentage of the submitted manuscripts is accepted, as long as their standard is deemed tolerable. Now, some (of course not all) of these additional, 'second-best' papers would have been suppressed, i.e. cancelled by the author, had they been rejected.

(d) This mechanism is aggravated by the fact that some of the journals employing this policy do not advertise it. The impecunious author is led to believe that the page charges are obligatory (only the more brazen author finds out differently); he may in the first place turn to another journal, such as one published by a commercial publisher (he then has the additional if doubtful benefit, that this journal is often less selective). So what? So the manuscript pressure on the learned society journal is reduced, its selectivity further lowered, and the number of expendable manuscripts it publishes, further increased.

(e) Second, some of the authors faced by page charges of the order of magnitude of US$ 50.- per page, discover that they can publish their paper, or maybe monograph, privately at a much lower cost.

(f) Then, of course, there are the journals in which page charges are truly obligatory. Here the same two effects will occur, only more frequently.

I believe that through all of the above or other mechanisms, in the long run the persistence of page charge systems will lead to a situation that authors simply publish (anything and only) what they pay for. Another and preferable solution to the economic quandary of the journals may lie in direct support from the funding agencies, rather than channelling the same monies through individual research grants.

CONCLUSION

We have seen some of the ways in which editors, through their role in editor-author interactions, aggravate and catalyze the publication explosion of primary scientific papers. Why do editors not refrain from taking such undesirable courses? The full answer may make up yet another article... Suffice it to say here, that science editors should more deeply appreciate their power and grave responsibility in affecting the publication explosion; then, draw some conclusions and apply them. To begin with, they should, in parallel to what is requested of authors, learn to derive personal satisfaction not so much from the number of pages they edit and publish but, rather, from the level of importance of each page.

-"Sure you don't have a peach the size of a melon?"
-"Nor do I have a melon that tastes like a peach!"

ACKNOWLEDGEMENTS

I am grateful to those participants in the First International Conference of Scientific Editors (Jerusalem, 24-29 April 1977) who, through their questions after the oral abbreviated presentation of this thesis, helped me towards a more nearly understandable written formulation and to Dr. J.A. Hopson for his comments on the manuscript. Preparation for publication was supported from a National Science Foundation, USA, grant to R.E. Lombard, University of Chicago (no. NSF DEB 76 18987).

REFERENCES

Ben-Eliahu, M. N. 1975a Polychaete cryptofauna from rims of similar intertidal vermetid reefs on the Mediterranean coast of Israel and in the Gulf of Elat: Sabellidae (Polychaeta Sedentaria). Israel J. Zool. 24: 54-70.

_______ 1975b Polychaete cryptofauna from rims of similar intertidal vermetid reefs on the Mediterranean coast of Israel and in the Gulf of Elat: Nereidae (Polychaeta Errantia). Israel J. Zool. 24:177-191

_______ 1976a Polychaete cryptofauna from rims of similar intertidal vermetid reefs on the Mediterranean coast of Israel and in the Gulf of Elat: Serpulidae (Polychaeta Sedentaria). Israel J. Zool. 25: 103-119.

_______ 1976b Polychaete cryptofauna from rims of similar intertidal vermetid reefs on the Mediterranean coast of Israel and in the Gulf of Elat: Sedentaria. Israel J. Zool. 25: 121-155.

_______ 1976c Errant polychaete cryptofauna (excluding Syllidae and Nereidae) from rims of similar intertidal vermetid reefs on the Mediterranean coast of Israel and in the Gulf of Elat. Israel J. Zool. 25: 156-177.

_______ 1977a Polychaete cryptofauna from rims of similar intertidal vermetid reefs on the Mediterranean coast of Israel and in the Gulf of Elat: Syllinae and Eusyllinae (Polychaeta Errantia : Syllidae). Israel J. Zool. 26: 1-58.

_______ 1977b Polychaete cryptofauna from rims of similar intertidal vermetid reefs on the Mediterranean coast of Israel and in the Gulf of Elat: Exogoninae and Autolytinae (Polychaeta Errantia: Syllidae). Israel J. Zool. 26: 59-99.

Bustard, H. R. 1968 The egg-shell of gekkonid lizards: a taxonomic adjunct. Copeia 1968: 162-164.

Haacke, W. D. 1975 The burrowing geckos of Southern Africa, 1 (Reptilia: Gekkonidae). Ann. Transvaal Mus. 29: 197-243, pls. 10-11.

_______ 1976a The burrowing geckos of Southern Africa, 2 (Reptilia: Gekkonidae). Ann. Transvaal Mus. 30: 13-28, pl. 1.

Haacke, W. D. The burrowing geckos of Southern Africa, 3 (Reptilia: Gekkonidae). Ann. Transvaal Mus. 30: 29-39, pls. 2-3.

Haas, G. 1972 Book review: Palaeozoic Fishes, by J. A. Moy-Thomas, Second edition revised by R. S. Miles. Israel J. Zool. 21: 53.

Lipkin, Y. 1975 Food of the Red Sea Dugong (Mammalia: Sirenia) from Sinai. Israel J. Zool. 24: 81-98.

Safriel, U. N. 1975 Reoccurrence of the red avadavat Amandava amandava (L.) (Aves: Estrildidae) in Egypt. Israel J. Zool. 24: 79.

Valen, L. V. 1975 Financial discrimination by journals - and what to do about it. Evol. Theory 1: 119-130.

Werner, Y. L. 1961 The vertebral column of the geckos (Gekkonoidea), with special consideration of the tail. Unpublished Ph.D. thesis, the Hebrew University of Jerusalem.

______ 1965 The comparative caudal osteology of some gekkonid lizards from Israel. Israel J. Zool. 14: 286-301.

______ 1967 Regeneration of the caudal axial skeleton in a gekkonid lizard (Hemidactylus) with particular reference to the 'latent' period. Acta Zoologica 48: 103-125.

______ 1968 Regeneration frequencies in geckos of two ecological types (Reptilia: Gekkonidae). Vie et Milieu Ser. C, 19: 199-221.

______ 1971a The ontogenic development of the vertebrae in some gekkonoid lizards. J. Morph. 133: 41-92.

______ 1971b Some suggestions for the standard expression of measurements. Syst. Zool. 20: 249-252.

______ 1972 Temperature effects on inner-ear sensitivity in six species of iguanid lizards. J. Herp. 6: 147-177.

______ 1976 Optimal temperatures for inner-ear performance in gekkonoid lizards. J. Exp. Zool. 195: 319-352.

______ 1977 Manual mapping of locality records - an efficient method J. Biogeogr. 4: 51-53.

Zoological Society of Israel 1976 Proceedings of the meeting of the Zoological Society of Israel (second series, fourteenth meeting). Israel J. Zool. 25: 199-219.

LOGOPOLLUTION AND WAYS TO PREVENT IT

Itamar Perath
The Geological Survey, Jerusalem, Israel

ABSTRACT
Logopollution is caused by the production of information that cannot reach its potential users. It is dangerous because it makes useful or vital information inavailable. Factors that influence logopollution include message Length, assimilation Time, message Efficiency and Availability, message Response, Merit and Dispersal. These concepts and their relationships are defined. The proposed information model implies that message packaging should assume new forms in order to reach the right recipient, and him only.

In order to cope with the rapidly-developing information crisis (the physical inability of experts to keep up with their subjects), the problem must be broken down into components and their dynamics examined. The following concepts are proposed: The *logouniverse*, which is the sum-total of verbal knowledge, and which consists of discrete *messages*. The *logosphere*, which is the sum-total of logouniverse information available to the single person. Its size is determined by:

$$(\text{opportunity}) \times (\text{capability})^{(\text{memory factor})}$$

Capability and memory are regarded as biological constants, the memory factor being a function of message-retention with time (i.e. smaller than 1). Opportunity, on the other hand, may be influenced by manipulation of the message flow.

Usable information consists of messages that are used to produce more messages (or, in fact, any artificial product), as opposed to unusable messages. The degree of utility is measured by *response*. The ratio unusable/usable messages, in either the logouniverse or logosphere, is the degree of *logopollution*.

The logouniverse expands by passive accumulation, the expansion rate of each logosphere is restricted by the time required for assimilating messages. When more messages are added to the logouniverse than can be assimilated in the sum of logospheres, they become unusable, and logopollution is increased. Information crisis may be the result:energy is squandered in the production and obtainment of information that cannot be used, and situations may arise where required usable information is not available. The condition is not academic: it may prove to be lethal in the fields of warfare, medicine and food production.

A model is proposed for evaluating the logopollution potential of messages (not including other properties, such as veracity, complexity, value, cost, etc.). The following factors are defined:

1. Message Size ($V \times T$), where V is message length (words & symbols) and T is assimilation time (reading, comprehending, recording).
2. Message Efficiency $= \frac{V}{T^2}$
3. Message Availability (N^Y), where N is the number of message recipients, and the exponent y expresses their potential of serving as secondary message sources or transponders.

Miriam Balaban (ed.), Scientific Information Transfer: The Editor's Role, 123-124. *All Rights Reserved.*

4. Message Response ($R = Q \times Z$), where Q is the number of quotations (explicit or implicit use of the message), and Z is the time lapsed from publication to latest quotation. (The Bible easily holds the record for Message Response).
5. Message Merit (M). This is the ratio of Response to Availability

$$M = \frac{R}{N^y} \text{ or } M = \frac{Q \times Z}{N^y}$$

6. Message Dispersal (D), a measure of the message's probability of entering a logosphere:

(Dispersal) = (Availability) × (Efficiency) × (Size)

$$D = (N^y) \times (E) \times (V \times T) = \frac{V^2 \times N^y}{T}$$

7. Logopollution $\frac{D}{M}$, the ratio of Dispersal to Merit, expressing the ability of a message to evoke response, i.e. its degree of utility.

$$\frac{D}{M} = \frac{\frac{V^2 \times N^y}{T}}{\frac{Q \times Z}{N^y}} = \frac{V^2 \times N^{2y}}{Q \times Z \times T} \quad \text{........... (a)}$$

or

$$\frac{D}{M} = \frac{E \times V \times T \times N^y}{\frac{Q \times Z}{N^y}} = \frac{E \times V \times T \times N^{2y}}{Q \times Z} \quad \text{.......... (b)}$$

Some conclusions can be drawn: 1. Logopollution is an inevitable by-product of publication. 2. Equation (b) shows that Efficiency increases logopollution by increasing the dispersal of low-Merit messages, but equation (a) shows that message length is far more pollutive than assimilation time. In other words: at similar efficiencies, the short message pollutes less than the long one (a suspect conclusion: do books pollute more than articles?). 3. Message Availability is the most polluting single factor. 4. Only increased Response can actively diminish pollution. 5. Logopollution is minimal in the case of personal communication.

The above model implies that, in order to avoid logopollution, future systems of publication should develop radically new forms of message packaging, guided along the following principles: a) The opportunities for personal communication should be enlarged, and given priority over other approaches; b) the personal archive should receive more institutional service, being the most efficient; c) editorial boards should encourage feedback response, and assume functions of information brokers; d) length of messages should be reduced, preferably by cutting nonessentials of documentation and style. The abbreviation can be aided by the coining and active usage of comprehensive neologisms.

The breakthrough will be achieved when the present modes of assimilation and storage of information in the individual brain are replaced by more efficient systems of information handling (e.g. message implantation, total recall, synoptic scanning, shared intelligence). Today this is still science fiction. Nevertheless, these are worthy and even obligatory targets for neuropsychology and cybernetics.

MODEL OF THE GROWTH OF SCIENCE ILLUSTRATES STRUCTURAL PROBLEM IN SCIENTIFIC PUBLISHING

A.A. Manten, *Senior Editor, Elsevier Scientific Publishing Company, Amsterdam, The Netherlands*

ABSTRACT

The total body of scientific knowledge may be represented, as a model, by a roughly spherical body, growing at its epidermis. The rate of growth is determined by various factors. Over the last three centuries, the number of active research workers and of research publications has kept pace with the growth of the epidermis or increased even faster. Very recently a reverse situation has set in, causing structural problems in the development of science and of scientific publishing.

INTRODUCTION

The formal transfer of original scientific information occurs to a very large extent through the conventional publishing of scientific documents (in particular journal articles, monographs and conference proceedings). Information transfer today meets with problems of different kinds: economical, technical, structural. These problems do not act independently; they interrelate in both their causes and consequences. In discussions of the present situation in scholarly publishing, emphasis is frequently placed on the economical problems and technical prospects, bypassing the structural difficulties which affect modern and future transfer of primary information. Methods need to be developed for a more integral examination of developments in this subject area.

GROWTH OF SCIENCE

Empirical observation has learned that the total amount of published scientific work between the middle of the seventeenth century and World War II underwent a continuously stronger growth. It had the character of an exponential expansion, with a doubling time of about 15 years (Price, 1963). After World War II this growth became even faster until in the early nineteen-seventies a retardation in the growth of science began to make itself felt.

For centuries the increase in the annual output of published scientific work kept pace with the velocity of growth of the total stock of published work; both doubled over a period of about 15 years. In the nineteen-sixties, however, the yearly production of published scientific publications showed a higher growth rate than the total body of knowledge. During this period the total volume of the body of scientific knowledge doubled in about 11 years, while over this same period the annual output in scientific publications increased by a factor of about 2.5. The reverse situation has set in during the nineteen-seventies.

Controlling factors

Many processes influence the velocity of the growth of science. These include:
(1) The social recognition of the importance of science, finding its expression in the percentage of the gross national product made available for

Miriam Balaban (ed.), Scientific Information Transfer: The Editor's Role, 125-127,

research and development.
(2) The growth of national and world economies.
(3) Population growth, particularly in highly developed countries.
(4) Democratisation of higher education, enabling a greater fraction of the population to potentially contribute to research and development.
(5) The self-slackening normally inherent to rapid growth processes (the doubling of the percentage of the population employed in research and development could easier be achieved when going from 0.1 to 0.2% than from 0.8 to 1.6%, whereas a doubling from 2 to 4% is extremely difficult to achieve).

Changes in these controlling factors are often reflected with some delay in the extent of scientific activity.

Developments in scientific publishing have roughly kept pace with the growth of science, although some deviations have occurred, such as relative over-investments (nineteen-sixties) or relative under-consumption (nineteen--seventies).

MODEL OF THE GROWTH OF SCIENCE

It has been suggested that science is a body of knowledge which grows on its epidermis (Price, 1969). This means that science would grow along the research fronts, based on knowledge acquired by earlier research.

An analogy

A very rough analogy that could be used is that of the growth of a crystal placed in a saturated solution: as the crystal increases in size, the new material is deposited on the surface, so that the rate of increase of the total weight of the crystal per unit time is proportional to the amount of surface area (Moravcsik, 1975).

If one is willing to make a few specific assumptions, it can be demonstrated that a near-exponential increase of the body of scientific knowledge is a consequence of the dynamics of the epidermal growth process (Moravcsik, 1975).

So far, the analogy is approximately consistent with the observed rapid growth in scientific knowledge that took place up to recent decades.

Spherical model of the body of scientific knowledge

The analogy induced me to experiment with a growth model of scientific knowledge, in which the latter is represented by a roughly spherical body, expanding at its outside. The velocity of growth is determined by factors as mentioned above.

Moravcsik (1975), when elaborating on the crystal analogy, wondered whether the research front along which science grows is of infinite thinness or rather has a finite thickness. It appears from his calculations that if we assume a finite but fixed thickness, the resulting growth function will be essentially identical to that of a strictly surface growth. If, on the other hand, we assume that the thickness of the research front is always the same fraction of the radius of the total body, we get a growth curve which is essentially identical with that for the volume deposition of new material, namely an exponential curve.

Empirical editorial observation suggests that the epidermis of the spherical body of scientific knowledge has indeed a certain thickness. Only a modest part of primary scientific publications report work that notably shifts the outer boundary of the epidermis. Much more work rather serves to fill-in interstices in the epidermis.

If the contents of a spherical body doubles, its surface area increases by a factor of approx. 1.6. This is a critical figure if the epidermis is of a fixed thickness. Should the thickness of the epidermis be a constant fraction of the radius of the sphere, this critical figure is 2. The actual factor is somewhere between these two extremes, presumably between 1.9 and 2.0, but a concrete figure has not yet been determined.

Predictions for the near future

Whereas the doubling time for the volume of the total body of published scientific knowledge dropped from a rather constant 15 years to about 11 years during the late nineteen-fifties and most of the nineteen-sixties, this period is now increasing again. If the factors controlling the growth of science continue to operate more or less as they do now, the volume-doubling period is likely to become approx. 13 years during 1968-1980 or approx. 14 years for the period 1970-1983. More spectacular will be the reduction in the growth of the annual production of primary scientific publications. This production is likely to increase only by a mathematical factor of approx. 1.5 over the period 1969-1981, and by a factor of no more than 1.4 over the period 1973-1986; again assuming that present-day conditions persist. This would imply that for the first time in centuries this factor would drop below the critical figures of 1.9 - 2.0.

Similar figures appear to apply to the world total of active scientific research workers.

This observation reflects the structural problem that is facing the future development of science and of scientific information transfer.

Expressions of change

The epidermis of the spherical body may be assumed to consist of unit cells. For our purpose a unit cell may be defined as the average volume of scientific knowledge which one research worker, or a group of workers with the same specialisation, can reasonably command. Space limitations prevent me from discussing this matter more thoroughly. The number of unit cells increases in proportion with the growth of the epidermis. The growth model indicates that at present the number of unit cells in the epidermis increases faster than the number of research scientists or the number of published research papers.

What we see in practice is that more and more research institutions, both those of universities and others, have to give up attempts to cover all specialisms within their scientific field. Collaboration between institutions increases, with a trend to concentrate on complementary subject areas rather than on duplication. This tendency is strongest at the level of specialisms, but also occurs at that of subdisciplines, and there are already examples where universities have merged their activities in a whole discipline and have concentrated these activities entirely with one of the collaborating universities.

Reactions to the present-day situation in the world of scientific publishing are less clear. Changes in the volume and nature of literature accepted for publication seem to be generally based on ad hoc decisions rather than on in-depth analyses. The time has come when we need to thoroughly re-asses the foundations of publishing monographs and highly specialised research journals.

REFERENCES

Moravcsik, M.J., 1975. Phenomenology and models of the growth of science. *Research Policy,* 4(1):80-86.

Price, D. de Solla, 1963. *Little science, big science.* Columbia Univ. Press, New York, N.Y.

Price, D. de Solla, 1969. The structure of publication in science and technology. In: W. Gruber and D. Marquis (Editors). *Factors in the transfer of technology.* MIT Press, Cambridge, Mass.

PUBLICATION-CENTRED SYMPOSIUM: AN EXPERIMENT

A.A. Manten, *Senior Editor, Elsevier Scientific Publishing Company, Amsterdam, The Netherlands*

ABSTRACT

Increased dissatisfaction in many circles with traditional symposium proceedings induced an experiment in which the submitted information and discussion results have all been integrated into one monograph-like publication.

INTRODUCTION

Symposium proceedings have been the subject of increased criticism over the past decade or so. Negative views have been expressed by authors (poor editing, delay in publication, limited distribution), readers (information of mixed quality, often rather dated upon its appearance in print), publishers (average sales have dropped faster than for other printed learned works), librarians (too many such volumes, too expensive), documentalists (bibliographically difficult to identify, harder to retrieve) (Manten, 1976).

BACKGROUND

When I began to make an analysis of current criticism about symposium proceedings and to search for possible improvements, it soon became clear that this could not be done well in isolation. The entire subject of symposium planning, organization and publishing had to be examined. This can be done by interviewing organizers and participants, and by written enquiries. Possible improvements and alternatives which arise from such studies can only be tested when tried out in practice. It was in this context that in 1976 an experimental symposium was held. The subject of the meeting was chosen in consultation with a co-organizer of this experiment, the Royal Netherlands Land Development Society, and was: the cycling of mineral nutrients in agricultural ecosystems. For details the reader is referred to other papers (Manten, 1977, 1978).

The entire course of the symposium is still being examined and has also given rise to other studies. Another experimental symposium is planned for 1979.

Since space is limited and editors are likely to be interested most of all in the editing and publishing of proceedings, this contribution will centre around that segment of the study project.

THE TERM "SYMPOSIUM"

I have opted for the term "symposium", to label the meeting under discussion, but several other terms could equally well have been used (Manten, 1976, pp.2-5). Among scientists the term "symposium" is used both for a conference and for a proceedings volume. Both terms are the subject of criticism. Moreover, a symposium can be a "convivial meeting for drinking, conversation, and intellectual entertainment" (the original meaning of the word), but also an arid presentation of an abundance of scientific information. Thus it is clear that the very complexity of the problem field that I am studying is even met when thinking about the term "symposium".

Miriam Balaban (ed.), Scientific Information Transfer: The Editor's Role, 129-132.

IMPORTANCE OF DISCUSSIONS

Consultations with symposium participants make it evident that they regard the discussions taking place at the meeting as the most valuable element in scientific conferences. Provided that they are clearly instructed well beforehand that pre-circulated information will not be publicly read at the symposium, but will be immediately subjected to thorough discussion, participants generally do take the trouble to take good note of it before arrival.

Discussions in small groups are preferred to discussions in plenary sessions. The latter should rather be used as a forum to test whether conclusions reached in the groups meet with general approval or give rise to further questions which require examination, and also to integrate the conclusions from the various groups. Group discussions further encourage informal discussions between symposium participants, and in reverse these informal discussions are beneficial to the outcome of group meetings.

Readers of symposium literature indicate that they generally prefer to see research papers published in primary journals. They feel that proceedings volumes and symposium reports have a value if these present some sort of state-of-the-art picture of the symposium theme and are published with no excessive delay. They normally wish that more information could be included about advances made at the meeting itself as a result of bringing people with differing backgrounds, experiences, and opinions together. It would be useful to hear whether people deepened or altered published views, or the interpretation of research, as a consequence of discussions at the symposium.

CHARACTER OF THE SYMPOSIUM PROCEEDINGS

The above considerations were thoroughly considered when planning of the publication that would result from the experimental symposium was done.

Authors were invited to prepare a review of knowledge relevant to the symposium theme that had been published in a particular language area of the world or by workers in a specific specialisation. They were requested to use a prescribed series of subheadings for their text and a standard table for numerical data.

Contributions had to be submitted about 6 months prior to the symposium. They were integrated by the symposium editor into one draft manuscript, which became the basic document for the symposium discussions and was sent to all participants about 2 months prior to the meeting. To this was added a list of questions and conflicting data or views, which had been compiled by the editor while working up the individual contributions. Further questions for discussion at the symposium were derived from a brainstorming session held by the organising committee together with most of the Dutch symposium participants, a week after they had received the draft of the integrated manuscript.

At the 5-day symposium itself the groups had to prepare reports which were subsequently typed and distributed to all symposium participants. Plenary discussions were recorded on sound tapes. All discussion information was made available to the editor after the meeting for incorporation of the relevant parts into the integrated manuscript. This was done at the logical places within the text.

The final symposium proceedings thus obtained the character of a monograph. All contributors are collectively listed as authors, the symposium editor being mentioned as the first author. Where portions of the text had been contributed entirely by one person, his authorship of that portion is acknowledged in the text.

EVALUATION OF THE INTEGRATED PROCEEDINGS

Reactions to the draft of the integrated manuscript were tested by means of a questionnaire which participants had to complete during the last symposium day. Statistical data on opinions about the final printed version are at this moment not yet available.

Readers' judgements about the draft manuscript were positive by a distinct majority.

Authors often took a more reserved attitude. This reserved attitude could partly be traced back to an unfamiliarity with the procedure followed. There also was some reluctance about the ultimate content and quality of the proceedings, which under the given circumstances was not fully under their control. A few contributors complained that bits of information they had supplied had not been used by the editor or had become too loosely formulated when rewritten or combined with material from other authors. Although this was not explicitly stated it may well be that some also felt that their incorporation in a large authors' team might add less worth to their credit list of published papers than if their contribution has been published separately under their own name alone.

Finally it must be admitted that both the organizers and the editor had underestimated the task given under this regime to the symposium editor. This and the functioning of the authors are two of the items requiring thorough re-examination.

REFERENCES

Manten, A.A., 1976. *Symposia and symposium publications. A guide for organisers, lecturers and editors of scientific meetings*. Elsevier, Amsterdam, 155 pp.

Manten, A.A., 1977. Note on the experimental nature and organization of the symposium. *Agro-Ecosystems,* 4:347-349.

Manten, A.A., 1978. Experimental publication-centred symposium. *Journal of Research Communication Studies,* in prep.

DISCUSSION on "Publication-Centred Symposium: An Experiment"
by A.A. Manten

RONALD SWARTZ - There seems to be an implied assumption that somehow the proceedings of a conference have lower standards than a journal. Since I don't know the standards for proceedings or for journals, I'd like to know whether that assumption is correct and what about it?

BRUCE BATTRICK - I'd like Mr. Manten to expand a little on what he calls integration. Could he tell us something about any sort of refereeing procedure that he could carry out in advance before embarking on six months of preparatory work.

ARIE MANTEN - I find myself in a difficult position, Mr. Chairman. Because I have to choose between saving time and portraying myself as a vain character, but answering the question on the quality of proceedings as compared to journals, I could refer to a book I published last year, "Symposia and symposium publications", which is on display on this building and that would save a long story. The major problem is that in proceedings, editors apply criteria other than pure scientific ones. If you have worked hard to get a big shot lecturer at your meeting you can hardly turn down his paper for the proceedings. Moreover, papers for a conference have to be prepared at a given moment of time, while the research can either have been completed a long time ago and it's a rehash, or it is premature and the appropriate conclusions cannot yet be drawn. This also has an effect on quality. These are just two examples. There are several more that could be mentioned. The definition of what is an integrated manuscript. An integrated manuscript, as we use the term, is one in which all the information brought together by the various authors is integrated into a single coherent story, not a series of papers. It is one paper subdivided into chapters, and in most cases the contributions of the individual participants can no longer be recognized. It reads as if it were a book, which has the great advantage that you get a balanced work in which discussion and confrontation of information from various parts of the world and from various authors are knit closely together thematically.

Refereeing: The organizing committee was set up to comprise representatives of learned societies so that we had a very thorough discussion on the development of the subject chosen and the subdivision of the subject. The editor did all the integration work and then, of course, the symposium itself was a kind of collective refereeing through all the discussions that took place. After rewriting the integrated manuscript and integrating all the discussion information, the manuscript was again circulated to all contributing authors to collect their opinions and criticism which, in a second round, was again taken up by the editor. Thus we have a very thoroughly discussed final product.

IN DEFENSE OF STANDARDIZED ON-DEMAND PUBLICATION

Joseph Agassi, *Department of Philosophy, Boston University, Boston, Massachusetts 02215 U.S.A.* and *Department of Philosophy, Tel Aviv University, Tel Aviv, Israel*

Due to current myths and to publication pressure, science is often pictured as a reservoir of knowledge, where all published material is accumulated and processed. This picture is sheer myth. Scientific literature is chiefly a means of communication, not a reservoir but a channel-system; and the channels are now clogged.
This paper has four sections. I first (1) described some of the most obvious facts of the matter; then (2) the view in defense of the status quo; then (3) the limits to improvements due to ambiguity of the importance or otherwise of new items and the reasons the scientific leadership uses for enhancing the ambiguity and slowing down progress in the name of playing it safe. Finally (4) I make some simple proposals for radical reform that may easily overcome the worst obstacles. In particular, I recommend the establishment of university and learned society clearing houses which will publish lists containing only quality items to be deemed as published, copies of which items should be available upon demand. Such an arrangement will be cheaper, more efficient, and less harmful, than pseudo-publications of unread journals.
That scientific knowledge is both stored and transmitted for all sorts of use is an accepted fact. The popular view of science as a storage or deposit of knowledge leads one to the problem of retrieval and transmission of stored knowledge -- i.e., to the problem of scientific communication. The view of science as a communication network, on the contrary, raises the routine problem of storage, i.e. of temporary storage of scientific pooling of information to be transmitted at will. The view advocated here is of science as a communication system, and so it raises the question how should knowledge be stored so as to become most readily available at will. Indeed, the claim from which I begin is that publishing books in the traditional manner is a poor mode of publication just because it relies on old modes of storage.

1. Science as a System of Communications

The problem of communication each person faces is technical: he is a producer or consumer of information in search of both channel and counterpart, where counterpart to producer is consumer. As usual, the search for a channel and the counterpart are intertwined since consumers are all too often identified with a channel, e.g. readers of Science and readers of Scientific American and readers of Nuclear Physics and readers of the Bulletin of Atomic Physicists, readers of government daily papers

Miriam Balaban (ed.), Scientific Information Transfer: The Editor's Role, 133-139.

and of the official organ of this or that party, of the leading independent paper in the country and so on and so forth.

Traditionally, control of the channels was political power, but not an unlimited one since when the demand for a channel is strong enough yet ignored by officialdom, new channels are created. The new channels may be illegal or semi-legal, despised or not respectable. But this is the struggle of a channel for survival.

A struggle indicates a conflict - of interest or of opinions about them.

The fact that authors have limited access to respectable scientific literature should be obvious from the very consideration of the employment benefit accrued to any author who gets published and from the fact that all established intellectual leaders, except Ludwig Wittgenstein and Don Juan and a few other gurus, are published first and established later and become leaders still later on.

Some material, however, comprising much scientific literature, has leadership which may be more easily accessible through personal communications. That is to say, much material published in the scientific press that may be science proper, and which gains the status of science proper from being published in the scientific press, is very cumbersome as communication. What should be done about it is to find a new way of according it scientific status, while employing better means of communication than printing it: status and printing need not go together.

Also, some material which is genuine communication is barred from the scientific press; both rightly and wrongly; to wit both on correct and incorrect judgment of lack of scientific merit. The question here is, who judges the judges. But a deeper question is here, how can we improve the criteria, the rules of judgment, so as to improve matters.

There is, also, the trade press: no press is an island. The interaction of the trade press with the academic world is complex. At times a publisher voluntarily submits his material to the profession's censorship and thus he raises again the problem of critic and of criteria. Also, at times the trade press publishes material of obviously low quality with high pretense. At times, the academics try to resist this, but they cannot possibly win. At times, the most trashy trade literature gains recognition from the learned press, again at times rightly, at times not.

But on the whole it is much harder to get published at trade channels because their readerships are less definite and so, for purely commercial reasons, they can only cater to mammoth readership - with the exception of the vanity presses, including most university presses, of course, which print books not for any distribution, i.e. not in order to communicate, and except for trade series that are deemed a must for any self-respecting library whose publishers print them only after they have presold enough copies to cover all costs and mark-up to boot.

It seems that the open market has its own rules, and that planners of improvements of learned communications should let them be. But this ignores the interaction between the trade and the learned press, or, more precisely, the institutionalized side of it. This includes, to begin with, standard text-books, standard monographs that have been essential for any given profession, and, the complex process of book reviewing in the learned press. This is far from being all. Anyone who has tried to publish a kudos volume, whether a collection of essays of one author or a collection by different authors in honor of one, or a proceeding of a humble conference, must have noted that there are unwritten unarticulated taboos. The first question put to a referee is, is the book up to standard and he judges this not by content but by context: is the kudos publication endows quite warranted? The second question is concerning market ability; and referees make it dependent on the first - as is at times true, but at times patently false, e.g., in cases of popular pseudo-science.

The question here, then, is how can the institutional attitude to trade publishing improve the service of the free market to the scholarly community. There are other obvious means like subsidizing and like institutionalized guidelines for controlling purchase in university libraries and even university bookstores. But these are refinements.

Thus far, I have approached the problems with minimal initial information and repeated, in their context, the standard socio-political questions about standards, modes of applying them, of checking the application, of testing for efficiency and for quality control. This hardly scratches the surface.

2. Science as the Best of All Possible Worlds

The idea of the owrld of science as a utopia here and now is very much in the air. Again and again one hears that many scientists consider Fred Hoyle's charming Black Cloud the way a Catholic may consider Lives of the Saints or a member of the counter-culture consider Hermann Hesse's Sidharta. It describes a colony of scientists that quite of necessity becomes a utopia when it had to govern itself. Of course, other science fiction stories describe scientific dictatorships far from perfect and the stories are neither here nor there; what matters is the widespread readiness to endorse a simplistic utopian view of the sociology and politics of science.

To apply this to communication, I have tried a small experiment. I have expressed a certain highly naive view to a number of natural scientists. Some accepted it, some confessed their inability to reject it despite grave doubts. The view is this: science is a body of knowledge and publication of worthy material is the addition of that material to that body. As long as this is taken as a myth it seems unanalyzable. But the current Levi-Strauss theory that is so much in vogue these days is just the means for analyzing a myth from the outside in terms of communications.

In terms of communications, the myth is that science is the best of all systems of communication. Best may mean most excellent conceivable and so better than very good and it may mean much inferior to very good but can hardly be improved upon under present circumstances. The first reading permits complacency, the second permits and even encourages reform from the inside: we can try to alter circumstances so as to permit improvement and the Establishment should be enlisted to effect the reform; true enough, the Establishment will not incite for any revolutionary alterations, but these are not called for anyway. Thus, the two views are the quietest and the moderate traditionalist. The quietist view vis-a-vis scientific communication is openly and vociferously advocated by Derek J. De Solla Price and by Joseph Ben David. The moderate conservative view is advocated by Michael Polanyi and by Sir Karl Popper, though the latter is a bit more reform conscious. Thomas S. Kuhn holds a position somewhere between the quietist and the moderate traditionalist; perhaps he holds both, but this, to employ Levi-Strauss' idea again, means a middle position all the same.

Let me state my own position. I take science to be a conduit, not a reservoir, of contributions of any import. Science is a social institution, a tradition, a culture; the declared canons of science, of the requirements for explanation, empirical evidence, probability, etc. These declared canons are, first and foremost, not necessarily the real canons, as noticed by Duhem, Freud, and Malinowski. Secondly, any canon is not one universally observed. Hence, whatever is the true canon, characteristic trademark, or touchstone, of science, the scientific tradition also embraces, perhaps only marginally, a sub-tradition that violates it.

I take the communication of scientific material to be the prime factor of the integration and rapid growth of modern science. I take it that some scientific communications channels are more institutionalized than others, that some scientific items of information are communicated more as an institutional factor - e.g. scientific text-book material - than other items - e.g. scientific gossip.

The communication system is, as most systems, partly a natural growth, partly a matter of design and legislation. For example, the idea of a scientific essay, of a standard style of one, belongs to Francis Bacon and Robert Boyle. The idea of a book review belongs to the Royal Society, perhaps to others. The idea of surveys belongs to Francis Bacon and Bishop Spratt. Yet Spratt's execution, as well as Joseph Glanville's, were so poor that the idea had to be resurrected in the 18th century by various historians of science, such as Joseph Priestley. The 1812 History of the Royal Society of Dr. Thomas Thomson is a survey and has already an old fashioned ring. Surveys of recent scientific studies, such as Faraday's anonymous survey of the electro-magnetic literature of 1821 and J. J. Thomson's of dielectricity of the 1860's, stand out, but the diverse encyclopedias of the early nineteenth century made surveys a fixture, as did Thomas Young's medical dictionary and similar works, not to mention the rise of the review articles in the learned press, review journals, and monographs like de la Rive's Treatise on Electricity of the mid-century.

I cannot here discuss the canons of scientific publication as implemented by the Royal Society, and further evolved through the ages. Let me only say that Robert Boyle insisted that scientific work is chiefly a communication - and he put all else as secondary to this single consideration. He recommended the essay form as a means for quick communication and his book of essays, Certain Physiological Essays of 1661 was a model for the scientific journals that started appearing at that time and it remained a model for generations of writers, including Joseph Priestley, whose style was praised by Goethe who endorsed verbatim the praise lavished on him by Lavoisier: his writing, they said, was facts uninterrupted by thought. For what we wish communicated most, says Boyle in his proëmial essay to his Certain Physiological Essays, are facts, not conjectures.

So much for science as the best system of communication to date. But the picture is more complicated. Some facts are deemed too uninteresting to publish and even the most inductivist editor will reject papers reporting them - at times on the pretext that the papers in question are too long - and some conjectures are very seminal and should be published, as even Boyle and Priestly noted. What is the guideline? Is there one at all and can there be one?

But, no doubt, the chief end Boyle had in mind is now severely frustrated. Papers are refereed months and years, and then shamefacedly rejected. Even when finally accepted, they are on a waiting list. (A paper of mine was four years on a waiting list before I protested and got it published after a few more months.) The reason is that people now publish less in order to communicate and move as a result of publication-pressure. What is to be done?

3. The Inherent Ambiguity in the System

The chief argument that traditionalists can offer against all radical reform proposals is that radical reform is based on standards which are untested and may be erroneous and harmful. This argument was forcefully presented by Michael Polany, in his classic Personal Knowledge and elsewhere. Yet he is evidently in error: the scientific tradition is not one organically evolved (if there ever exists any that is), but one developed by the rational growth of rules - rules that were often found erroneous and in need of mild or of radical modification.

Nevertheless, clearly, we must have lax rules, and we must have rules to take care of the fallibility of our rules as a system: constitutional rules concerning constitutional amendments.

Not only do we need constitutional amendments; we also need much leeway and the benefit of doubt. Admittedly, the case of great ideas overlooked by the scientific establishment is less damaging than the same case in other quarters, such as politics, the arts, and religion. These days even religious Establishments are not totally blind to new ideas. The scientific press is only one small part of the Establishment, and it has its share in the game played by the Establishment, the game of increasing ambiguity. Let me explain.

The well-known fact is that some contributions - to science, art, politics, religion - are direct answers to prayers and seem so at the very first blush. They may be theoretical - like quantum statistics - or empirical - like Eddington's ecclipse observation - but they fit certain specifications that are well accepted by all students of the relevant questions. Other contributions, though unexpected, are noticed as important the moment they appear. Oersted's discovery of currents' deflection of a needle is the paradigm here. It took Einstein to put his authority behind the electron wave of Davison and Germer, but once the idea clicked its instant success was assured, Einstein or no Einstein. But some contributors are discouraged from the start - Mössbauer is a paradigm. And, at times, they are subject to a mini-witch-hunts for decades before they make it - Oersted is a paradigm again - or even after they should be seen as important discoveries - Brownion motion, for example.

The citation index that Derek Price has devised cannot but prove success and ignore neglect of what should count as success: the ideas that are neglected are simply not cited. Citation is traditionally - since Bacon and Boyle made it so laudatory - a means of stimulating interest in the game. But the citation index can also be used to show that the Establishment's game is hardly ever right: they neither back the right horse, nor do they back no horse while letting the better one win; the citation index shows that the game is, only if you can't beat them, then you must join them. The result is a constant sluggishness of progress that is necessary for members of the Establishment to readjust: unlike the public speakers in Orwell's 1984 they cannot change sides in mid-speech, and unlike the scientists in the science fiction literature of Asimov, Merton, Popper and Bunge, they are unable to say what I thought yesterday to be true, turns out to be false. When forced to say so, at gun point, they say of a refuted theory that it was true yesterday, or that times have changed but the theory - especially Newton's theory - is still all right. Here it is that the worst and most reactionary philosophical rubbish, relativism and/or instrumentalism, comes handy. But only at gun-point. In less trying times even the scientific establishment is not as reactionary as Polany and Kuhn.

The rationale for the sluggishness, I say, is the maintenance of the scientific leadership. When it is not a sufficient means for maintaining control, even while crowning a young turk the king of the day, then specialism splits to sub-specialisms so that the leaders have less and less to learn. The effect is that any study of sub-specialism - one which penetrates the area of sub-specialism - two is handled with extra care and caution, i.e. with extra sluggishness.

There is a limit to sluggishness, though, imposed by the non-academic trade press. The paradigm still is Helmholtz' Erheltung der Kraft of 1847, rejected by a journal editor but well sold in the open market. It does not always work. John Newland used the trade press to publish his Law of Octaves, yet was beated by Mendeleev. Some will say, Newland did get the table of elements right; the answer is that Mendeleev did not either; that if anyone did get it right, then it was Niels Bohr and no one before him.

Nor do we know whether to honor a by-passed innovator in retrospect or not; Newland was not honored, but the celebrated Mendel was. It was a touch-and-go in his case too, but this is a different story. It was for generations quite customary to blame Mendel's contemporaries for blindness and prejudice. This attitude is now passé: we now fully recognize the difficulty of recognizing the import of a novelty. And reactionary philosophers of science such as Thomas S. Kuhn and Imre Lakatos declared as a general law that all recognition must come too late. Lakatos even cited on this the Great Reactionary Philosopher, Georg Wilhelm Friedrich Hegel: The Owl of Minerva, he said, flies in the dusk. All this means is, clear sight is hind-sight and in the twilight of the day of glory and achievement!

I wish to stress that ambiguity is always there. Even when we all know what we want and recognize it at a glance, this is so because we all accept a certain intellectual framework that may soon be overthrown and this overthrow may totally alter the picture. Remember that in the mid-19th Century action-at-a-distance won over field-theory, and when later field theory won text-books and histories were rewritten.

4. Some Obvious Proposals

The system of scientific communication, to recapitulate, is the best we have, but it is not perfect. Partly it's defective because its responses to innovations, such as publication pressures and research grants, were more spontaneous than deliberate, partly because the deliberate part of it is out of date. Yet there are clearly defects that can be overcome, especially the evil current mythology of science as perfect; and those that cannot be overcome, especially the problematic nature of many possibly important, possibly worthless contributions. Yet the sluggishness caused by the ambiguity need not be enhanced as it nowadays is. Hence, radical reform is possible and may be tried out.

First and foremost, comes the desire to keep the scientific press as a means of communication plain and simple. Of course, it is an ideal, but one worth struggling for. Since publication pressure cannot for the time being disappear - until someone has a bright suggestion - we may designate publication for kudos purposes one way and publication for communication another way. I observe here an existing practice: all U.S. doctoral dissertations we now already deposited in a certain clearing house in the U.S. and are thereby deemed publications and photo-copies of them are made on demand. Two things are wrong with this, fundamentally excellent idea. First, the clearing house receives dissertations that can be exposed as pure charlantanry at a glance, as I can testify in court of law. Second, it is confined to doctoral dissertation. We should have more such clearing houses, and they should make money by publishing lists of available items and providing efficiently photocopies or computer reprints of these on demand, and send to publishers and/or distributors items for which demand is sufficiently high.

Also, clearing houses may publish their policies, debate them publicly, etc. This would be a factor highly democratizing, diversifying and encouraging. And if the clearing houses will be of learned societies, then debating their policies will increase participatory democracy in them yet, on the whole, this is a long term project. The short term project is to make scientific publication more frankly a reservoir so as to facilitate possible flow to any conduit as desired.

Not only that. Publication pressure leads to much repetition and to minute studios of all sorts of insignificant detail for which there is almost no demand. It would be much better if instead we encourage the publication of book reviews, surveys, proposals for institutional reform of the organization of the sciences and their institutions, etc. Rather than see an ex-scientist desperately try to pretend to be keeping up-to-date, I would suggest he write essays on his experience as a scientific administrator.

There is more that can be instituted, but I am more anxious to discuss implementation now. When I put this proposal before the public, it may pass unnoticed, but it may draw attention since it deals with a problem painfully felt by many. If it be noticed it will surely draw criticism, and to the extent that criticism may be just, it may have to be modified, or be replaced by better alternative proposals. Anyway, sooner or later a reform will have to be tested in the field.

For this what now seems to be needed is a number of leading members of the commonwealth of learning who will be ready to lend the project their full moral support. This means that the venture will have some limited currency. If it succeeds - and in modifications prescribed by new experiences - then it will both increase its limits of validity and be emulated to varying degrees. This will lead to the proliferation of competing standard, competing intellectual framework and more.

It will be a step towards making science more democratic, more open, more encouraging, more offering options to all sorts of oddballs.

But the major objection to the proposal made here is the objection to all radical reform: science functions very well and much better than any other aspect of human society. As Bacon already stated, the conservative says, things could be much worse and the reformer says, and they can be much better. The troubles people who use the academic press for communication encounter are real. Some of my younger colleagues are complaining and with justice. I plead with those who reject my proposals to remedy the situation to come up with alternatives, not to defend the status quo, not to overlook just complaints. I admit, leaving the status quo could mean deterioration; but it can also be progress.

One final technical point. Vanity publication is an abortive form of publication geared to authors' demand. Like in movies, nowadays, production is a minor venture as compared with marketing and distribution. The reason private vanity in publishing (and movie making) fails is the absence of large scale clearing houses. University vanity publications channel material through clearing houses or have clearance sales just like remainder clearing houses of the tradesmen. The trade presses and the trade clearing houses are the middle man to be cut: all we need is the private or vanity publisher who is thus far inefficient because of lack of distributors, plus the distributors in forms of influential advertising clearing houses which can cater ad hoc to increased demand. Whereas, today a trade clearing house will sell university press remainders but not vanity press remainders, thus cashing in on the reputation of the university, it is better that the university not have a university press but a university clearing house. In this way it will be able to transcend the primitive printing press and use more modern means of duplication, better employed to fit demand.

THE PUBLICATION GAME AND THE EDUCATION OF FUTURE SCHOLARS

Ronald Swartz, *Assistant Professor of Education,*
Oakland University, Rochester, Michigan 48063 U.S.A.

ABSTRACT

This papers centers in on the following two questions: 1) Should teachers encourage students to write in clear and direct ways, or is it wise to school people so that they can write in "a jargon only to be understood by the erudite few?" 2) Should editors of academic journals expect professional scholars to write in clear and direct ways? The author's answer to these questions can be summarized as follows: All scholars, both students and professors, should immediately begin to write about their problems as clearly and directly as they can. Whenever possible, scholars should consult with others who have written about the problems under consideration; but when consulting the literature associated with one's academic interest, it is always a mistake to learn the language of the professional for its own sake. And furthermore, students should seek out teachers who will help them become more clear about their academic interests, and professional scholars should submit their essays to editors who understand that scholarship may be worthwhile even if it is confused. In arguing for the toleration of some confusion in academic work, the author attempts to apply and interpret some of the ideas Sir Karl R. Popper has developed about language and the growth of human knowledge. Specifically, Popper's philosophy of science is used as the basis for an educational philosophy that encourages teachers to help their students become less confused and more clear. In addition, Popper's ideas on science and theory development suggest that it may be desirable to reform the way editors evaluate academic work. It is claimed that Popper's views on the inability to avoid confusion suggest that editorial policies need to be liberalized.

We all have the pleasures of the reader, but the writer also has the pleasure and the task of writing. This is not only a strange but a rewarding experience. We owe all young writers the opportunity of getting together, of agreeing or disagreeing, and finally of achieving the craft of writing.[1]

I. Preliminary Note - Language and Learning as Common Factors Which Affect All Academic Communities

Since I am a philosopher of education, I consider it desirable to start this paper with a brief word about why I wish to speak to scientific editors.[2]

To begin with, it is often unnecessary and undesirable for academic communities to isolate themselves from one another. In other words, the isolationism that exists between individuals who work in different academic disciplines is at times the result of convention, rather than necessity. And

Miriam Balaban (ed.), Scientific Information Transfer: The Editor's Role, 141-155. *All Rights Reserved.*

furthermore, communication between different kinds of scholars might greatly aid our desire to advance knowledge because there are some problems which all academic communities need to solve.[3] Specifically, all scholars must use human *language* as the vehicle by which to communicate their ideas. Also, every academic community is constantly faced with the problem of creating a satisfactory *learning* environment that will allow for personal growth and the improvement of human knowledge. Thus, as I see matters, all academic groups can benefit from discussions about the effects that language and learning might have on scholarship.

What I am trying to get at here is that individuals such as poets and physicists have more in common than is usually realized. It is a fact of life that all scholars must use language as the means to communicate ideas. In a similar sense, the learning situations which are created for both physicists and poets may have much in common because all human attempts to understand a problem or a question seem to have many similarities.[4] The poet who is struggling to find the best way to describe an emotion or a feeling is *not* that different from the physicist who wishes to articulate a new or innovative theory that may eventually lead to new scientific information. Of course, the subject matter of the poet and the physicist may be very different, but all scholars share much in common if we center in on problems related to human language and learning. When it comes to discussing language and learning, scholars do *not* have distinctly different problems, but instead there is merely a difference in the way to view the same or similar problems. In regard to how the language of science is both similar and different from other forms of communication, Einstein once wrote the following:

> The scientific way of forming concepts differs from that which we use in our daily life, not basically, but merely in the more precise definition of concepts and conclusions; more painstaking and systematic choice of experimental material; and greater logical economy.[5]

Einstein's views on the similarities between scientific activities and daily life experiences help us see that all academic communities are faced with problems associated with using language. Regretfully, I have come to the conclusion that scholarly groups often hinder personal growth and the advancement of knowledge partly because they place too great an emphasis on learning the specialized language of an academic discipline. As I see matters, the educational practices in most scholarly communities leave much to be desired and meaningful learning is often replaced with the pseudo-learning of academic jargon.

Later in this paper I will return to issues related to how scholarly communities inappropriately teach students academic jargon. But given all that I am saying here about language and learning, it is perhaps a bit easier to understand my desire to speak before a group of scientific editors. In particular, although scientists have been somewhat successful in achieving an understanding of the physical world, I do *not* view the success of science as an indication that scientists are now being educated in the most efficient or effective way. On the contrary, much of the past success of science is due to the fact that some scientists such as Einstein were able to overcome their educational training.[6] Furthermore, the education of scientists, and the distribution of new and potentially worthwhile ideas, could be aided by better understanding how individuals and groups of people process information. Thus, whereas scientists and poets do *not* always share the same problem, when it comes to understanding human learning and human language, it does seem that all scholars have much in common.

In short, it is the common concerns of learning and language which I wish to discuss here. And my hope is that editors of scientific journals will see that my ideas are indeed relevant for those who work in the scientific disciplines. Hopefully, my ideas on learning and language will challenge those who work in scientific communities to reconsider the existing educational and editorial policies which now affect the way scholarship is evaluated, viewed, and distributed.

II. Russell's Advice to Young Professors and Two Problems Associated with Clarity

During my years as a student and a teacher I have found it difficult to acquire good advice about how to write essays for scholarly communities. One of the few pieces of advice that I have found useful for those who work in existing academic disciplines is the advice that Bertrand Russell has offered to young scholars. In regard to how a person should do academic work, Russell once said the following:

> . . . a word of advice to such of my readers as may happen to be professors. . . . I suggest to young professors that their first work should be written in a jargon only to be understood by the erudite few. With that behind them, they can ever after say what they have to say in a language "understanded of the people." In these days, when our lives are at the mercy of the professors, I cannot but think that they would deserve our gratitude if they adopted my advice.[7]

Throughout this paper I will refer to the above comments as *Russell's advice to young scholars*. As we will see, this advice, although it has much merit for those who work in academic communities, needs to be modified. But before I discuss Russell's views on scholarship in detail, I would first like to explain how Russell's advice has led me to understand specific problems related to educational and editorial policies.

Russell's ideas about how to do academic work first came to my attention during the week in which I successfully defended my Ph.D. dissertation. With the completion of my doctorate degree, the university I worked for immediately promoted me to the rank of assistant professor. Thus, I really was one of the people Russell was talking to when he recommended that scholars write in a language "understanded of the people." And with my professorship in hand, I decided that now I would *stop* using academic jargon as much as possible.

I have recorded the above autobiographical information because I wish to use this essay to discuss two problems that have been suggested to me from my efforts to follow Russell's advice to young scholars. Specifically, the two questions I wish to deal with here are the following: 1) Should teachers encourage students to write in clear and direct ways, or is it wise to continue to school people so that they can write in the specialized language used by professional scholars? 2) Should editors of academic journals expect professional scholars to write in clear and direct ways? For convenience I will label the first question *the educational problem of clarity* and the second question can be referred to as *the editorial problem of clarity.*

The two above problems are an outgrowth of my interest in Russell's advice to young scholars. Unfortunately, I must report that I have *not* always been able to follow Russell's recommendations as much as I would have liked. My failure to act in accord with Russell's advice has led me to consider many issues associated with educational reform, writing, and editorial policies. Upon reflection, I now find that I can no longer wholeheartedly endorse Russell's views about how to do academic work. Nevertheless, I still consider Russell's advice to be reasonable and somewhat suitable for individuals who wish to survive in today's academic communities. But before I discuss the specific areas in which I agree and disagree with Russell's ideas on scholarship, I would further like to explain Russell's views on clarity, schooling, and editorial policies.

III. An Interpretation of Russell's Advice to Young Scholars

From the advice Russell offers to young scholars, it is difficult to figure out how he would answer the educational problem of clarity (i.e. Should teachers encourage students to write in clear and direct ways, or is it wise to continue to school people so that they can write in the specialized language used by professional scholars?). Russell's advice, although it appears to be against training students to speak in academic jargon, both condones and perhaps implies that it is desirable for students to learn the language of the professional scholar. As many teachers know, academic jargon, despite the fact that it is usually difficult to learn, may at times be very helpful; since Russell used jargon in many of his early and later works, we can assume that he may have found something valuable in the idea of having students learn the language of professional scholars.

Whereas Russell probably knew that some good could come from having students learn the jargon of their teachers, he also was aware of the fact that too much academic jargon presents a danger to writing essays in a clear and direct way. And since Russell valued clarity so highly, he did not consider it wise for students to become too steeped in academic gobbledygook.[8] For Russell, the language of the professional scholar should be used minimally; when possible Russell prefers that scholars, and perhaps even students, should write in a clear and direct fashion. However, in regard to having students write clearly, Russell knew that teachers would probably think that their students did not know the subject matter of an academic discipline unless one could competently use the language of the professional scholar. Thus, it appears to me that Russell recommends that students should learn academic jargon for pragmatic reasons. That is, in order to pass tests and get academic degrees, Russell knew that students needed to learn how to speak like their teachers. But, when one finally is viewed as a full-fledged member of an academic community, Russell suggests that a person should reverse gears and use as little jargon as possible.

Why is Russell willing to allow students to be schooled in the language of the academic disciplines while at the same time he recommends that one should *not* use academic jargon once it is learned? The answer to this question seems to be that Russell is willing to tolerate the bad educational policy of teaching jargon to students because he is aware of the fact that when individuals learn the language of the professional scholar, they are likely to become familiar with the important questions and ideas in a scholarly discipline. Thus, it appears to me that Russell is willing to tolerate a bad educational policy because such policies may be able to foster some desirable consequences. In other words, when students learn the jargon of their teachers, they will come in contact with many worthwhile ideas. Furthermore, if students learn their lessons well, then they will be given academic degrees and honors which will allow a person to do worthwhile scholarship. But remember, Russell tells us that good scholarship should be written with as little jargon as possible.

As I see matters, Russell is mistaken if he indeed assumed that many young scholars would use academic jargon sparingly once they had achieved a modest amount of social success in their chosen fields of study. On the contrary, my experience in the academic world has shown me that most scholars *continue* to write for the "erudite few" throughout their professional careers. This should *not* be surprising because success and survival in academic communities is partially dependent upon getting published in scholarly journals. And since those who grant Ph.D.'s are usually the same people who judge papers acceptable or unacceptable for publication, we should expect that the standards for academic work will be somewhat similar to the standards that teachers use to judge the work of their students. Of course, it is *not* necessary to teachers and editors to evaluate academic achievements by the same standards because people could decide to evaluate the efforts of students by entirely different standards than those which are used to judge the efforts of professional scholars.

It is interesting to note that many individuals seem "naturally" to assume that students and professional scholars should be evaluated by the same or similar standards. Editors of academic journals, like teachers in graduate school, expect people to write in the specialized language of an intellectual discipline. A major difference between the standards for evaluating a student's achievements and the work of a professional scholar is that scholars are expected to use academic jargon properly while students merely need to demonstrate that they are in the process of learning the jargon that their teachers value. Thus, if it is correct to assume that editors and teachers do indeed use similar standards to evaluate academic accomplishments, then Russell's recommendations that young scholars write in a language "understanded of the people" is almost a surefire way for a person to become a failure and an outcast in today's academic communities. Unfortunately, if individuals do not continually demonstrate competency in the jargon of their chosen disciplines, then most likely they will never have their essays accepted for publication.

Russell's advice to the young scholar will *not* lead to failure if editors of scholarly publications offer an affirmative answer to the editorial problem of clarity (i.e. Should editors of academic journals expect professional scholars to write in clear and direct ways?). As it turns out, Russell's advice to young scholars is *also* advice to the editors and publishers of academic works. In particular, Russell's advice to young scholars should be seen as an attempt to reform and change the standards used to evaluate scholarly efforts; in his advice to young scholars, Russell tells those who control the publication of academic essays that they should *stop* publishing works that are loaded with the jargon of the professional scholar. Rather than publish essays that contain a great deal of specialized language, Russell seems to suggest that editors should immediately begin to publish works that are written in a clear and direct fashion.

One of the most intriguing aspects of Russell's advice to young scholars is that it *separates* two problems which people often confuse or identify as one. Specifically, Russell seems to suggest that it is possible to solve the educational problem of clarity with an entirely different solution than the one used to solve the editorial problem of clarity. These two problems related to clarity are solved in unique ways because Russell recommends the following: 1) Students should learn how to write essays in the language used by the professional scholars. 2) Editors should publish works which are written with a minimum amount of academic jargon. Of course, as I noted earlier, Russell makes the first recommendation somewhat reluctantly. Also, he makes the second recommendation indirectly. But whatever the case, Russell has helped us see that teachers and editors can judge intellectual accomplishments by different standards.

IV. On the Failure to be as Clear as Russell Envisioned and Clarity as a Worthy Unattainable Goal for Imperfectionists

With a working interpretation of Russell's advice to young scholars in hand, I would now like to begin a critical analysis of the views I have attributed to Russell. At this time I would like to explain that the goal of writing clearly is not as easily achieved as Russell envisioned. In fact, the kind of clarity that Russell hoped for seems to be a worthy *unattainable* goal. But let me hold off for a while about the quest for unattainable human aims. Let me begin this section with an account of Russell's failure to write as clearly as he would have liked.

Few philosophers of the twentieth century have achieved the fame of Bertrand Russell. Russell was not only a great philosopher of mathematics and science, but he also was a great social philosopher who wrote about many political, economic, and educational issues. When I go to the library to survey Russell's work I find it difficult to believe that one person could do so much writing in a lifetime. Furthermore, not only was Russell a prolific writer, but he also was a great thinker and I marvel at the fact that he could write so clearly and eloquently about various kinds of problems. However, despite Russell's efforts to be perfectly clear, I do not think that he or any writer can avoid being confused and muddled at some times. As it turns out, it may be impossible for anyone to achieve the kind of clarity that Russell expected. It is interesting to note that in the volume in the Library of Living Philosophers devoted to Russell's intellectual achievements, Paul Schilpp has reported the following:

> . . .Mr. Russell intimated that his greatest surprise, in the reading of the twenty-one contributed essays, had come from the discovery that "over half of their authors had *not* understood" him (i.e., Russell). This fact amazed Mr. Russell all the more because he always thought that he had been making every effort to write clearly and to express his ideas in the briefest possible and most direct way.[9]

I am *not* surprised that Russell was misunderstood in spite of the fact that he tried to, and in fact often did, write in a fairly clear and direct fashion. As I see matters, all authors should expect to be misunderstood by many of their readers. Regretfully, it is impossible to achieve the kind of clarity that Russell strived for throughout his life. A major reason why perfect clarity is outside of our reach is that every human sentence or thought implies an infinite number of ideas and concepts. In other words, perfect clarity and perfect communication are unattainable because human beings can never say or write everything that can be said about any single thought. In regard to the notion that all sentences and theories have an infinite number of ideas associated with them, Karl Popper once stated the following:

> . . . when we propose a theory, or try to understand a theory, we also propose, or try to understand, its logical implications; that is, all those statements which follow from it. But this . . . is a hopeless task: *there is an infinity of unforeseeable nontrivial statements belonging to the informative content of any theory*, and an exactly corresponding infinity of statements belonging to its logical content. We can therefore never know or understand all the implications of any theory, or its full significance.[10]

Popper is not the only contemporary philosopher who has realized that human language cannot allow people to express ideas in a simple and perfectly clear manner. Another recent example of a writer who recognizes that all human utterings have an infinite number of ideas associated with them is Jorge Luis Borges, the Argentinian poet and short story writer. In speaking about his attempts to write in a clear and direct way, Borges has noted the following:

> I have done my best--I don't know with what success--to write straightforward stories. I do not dare state that they are simple; there isn't anywhere on earth a single page or a single word that is, since each thing implies the universe, whose most obvious trait is complexity.[11]

Popper and Borges are two contemporary philosophers who endorse the notion that communication between people must always be imperfect and somewhat ambiguous because of the structure of language.[12] That is, for Popper and Borges, all human languages are imperfect tools for communicating our thoughts because no language can be created which can perfectly translate any human idea into a verbal context. For convenience I would like to refer to Popper's and Borges' ideas about language being an inadequate tool for expressing a human thought as *the thesis of imperfect communication*. As it turns out, this thesis has many implications for the way in which we should view such goals as clarity. Specifically, if one endorses the thesis of imperfect communication, then it is possible to see that Russell's demand for a great deal of clarity is really an unattainable goal for human beings. In other words, if every idea has an infinite number of statements associated with it, then human attempts to be perfectly clear are doomed to fail because verbal utterances only communicate an insignificant amount of all the information that is implied by any single thought.

To suggest that a thing is unattainable does not necessarily imply that it is an unworthy goal. On the contrary, at times human beings may decide that unattainable goals are worthy to strive for because in the process of approaching an unattainable goal we may learn something valuable along the way. In relationship to striving for unattainable goals, people such as Popper and Russell have suggested that absolute truth can be viewed as a worthy unattainable goal of scientific inquiry.[13] In a similar sense, it is possible to say that perfect clarity, despite the fact that it is an unattainable goal, is a worthy aim that human beings should attempt to approach.

Human beings may find that it is undesirable to spend an entire lifetime striving for only unattainable goals. At times people may desire to achieve their aims. It is a difficult matter to decide when and how we should strive for the impossible, but if we wish to understand our own thoughts and those of others, then we may find that some degree of clarity is desirable even if we can never be perfectly clear. That is, whenever we try to understand human ideas, any amount of clarity may prove to be more desirable than total confusion. The problem with working at being confused is that the more muddled we attempt to be the more difficult it is to have meaningful human communication. For those of us who want to share our thoughts verbally with others, any amount of clarity is desirable because our attempts to be clear can greatly facilitate communication.

Much of what I have said about clarity might have been known to Russell. However, Russell does *not* seem to be content with accepting the idea that all human communication is a highly imperfect process. Russell's disappointment with his own attempts to write clearly might not have existed if he had sufficiently understood the implications of the thesis of imperfect communication. Nevertheless, despite Russell's failure to comprehend that he was doomed to be misunderstood by others, this remarkable man continued to do academic work for over seven decades. As I see matters, Russell's work can help inspire all of us to seek such worthwhile unattainable aims as clarity. Russell's academic accomplishments make it possible to see that an imperfect and fallible person can have a meaningful life striving for some impossible goals. What is so sad about Russell's quest for the impossible aim of writing clearly is that Russell seemed to expect far too much from himself and the rest of us. As it turns out, Russell has done much to clarify important problems even though he failed to be perfectly clear.

The arguments in this section have *not* tried to suggest that human beings should give up the impossible quest for clarity. On the contrary, I have attempted to point out that the arguments put forth by imperfectionists such as Popper suggest that we must now view the aim of being clear somewhat differently than Russell envisioned it. Specifically, we need to realize that in spite of our attempts to be clear others will undoubtedly find that much of what we say is confusing. But what I wish to emphasize here is that it is all right to be confused and none of us should expect to be perfectly understood by ourselves or others. And furthermore, confusion does not have to be an insurmountable obstacle which totally prevents communication; at times confused individuals can share their thoughts with one another. Also, we may even become less muddled if we can find people who will criticize our ideas. Thus, whereas perfect clarity is *not* humanly possible, the problem of trying to be clear still remains with us because it seems that some degree of clarity is better than total confusion. In regard to how an imperfectionist might attempt to be clear, Popper has said the following:

> . . . .what should we do in order to make our meaning clearer, if greater clarity is needed, or to make it more precise, if greater precision is needed? . . .the main answer to this question is: any move to increase clarity or precision must be *ad hoc* or "piecemeal". . .reformulate your formulations *ad hoc*, with a view to avoiding misunderstandings which have arisen or which you can foresee. And always remember that *it is impossible to speak in such a way that cannot be misunderstood:* there will always be some who misunderstand you.[14]

Later in this essay I will return to some of the ideas associated with an imperfectionist view of human communication, but first I would like to return once again to Russell's advice to young scholars.

V. Academic Jargon as an Obstacle in Education and the Need for Teachers to Help Confused Students Become Clearer

Although I still agree with much that Russell has said about how to write academic essays, I have come to see that his advice to young scholars has to be modified and at times even rejected. In particular, I think Russell expects too much clarity from young professors and not enough clarity from students. That is, as I tried to demonstrate in the previous section, it is a mistake to assume that any scholar can be as clear as Russell would have liked. Also, I think Russell is wrong to assume that we should train students to write papers in the jargon of the various academic disciplines. As I see matters, the educational problem of clarity (i.e. Should teachers encourage students to write in clear and direct ways, or is it wise to continue to school people so that they can write in a "jargon only to be understood by the erudite few?")needs to be answered as follows: All students, no matter what grade they happen to be in, should be encouraged to write as clearly and directly as is humanly possible. Let me now say something about my answer to the educational problem of clarity.

It is not usually realized that many teachers, and particularly college professors, help their students become more, rather than less, confused. My experience with teachers has shown me that they usually like to see their students use the language of the professional scholar; teachers emphasize the language of the scholar because they think that it is more powerful and precise than ordinary language. In addition, my guess is that teachers assume that academic language will eventually help their students be less confused and clearer. But contrary to what teachers think might happen, the language of the professional scholar can become a major source of confusion because students are likely to learn the jargon of an academic discipline without any understanding of the problems or questions which have led scholars to develop very sophisticated language systems.[15]

What I am trying to get at here is that many students learn academic jargon for its own sake, rather than as a tool to help solve worthwhile or significant problems. In a similar sense, I have come to realize that teachers and professional scholars may at times lose contact with the fact that academic language has been created to help people better solve and understand problems. Of course, others before me have noticed that academic communities often develop elaborate language systems that have allowed individuals to lose contact with meaningful questions.[16] However, in spite of the fact that many teachers are aware of the idea that academic jargon might be an obstacle for learning, students are regularly expected to learn how to speak and write as professional scholars; as I see matters, students should *not* learn academic language until they are able to articulate a problem to be solved.

My impatience with the way students are inappropriately taught has led me to become an educational reformer.[17] Unlike Russell in his advice to young scholars, I do not recommend that beginning professors write their first work in a "jargon only to be understood by the erudite few." On the contrary, I think all intellectual work needs to be written as clearly and directly as possible because individuals should always do their best to communicate with others. As I see matters, students should be encouraged to write and think clearly as *early* as possible. I even recommend that kindergarten teachers should help their students to speak and think clearly because the process of learning how to be clear is a life-long endeavor which is never completed; at all times people need to struggle to become less confused.

The age or grade that we should begin to encourage students to be clear is an educational matter which Russell briefly alludes to in his advice to young scholars. And although Russell wrote about some very daring educational reforms for the lower levels of schooling, he seems to be a bit hesitant to advocate innovative learning practices in regard to the higher levels of education.[18] As I suggested earlier, Russell seems to tolerate conventional educational training at universities partly because he hoped that while students were trying to master academic jargon they would also tacitly learn the important problems and concepts of a field of study. However, learning the language of the scholar can often become an *obstacle* for understanding the problem situations of a field of study because academic jargon, when it is first being learned, requires so much attention that it is unlikely that students will have time to figure out why scholars have developed sophisticated language systems. In other words, the mastery of jargon usually becomes a full-time job and students are regularly trained in such a way that they do not learn that a specialized language has been developed to serve a particular purpose (i.e. help people better understand specific problem situations).

Despite Russell's general dislike for academic jargon, he undoubtedly knew that the language of a scholar has the potential to be a powerful tool if it was used correctly and sparingly. Regretfully, teachers and professional scholars do not usually follow Russell's advice to minimize the use of jargon. On the contrary, jargon is often used as a sign of being associated with a specific intellectual group and part of a student's initiation into an academic community entails learning the specialized language of the scholar. But teachers who teach jargon as part of the ritual of being initiated into a field of study may not have an understanding of the important problems of their discipline. As it turns out, many professional scholars may not have a grasp of the problems which their fields of study are trying to solve; it may be the case that teachers are often confused and divided about how and when to use the jargon they know so well. Thus, teachers are not always in a position to teach both academic jargon and the important problems of their field of study. In fact, it is highly likely that many teachers are as confused as their students! But there is a difference. Teachers have been certified as social and intellectual authorities and, in the eyes of their students, teachers possess "knowledge" that must be mastered.

What I am trying to get at here is that those who teach students the jargon of a field of study may be quite confused themselves. Since this is so, students are often better off if they ignore much of what their teachers have to say. Nevertheless, it is *not* the case that teachers have nothing to offer their students. Teachers can help students become less confused about the problems and questions that are studied in school. One way teachers can begin the process of helping students become less confused is that teachers themselves should try to use less academic jargon; teachers and professional scholars should follow Russell's recommendations about trying to be much clearer and more direct.

The kind of teacher I have briefly outlined in the above has much in common with the Socrates who was portrayed in Plato's *Apology*.[19] But it would be inappropriate for me to use this paper to provide a detailed account of

Socratic teachers. Much work needs to be done in order to understand how Socratic teachers can help individuals become less confused. However, despite my sketchy description of Socratic teachers, it should be evident that these people never use academic jargon for its own sake. On the contrary, a Socratic teacher will only use the language of a scholar to help a person solve a problem or answer a question. And Socratic teachers endorse the idea that human beings can never communicate with perfect clarity. For a Socratic teacher, confusion can never be entirely eliminated, but it can at times be diminished; teachers who follow in the tradition of Socrates recognize that the elimination of some confusion allows people to approach the unattainable goal of perfect clarity.

VI. Clarity as an Obstacle for the Growth of Knowledge and the Need for Editors to Tolerate More Confusion

I would now like to shift the discussion to issues related to the editorial problem of clarity (i.e. Should editors of academic journals expect professional scholars to write in clear and direct ways?). My solution to this problem is as follows: Editors of scholarly works should not expect a great deal of clarity from writers, but they should desire that authors write as clearly and as directly as may be possible. In addition, editors should learn to tolerate some confusion in the essays that they publish because if we ask writers to be too clear, then we may prevent worthwhile ideas from being publicly discussed. Whereas it seems a bit strange to recommend that editors publish works that are admittedly confused, it is worthwhile to note here that confused essays may contain some valuable ideas and arguments; if editors ask for too much clarity from authors, then they may *unintentionally* hamper the advancement of human knowledge.[20]

The above solution to the editorial problem of clarity might be viewed as unsatisfactory from the point of view of both editors and scholars. People who work in academic communities do not like to think that the essays which appear in scholarly journals are confused and muddled; it is usually assumed that when an essay is accepted for publication it is written in such a way that it can be understood by one's colleagues. However, if one endorses the notion that all languages are imperfect tools for communicating human thoughts then it is possible to claim that there is much self-deception going on in the academic world. Specifically, both editors and scholars are likely to think that they have a clear understanding of what is being said in a published essay, but in reality everyone usually has a very incomplete and inadequate comprehension of what most authors are trying to say. As it turns out, it often takes many years and many interpretative essays in order to have a beginning grasp of what an author has attempted to write about in a published work.

Scholarly achievements must always remain somewhat unclear because, as I noted earlier, clarity is an unattainable goal which human beings can only approach.[21] Thus, editorial policies which demand a great deal of clarity may be too restrictive because they expect too much from fallible human beings. Rather than expect writers to be clear, editors need to learn how to accept the idea that all authors are at times confused about what they actually want to say. In addition, although editors may still desire some amount of clarity from scholars, it may be worthwhile for academic communities to publish and circulate works that are judged to be a bit confused and unclear; the publication of unclear essays is desirable because these works may at times challenge an academic group to rethink and reconsider the questions and problems which scholars are attempting to solve.

It is indeed a bit radical to recommend that editors should start to publish essays that are admittedly somewhat confused and unclear. If editors endorsed this recommendation, then we would undoubtedly have a number of significant changes in the standards which are used to judge essays acceptable or unacceptable for publication; as was the case with Russell, I too have a desire to change the standards which editors use to evaluate scholarly achievements. However, unlike Russell, I do not think that clarity should be a major criterion for determining which essays appear in print. I think editors should *liberalize* the standards used to evaluate academic work. As I see matters, the standards used to determine acceptable forms of scholarship need to become less restrictive and academic communities should support the publication of works that are admittedly somewhat unclear.

If we assume that editors do indeed look unfavorably upon essays that are unclear, then my recommendation about editors learning to tolerate and even publish confused essays should be seen as an attempt to change existing editorial policies. And furthermore, my suggestion that confused essays should be published needs to be understood within the context of many of the things I have said about clarity. In particular, it is helpful to remember that academic communities should not be satisfied with confusion for its own sake. On the contrary, as with Russell, I recommend that all authors should try to be as clear and direct as possible. But, unlike Russell, I do not expect much clarity from writers. That is, I recognize that in order to develop a better understanding of a problem situation scholars usually *begin* with a confused state of affairs; it is only *after* an academic community is exposed to a confusion that we can hope to make a problem situation somewhat clearer and more understandable.[22]

Confusion for its own sake should not be tolerated by editors and scholars. Nevertheless, if we endorse the notion that some confusions may eventually lead to the advancement of human knowledge, then it is possible to see why confusion must at times be tolerated and even encouraged. As it turns out, it is very difficult to contribute to the growth and development of new ideas. And for those who wish to go beyond the existing state of knowledge in their chosen academic fields, it is usually necessary to work through one's confusion, rather than totally avoid being confused; the growth and advancement of an idea is partially dependent upon a person's willingness to be confused in the early stages of theory development. If editors would tolerate an author's confused early works, then academic communities might be able to advance knowledge quicker and more efficiently because scholars who are developing confused but potentially worthy ideas might be able to get help from others if they could get their essays published. Regretfully, many scholars have been forced to work in isolation because their ideas appeared to be confused and unorthodox.

There are numerous examples of individuals who have been ostracized from the academic world because they attempted to develop unconventional ideas which appeared to be quite muddled. In the field of the philosophy of education I have come to realize that a great teacher and thinker such as Socrates was isolated from the teaching profession of ancient Greece partly because his ideas were different from those of his contemporaries. And since being different is often translated to mean that one is confused, it should not surprise us that Socrates was usually viewed as a muddled scholar who did not know what he was saying.[23] Of course, Socrates was somewhat confused about many educational issues, but his ideas are nevertheless worthwhile and they merit our attention because they are so challenging and rich in thought.

The histories of most fields of study probably contain numerous stories about scholars who were isolated from an academic community because their contemporaries viewed them as confused. And as fate would have it, many so-called "confused" scholars have been credited with initiating intellectual revolutions. Of course, every confused scholar is not a genius and many muddled ideas will lead nowhere. But if we are aware of the dangers of romanticizing confusion, then it is possible to see that an emphasis on clarity may allow scholarly groups to ignore individuals who are developing worthwhile ideas that usually begin as confused utterings which need to be clarified and criticized; a toleration of confusion may allow us to advance knowledge because many potentially worthwhile theories have to go through a confused stage before they can be adequately understood by a community of scholars.

If academic groups begin to recognize that it is at times desirable to publish, circulate, and criticize ideas which are openly labeled confused, then the scholarly world is probably in store for numerous changes. Unfortunately, I cannot discuss the numerous consequences associated with having liberalized editorial policies which would allow for the publication of confused ideas. However, I would like to conclude this section with some questions that editors and scholars may wish to consider. Specifically, the questions I have in mind are the following: 1) Can we distinguish between those confused works that might eventually lead to improvement and those essays that are confused beyond hope? 2) Should editorial boards reject clarity as a criterion for accepting an essay for print? 3) What might happen to academic journals if editors began to publish essays that were admittedly confused? 4) How might academic communities begin the process of learning to tolerate more confusion from scholars? 5) Is it correct to assume that confusion does indeed have the potential to lead to the advancement of human knowledge? These questions are only a few that can be inferred from the arguments in this essay. As I see matters, all academic communities could improve existing editorial policies is they would discuss these and other questions related to clarity, confusion, and the growth of human knowledge.

VII. On the Imperfectionist's Rejection of Clarity as a Criterion for Evaluating Intellectual Work and Russell's Advice Revisited

In light of the arguments in this paper, it should be obvious that I am of two minds about the issue of clarity. That is, on one hand, I am willing to tolerate a great deal of confusion from both students and scholars because I endorse the notion that confused ideas may eventually lead to the advancement of human knowledge. But on the other hand, I view clarity as an unattainable human goal which is worthwhile to strive for because I think that people can improve their understanding of a problem situation by attempting to be as clear as possible. Thus, whereas I am of two minds about clarity, I do not think my ideas on this issue are contradictory because I view clarity as an unattainable goal. For me, the process of becoming clearer is viewed as the same as the process of becoming less confused. And furthermore, the process of becoming less confused *never* ends and people often spend a lifetime trying to approach the unattainable goal of perfect clarity.

Much can be said for or against my views on clarity. It is my hope that this essay will eventually be taken seriously enough to be criticized by those who are interested in developing adequate standards to evaluate academic accomplishments. However, until I receive some criticism for the arguments in this paper, I must remain under the illusion that what I have said about clarity does indeed make some sense. In addition, in order to better understand my views on clarity, I would now like to explain how an imperfectionist might modify Russell's advice to young scholars.

An imperfectionist might recommend that scholars do the following: All scholars, both students and teachers of every age, should immediately begin to write about their problems as clearly and directly as they can. Whenever possible, scholars should consult with others who have written about their problems. Nevertheless, when one goes to the literature associated with an academic interest, it is always a mistake to learn the specialized language of a scholarly community for its own sake; only learn academic jargon when it helps you better understand the problems you wish to solve. Furthermore, when writing or speaking about your problems, always try to be as clear as you can, but never expect to be perfectly clear; when your thoughts are confused in the essays you write, try to write through your confusion, rather than give up because you are unclear about what you are saying. And finally, students should seek out teachers who will help them to become clearer about their academic interests, and professional scholars should submit their papers to editors who understand that academic work may be worthwhile even if it is somewhat confused.

Since the above recommendations incorporate many of the ideas associated with the thesis of imperfect communication, I would like to call these recommendations *the imperfectionist's advice to scholars*.[24] This advice modifies Russell's views on scholarship in at least the following four ways: 1) The imperfectionist's advice is meant for all students and teachers, rather than just those who are beginning professors. 2) The imperfectionist's advice suggests that not only should young professors attempt to be clear, but all individuals should attempt to be as clear as possible during their school days and throughout their entire academic careers. 3) The imperfectionist's advice challenges teachers to try to help their students become less confused and clearer about the problems that are studied in school. 4) The imperfectionist's advice challenges editors to develop editorial policies which will tolerate more confusion in the essays that are accepted for publication.

VIII. A Concluding Remark

I would like to end this paper with both a warning and an observation about how the imperfectionist's advice to scholars might affect an individual's life. My warning is that those who follow the imperfectionist's advice may become failures rather than social successes. My observation is that contemporary scholars should have a skeptical attitude towards being successful. Although I would readily admit that a modest amount of social success may make it easier for scholars to do their work, I think it is important to realize that there are dangers associated with being successful in many of today's academic communities. Let me now conclude with a brief word about the warning and observation I have made here.

The imperfectionist's advice to scholars may help people become failures rather than successes in the academic communities they wish to join because students who decide to write in clear and direct ways will only be able to survive in the academic world if they can find teachers who allow them to write essays in a language "understanded of the people." In a similar sense, professional scholars who have the courage to submit confused essays which contain their latest and best efforts to solve a serious problem may not be able to find editors and journals which will publish their ideas. Thus, if one is looking for a surefire way to be socially successful, it may be best to ignore the imperfectionist's advice to scholars.

In regard to the observation that success may be dangerous for scholars, it is important to remember that the praise of one's peers does not necessarily indicate that a person is indeed doing worthwhile and meaningful work. On the contrary, success in academic communities may mean that an individual is merely satisfying public opinion. And what the public demands of us is often base and of little value.[25] Given this state of affairs, we should all be a bit wary of being socially successful. Unfortunately, the desire to be successful may be an obstacle in our quest for clarity, truth, beauty, and justice. Of course, all of these worthy goals may be unattainable. But if we wish to make our lives more meaningful, then it may at times be desirable to seek unattainable goals, rather than social success. In relationship to the quest for impossible aims, Russell once noted the following:

> Let us preserve our respect for truth, for beauty, for the ideal of perfection which life does not permit us to attain. . . . In action, in desire, we must submit perpetually to the tyranny of outside forces; but in thought, in aspiration, we are free, free from our fellowmen, free from the petty planet on which our bodies impotently crawl, free even, while we live, from the tyranny of death.[26]

There is much more to say about the issue of unattainable goals. Regretfully, detailed commentary on this important subject must wait for another time.

FOOTNOTES

1. Jorge Luis Borges, "The Writer's Apprenticeship," *Borges on Writing*, eds. Norman Thomas de Giovanni, Daniel Halpern, and Frank MacShane (New York: E. P. Dutton and Co., Inc., 1973), p. 165.

2. At this time I would like to thank Joseph Agassi and Miriam Balaban for the suggestions and encouragement they have given me in regard to my ideas on the connection between schooling and the publication game. Whereas neither of these individuals necessarily agrees with the arguments in this paper, I nevertheless am very grateful for the help these people have given me. Also, I would like to note at the outset of this paper that my use of the word "game" in the title of this essay is not meant to depreciate the activity of publishing scholarly works. On the contrary, the word "game" is merely used to suggest that the publication of academic works is a human activity that is governed by a set of rules which guide and somewhat determine human behavior. Finally, I would like to thank Oakland University for providing me with financial support which made it possible for me to attend the First International Conference of Scientific Editors.

3. For an explanation about how the distinctions between academic disciplines are a matter of convention see Karl R. Popper, *Conjectures and Refutations* (New York: Basic Books, Publishers, 1962), p. 67. Also, my early ideas about how to view the study of problems can be found in Ronald Swartz, "Mistakes as an Important Part of the Learning Process," *The High School Journal*, Vol. 59, No. 6, (March, 1976), pp. 246-257.

4. See the references in footnotes three and six.

5. Albert Einstein, *Out of My Later Years* (New York: Philosophical Library, Inc., 1950), p. 75.

6. For an account of Einstein's views on his scientific training see Albert Einstein, "Autobiographical Notes," *Albert Einstein: Philosopher-Scientist*, ed. Paul Arthur Schilpp (New York: Harper and Row, Publishers, 1969), p. 17.

7. Bertrand Russell, *Portraits From Memory* (New York: Simon and Schuster, 1969), pp. 213-214.

8. For references which clearly explain how Russell wanted to grant academic freedom to students see the following: Bertrand Russell, *Education and the Good Life* (Boni and Liveright, Inc., 1926), p. 247; Bertrand Russell *Education and the Modern World* (New York: W. W. Norton and Co., Inc., 1932), pp. 28-42; Bertrand Russell, *Sceptical Essays* (New York: Barnes and Noble, Inc., 1963), pp. 127-138.

9. Paul Arthur Schilpp, *The Philosophy of Bertrand Russell* (New York: Tudor Publishing Company, 1951), pp. xiii-xiv.

10. Karl R. Popper, "Intellectual Autobiography," *The Philosophy of Karl Popper*, ed. Paul Arthur Schilpp (LaSalle, Illinois: The Open Court Publishing Co., 1974), pp. 19-20.

11. Jorge Luis Borges, *Doctor Brodie's Report* (New York: E.P. Dutton and Company, Inc., 1973), pp. xi-xii.

12. Many other philosophers beside Popper and Borges have endorsed the notion that human language cannot verbally communicate all that people know. For other examples of people who recognize the limitation placed upon

human communication see the following: William James, *Pragmatism and Other Essays* (New York: Washington Square Press, Inc., 1968), pp. 5-6; Michael Polanyi, *The Tacit Dimension* (Garden City, New York: Doubleday and Co., Inc., 1967), pp. 4-10.

13. For Russell's and Popper's views about the quest for unattainable goals see the following: Bertrand Russell, *A History of Western Philosophy* (New York: Simon and Schuster, 1945), pp. 819-828; Karl R. Popper, *Conjectures and Refutations*, pp. 228-229.

14. Karl R. Popper, "Intellectual Autobiography," pp. 22-23.

15. See the references in footnotes three, six, and eight. Furthermore, for my early ideas about how to foster meaningful learning see Ronald Swartz, "Education as Entertainment and Irresponsibility in the Classroom," *Science Education*, Vol. 58, No. 1 (Jan.-Mar., 1974), pp. 119-125.

16. For an example of a scholar who is very much aware of the fact that social scientists often create and use a great deal of unnecessary jargon see C. Wright Mills, *The Sociological Imagination* (London: Oxford University Press, 1967), pp. 3-75.

17. Refer to the works cited in footnotes three and fifteen.

18. For an account of Russell's views on higher education see Bertrand Russell, *Education and the Good Life*, pp. 301-313. Also, see the references in footnotes seven and eight.

19. An account of my early ideas about how to have a Socratic teaching method can be found in the reference cited in footnote fifteen.

20. Whereas it seems strange to say that too great an emphasis on clarity might hamper the growth of human knowledge, this idea seems to follow from what Popper and Borges have said about language being an imperfect tool for human communication. Also, for an introduction to my views on editorial policies see Ronald Swartz, "On How Different Philosophies of Education Might Affect Publication Standards," *Philosophy of Education 1976: Proceedings of the Thirty-second Annual Meeting of the Philosophy of Education Society*, ed. Kenneth A. Strike (Worcester, Massachusetts: Heffernan Press, 1976), pp. 169-180.

21. See the references in footnote thirteen.

22. Refer to the works cited in footnotes three, thirteen, fourteen, and nineteen.

23. For a contemporary account of the tragic circumstances associated with how Socrates was treated by his peers see Karl R. Popper, *The Open Society and Its Enemies* (New York: Harper and Row, Publishers, 1962), pp. 189-201.

24. It is important to note here that Russell often advocated imperfectionist ideas. However, when it came to the issue of clarity, Russell seemed to lean towards perfectionist views. For a reference which clearly shows Russell's imperfectionist leanings see the work cited in footnote twenty-six.

25. For an introduction to some of the dangers associated with basing decisions on "public opinion" see Karl R. Popper, *Conjectures and Refutations*, pp. 347-354.

26. Bertrand Russell, *Mysticism and Logic* (New York: Doubleday and Co., Inc., 1957), p. 48.

EDITORS' IMPACT ON SCIENCE POLICIES AND POLITICS

Dimitri Viza, *Editor, Differentiation*
Laboratoire d'Immunobiologie, Service d'Hématologie
15, rue de l'Ecole de Médecine, 75006 Paris, France

ABSTRACT

In the present era of conservative scientism, where the number of papers counts more than their content and the number of experiments more than the ideas behind them, where security is appraised and intellectual boldness frowned upon, the editors of scientific journals, by the mere selection of papers can not only help to re-establish the circulation in the clogged information channels, but also change the research policies and the established habits, so that scientists may resume their individuality and counterbalance the *efficiency* conceived by the politicians and their technocrats.

From the technological revolution of the 19th century the scientist emerged as the modern high priest exorcising the demons of both nature and the human soul. Until his magic disappeared in an atomic explosion, followed by a population one. And the results of science have since been regarded as the work of either irresponsibility or genius, depending on the mood, the political or religious point of view and the psychological make-up of the observer. The reality may be different. However, through worship or criticism, through encouragement or ban on his freedom, through infatuation or contempt, contemporary society is transforming the scientist into a soulless monster.

Crushed under personal and professional obligations, miserable salaries and miserable grants, the modern scientist's career is worthy of a Kafka plot, and he is slowly becoming the obedient servant of a system gradually developed to suppress individuality and to promote mediocrity, an anonymous servant of anonymous masters of socio-economical systems obeying their own laws and defying the analyses of the so-called experts.

Scientific research today is far from being the creative work of a personality, or, to paraphrase Bronowski, *scientific creation is far from flowing from a personality* and *being alive and charged with individuality.* Rather, it looks like the mechanical drudgery of a mole's digging and modern research investigators are too concerned with their tenure, their career - securing them in some places six weeks' holiday and a research bonus, not a discovery one - to bother about individual creativity. With the methodical unimaginative data collection and accumulation, with too few hypotheses and synthetic views in comparison with the avalanche of descriptive reports, scientific research became a butterfly collection devoid of any joy of creativity. And those scientists with the extra energy, whom the system has not as yet subdued, will spend it in the endless search for money - on which scientific empires are constructed - in "scientific tea parties" called scientific conferences, symposia or congresses, suffering from the jet lag of the new jet scientific society, and in occasional appearances on radio or TV and press conferences to fool the lay public, after they have fooled

Miriam Balaban (ed.), Scientific Information Transfer: The Editor's Role, 157-160.

themselves.

The absurdity of the present structure of scientific research is well-known. It is generally accepted without any protest that new, off-the-beaten-track ideas for original research can not be serious and should not be funded by serious organizations - and they are not; that only conservative projects based entirely on evidence from existing data can be presented without the risk of wasting time and paper for the applicant; that backstairs influence is quite useful whether one calls it recommendation, acquaintance, friendship, or bribery; that those who control grants proliferate more rapidly than scientists; that technocrats are highly prized by governments, whereas young fundamentalists are treated with utter contempt, left only with the chance of hitting the jackpot, by inventing the earth blasting bomb and winning the nobel prize for the glory of their government, which will then proudly discover them to prove to the others how highly civilized it is.

It is amusing and it should be mentioned here, so that one gets the whole picture of the Kafka's plot of modern scientific research, that in order to get funds for a research project one has to state what assets one already possesses in the form of equipment, personnel and grants from other sources, and then, after an incredible waste of time and money in bureaucratic red tape, he finds out that his grant application would not have been accepted anyway as his competitor, enemy, rival or a jealous moron was sitting on the committee. And if he complains about injustice, bad faith, sclerosis or neurosis, he will be accused of making slanderous allegations prejudicing the honour of the committee, in particular, and the scientific community in general. And so it goes on, for the policy regarding modern science evolves around the fundamental principle: *never take the risk to create, it is an expensive hobby, it fails often, it disturbs always*. It has always been easier and safer to flog dead horses.

We live in a technological era, where true science, original and unorthodox ideas are considered 'anarchistic', with all the fear the connotation of this word implies. The safest way to achieve security is through technology and applied research, the development of what has already been developed. No revolutionary movement is in sight to save science from its lethargy by destroying the limitations imposed by stupidity, technocracy and bureaucracy, at the service of politicians obsessed by ancestral fears, against which their defence takes the form of aggression. The scientist has thus no alternative but to accept the role of a puppet in the bureaucratic society with the remote hope that a society of scientific meritocracy will one day exist.

This is not an imaginary possibility, but a real one, appealing, however, to few laymen and even fewer scientists. For it might have seemed logical to give the power to those who derive it from nature, to shape the form of future civilizations, if it had not become obvious that psychological and emotional development do not keep pace with technological advances. One may judge from the scientists who have become administrators, and the nightmarish image of a scientist-controlled society immediately appears before us in all its horror. Whether it is scientists or technocrats who control the lives of this potential society is of little importance. The technological and bureaucratic ritual will become dominant in the short term, destroying the remnants of all creativity.

Scientists have failed to control their own affairs and have demonstrated better than any other profession that no one can "be on top and on tap"! However, leaving the science policy-making in the hands of clerks, whether they are government technocrats or trained scientists short of imagination, has been calamitous and will continue to have catastrophic results destroying all creativity. Allowing lay public interference would be deleterious and the taking over of policy-making by scientists has also proven to be disastrous; for the system beats and perverts everybody it encompasses. A solution must, however, be found, if technology is not to completely destroy creativity in science.

In moulding the scientific community and the future of scientific research, the editors of scientific journals have a predominant role to play in the evolution of scientific research and this role has not as yet been assumed.

No definition of an editor's occupation is possible, since anybody without any specific training is a potential editor, his attributions being highly variable and broad, depending on the publisher and/or the financial or scientific body behind his publication. One thing, though, is certain about the editors' occupation: it is highly subjective, if not highly specialized. But subjectivity is more than often associated with creativity.

There is little doubt that the impact of editors' decisions on the scientific community can be enormous. The mere selection, acceptance or rejection, of papers can influence the evolution of science as a whole as well as the scientist's personal career. Scientific reports are indeed the only means of expression as well as recognition for a scientist, recognition which leads to promotion, availability of facilities for research and, eventually, freedom to impose new and revolutionary ideas - if there are any still remaining when one reaches the age of recognition. But scientific publications are also supposed to protect the freedom of scientific thought and new and revolutionary ideas, which come mainly from the young and unknown scientist, against the stifling and suppressive reaction of the establishment's sclerosis. The editors' responsibility is thus beyond their own journal: it is the maintenance of the freedom of scientific communication. One may wonder how true this is today and how many editors at present would be prepared to take the risk of publishing reports unwarranted by their highly respected referees' criteria. One may wager that the number of journals which would today accept reports confirming, for example, some of Summerlin's faked observations, would be as few as those which would accept reports in favour of Lamarkism in the 1930's.

If it is certain and obvious that editors hardly play any significant role today in the promotion of novel and original lines of research outside the beaten track established by government committees, administrative executives with no scientific knowledge and research background, respectable scientific sub-committees and financial controllers, it is also certain that the possibility exists and is within our reach.

Influencing research policy-making is precisely the key point of expression of editors' power. From the selection of research reports on independently established research priorities and scientific criteria, to the publication of editorials and dissemination of information outside the specialists' domain, even to the layman, the editors hold the possibility of imposing new lines of research and eliminating existing fossilized ones. Who, for example, would like to continue to flog the dying horse of BCG and non specific cancer immunotherapy, as conceived today, if the reports on the subject received a low publication priority from all biological journals? Instead of this, new journals are created just to accommodate the blizzard of data on the subject and the lot will be added to the heap of meaningless facts and figures, result of the *well-planned and directed cancer research*. And this, whilst genuine fundamental research, aiming at the understanding of the basic problems of cancer, is lacking the necessary funds, and fundamental revolutionary observations are buried under this meaningless stamp collection.

What I propose here, is the possibility for the editors of scientific journals to become a new force, selecting, and even imposing, scientific research policies by the control they may exercise on the publication of research reports. This creates a new responsibility for them since they may thus intervene in the future of not only science but of evolution of human civilization. However, when a power of such a fundamental importance is available, the duty for using it becomes as compulsory as the sense of responsibility associated with it.

This conference could thus be more than a technical exchange meeting. It offers a unique opportunity to lay the foundations of an organization which would be a new force on the chessboard of science politics. The responsibility of scientific policy-making, which is enormous, has not as yet been assumed by those who are the first to be concerned: the scientists. Scientific editors should play the role which is theirs, if they have the courage, the energy, the enthusiasm and the sense of responsibility to accept the challenge and create such a structure beyond language or discipline barriers, financial or academic personal interests. This international body, of which we are all thinking in different ways and for different purposes, may become a reality as from today. Let's state clearly that it will not be a substitute for scientists' own organizations, rather, it should be viewed as the crucible for such organizations, which are lacking more than ever. And this lack plays a role in the slackening of social conscience and responsibility of the modern scientist, who is becoming accomplice to the system of lethargy and indifference and technocrats' exploitation, which governs today's scientific research policies.

If such an international body of scientific editors were to be created, and because of the power which might be derived, it is necessary to envisage from the date of its creation a code of ethics, which far from crystallising it, should preserve it from exploitation by equivoqual personalities. This should help to maintain the integrity and independance of editors from all sorts of directive forces, whether they call themselves "government agencies", "publishers" or even "learned societies", and it will be the guarantee of maintaining high conscience and moral standards in this new structure, which should be seen as a diversification of the existing powers.

A diversification of forces moulding a society is the best guarantee for its freedom.

THE EDITOR AS MOVER OR RETARDER OF SCIENTIFIC IDEAS

Artur Isenberg
Editor, Kidma, Israel Journal of Development
P.O.B. 13130, Jerusalem, Israel

ABSTRACT

The conscientious editor of a scientific publication is — or at any rate should be -- haunted by two fears: denying the dignity of print to an article which, however heretical at first blush, subsequently proves to have come nearer the truth than the reigning orthodoxy which it challenges; and -- the reverse side of the coin -- publishing unscientific rubbish, pseudo-science masquerading as the real thing behind a cloak of neatly tabulated "data", trailing fringes of footnotes, ibids. and op.cits. Happy the editor who can truthfully assert that he has always been able to avoid that Scylla and that Charybdis!

By "editor" I mean not only the person properly so designated, but the entire editorial process, including the referees and, where he or she decisively interferes in the process, the publisher as well.

The editor's position is not always enviable, for there are no absolute criteria for assaying the veridity of a new theory, and the pitfalls are many: results have been known to have been forged, sources misquoted or even freely invented, and a stately theoretical structure may involve a concealed logical fallacy -- obvious once found but sometimes quite easily overlooked. The full-fledged crackpot is easily spotted; not so the borderline case. And sometimes an authentic genius is mistaken for just another case of psychoceramics. Thus editorial fiat denied Gregory Mendel access to the leading botanical journal of his day -- and his genetic work languished in an obscure periodical until it was re-discovered more than a generation later, a generation during which much progress might have been made. Less well known perhaps is the case of J. J. Waterston whose paper proposing a molecular theory of gases, partly anticipating Maxwell, was denied publication in 1845 after a Royal Society referee had described it as "nothing but nonsense". The paper was found in 1890. As W. I. B. Beveridge says (in The Art of Scientific Investigations; London; Heinemann, 1950; p.108) in commenting on the Waterston tragedy: "Many discoveries must have thus been stillborn or smothered at birth."

In point of fact, the editor has no choice but to accept a criterion which I once heard Henry Kissinger define as that of "being reasonably arbitrary". What is troubling is the fact that the editor has to act as absolute arbiter, as if he or she were omniscient, even while remaining mindful of the fact that there is no omniscience outside the theological sphere. This inescapable state of affairs ought to engender a certain humility.

Miriam Balaban (ed.), Scientific Information Transfer: The Editors' Role, 161-165. ***All Rights Reserved.***

It may help to be aware of certain tendencies, including one noted by Francis Bacon (Novum Organum; Book 1; Aphorism 46):

> "The human understanding, when any proposition has been once laid down, either from general admission and belief, or from the pleasure it affords, forces everything else to add fresh support and confirmation; and although most cogent and abundant instances may exist to the contrary, yet it either does not observe or despises them, or gets rid of and rejects them by some distinction, with violent and injurious prejudice, rather than sacrifice the authority of its first conclusions."

Of the two evils -- excluding true new scientific insights and including pseudo-science -- the latter is surely much the lesser: if a theory is wrong, publication in a scientific journal is most likely to expose its error most efficiently by submitting it to the normal -- and normally unemotional -- critique and tests of other scientists; while non-publication in scientific media is apt to encourage cultism and faddism. On the other hand, if the innovative theory turns out to be correct, its non-publication is at the very least a matter for regret and may be a real tragedy. What is of course entirely unforgivable -- but not alas! unknown -- is to bar a scholar's access to scientific publications, then castigate him for publishing his work in lay journals.

Scientific censorship of a theory becomes the more untenable the more evidence accumulates in its favour as time goes by.

II

All this is strikingly illustrated by what has been called the Velikovsky Affair. I am quite prepared -- if requested and permitted to do so -- here and now to list more than three dozen anticipations made by Dr. Immanuel Velikovsky, all said to be preposterous and even "crazy" when first made by him and today accepted if not in every instance by scientific orthodoxy then by a growing number of "orthodox" scientists.

Instead, let me just cite a handful of specific instances of correct prognostications -- not predictions, but necessary inferences from one or two basic theories.

Thus Dr. Velikovsky anticipated and stated that electromagnetic phenomena must play a vital role in the physics of the solar system and the universe at large. A heresy in 1950 when he made this advance claim, it is today accepted scientific orthodoxy. He similarly anticipated that Venus must be very hot; and that, too, is orthodoxy today. Jupiter emits radio noises: stated by Velikovsky at a lecture at Princeton College in October 1953; confirmed by B. F. Burke and K. L. Franklin of the Carnegie Institute and announced on 5 April 1955. Evidence of petroleum hydrocarbons will be found on the moon: Evidence found by Apollo XI mission -- see C. Ponnamparuma et al.: Search for Organic Compounds in the Lunar Dust from the Sea of Tranquility. Lunar rocks will reveal remanent magnetism despite the fact that the moon itself has an exceedingly weak magnetic field: so stated in Velikovsky's Memorandum to Space Science Board, National Academy of Science, 18 May 1969. Strong remanent magnetism was found in lunar rocks by Apollo XI and XII missions. The earth has a magnetosphere reaching at least as far as the moon: memorandum by Dr. Velikovsky to U.S. National Committee for the International Geophysical Year, dated

5 December 1956. Discovered by J. A. Van Allen in 1958 ("Van Allen Belt") and at first thought to extend only a few earth diameters; later found to reach well past the lunar orbit.

Let me repeat that these are just a very few instances taken at random and that I am prepared to document several additional dozens of similarly important and subsequently verified anticipations made by Dr. Velikovsky.

In this context, I am not interested in whether Dr. Velikovsky is right or wrong. My interest in the present context lies in the proper role of editors of scientific journals many of whom continue to close their columns to an authentic account of his theories even while giving the widest latitude to his critics, scientists and non-scientists alike, thus substituting ex parte or Kangaroo Court proceedings for free scientific discussion.

These are serious charges. They are, however, fully borne out by any impartial examination of the record. For one such scientific study of what I have said, see The Velikovsky Affair — Scientism vs. Science, edited by Alfred de Grazia, Ralph E. Juergens and Livio C. Stecchini, originally published as a special issue of The American Behavioral Scientist and thereafter in book form by University Books, Inc. (1966). There you will find documentation — not refuted to this day — of such strange matters as scientists reviewing Velikovsky's books before they were published; branding one of his books "lies! all lies!" while admitting, in the same paragraph, that "no, I have not read the book...and I do not intend to waste my time reading it." (So Dean MacLaughlin, University of Michigan astronomer in a letter to G. P. Brett, Jr., then President, The Macmillan Company, New York.) Or compelling a publisher (Macmillan) to stop sales of a best-selling book and to transfer it to another publisher (Doubleday & Co.) after threatening an all-out boycott of Macmillan's textbook business, the threat being spearheaded by a group of academic astronomers.

Among scientists who did not think that Velikovsky was a charlatan, a faddist, a fraud, but who regarded him as a serious scholar, let me just mention a few: Albert Einstein; H. H. Hess; Horace Kallen; Cyrus Gordon — the list can easily be lengthened for today the number of such scientists runs into the hundreds. In fact, courses on Velikovsky's theories are today offered in scores of U.S. and other universities. That, however, did not prevent a highly respected U.S. science publisher from writing: "We have not encountered a single scientist working in any of the many fields from archaeology to astrophysics on which Velikovsky touches who finds any interest whatever in anything he has to say. That is why you have not seen any account of Velikovsky in our pages."

Small wonder that the journal published by that particular publisher will not accept advertisements of books by or about Velikovsky although most U.S. scientific journals which once also refused to accept such advertisements, more recently have begun to accept them. The list includes SCIENCE, SCIENCE NEWS, SCIENCE AND PUBLIC AFFAIRS, PHYSICS TODAY and many others.

The matter of advertisements is entirely germane to my topic, for it too involves the editorial domain. Thus when PENSÉE submitted an advertisement on a Velikovskian theme (an ad accepted by many other scientific journals), the advertising representative of THE AMERICAN SCIENTIST wrote to PENSÉE that "although the ad meets the requirement set forth in my previous correspondence, the editorial board has exercised its prerogative of refusing your announcement." Whatever policy such a decision reflects, it can hardly be described as an example of the "democracy of the intellect"...

Equally to the point is the statement made by Charles Federer Jr., editor of SKY AND TELESCOPE, to PENSÉE editor Steve Talbott: "We absolutely wouldn't touch Velikovsky."

Against such a background it should occasion no surprise that we find the same leitmotif when it comes to reviews of Velikovsky's books by certain science editors. His latest book, Peoples of the Sea, which deals exclusively with a twohundred-year period in Egyptian history was reviewed by the physical sciences editor of SCIENCE NEWS, in the NEW YORK TIMES. He dismisses Velikovsky's reasoning as coming "close to cabbalism", accusing him of straining "at gnats in order to swallow a camel." Compare this with the view of Dr. David Lorton, an Egyptologist who teaches at Johns Hopkins University and who after minutely examining the book writes: "On the whole, then, I have a very positive reaction to the book...I can testify that anyone wishing to attack it will have a very difficult job."

Those of us who are interested in Dr. Velikovsky's theories do not capitalize the personal pronouns when they relate to him, nor do we consider him to be at all infallible. We do not want science editors to endorse his or any other scientist's theories. What I do urge and advocate is something quite different: let the editors concerned make the relevant scientific channels of communication available to Dr. Velikovsky's theories so that they can be tested, examined, discussed — dispassionately, objectively, thoroughly, rigorously: no more, no less. Professor Lloyd Motz of Columbia University has put it very well indeed:

> "I do not support Velikovsky's theory but I do support his right to present his ideas and to have these ideas considered by responsible scholars and scientists as the creation of a serious and dedicated investigator and not the concoctions of a charlatan seeking notoriety... Inspite of my rejection of Velikovsky's concepts, however, I believe that his predictions should be recognized and that his writing should be carefully studied and analyzed because they are the product of an extraordinary and brilliant mind, and are based on some of the most concentrated scholarship and research of our period."

III

Editors of scientific publications undoubtedly have much to which they can point with justifiable pride: in many instances they have contributed handsomely to the advance of knowledge. But to the student of the history of science familiar with the details of a record which I have been able to sketch only very superficially, in the barest outline, there cannot be any doubt whatever that in the case of Dr. Immanuel Velikovsky, too many science editors have acted as retarders of scientific progress: discussion of his theories even if wrong, would have shed important new light on scores of anomalies noted by Velikovsky in astronomy, archaeology, history, geology and still other disciplines. And this is true even more so if, as I personally believe, Velikovsky is basically correct in his reconstruction of the history of the solar system and of the history of the Middle East from 1500 to 300 B.C.

Is the Velikovsky Affair an isolated event or does it hint at a more pervasive problem in the editing of scientific publications? There are some indications that I am not alone in thinking that the problem is far more general, the Velikovsky Affair being merely an unusually striking example. I venture to suggest that a poll conducted in June 1973 is worthy of some real editorial soul-searching. The poll in question was conducted by the editors of a leading United States technological publication, INDUSTRIAL RESEARCH, circulated, it is stated, "among qualified scientists, engineers and administrators in technically-oriented organizations". Normal responses to its polls vary from 1,300 to 1,800. The poll in question elicited 2,000 replies. To the question, "Do you yourself believe in Velikovsky's theory of catastrophic upheavals in Man's past?" 81% replied either "yes" (19%) or "to some extent" (62%), with only 19% replying in the negative. Further, noting that Velikovsky's wiritings have not been published in technical journals, the editors went on to ask, "Does this represent a basic flaw in the concept and functioning of these journals?", 12% said "no"; 24% replied that they were not sure. 64% said "yes".

But the most interesting reply and the one I especially commend to your attention was that to yet another question: "Are the technical journals basically too conservative?" 29% replied in the negative. <u>71% answered affirmatively</u> -- more than seven out of ten of the journal's readers, "qualified scientists, engineers and administrators", said thay considered technical journals to be basically too conservative.

In other words: there seems to be ample scope for editors to expand their role as communicators or movers of new scientific ideas.

DISCUSSION on "In Defense of Standardized On-Demand Publication" by Joseph Agassi

"The Publication Game and the Education of Future Scholars" by Ronald Swartz

"Editors' Impact on Science Policies and Politics" by Dimitri Viza

"The Editor as Mover or Retarder of Scientific Ideas" by Artur Isenberg

ITAMAR PERATH - I'd like to take up a point which Dr. Swartz raised, and to consider another aspect of the use of jargon. Under the term jargon I understand not what Dr. Kohn called gobbledygook (of which he gave some examples) but something more in the sense of what George Bernard Shaw called a "professional terminology." Every professional terminology, he said is a conspiracy against the layman. To which as a professional, I would respond: so, what did the layman ever do for us?

But the serious point is, there are things about jargon which are important. The first point is that it is something which I regard as inevitable. It is part of the mental shorthand which thinking people tend to make and this results in a verbal shorthand which finds its way into the written shorthand. And if we look at the verbal explosion of written information this form of condensation is imperative and it should be encouraged as such. I would even make a further point and suggest that jargon is a kind of antidote to some of the ills which open publication imposes upon us. As has been said here, very few people read an entire journal. Much effort of publication is devoted to spreading information which is really aimed at only very few readers. In such information, the use of jargon is a kind of warning sign saying: don't read this, it wastes your time. On the other hand, if you are familiar with the jargon, it is a sign: read this, this is aimed at you. As such, jargon is a time saver and I think saving time is today the only thing which can help us out of this information crisis, wherein the constricting factor is the amount of time which any scientist can devote to any subject.

DIMITRI VIZA - The original meaning of jargon is in fact a precise terminology of a precise use of words. What we're talking about is the non-language of the gobbledygook that comes out. We have the unsolicited application for a job in our research department for someone who is working on language process using a computer from a University in the United States. It looked marvelous until you actually tried to look at what it really meant. There was no sentence structure in it at all. The second point is the bookshops. Comparing on-demand of books - there may be as much effort to get an on-demand publication as there is to get a book, but journals do not require any effort to receive them in the library once they are ordered. They arrive every month or every quarter in the library, on the desk whereas in on-demand you have to make the effort to actually get the material. These are the sort of things that are eating up the journals by bits. Because it takes a positive action to cancel them.

One last point which may partly answer your question concerning jargon. We publish a multi-lingual journal which is transdisciplinary and we eliminate jargon for the simple reason that jargon does not translate and it does not translate economically. And the device by which we eliminate jargon is by rewrite.

PETER LENGYEL - I'd like to say one thing from the point of view of the social sciences, where, undoubtedly, jargon is used as a protective device. If you remove it you remove the emperor's clothes. As an editor, therefore, you are taking a heavy risk in removing jargon because you thereby tend to reduce the author to the everyday language which he precisely wishes to avoid, for a variety of legitimate and illegitimate reasons. I will give you one word, for instance, which I would not allow in my journal for many a long year: "Consumerism." I still don't know what consumerism means, other than acting as consumers, which we all do all the time. One might as well speak of "lovism" or "breathism". But there comes a point when one cannot control the situation anymore because a word has passed into every newspaper and magazine and it would be absurd not to publish "consumerism" anymore. Then you are overtaken by the jargon. And there is very little that an editor can do about it except stop, stare up for some time and finally submit to usage which is hardly a very creative way of dealing with redundant neologisms or jargon.

RON SWARTZ - I'm not against all jargon - minimum use rather than maximum use. I think we have maximized it more than necessary.

MYRA KAYE - It was wonderful to hear the golden combination - editors, teaching and scientists. Dr. Swartz, what would be your opinion that the teaching process should be at the graduate school or the undergraduate school level? That the scientist has to start learning to write papers before, and not after he is a researcher who has already something to publish.

FRANK PORTUGAL - We have a number of problems with jargon at the Proceedings of the National Academy of Sciences, and it is particularly important for us because we are a multidisciplinary publication. One of the ways that we handle this problem is to simply request a definition of the jargon term initially at the beginning of the paper when the term is first used. Then the term may be used throughout.

I'd also like to make a very trivial point with Dr. Isenberg. I think that Mendel's work was rediscovered by three people rather than two.

JEFFREY WATSON - I guess I took Dr. Swartz off the hook here by basically starting off with the false premise that it is editors who ask authors to write and at the same time to use jargon and there is nothing really being done about it. I think a lot is being done about it particularly CBE publications on editorial policy. And a lot has been written in the literature about the problem of why doesn't the university do the teaching first.

MICHAEL BOWEN - First a little comment to follow up on that. I would say that jargon, as Tony Woodward defines it and clarity are not necessarily comparable at all. They may be entirely compatible.

JEAN SACHS - I have two questions to Prof. Agassi. I think it is very understandable that if you are going to do a twenty volume series of the complete works of Berry or you have a ten volume state-of-the-art in history of infectious diseases that you might need a little seed money to get a start. But what about books of about three hundred and fifty or four hundred pages with a six hundred print number and no subsidy. There may be only four hundred people in the world who can read it. It is a very brilliant book. And it is one of those that didn't fall away, that wasn't lost. My question is, how does that jibe with what you're saying?

The second one is, I don't understand what you mean by administrative pressure. Are you talking about the director of the press, the administration of the university, the academics themselves?

JO AGASSI - I want to answer the question about what is the clearing-house, the on-demand, and how do I think one can finance excellent ventures of publishing first rate books which have no more than a few hundred readers. Publishing is a complex social institution. The Ancients had publishing houses where slaves copied works. The modern publishing houses as we know them, evolved out of printing houses. In the Middle Ages they only had summas, as professors exploited students to copy their lecture notes. Which is why medieval summas are so common. Nowadays professors still use students as slaves, if the press doesn't exist. That's another problem. There is no question that a book that has no more than a few hundred readers need not be published. If some well wisher wants to invest a few thousand dollars in order to have an expensive edition of a few copies, there can be no objection to it, but to have for this purpose the whole apparatus of a university press, for fifty eight percent of the expenditure, according to the Princeton University Press chief editor goes not to publishing at all but to bureaucracy, or even if it's only twenty seven percent, then it sounds terribly inefficient. What we need is a clearing house. I have two unpublished manuscripts in circulation, not as well circulated as Wittgenstein's colored books, but there are quite a few copies of each in circulation. Now one of them is my doctorate and is not advertised in any list, but some people heard about it and ask about it and they get permission from me and they write to the University of London and they get a microfilm of it and it costs them or their libraries a minor fortune, but they can get it. It doesn't cost the university press anything which is better. The other book I have is not circulated in any clearing house and I don't have it in as much circulation as I want. Moreover, every copy in circulation I have paid for. Publication pressure began by boards of trustees and presidents in American universities, for no good reason. And I think it is clear that we can do without publication pressure; at least we should debate the question in every university which is a self governing body or at least should be, rather than a university governed by administrators. This debate is raging at least in the United States and I cannot go into it here. But I hope that I have answered the question, how can we contribute to the lessening of the pressure. Had a reputable publisher published lists of books, like the list of "books received" that many journals publish, so as to inform the public that some very, very important editor thinks that your book is worth reading, and that if you write to him he will send you a copy or a microfilm for an appropriate price, then much financial loss will be avoided and many books will become available. This is all that I can suggest at the moment.

MICHAEL BOWEN - Since we've been talking about morals, ethics and responsibility, I'd like to pose a question especially to Dr. Isenberg. Suppose a scientific journal accepts advertising for books and other products that are compatible with the interests of its readers, should it accept as a paid advertisement a paper which has been turned down by the review process of the same journal?

ARTUR ISENBERG - In principle, yes, for the case does not differ from that of - say - *Nature* publishing an editorial in one issue and a letter from a reader in the next issue, vigorously attacking the editorial. In practice, however, each case should be decided by the editor on its merits. Thus if the rejected article had advocated genocide, I as editor would also refuse an advertisement for such an article. The final decision must rest with the editor. But, as I said, in principle I can see no reason why a paper which was rejected should not be advertised in the journal concerned.

CREATORS AND GATEKEEPERS – AUTHORS, EDITORS, REFEREES

Joseph Gillis, Chairman

DEMOCRACY OF THE INTELLECT

Gerard Piel, *Publisher, Scientific American*
415 Madison Ave., New York, N.Y. 10017

ABSTRACT

The welfare and the sanity of the human species hangs upon the integrity and vigor of its scientific enterprise. In the fostering of this enterprise the popularization of science has a central role.

This conference is blessed in having a muse whose call brings us from so many different points on earth. Miriam Balaban is also a muse with a needle and a pragmatic sense of how to run a meeting. She brings us each here to pool our respective trained incapacities.

My concern is with scientific information transfer to an audience in the penumbra of most participants here; that is: to the wider audience of people concerned with the advance and the uses of scientific knowledge. I should like to talk to you about the urgency of that part of our total enterprise, as distinguished from the technique for accomplishing this purpose. In other words, I am here to give you a sermon.

International meetings are always of great value because they show us that the concerns each of us brings with us -- concerns with retrograde movements and troubles at home -- are the concerns of all assembled. The movements that disturb our respective national cultures somehow or other move in concert. You will not be surprised, therefore, to learn that I come here today concerned about the recoil from technology and science at home in the U.S.A. The smoke plume from the factory chimney which, when I was a boy, gave promise of abundance and prosperity has become now the desecrator of the landscape. Nameless fears seizing our electorate have brought nearly to a halt the installation of nuclear power plants. Something worse than nameless fear, something more like panic, has seized our university communities over the subject of recombinant DNA, and there is dark talk about a supposed necessity to set a limit on scientific enquiry. This loss of confidence in science and technology has brought with it a widespread retreat from rationality to the comforts of absolutes founded on the mystic, the psychic and the occult.

All of us here at this meeting come from developed industrial countries, whose populations share the luxury of being able to complain about the external diseconomies of technology. At home in the U.S.A., at least, the retreat from rationality is

Miriam Balaban (ed.), Scientific Information Transfer: The Editor's Role, 173-177. *All Rights Reserved.*

associated with a callous disregard for the predicament of the peoples of the underdeveloped countries. Those peoples do not yet suffer from surfeit; they have a forthright commitment to the sciences and technologies that are going to answer the question with which they go to bed every night: "When do we eat?" It is to the people of the rich countries, however, that they must turn for instruction in the sciences and technology -- just at this ironic turn of history, when the rich are taking leave of their senses.

This meeting is not only well timed but well placed. Israel stands here in the Middle East as the example of modern industrial democracy, and a model demonstration of the great task of economic development that still lies ahead for the peoples of feudal, pastoral nations in the midst of which Israel is set. Tragically that example has been lost in the strife that still wells up from passions and prejudices and ancient history that is mostly irrelevant to the great task of industrial development that lies ahead.

It is no accident that industrialization and democracy occur together, for science and self-government are coeval in history. The celebration of the bicentennial of my country last year gave insufficient recognition to the fact that the authors of our political institutions were natural philosophers. Natural philosophy was what, in the 18th century, they called science. As early as 1743, Benjamin Franklin had founded the American Philosophical Society... "to be held in Philadelphia for promoting useful knowledge." Scientific knowledge is useful knowledge because it makes a difference, because it confers control. Benjamin Franklin fully comprehended that the pursuit of this kind of knowledge is a subversive activity. By enlarging human capacity, opening up new possibilities and choices to human life, the advance and the application of science force the restructuring of human values and the dismantling of old institutions. Thomas Jefferson, who was the third president of the American Philosophical Society as well as the third President of the United States, put the matter succinctly when he said: "So long as we may think as we will, and speak as we think, the condition of man will proceed in improvement."

The revolutionary proposition of the founders of the United States had built within it the means to its fulfillment. The promotion of useful knowledge, securing improvement in the condition of man, has finally brought all Americans into the full sovereignty of citizenship. It is worth recalling that, at the time of the founding of the country, a full 25 percent of its labor force was in chattel slavery. Universal white manhood suffrage did not become established in the new states until after 1800. Citizenship was not extended to black men until 1867, with the adoption of the 14th Amendment. And the United States did not finally admit the better half of its citizenship to full sovereignty until 1920, with the 20th Amendment that established female suffrage.

The sunshine patriots of my country might also have recalled that these ideas were not original with Americans. The sovereignty of the individual was first asserted in modern times in the Renaissance. Historians are now beginning to recognize that this seminal epoch in our history, hitherto romantically dedicated to the arts, was also the beginning of the scientific-industrial revolution. In fact, the dawn of that revolution in the sciences supplied many of the most inspiring themes of the Renaissance art.

The symbolic figure of this turn in events is Galileo. Against the absolute spiritual and temporal powers of the time, he asserted the autonomy of the individual. To them and to the world, he demonstrated that a scientist can accept no authority but his own judgment. He yielded his person, but not his conscience, to their power; he wrote his dialogue on the new sciences under house arrest. Its publication and wide circulation in Europe during the next century made the claims of the spiritual and temporal powers absurd. Galileo thereby set the foundation for recognition of the dignity and the sanctity of the individual that makes all men equal.

The ethic of self-government then rests on the same ground as does science; that is, on the autonomous individual. Now the individual of whom I speak is not that rugged individual of the laissez-faire economy whose object in life is to get out of the sight of the smoke of his neighbor's chimney. The autonomy of the self-governing citizen and of the scientist is asserted in the tightly drawn web of the interdependence of us all with all of us. The principle is best illustrated by the work of science. That enterprise imposes what J. Bronowski called the habit of truth. This is not a virtue that suffuses scientists from within, as if they were a special and a chosen priesthood. It is imposed upon them by the kind of work they do. A scientist cannot fake his results or edit them in favor even of his ethical prejudices; for he will certainly be caught. As Gunnar Myrdal speaking for the softer kinds of evidence with which social scientists work has put it, "Facts kick."

In this aspect, science is an intensely private undertaking. Percy Bridgman put it vividly in a memoir in which he said, "The process I want to call scientific is a process that involves the continual apprehension of meaning, the constant appraisal of significance, accompanied by the running act of checking to be sure that I am doing what I want to do, and of judging correctness or incorrectness. This checking and judging and accepting, that together constitute understanding, are done by me and can be done for me by no one else. . . . They are as private as my toothache, and without them science is dead."

Yet the scientist is engaged at the same time in a highly public enterprise in the full sense in which John Ziman has defined it. The work of a scientist will not be challenged or receive even the honor of disproof unless it significantly challenges the interest of his contemporaries. The importance of his work is embedded in the historical context. It may be measured by the degree to which it stresses that context and forces the reconstruction of ideas in a field; by its relevance to the work of the past from which it must be a departure; by its relevance to the work that follows which it must urge forward.

Scientists are thus able to address one another not only across national boundaries and even across generation gaps but also across the boundaries of their disciplines. General Leslie Groves, who was appointed by President Roosevelt to take charge of the Manhattan Project, attempted to keep the solemn purpose of the enterprise secret by compartmentalizing from one another the scientists engaged in various aspects of the work. He could not, however, compartmentalize the context. Scientists in that project -- wherever they were: in Chicago, in Los Alamos, in New York -- knew what was going on at all times because they were able to read each others' minds. Out of their common comprehension of the context in which they were working, the slightest clues tripped a whole line of deductive insight.

There was always a chance, as you know, that the first experiment at Alamogordo might have catalyzed the combustion of the nitrogen by the oxygen in the atmosphere. The official calculation of that possibility showed the first significant figure to be at many places to the right of the decimal point. But that was by no means the only calculation of that possibility. The same idea crossed many minds and, from calculations on the backs of envelopes, there came the general community assurance that the chance was infinitesimal.

Now the chance that some crazy cabal of scientists might try to ignite the atmosphere may be somewhat larger than the chance that it could be done. Nonetheless, we can place our confidence in the capacities of the living, self-governing community of scienists to evaluate and keep abreast of, and to police, the work of their contemporaries.

By contrast, it is the entire absence of context and relevance to any significant idea that shows the total worthlessness of the monstrous experiments conducted by the Nazi physicians in the concentration camps and convicts them of the bestiality of which they were guilty. The lack of context was the essence also of the Lysenko story. It was possible for external authority to blunder in and to destroy a generation's work in genetics and to create disaster for the generations of the Soviet Union that followed.

In contrast, again, to these violations of the context, is the historic story of "recombinant DNA." It was the scientists engaged in this work who first recognized the hazards, they who first recognized the ethical questions, and they who called, first of all, for a moratorium on the work until the questions had been more closely examined and, then, for a conference to establish the safeguards under which this kind of work can be carried on.

One cannot think of an agency that can be placed, in any way, in control of science that could be any more reliable and responsible than the self-governing polity by which this public enterprise is conducted. As Warren Weaver who was for many years the bursar for the sciences of the Rockefeller Foundation observed: "The sober record of experience shows that the trained human mind, if you give it free play and a congenial climate, turns to deep and significant enterprises. The rational approach to life is a successful and productive approach. The most imaginative and powerful movements in the history of science, have arisen not from plan, not from compulsion, but from the spontaneous enthusiasm and curiosity of capable individuals who had the freedom to think about things that they considered interesting."

The freedom and the autonomy of the scientist then is the same freedom that makes possible self-government and makes the citizen a sovereign. By the exercise of the freedom of enquiry the sovereign citizen comes into possession of knowledge that makes a difference. That knowledge has brought the entire populations of industrial societies into the political process. It has given them the pragmatic wherewithal to ask the interesting questions that follow upon the answer of the question that still besets two thirds of mankind: "When do we eat?" Only the children of the fortunate are asking: "Why do we eat?" In the end, again in the words of J. Bronowski: "We must be a democracy of the intellect. We must not perish by the distance between people and power that destroyed Nineveh and Alexandria and Rome . . .. That distance can be closed only if knowledge sits in the homes and heads of people and not up in the isolated seats of power."

The conduct of the tertiary enterprise of public information in the sciences is an essential part of the conduct of the total enterprise. This is an urgent undertaking because the values of science must be more widespread among self-governing peoples. It is that method of knowledge and understanding by which people may learn in the end to tell the difference between sense and nonsense. With a deeper understanding of science, there will be a wiser sponsorship of science by self-governing democracies, and the work of science will fulfill its promise for the liberation, ultimately, of the rest of mankind.

THE ROLE OF REVIEW JOURNALS IN SCIENTIFIC PUBLICATION

Norman Chigier, *Editor, Progress in Energy and Combustion Science, Department of Chemical Engineering and Fuel Technology, The University of Sheffield, Mappin Street, Sheffield S1 3JD, UK*

ABSTRACT

Editors of scientific journals normally require that papers submitted to their journals are both original and significant contributions to the field of specialty of that journal. Editors rely on the expert advice of referees in order to make a judgement as to whether the submitted paper satisfies these criteria. Many journals have an additional restraint of severely restricting the length and number of diagrams in a paper accepted for publication. The referee rejects material which is familiar to him, as well as material he considers to be of little significance. The net result of this process has been to encourage a very high degree of specialization so that there are very few subscribers to journals who are able to understand the language and basic concepts of all the articles in any particular issue. This trend towards specialization and the increasing costs of publication are leading to a decline in the number of individual subscribers to many scientific journals.

The student at postgraduate level and, to a lesser extent, at the undergraduate level, seeks information as background reading to amplify lecture notes. If there are no recent textbooks in the field, he has difficulty in finding the information in journals and will often find the article in a scientific journal incomprehensible to him. A scientist or engineer entering into a new field is faced with a similar difficulty. Review journals provide articles which introduce the subject, give a current state-of-the-art and describe the subject in a much more general manner than is found in a shorter, more specific article. The author, who is usually a leading expert in the field, provides an overview which enables the reader to see the subject as a whole in perspective. He can use such an article as study material and is supplied with a comprehensive set of carefully selected references. Review journals can play an important intermediary role between textbooks and scientific journals.

I should like to examine the specific role of review journals in scientific publications from the point of view of the requirements of undergraduate and graduate students and of research scientists in university. And let us look also at the requirements of those working in industrial and government research establishments.

We start first of all with undergraduates. Teachers of undergraduates would like to recommend textbooks to their students, but they find that because of rapid advancements in science, and because of the very long delays in publication of textbooks, they often cannot recommend a textbook unless they have written it themselves. So when a student comes to his professor and asks what he can look at, very often he refers him to the general scientific literature.

Miriam Balaban (ed.), Scientific Information Transfer: The Editor's Role, 179-182.

But there we come across the problem of specialization. The vast majority of articles published in the scientific journals is not addressed to the undergraduate. The reasons for this are that the advances in publication have been such that nowadays an author submitting an article to a scientific journal is required to prove that his contribution is both an original and a significant contribution. In the majority of cases the editor cannot make this judgement. If he is fortunate enough to find a referee who is sufficiently close to the subject, he prays and hopes that the referee will help him to make that judgement. The referee is asked by the editor of a scientific journal to make sure that the more irrelevant material is eliminated from the article, that it should be as concise as possible with the usual restrictions in the number of equations and the number of diagrams, and he's also on top of this asked to make this critical judgement as to whether this is an original and significant contribution. If he has been well chosen as a referee and he understands the subject, he then eliminates all material with which he is familiar, and he will only allow material which is new to him to be included in that article. We have now set up a dialogue between the author and the referee, and the editor very often is left out of this dialogue.

When we come to trying to follow this specialized literature, we find that even if we are specialist in our own subject, there is a very limited number of papers in any one journal which we can even understand. And this is due to the tremendous degree of specialization developed in science.

I think that this is one of the explanations why we have now a return to the review journal because it is at least beginning to offer some information to the people who are looking for this information. And is we look at the specific problems of the students at the university we have to understand that they are not yet specialized and even until they finish their doctorates they still have not achieved the degree of specialization which is required at the level of many of the articles in the more advanced scientific journals. And so in the review journals, and there is a wide spectrum of review journals, what we set out to do is to try and cater for this audience. I believe that this is probably the largest reader audience that we have in science - the students at the university. They have the time to read and they are eager to learn. In order to learn we have to provide them with guidance, we have to provide them with leadership, and we also have to provide them with explanation.

The ordinary scientific article offers none of these things because it has become so specific. If we couple it with the fact that we have these huge quantities of publications which we now find in the scientific literature any of which, as we have heard during this conference, have not been carefully screened so that some of them are either badly written of in fact are not particularily relevant. We thus find that the student can't get very much help from this general scientific literature. So those of us who are now working in these review journals feel that our specific aim should be addressed towards these people who are looking for guidance, who are looking for a state-of-the-art of the subject and who are looking for information.

These review journals are beginning to develop an intermediate role between the very specialized scientific journal and the textbook. And in fact as scientists become increasingly busy and do not have the time to devote their time to writing a complete textbook they are very willing to at least write a comprehensive review article. If we ask ourselves what we expect from an author of a review article and the way that I ask my authors, I ask them to at least put themselves in a book writing frame of mind as distinct from that of writing for a scientific journal. I asked them to bear in mind that though they are very familiar with their subject the vast majority of the audience will be unfamiliar. This doesn't mean that they have to go to the simplicity of a popular article but that when they're giving long equations they should give intermediate steps, and that they should explain complex diagrams.

The authors I invite to write for the review journals are those I know to be experienced writers. I feel that we have had too much of having totally inexperienced writers who are exercising their ability or trying to get their training at the expense of the audience. This has added a great deal to the confusion and unfortunately because we have had many lax editors or editors who have not taken the role which I believe they should have, they merely allow this indiscriminate amount of publication to floor the scientific literature. So these are the roles which a review journal can play today.

The criteria which I think are most important in the review journals are first of all the selectivity. If we look at our whole educational system, it is based entirely upon examination. We look for excellence in our educational system. I cannot understand why in the scientific literature we have not used the same criteria. Why we have to have certain publications which will allow almost anything to be published simply because the author wishes to write this. I do not accept that this is a criteria for publication. In fact we have heard even here at this conference that if somebody has got something to write, then he should be allowed to have it printed. I do not believe this at all.

I believe that if anyone wishes to have something printed that he has got to pass the examination and the person who is to make that judgement is the editor. People here have been asking, who is going to take on the responsibility and I say that that responsibility lies clearly and definitely in the hands of the editor. The author wishes to have things published and he is looking for somebody to make the judgement. The publisher says well how can he make the judgement, he doesn't understand and he turns to the editor and so when we have editors who themselves do not accept this responsibility then we have the chaos which we have at the present time. Now we have had some people here at this conference who talked about the enormous quantity, they talk about constipation in order to try and describe the situation. If I look at the situation at the university and the research group which I work with, we have solved the problem in various says. Where we cannot find textbooks, then we return to review journals for getting our general knowledge, but when it comes to the specific problem of finding out the most up-to-date information about the research which we are doing, then we find that the scientific journals do not help us. Because there is such a huge quantity that even with the most up-to-date data processing which is available in the libraries it is very seldom that I can go to a librarian and I can say: "give me the most up-to-date-information on the subject that I am working on to day" and that he can come up with a list of articles. Normally the librarian wants me to sit down for three hours in order that he will establish a profile of the subject which I am interested in, and then he provides me with a huge quantity of material and it is extremely difficult for him to clearly identify the specific article which I am looking for.

So the net result is that those of us who are active in research re-establish our own lines of communication through conferences and workshops to which a very selective number of people are invited to discuss a highly specific problem. And many of us are now resorting to correspondence and to setting up of small clubs of active people in a field. And these are the ways that we as animals in the system are beginning to find ways of working.

But what we have to consider here at this conference is what is the role of the general scientific journal. And I see that we have a few alternatives. The one is the review journal. The fact that it is spreading so rapidly and is being turned to so much by our general audience is a clear indication that we have become overspecialized in most of the journals which are coming out at the present time. And so I think we have to use the criteria which are used by the average editor of the review journal. The editor of the review journal is much more ruthless about the articles which he allows to be published in his journal than that of the average journal.

The editor of a review journal first of all seeks advice as to whom he should invite to write articles. The majority of these articles are invited contributions, and he will make this invitation only to a person who has satisfied the criteria required to be an experionced writer, a person who is an authority in his field. Once he has him, he then gives him a set of guidelines which clearly indicates that he must remember who is his audience and that he must explain things to them. Then if he is a good editor, he doesn't accept that because a person is a great authority in his field that he is entitled to write whatever he likes. He then turns to his referees and he gives very clear instructions that no matter how distinguished that scientist is, he still must pass this examination requiring that he has got to at least write clearly. And if anything which he writes is not easily understandable or it makes no sense, then he must be told so even if he should take this as an insult. And so I would like to suggest that this is the role that I see for the review journal today. I see that it will grow and that the biggest problem will be what happens to the general scientific journal which will be pushed aside as more and more of the general audience turns towards the revie journal for information. The one way in which we competed for the high degree of specialization is to try and be so specialized in a journal that one is trying to cater for the specific research interest of a very small group of people. But I do not think that that is the majority of our audiences. Therefore I see a very important and growing role for the review journal.

REPORTS ON PROGRESS IN PHYSICS

John M. Ziman, F.R.S., *Honorary Editor, Reports on Progress in Physics 1967-1976 , University of Bristol,*
H.H. Wills Physics Laboratory, Tyndall Avenue, Bristol BS8 1TL, UK

ABSTRACT

By critical re-assessment and synthesis of the primary literature, review journals play an essential role in the information system of science. Ideally, every active research field should be reviewed thoroughly as significant progress is perceived, not only for the benefit of specialists in the field but to inform a variety of wider audiences. The ideal review author is learned but unpedantic, tough-minded but tactful, critical but not unimaginative - an unlikely combination of talents.

In pursuit of these ideals, the editor of a major scientific review journal faces many special problems: getting expert advice on choice of topics for review; finding suitable authors; following up commissions; assessing the quality of completed articles; ensuring adequate revision of obscure writing; maintaining a steady flow of papers to the publisher. These problems are discussed in the light of practical experience.

I'm going to take up the current from Norman Chigier by talking about being the editor of a review journal. After all, as Gerard Piel was telling us, the scientific community has to manage its own affairs, and one of the ways of managing its own affairs is running its own press and in a certain sense, this is what the review journals are.

Now I was talking about my personal experience with two journals, *Science Progress* and *Reports on Progress in Physics*. *Science Progress* is I think some seventy five years old and *Reports on Progress in Physics* is forty years old, so it's not quite a new institution, and also notice the word "*Progress*" in them. We are still optimistic, Mr. Piel. We still want to believe that we can do some good.

Now what do we want. Let's first talk about readers. They're the people who are going to pay for it. They want it, so they're the most important people. And what do they want? They want reviews that are timely, the right subjects at the right moment. They want them to be intelligible. They want them to be authoritative, true, comprehensive. We mustn't be missing the important points. On the other hand, they have got to be critical and concise which is not always quite so easy to balance. So that's what the readers want and that's what we have to try and produce for them.

Then we have authors because these things are not going to be written by machines. I think we can wait a few decades before we can have that done for us. So we are going to have human authors, and that is a problem because they have to be very industrious and reliable, and they are not so industrious and reliable as IBM machines running night and day, and they can be a little slow

Miriam Balaban (ed.), Scientific Information Transfer: The Editor's Role, 183-187.

in responding. "Well it s not the right moment", "Next year perhaps", "I have to go to a conference in Jerusalem", all the different reasons why people don't do what they ought to be doing.

Then they have to be very clear-minded, which all scientists are by definition. But some people's clarity is not quite the same as others. They have to be sympathetic and I really use these two words to cover the need to be readable. This ability really depends upon the combination of a clear mind and a sympathetic spirit in relation to what other people are going to take in from the work. These are important. Of course they have to be learned and that is not quite the same thing as expert. People can be learned without being expert; they can be expert without being learned, but for a review library he has to be both. And then he has to be very tough and tactful because they are going to be critical, and being critical in the scientific community means that you have to be very tough about some people's work. And on the other hand you have to be very tactful. You know the phraseology that's necessary under the circumstances. Finally I would just add one word. They must be undramatic because nothing makes for less readability or intellegibility than certain forms of pedantry. And that's precisely what we want from the authors.

Now the job of the editor of such a journal is to get this lot to do this sort of thing. And I want to tell you my experience and about what we've been doing over some years now in the Institute of Physics with *Science Progress* which is published by a commercial publisher and we have our scientific editor.

First of all, you've got to try and decide the coverage. What topics are you going to talk about in your review? That's not such an easy thing to decide. You can look around for gaps in the existing material. But then you have to decide how wide a scope should the subject have? How big is the gap really? How specialized should your particular article be? The editor cannot do everything himself. When you are dealing with a review journal, editorial advisors are an absolute essential. You must have discerning scientific advisors who can tell you where the gaps are in their particular, specialized part of the subject.

One of the things which is very important, and I've underlined it particularly here is that it is very easy to fall into the mistake of just doing the fashionable subject. Now a year or two ago all my friends in theoretical physics were saying, "Oh, phase transitions, critical phenomena, the renormalization groups, that's the thing." And indeed it was the thing. I'm not denying that. So you must have a review on that, but of course this was happening in each of the six other major review journals in physics which all came out with reviews on that particular subject. This of course overburdene the people who could write for it and ended in a chaos, leaving big gaps.

So one of the tricks of the trade is to be a little bit counterfashionable. Think of what will be important next year and not just what was fashionable last year. And that's where getting hold of your advisors and pulling it out of them can be very important. Also, subjects must be reviewe again every few years. You just can't rely on the review of ten or fifteen years ago. You must take a systematic look at the coverage.

Also in many cases the pattern of subjects is changing, and you have to see where the overlaps are between disciplines or between different specialties. One of the tricks, one of the nice things to do in building up a journal, is to try to get somebody to write in a way that a little bit covers two subjects previously regarded as separate. That sort of review paper is extra ordinarily valuable to everybody concerned. This is a job which really can b done by an editorial board.

In *Reports on Progress in Physics* it has been a matter of having a committee of ten or fifteen people meeting three or four times a year. The job of the editor at that meeting is to sit them down and say "You must produce at least two or three suggestions in your branch of the subject at this meeting". And we go around the table where we have all the different subjects of physics represented and try to bring forward ideas and make appropriate suggestions for authors.

The job of the editor is to give the right briefing, the right guidelines which consist of a number of factors. And the fundamental point is the reader Who is it for? *Reports on Progress in Physics* is for the physicist, the graduate physicist probably involved in research, the graduate student perhaps starting up in a new subject or the experienced physicist in another subject who wishes to inform himself rapidly of the situation in a field with which he is unfamiliar. You've got some idea of who it is addressed to.

In *Reports on Progress in Physics* we look at a slightly broader audience - students, undergraduate students, in a particular field. Then you've got to tell them the length to go for, the style, the conventions. These are all the things that have got to be worked out and kept stable over a long period. Remember that a review journal is not a fly-by-night institution. It is something that lasts for decades and has a very long time scale. I'll come to that in a moment. So it has got to be stable and organized in that way. Then you must commission your papers. And these review papers have to be commissioned. There are a number of papers that come in unsolicited. A few are worth publishing, but there are very few and basically you couldn't get anywhere on that basis. You must ask people to write. And the job of your editorial advisors is to tell you who is the obvious authority on the subject. Who's the chap who knows the subject first rate, who is the scholar on the subject, and who will write reasonably, sympathetically and so on. In other words, whom they would choose and whom you should approach in the first instance. But of course you don't get that chap in the first instance. The chap you write to in the first instance tells you that the subject in undergoing such changes at the moment that it would be untimely to write a review and should wait a year or two. And then somebody else you write to will say, well unfortunately there hasn't been very much change in the subject recently and perhaps something might happen in a year or two.

These are the two poles you see, and unfortunately the region inbetween is a set of measure zero. And then somebody will say that this is the moment to write the review, because things are changing and because there are also things which have been established. So what you've got to do is very often to go from the obvious person to other less obvious choices and the advice on this of your first choice is very often extremely valuable. He knows the subject better than anybody else, and he is the man to adivse you. My experience is that very often some of the best reviews are written by quite young and inexperienced authors perhaps under the eye of a very experienced leader. Probably the best pattern you can find is a very very bright chap indeed sort of a few years post doctoral, getting into the subject, is enthusiastic about it but with a tough professor whom he can go to for advice for getting the balance right. That's the best combination. After all Pauli wrote his great review on relativity when he was twenty one, and we still read that even now. So that's important there.

You must think about how much you pay people for this. It's worth paying because then you can reject. If you commission a paper from somebody and you don't pay him, how can you reject it? You think about that. If you are paying him, then you can say "well all right you have got your fee and we're not going to publish it", and you're in fact able to do that, and this is very important. I don't think the fees come to very much in getting people to write but they provide a certain committment. What you've got to allow for is that overall there is going to be a considerable failure rate.

Regarding the number of places where you think you've found gaps and you've got a subject, in fact you'll finish up with at best fifty percent of the papers that you thought you would. But in *Reports on Progress in Physics* we always did very well because Dorothy Fisher who is the managing editor of the Institute of Physics did all the real hard office work - this is rather taken for granted here - but it is really important. She had such charm in her letters that she managed to get them to do things that otherwise wouldn't have gotten done. Now you get the papers in and then you've got to referee them generally, and there they have to be refereed for content. Here your editorial advisors are people who can tell you, probably give a first opinion. Is this paper basically all right in the framework of the subject they are familiar with, but they may need to go to much more specialized referees to be quite sure of that.

One of the things you've got be very careful about indeed in the review is idiosyncracy. There are scholars alas alas - you would think they would know better, but when you get a review paper from them it turns out to be all their own work - all their wonderful things. The work of those whom they regard as enemies - their rivals - somehow or other are not given quite sufficient balance. You have to be very careful indeed to make sure that you are not making that mistake. That sort of idiosyncracy is very dangerous and must be got rid of. And then you have to look at the style. You may have to say it is too long and should be cut or that you've got to expand it, or it may be too cryptic. You may have to say revise, you may have to say reject, and here again the fact that the editorial advisors are a board behind which the honorary editor can hide when he is rejecting is immensely valuable. I must say that an institutional framework, there for anonymity, is a very important garment for any editor to wear.

There must be times when you reject in order to keep the standards up because I think that the quality of a journal is judged by its first class contributions. The first class ones people remember, and the third class ones they remember; you've got to cut the third class ones it you possibly can. Well now you have the framework.

Now what happens? Well you get the things in and you try to deal with them. It's no good expecting to work deadlines into this. You cannot run a journal of this kind on the basis that when he says he is going to have his paper in September, he is going to come in September, ready to be published for the January issue, run through the machine and so on. You are not going to get that. People are not like that. They go to Jerusalem to conferences of editors and have all sorts of problems. And it turns out to be December before it comes in and your editorial policy is wrong. So the only way to do it, is to have a lot of reminders. I have two sorts of reminders for *Reports on Science Progress* papers that are overdue. One of these is a letter which is very dove-like, which says, "You know I just hate to disturb you but we would just like to know what your are doing so that we can get some idea of when we are going to get your paper". And the other one says, "You stinker why haven't you turned up with your paper". And I keep forgetting to tell my secretary which year we are in - whether we are doing the dove-like ones or the hawk-like ones.

What I am trying to make clear to you is that this is not an operation that can be pre-programmed very sharply. You have to assume that what really happens is that you commission enough and you get a random flow of papers coming in, with peaks depending on people doing their writing on airplanes when they are coming to and from conferences as this gives them a chance to write the papers that otherwise they cannot do because of their research.

And so these come in rather randomly. And you see the point is that there is a very long time scale involved in this. It is several years between commissioning and getting your papers. I haven't got the statistics with me of the exponential rates of decay, but basically what you've got to say to yourself is: you send out a hundred letters each year shall we say. And over the next two or three years the reviews come in at a rate of twenty five, or thirty, or fifty of those spread over the time. And you make up your journal accordingly. And that's how a review journal is made.

Behind review journals there is a tremendous lot of very devoted and efficient work by publishers, by the learned societies' professional editors and so on which is actually essential to their making. When we're talking about the self governing scholarly community, we ourselves as scholars have to do this sort of thing, and I'm trying to show you what it actually implies in practice.

DISCUSSION on "Democracy of the Intellect" by Gerard Piel

"The Role of Review Journals in Scientific Publication" by Norman Chigier

"Reports on Progress in Physics" by John Ziman

MICHAEL BOWEN - I have a quick question for Prof. Chigier. You mentioned several times a spread in review journals. Do you have any numbers which indicate the number of journals that have come out in recent years, the number of articles, etc.

NORMAN CHIGIER - I don't have the statistics, but Pergamon Press has brought out a large number in the past few years under different titles such as *Progress* and *Advances*. I have the impression that other publishers have also increased the number of new journals in this category.

URI BLOCH - I wonder why it took three hours to prepare a profile. It usually takes about half an hour to prepare a profile; if we sit for an hour, it is because we are also chatting about the joint and not because we have difficulties with the profile.

I do not believe that the way to find the proverbial needle in the haystack is to bypass the whole publication system and go to the grapevine. But if you really think the closed shop is the right way, then I would like to ask Mr. Piel whether he doesn't think that this closed shop somehow violates his principles of democracy.

GERARD PIEL - I, for sure, think there are dangers in the isolation of the invisible colleges one from the other. The danger is that the invisible colleges can hive-off in devotion to their self-selected interests without response to the wider community around them. We have a splendid case in the peculiarly American school of psychologists who have been captured by the non-question proposed by Arthur Jensen. You correctly identify his name, if you know it at all, with the heritability of intelligence and the identification of that population statistic with genes. Now, so long as they are simply reviewing one another's papers, this literature can flourish and proliferate. They finally came under serious external scrutiny, however, when Jensen surfaced with his ideas in the *Harvard Educational Review*. The ensuing public controversy brought the geneticists to attention, and they have been instructing the public as to Jensen's total misunderstanding of the concept of heritability. We haven't put the fire out yet, but I do still count upon the fluid boundary between disciplines to bring monitoring of one set of scholars by another.

CHARLES LANG - I'd like to put a question to Mr. Piel. If I understood you correctly, you implied that when responsible scientists were working in context - and I'm thinking of the example you gave on recombinant DNA - that they would be able to control any maleffects of the application of the scientific knowledge. Yet I think, as Walter Rosenblith at MIT pointed out, if scientific knowledge is available it would be used for good or for bad. The example he gave was, "suppose somebody came to the President of the USA and said: 'I could provide a transport system for getting anybody from anywhere to anywhere at any time of the day or night, but the cost of this would be fifty thousand killed a year and over two hundred thousand injured.' The President would say 'no'. Yet Henry Ford introduced the motor car." I'd be interested in your comment on this.

GERARD PIEL - We don't yet have to contend with a technology issuing from the recombinant DNA research; I was talking about research and not about the uses of the knowledge. I was suggesting that the engagement of citizens outside of the scientific community becomes appropriate even when it comes to research questions. The licensing of the scholar to pursue questions that interest him does not license him to endanger the community. The panic that has seized the university communities in the United States may yet serve a good purpose. Remarkably, when panels of citizens from outside of the sciences have been appointed to review the question - as for example in Cambridge, Massachusetts, and even though they were selected by a hostile, highly political, demagogic mayor - they have come to sensible conclusions and accepted the guidelines of the National Institutes of Health. Those citizen panels have thus closed the loop by accepting the recommendations put forward in the first place by the scientists who were engaged in the work.

The use of scientific knowledge is even more appropriately a community enterprise and invites participation from far beyond the individual college from which it issues.

ARIE MANTEN - To Prof. Chigier. I notice that you are editing a review journal on energy and combustion science which has a very heavy applied aspect. It is my impression that the requirement for a review article is greater among workers in applied sciences, particularly in industry and in learned professions as compared to scientific institutions. It is also the kind of review article demanded by those people whom you have described, and I wonder in how far you have confirmation of this by the distribution of subscriptions to your journal in terms of the different sectors of employment.

NORMAN CHIGIER - There is a point you make about applied science that I didn't emphasize. There are some special problems of the applied scientist. Certainly the engineer in industry has less time to browse in libraries. He is looking for much more concise and specific information than perhaps the pure scientist and certainly the non-scientist. So there is possibly more in applied science. I'm afraid I can't give you actual numbers, but certainly the number of commercial publishers who are entering this field must be a clear indication that they see a demand.

Since both inside and outside universities the applied scientists has less time to browse, the librarian who wishes to cater for the busy applied scientist has to find a way of satisfying him more quickly than at present. The librarian can offer us excellent service if we ask him for a specific publication, but when it comes to a more general question, even a highly qualified librarian has considerable difficulty in providing answers. The review journals can thus help.

URI BLOCH - It is not really a question, but a reply to Prof. Chigier. You mention that librarians have found a way to provide you with a paper when you know which paper you want. I can tell you that they wildly desire to find the stuff when you have only a general question, and I think that it is you the editors that could help tremendously with some very simple gadgets. Why don't you make titles meaningful and comprehensive? What do I make out of a title that says "Alice in Wonderland"? What do you that a paper on "How to divide a Cake" is about? (Actually statistics). Why do you take the author's abstract as it is because it's more convenient? You're publishing the paper for the whole scientific community, and they need an abstract that is objectively written; make it short, but not too short, meaningful, comprehensive and indicative of what your paper is really about. Use standard terminology; it makes life so much easier or would enable you to get the papers through the general question you believe the librarian cannot solve now.

I would very much like that during this conference some germination to solve these problems would start. Try to do something in this field, and I think it will be much easier to find a general paper outside a closed shop.

JOHN ZIMAN - I would like to make this point very strongly - that there are in very many different constituencies, readerships available whose needs are not being met. I do not think that applied scientists need it more than pure scientists. I think there are students, there are applied scientists, there are very hard, practical technologists, there are very refined pure scientists of one sort or another. All of them apparently need to have their eyes opened to the broader issues of science that are relevant to them. The *Scientific American* plays a very important role for many of us as a primary main review journal on all aspects of science with which we are not closely familiar, but on the whole, the amount of writing and publishing of this kind is inadequate.

The figures we got recently calculating the amount of time actually spent by science on writing reviews was 0.1% - an idiotically small part of the time spent on doing what is a primary task of the scholar. So it's not enough to say, well the applied chap needs it. We all need it, on all levels and all different planes.

ARTUR ISENBERG - I have a question for Mr. Piel. While I share your high regard for Science as a form of human enterprise, I am somewhat puzzled by your criterion of "adequate context". You rightly castigated the experiments conducted in Nazi concentration camps on human beings, on the grounds that they had no context. I imagine that if the United States had lost World War II and we had this conference in Tokyo, there might be a Japanese scientist telling us about the lack of context of the Manhattan project: did it provide "useful knowledge"? Is your criterion sufficient?

I believe that Science is essentially - with one possible exception: respect for truth - a value-free system. I know of no project for which it would be impossible to solicit the cooperation of some scientists - be it chemical warfare or anything else.

In other words, I do not believe that the situation is quite as idyllic as you indicated. Science being essentially value-free, it is not enough, in my view, to deplore a lack of context: such a lack is very frequently purely in the mind of the beholder. I think you need another safeguard permitting scientists to participate as individuals, a safeguard grounded in ethics and some kind of public morality - a sphere in which, I submit, the scientific community has no monopoly and to which, *per se*, Science has nothing to offer to the rest of humankind.

GERARD PIEL - I rise to respond to this question which was not quite in the form of a question. The proposition that science is value-free deserves close re-examination. The decision to enter into the work of science is in the first place an ethical act. The discipline of truth is imposed by the nature of the work.

Science has still another relation to value of a deeply important kind. This resides in the possibilities for human existence that result from the application of scientific knowledge. The bomb is not the only product of issue from work in fundamental physics; out of the same work has come the possibility of infinite supplies of energy from the fission or the fusion route. That development changes radically the human condition.

The bomb itself lends a might compulsion to the Golden Rule. It may be added that it was the sainted Albert Einstein who signed the letter to Franklin D. Roosevelt that launched the Manhattan Project.

EARL TAFT - I want to reinforce what has been said about control of scientific activity by people other than scientists. I have been personally involved in recent months with the controversy going on in Princeton over recombinant DNA research. I am speaking from the point of view of a member of the public, by the way, not as a scientist - so I would just like to offer two short examples.

One I think is that most of the scientists who have objected to the carrying on of this research at the current time are people who have already won the Nobel prize in science. Most of those eager to get on with it are younger people who are perhaps hoping to get it later on. The other example that I would offer that you cannot regard as a completely altruistic motive of all scientists is a technical one.

E. coli is the strain of bacteria currently being used. There are some dangers in this since it exists in the human body and and enters into it very easily. We all get billions of them every day. It seems fairly clear from research going on at Rochester and various places that perhaps in five or ten years, there will be a safer strain of bacteria developed.

Most of the scientists now doing the research would probably admit that, but they are not willing to wait five or ten years. They want to get on with it now. So I would just sum this up by saying that just as you would not perhaps leave the conduct of war to generals, it might not be completely safe to leave the conduct of scientific activity to the scientists who want to do it.

REFEREEING AND JUDGMENT

EDITORIAL JUDGMENT IN SCIENTIFIC PERIODICALS.

Mark Keller
Editor, Journal of Studies on Alcohol
Rutgers University, New Brunswick, N.J., U.S.A.

ABSTRACT
The kinds of judgments that editors of scientific journals exercise are outlined and discussed. The most important judgment is made on every paper submitted for editorial scrutiny: to publish or not to publish it. The principal focus of judgment on reports of observations made in experiments or field work concerns the criteria of science: the quality of design, methodology, execution and analysis, and the propriety of inferences and conclusions. Here strictness is appropriate. In heuristic essays the editorial focus is on relevance and rationality, and judgment should be liberal. Since editors are not omniscient and frequently rely on the judgment of referees, the choice of reviewers for each article is a further critical area of judgment by the editor. Finally, the role of the editor as ultimate arbiter requires him to be sensitive to bias, to arbitrate between disagreeing reviewers, and to judge when to overrule even unanimous referee opinion. The several kinds and areas of judgment are illustrated with case histories.

In undertaking to address the problem of editorial judgment in scientific publication I was unwittingly indulging an illusion. Editorial judgment, it appears after hard consideration, is a fiction, probably spawned in the imagination of editors--including this editor--who would like to think they exercise a distinct art. (If any imagine it is a science, that is sheer delusion.) Actually there is only editors' judgment, idiosyncratic and inevitably fallible, and the growth of democracy in science has tended to dilute it, as committees, associates and referees increasingly share the editorial power along with the responsibility.

If in fact there is no science of editorial judgment, nor even a distinct art, nevertheless a skill exists--perhaps deserving to be called art--of editing and editors' judgment. It can be recognized in the distinctive characteristics and in the quality of some scientific periodicals. A lifelong attempt to practice that skill may justify the present attempt to delineate some of its problems and illustrate some principles. For convenience in expression I shall refer to it sometimes as editorial judgment, and I shall assume that we are talking about periodicals in which the editor is responsible for the final decision on each publication. (This excludes periodicals limited by obligatory publication of, for example, society proceedings, or those in which the decision of owners, committees or referees overrides the editor's opinion.)

Miriam Balaban (ed.), Scientific Information Transfer: The Editor's Role, 195-201. *All Rights Reserved.*

An editor's most important judgment is made on every document he is required to consider for his journal: to publish or not to publish. The science-oriented foci of evaluation are obvious and will only be mentioned: Does it belong in the particular journal? Does the content--apart from language and organization--meet the criteria of scientific reporting? If it reports an experiment, were the design, the methodology, the execution and the analysis appropriate and adequate? And are the inferences or conclusions reasonably warranted? On these issues an editor is justifiably strict. Curiously, experimenters--more often than one would have thought--can misinterpret their data and from a good experiment derive absurd conclusions. An editor can save the useful data by discussing a revision of their interpretation with the author. The effort will not always succeed. An experimenter who misinterprets the meaning of what he observed probably has some emotional investment in the outcome being what he supposes, and he may resist the corrective critique. The editor then has to exercise an exquisite judgment. He may decline to publish the paper; but instead, he may decide to publish it--arranging to plant the necessary critique of the faulty interpretation. This is not difficult for an editor who has good friends--as every editor has. There are at least two sound reasons for accepting the paper with the misinterpretation. First, the experiment may be original and excellent, and it would be a pity to miss the opportunity of publishing it; second, it might be a pity to let it be published elsewhere--without the needed corrective. The latter kind of reasoning exemplifies the more subtle aspects of editor judgment.

If the paper is a think-piece, a different sort of judgment is appropriate--liberal rather than strict. Experiment is the life-blood of science. Theory is the inspiration. If I could choose, I would rather publish a daring theory than a safe experiment. But a scientific journal should accomodate both kinds of article.

So much for generalities of editorial policy. I propose to illustrate with concrete examples the applications of editor judgment where problems predominate.

The Invaluable Referees. From a mass of patients with a specific diagnosis, two ambitious medical scientists collected practically a river of urine. Day after day they analyzed multiliters of the fluid by a new method recently described in one of the biochemical journals. And they found positive results which, triumphantly, they detailed in multitudinous tables with an accompanying text. The editor of the journal to which their paper was submitted declined it. Regretfully he called the authors' attention to a fact pointed out by the very first of his editorial referees: The inventor of that test had, in the issue immediately following the one in which it was published, withdrawn it--because, as he had belatedly discovered, it was hopelessly flawed and useless. Some months later the paper thus declined appeared in a respectable medical journal. Perhaps the authors felt that the second opinion of the inventor of the test they used had been published precipitously; he should have waited at least until the first paper based on it was published. Or else, they thought that after having tested a whole river of urine they were entitled to at least one publication out of it. In any case, it is obvious that the editor of the journal in which that paper appeared was inadequately provided with knowledgeable referees.

Blessed with knowledgeable referees was the editor who received the beautifully prepared report, by a highly qualified physiologist, on the results of an experimental induction of hangover. This condition, a sequel usually on the morning after an evening of overindulgence in alcoholic beverages, is remarkable for the wealth of cures recommended--anciently in Egyptian medical papyri and in the Ayurveda, currently in every western newspaper around New Years. An overabundance of cures suggests an abundant ineffectiveness; thus far physiologists have not been able to establish exactly what causes the hangover symptoms. In his attempt, the present experimenter induced human volunteers to drink copiously of a variety of spirits ranging from vodka to aged brandy. There has never been a lack of volunteers, all fully informed and cheerfully consenting, for experiments involving alcohol ingestion. The vodka --mirabile dictu--produced hardly any hangover symptoms. Unfortunately for the experimenter, the editorial referees noted that his design was such that the outcome was inevitable, and one referee felt sure the experiment was deliberately rigged to produce the result. Not without regret the editor declined the paper, with a frank explanation based on his agreement with the view that it was a rigged experiment. (Incidentally, all the reviewers were sure that an unrigged experiment would have yielded the same result; and indeed the same journal subsequently published similar results.) The author did not protest the decision. Instead he submitted his paper to another journal where, in due time, it was published. The other journal was specialized, as its title indicated, in proctology. Inasmuch as the commonest symptom of hangover is headache, the editor who had declined the paper consoled himself with the thought that, after all, it has found its appropriate place.

The moral of the two preceding examples seems to be that a minigram of editorial referree knowledge may be worth more than a maxigram of editorial judgment. The symbiotic role of the referee-reviewers in the function of the scientific editor is invaluable. Most of the time they provide not only the basis for a decision pro or con, but reasons the editor may not have thought of, and--above all--critique and guidance for the improvement of papers that are not quite fit but ought to be published. Unless a journal is focused on an extremely narrow subject, no editor can be expert enough to evaluate all papers confidently. The cultivation of a corps of referees, and the selection of the appropriate reviewers for each paper, are thus a prime area of editorial judgment. Referees nevertheless can be wrong in their conclusions, as they sometimes must be, and an editor must have the courage sometimes to go against their advice. Examples follow:

Negative Results. The prejudice against publishing negative results is nearly universal. It is understandable: The editor is saying, and saying it on behalf of his readers, if you didn't find out anything, who wants to read about it? Unfortunately this evaluation of "negative results" is often misleading, since the failure to get positive results may provide more knowledge than many "positive" findings. This is especially pertinent if the experiment is the sort that may be repeated in many laboratories--each of which has to discover anew that it leads to negative results.

Two psychologists carried out an acute experiment in which they studied the effects of a psychotropic drug on a behavioral response in rats. For convenience, these scientists divided the work in a very practical way: each experimenter handled, throughout, the same half-group of experimental animals and an equal half-group of controls. In the end their sophisticated computer-aided analyses yielded no positive results. Referees said, in effect, they got nowhere--their paper isn't worth publishing. But there was one curious statistically significant difference--not between experimental and control animals but between the groups handled by Experimenter A and those handled by Experimenter B. The authors had no explanation of this outcome. They thought they had done everything exactly the same, in the same time and place and with like animals. The editor wondered: Is it possible that rats can feel differently toward two psychologists? And give them different responses to the same treatment? The paper was therefore published. Thus far, no one has followed up on the hint of this positive finding. The authors' negative results, however, have been cited several times in the literature. Experimenters should resist feeling embarrassed to publish negative results and editors of scientific journals should judge them more kindly.[1] A negative result may be an important positive truth.

Animals as Subjects. If rats may show feelings toward the personalities of human experimenters, one would surely expect it of dogs. I have seen dogs eagerly jump up onto the table where they knew they were to be subjected to surgery. And I have seen them struggling to escape from the arms of an experimenter carrying them gently merely to their daily weighing. (Him they don't like.) The amount of inattention to this factor, and its possible influences, by experimenters and by the editors who publish them, is phenomenal. A famous Pavlovian physiologist studied the effects of alcohol on sexual reflexes in dogs; he tested "time to ejaculation." Two referees severely critisized the paper that came forth because the data seemed incomplete and, therefore, not confidently interpretable. The author, responding to a regretful editorial declination,[2] explained the defect in the data: it was owing to the circumstance that some of the dogs had ejaculated on the threshold of the experimental room! The editor invited the author to insert this explanation in a footnote, and with that, happily published the paper. The defects in the data seemed trivial compared to the dramatic evidence of the eagerness of his dogs to give the experimenter what he wanted. The objections of some self-styled animal-lovers to the use of dogs in experiments may be irrational,[3] but scientists--and the editors who publish them--should wonder about the objectivity especially of the dog--but of other animals as well.[4] Editors cannot question the legitimacy of findings made with experimental animals, but they can urge authors to be a bit less confident, a trifle more modest, in their interpretations and extropolations.

Serendipity versus Design. People sometimes make good experimental subjects. A psychologist who wished to study the effect of alcohol on verbal fluency gave his subjects a rather stiff drink and then asked them to perform such tasks as naming trees, speaking words beginning with a specific letter of the alphabet, and so forth. The results, as usual, seemed to confirm the folk wisdom--alcohol loosens the tongue. Unfortunately, the psychologist had neglected to consult anyone who knew about alcohol metabolism before designing his experiment. There was a serious technical defect: one could not really know how much alcohol did what. Two strict referees thought the paper unpublishable. The editor nevertheless published it. The experimenter fortunately had reported not only the statistical outcome but also illustrated the findings with examples. And it turned out that when, for example, subjects were asked to Name Trees, without alcohol they named such trees as birch, pine, apple, pear; with alcohol they named the sort of trees, but also, in addition, such trees as family tree, shoe tree, Christmas tree....

The preceding examples may suggest that editorial judgment is especially directed to overruling negative judgments by editorial referees. Indeed this is a vital function. The science editor stands on the threshold of Publication Paradise. He is the angel with the flaming sword, keeping the way. But he has the privilege of letting by the imperfect who nevertheless have something to teach--especially those who bring something new, refreshing, potential, inspired, which--given a chance to take root in the garden--may produce the most splendid offshoots.

Yet the danger of letting by the unworthy is great and eternal editorial vigilance is the price of scientific integrity.

The editorial life runs not smoothly, and a wise physician once advised that no one with an ulcerous personality should be an editor. Those incomparably competent and knowledgeable editorial referees don't always agree among themselves. Every editor experiences that Referee A picks on item X as needing revision, while Referee B has no complaint about X--it's item Y that he finds erroneous. An editor can decide that they are both right, or one is right, or neither is right. The hard time comes when Referee A enthusiastically recommends acceptance as is, but Referee B is equally sure the author is practically a charlatan and firmly recommends an unconditional rejection. The number of times two reviewers equally competent in their spheres of expertise disagree totally about a paper is amazing and might shake the confidence of an editor in his own judgment. It should not. Depending on the nature of the disagreement, the advice of additional referees might be sought. For example, if the negative referee's reason is that the author used the wrong statistic, and had he used instead the named right analysis his significance values would disappear, it is surely occasion for going to the best possible statistical consultant. But the difference between the referees may be philosophical. For example the negative referee may be turned off because he thinks the source of support for the study raises a suspicion of "bought science." Where expertise cannot be brought to bear on the issue, an editor must have the strength to exercise his own judgment. He is the referee of referees.

Semantics and Style. To print under an author's name anything he did not authorize is impermissible. Therefore editorial corrections or changes should be shown to the author--be it on galley or page proof--before publication. This necessity may discourage editors from correcting text--especially from improving it. Alas for our time! Too many scientific workers who get to write papers that need to be published are linguistically handicapped. (That may be a euphemism for illiterate.) They often fail to say what they mean, and sometimes manage to say what they don't mean. Fortunately most authors are content--many are delighted--to accept editorial improvements of their faulty expressions. When an author objects, however, every possible concession to his preference must be made--short of allowing outright misstatement or violations of style. Some of the problems may be most briefly illustrated in the form of imaginary dialogues:

Editor: The mice in your Group A did indeed drink a greater proportion of their fluid from the tube containing the 10-per-cent alcohol solution than from the one containing tap water. But why do you refer to this as an alcohol preference? Surely we can't read the mind of the mice? Perhaps they only disliked the tap water more? (Didn't a sanitary engineer once say that the best treatment for tap-water taste was a shot of Scotch?)

Author A: It's my opinion that if they drank more of the alcohol solution, that indicates a preference.

The author's opinion may not be good science, but his experiment is, and the possible defect of language is not grossly harmful; the editor should let the author win this argument.

Editor: You write, "Smith has proved that horse serum can cure hangnails." Actually Smith made that claim, but it has not been confirmed or generally accepted as a fact.

Author B: But I believe Smiths' work proves it.

Editor: You can say that Smith believes he proved it; or that you believe it; or that Smith believes it and you agree with him; but not that Smith proved it--implying that it is a generally accepted fact.

Author B: You are trying to put words in my mouth. I insist on saying that Smith proved it.

The editor must have the courage to insist on no misrepresentation of fact. In this case he has adequately guided the author toward the proper statement and he should decline to publish the paper rather than allow the gross misstatement.

Among his other functions, the editor determines the literary quality of a scientific periodical. There may be some who consider literacy inconsequential in scientific reporting. But illiteracy, and indifference to precision in expression, go with ambiguity and misapprehension.

Author C: You changed skid road to Skid Row. That's wrong. Here in Seattle we invented the skid road, and I insist on changing it back.

Editor: Yes, you Seattlers invented and named it first, and probably it should be skid road. But it has become institutionalized in most of the world as Skid Row, and this Journal has adopted that form as its style. To avoid hodge-podge in our pages, and without claiming any correctness for Skid Row, we have to maintain the form dictated by our style.

Author D: We Canadians (We English, etc.) spell it labour and counselling. You Americans must think you are being economical by changing it to labor and counseling, but it will seem strange and illiterate to my countrymen if these misspellings appear in the reprints of my article as it is set up in type in your proofs. Please change my text back to the British orthography.

Editor: We don't think American orthography is more economical, more literate or in any way superior to the British. However, this Journal adopted as its style authority for orthography the Webster II, and we are stuck with it. So if your article is to be published in this Journal--as we desire--you will have to tolerate our quirks of style. We would look hodgepodgy if we made non-style exceptions for authors' nationality or literary preferences.

An editor who does not have firm style rules or does not uphold them will eventually wish he had.

Ethics. An interdisciplinary group of experimenters gave 64 medical students on four occasions four doses of whisky--the largest, 180 ml; tested was their ability to solve problems in Boolean logic. No one questioned the ethics of the experiment, or of its publication. A psychiatric resident gave to uninformed but nevertheless consenting fellow resident psychiatrists an intravenous injection of an unnamed drug (it was alcohol, and some subjects guessed correctly what they were getting). No one questioned the ethics. But when in a study of addiction a psychiatrist gave intoxicating amounts of alcohol daily for several weeks to a group of prisoners, a priest wrote to protest that both the experiment and its publication were unethical, although the subjects, all former opiate addicts, had volunteered after being fully informed of the nature of the experiment (they were promised, as reward for participating, a reduction in their term) and those who regretted participating were permitted to withdraw. In another experiment a psychiatrist-neurophysiologist gave to ten prisoners who were confirmed alcohol addicts about a liter of whisky (at about 40 per cent alcohol) daily for three weeks to test addictiveness as well as numerous other parameters. All the subjects were fully informed and consented. Ten papers were published and no one challenged the ethics at the time; but subsequently the question was raised in discussion. One defender of the experiment noted that the subjects were all men who, on their release, would indulge in just such drinking--until they became disabled; and that they were better off drinking under ideal metabolic-laboratory conditions with protective nutrition and 24-hours-a-day nursing and medical supervision. A critic argued, however, that giving alcohol to alcohol addicts was never justified under any circumstances--even for the purpose of attempting to train them to drink moderately--and that editors should discourage such experimentation by refusing to publish their report.

The problems of ethics in experimentation--social and psychological as well as biological--have recently proliferated and they challenge the judgment of editors as well as the conduct of the experimenting scientists. Civil libertarians challenge the ethics of asking certain types of personal questions in demographic surveys--even in census taking. Animal-loving humanitarians challenge the ethics of inflicting pain on animals, and soon will challenge the ethics of imprisoning them at all. An anthropologist who desires to observe the behavior of the patrons in various types of pubs is told it is spying and an invasion of privacy. He argues that he is doing nothing different from what the objectors are doing--watching people--the only difference being that they are casual observers while he is systematic, but it is not easy to win an argument with earnest civil libertarians or fanatical antisciencists. It has come to the point that the authors of a new book on scientific editing[5] recommend that editors should require, with each article they publish which reports an in vivo experiment, a statement by the authors certifying that humane and ethical standards had been maintained.

Editors are perhaps in an easier position than scientists; they can argue (even from the perspective of freedom of the press) that their duty is to publish the report of what has been done, once it has been done. But it will be poor judgment to stand apart from the scientists. In the case of the report of a psychologist who placed wine on the dining table before children in an institution for the mentally retarded to determine how much they would choose, and allowed them to drink as much as they wanted, until some got drunk and sick, the question of ethics hardly arises; it was an indecent experiment. Nevertheless the editor who published the report was right to do so; but in my judgment he should have appended an editorial note condemning the experiment.

As for the proposed universal certification of innocence, I would hope that editors of scientific journals will not only not require it but will refuse to publish it if submitted by an author. The idea of a certification of decency in the reporting of each in vivo experiment implies that scientists are expectedly indecent and inhumane unless strictly controlled. To a scientist who would insist on publishing the certification of his innocence an editor should quote, "The lady doth protest too much, methinks." In any case, it is not for science editors to set themselves up as the ultimate judges of the morals of scientists, any more than they should be the censors of unsavory ideas. Their judgment should be reserved for the proper realm of their function--which is to publish what scientists know and think.

NOTES

1 An old suggestion, perhaps original with Dr. Howard W. Haggard, then director of the Laboratory of Applied Physiology at Yale University, was to publish The Journal of Negative Results, which might achieve the largest circulation in the realm of modern science.

2 The illustrative cases in the present paper are mostly from the Journal of Studies on Alcohol--advantageously, for the present purpose, an interdisciplinary periodical. The policy of that Journal is never to reject a paper; but about two-thirds are, regretfully, declined.

3 Concerning impassioned animal lovers an insightful Prophet once observed: "They that sacrifice men, kiss calves!" (Hosea 13: 2.)

4 As one animal-lover has remarked: "Dogs are crazy. They love people more than dogs."

5 DeBakey, L. et al. The Scientific Journal: Editorial Policies and Practices. St. Louis; Mosby; 1976; p. 35.

IMPROVING REFEREE-SELECTION AND MANUSCRIPT EVALUATION

Manfred Kochen and Barbara Perkel
University of Michigan

Abstract

To maximize the chance that manuscripts will be evaluated by the most competent referees, four systematic procedures are proposed. One expands the pool of potential referees. Another selects a set of people from that pool to evaluate a given manuscript. The third solicits from them judgments along five scales. The fourth procedure combines these judgments into a form of consensus to help the editor make a decision. The five scales involve measures of relevance to the journal's objectives, significance of the problem, validity of the results, novelty and clarity. Limited experiments with field trials over the past five years with the Journal of the Association for Computing Machines show that implementations of some aspects of these procedures help to improve the quality of accepted papers. The quality of a paper is interpreted primarily as a likelihood of its being appropriately cited. Analysis of these algorithms leads to interesting research problems, such as the possibility of non-subjective measures of quality by the use of modified co-citation counts. The use of computer conferencing to aid communication among referees, and among them, the author and editors, is analyzed for its potentialities and constraints.

1. Introduction

Derek de Solla Price (24) has suggested that the number of published scientific articles that are "good" increases only as the square root of the total number of published articles. This implies that if total publication volume continues to increase, the ratio of those that are "good" decreases. The burden on readers to sort "wheat from chaff" would increase. This might increase the need for lower acceptance rates, resulting in an increased quality of the manuscripts accepted for publication. The present acceptance rate is about 55% for journals receiving at least 300 manuscripts per year and about 70% for "smaller" journals (13). The acceptance rate is also higher for papers in the "hard" (e.g., physical) sciences than in the "soft" ones. It can be as low as 8% for some social science journals (21) and as high as 90% in mathematics, logic, and physics.

Responses from a third of some of the referees of the JACM who were recently sent a questionnaire indicated that they are on the whole satisfied with the manuscripts published by that journal. If this is representative of most scientific publications and/or readers, then perhaps there is no real need to improve the proportion of publications that are high quality for such journals as JACM.

One might even argue that all submitted manuscripts should be published in some journal without costly attempts at quality control or "birth control," leaving it to readers and reviewers to discriminate for themselves. This would raise questions about whether the resulting literature is worth reviewing (4). Discerning authors and readers might simply stop submitting or subscribing to journals with a high proportion of low quality papers. Journals known to publish high-quality articles would tend to attract more manuscripts of high quality than others. Publishers would incur excessive

Miriam Balaban (ed.), Scientific Information Transfer: The Editor's Role, 203-229.

costs if they accepted all submitted manuscripts and there would be many more low-quality journals. Some, possibly an optimal, degree of quality control seems desirable.

One way to improve the quality of what is accepted for publication is to improve the mechanisms of screening and evaluating the manuscripts that are submitted. One of the most powerful aspects of science communication, one that makes science self-correcting, is its peer-review evaluation system. Responses to the above mentioned questionnaire showed consensus that peer review is preferable to alternative systems, even if it has faults such as inconsistencies due to differing standards. "The scientific paper that has been refereed and published in a journal not only conveys information, but more importantly, it is a statement of knowledge that has been evaluated and declared acceptable by the scientists' peers" (29). A work of science is not complete until it is reported in a scientific journal, where it is freely citable. The scientific paper is the most important source of new ideas as well as new findings and methods (11).

Several studies dealing with the social structure of science have investigated factors in the "quality control" of the scientific literature (5, 6, 23, 30, 31). Journal refereeing practices have been studied, though not extensively enough to provide a comprehensive picture (13, 7, 12). In a state-of-the-art survey, Manheim (20) concluded that "practices that tend to reduce effectiveness or referees and the quality of scientific periodicals are gaining ground". The Juhasz survey showed that the editor of a refereed journal rejects 10% of the manuscripts and sends the other 90% to referees, more often to ad hoc referees than to those who have formal and permanent agreements with the editor.

If we consider that very few papers in refereed journals are never cited--a very important determinant of quality--then the refereeing process must be remarkably effective. There are nevertheless opportunities for increasing both its effectiveness and efficiency. For obviously bad or very good papers, the rejection or acceptance of manuscripts is not unusually sensitive to who does the refereeing. However, for the majority of papers in the middle, the choice of referees becomes quite important. Most scientists are aware of the fact that many mediocre papers are published while some slightly less mediocre ones are rejected, and that whether a published article is cited depends not only on the quality of the work it reports, but on how well it is presented and some chance elements.

The mechanisms of screening and evaluating manuscripts could be improved by a set of procedures or algorithms that: (1) expand the pool of potential referees in the direction of including the well qualified peers of manuscripts' authors and the most competent experts in the field that can be recruited, especially from communities of younger scientists who are less well known, and yet may be very good referees; (2) selecting the best possible set of referees from this pool to evaluate a given manuscript; (3) eliciting from them the best possible judgments about that manuscript; and (4) combining these judgments into a good editorial decision.

In this paper we describe ideas for improving such procedures and analyze when it pays to implement them partially by computer. To demonstrate that use of the algorithms improves referee-selection and manuscript evaluation, we must first clarify what we mean by "well-qualified peers," the "best possible referees," the "best possible judgments," and "good editorial decisions." We then provide logical arguments why the algorithms, combined with a variety of newer practical techniques such as computer conferencing, are likely, through a number of stages of development, to lead to the needed improvements; these arguments are supplemented by data and insights from six years of experience with a practical earlier version of these procedures and reported earlier (16).

2. Criteria for Evaluating Manuscripts

We propose to call a manuscript useful to a reader R in connection with an issue of question Q if:

(1) it is read and clearly *understood* by R;
(2) if it is *new* to R;
(3) R judges it *relevant* to Q;
(4) R is *persuaded* by the arguments, inferences, discussions and conclusions presented in the manuscript in support of a claim to have resolved Q or shed light on Q; and
(5) R regards both Q and the paper's discussion of Q *significant* enough to probably cite the paper.

Let us examine these in turn. Obviously a manuscript must be read before it can be evaluated. For this to happen, it must be able to hold the reader's attention and interest to at least a minimal degree. At the very least, it must employ language--including technical terms--in which the reader is fluent. Requesting a response to the first item in the questionnaire of Appendix A is one way to measure a reader's fluency with, or clarity of, a given paper for the reader (the two are relative).

A plausible hypothesis is that a reader who marked a manuscript well above the midpoint of this clarity scale is likely to cite that paper in some future publication of his own, assuming that the manuscript is citable (i.e., published). To test this hypothesis we would have to search at least four years of *Science Citation Index* for all the manuscripts that are published and calculate the correlation between the number of citations, especially by that referee, and the clarity score he assigned to it 5-6 years earlier.

If we give the same manuscript to several readers, we will obtain a distribution for the degree of clarity they assigned to it. If the paper contained terms that some readers did not know, they should give it lower scores. Alternatively, if a reader assigned a low clarity score to a paper that differed from another paper only in that it did not use certain key technical terms, and the same reader assigned a higher clarity score to the second paper, perhaps we can infer that this reader was not fluent in these technical terms. Manuscripts with some technical terms should not be sent to such a referee until his fluency score increases.

Some referees to whom we send manuscripts submitted to *JACM* return it, stating that they are not qualified to review it. That may be due to insufficient fluency with the language or concern or competency with the issue or techniques, or it may be a cover for other reasons for not wanting to judge the paper.

If a reader assigns a low clarity score to a manuscript, how can we tell whether the low score is due to the lack of clarity in the manuscript, or the lack of fluency by the reader, or both? A workable procedure may be to determine a "fluency profile" for each reader based on the clarity scores he assigned to each paper in a "standard sample" of manuscripts. The editor, possibly aided by a computer program, then judges the similarity of the language in a given manuscript with each paper in the standard sample. He finds the paper most similar to the manuscript and then searches all referees' fluency profiles for one that has the highest score on that paper.

The entire question of estimating the similarity of two papers in their language is far more complex than suggested here. Commonalities in technical vocabulary is but a crude first approximation.

The second criterion for evaluating manuscripts is its novelty for R. This is clearly not independent of the clarity judgment. If it is not new to R, R may not find it interesting enough to read. If it is too new, he may not understand it. Novelty judgments depend on how much R already knows and understands. To measure novelty we could examine the references used in R's own publications as an indication of what articles he has read and cited. We could determine a "newness profile" for R based on the newness scores he assigns to a "standard set" of manuscripts. The editor assigns R to a manuscript only if he judges the manuscript to be sufficiently close in novelty to a manuscript in the standard sample to which R is assigned the "best" newness score. We could define two papers as close, with respect to novelty, according to the fraction of the references in their bibliographies that they have in common (14). An alternate technique is to count the number of references common to the manuscript and the union of all papers ever cited by R's publication, then select the R for which that number is greatest.

R's judgment of relevance to a question or issue Q that concerns R depends on how closely the author's representation of Q, as expressed in the manuscript, corresponds to R's representation of Q. The editor should try to select readers whose representation of Q matches that of the author. Alternatively, he could give the author a chance to select the reviewers. One way to help the editor to do that is to examine the commonalities between the set of index terms assigned to a manuscript (by the author for JACM) and the referees' interest profiles, as recorded for each referee, based on how he expressed these in his own words. If exact matches in the phraseology of authors and referees do not occur, the use of an "adaptive thesaurus" may be employed.

A reader R is well-qualified to judge a manuscript according to these criteria if the manuscript is similar: (1) in its language to a paper P_1 in the standard set of papers to which R assigned to a high clarity score; (2) in its references to a paper P_2 in the standard set to which R assigned a reasonable novelty score; and (3) in its representation of the key question, issue or claim Q to a paper that R has written or that R says he is strongly concerned with and competent in.

The best possible referees for a paper, among those who are qualified, are those for whom the "distance" to the manuscript with regard to language, question-representation and knowledgeability is least. The "distance" measure combines the similarity of the manuscript to the closest "test" papers and the scores assigned by R to those test papers.

It is useful to formulate possible referee-selection criteria mathematically. Let $R = (R_1,\ldots,R_n)$ denote the pool of referee-candidates and $P = (P_1\ldots,P_N)$ a standard sample of test papers. Let M denote a new manuscript to which referees are to be assigned, and suppose that the following 3Nn numbers are on record:

f_{ij}: the clarity score referee R_i assigned to test paper P_j;
ν_{ij}: the novelty score referee R_i assigned to test paper P_j;
r_{ij}: the relevance score referee R_i assigned to test paper P_j;
$i = 1,\ldots,n;\ j = 1,\ldots,N.$

Suppose further that the following 3N numbers can be easily computed for M:

s_{Mj}: the similarity of the language in M to that of P_j;

ρ_{Mj}: the overlap in the references of M and those of P_j;

d_{Mj}: the overlap in the "index terms" or phrases expressing key questions or issues assigned by the author or professional indexers to M and to P_j; for $j = 1,\ldots,N$.

Given M, the s's, ρ's and d's are computed, and up to three papers in P are selected, the first, say P_j, so that it has a value, s_{Mj}, larger than a minimal threshold; the second (which may be the same as the first) so that its value exceeds some threshold; and the third so that its d value exceeds some minimum as well. If three papers in P, close enough to M in this sense, cannot be found, then an alternate procedure, e.g., the conventional one, using the editor's judgment alone, is used. If three such papers and the corresponding "distances," s_{Mj}, $\rho_{Mj'}$, $d_{Mj''}$ are determined, we could then select that referee R_i for whom the weighted sum (a linear combination) $af_{ij} + b\nu_{ij'} + cr_{ij''}$ is the least for all values of i; here a, b, c, are the relative degrees of importance attached to clarity, novelty, and relevance, respectively.

Whether R is persuaded by a manuscript depends on many factors. First and foremost is conformity by the authors to consensual scientific criteria for valid logical and statistical inference. Secondly is the author's style of proof. Thirdly is R's degree of skepticism, the extent to which R is demanding and critical in what he accepts as proof. Most important is R's competence about Q and methods used for its analysis. It is not necessary, though it is usually the case, for R to be a creative and productive contributor to the field, though criticisms may be much more readily accepted from a judge who has earned his eminence or qualification to criticize by exemplary work of his own. Our experience has shown that to a large extent the more eminent a judge is, the less time he seems willing to spend reviewing manuscripts. The younger, less well-known readers often are the most prompt, conscientious, thorough and objective referees.

Whether R judges a manuscript on Q as <u>significant</u> depends on how many other issues Q', Q", ... he considers it to shed light on, as well as on R's evaluation of the significance of Q, Q', Q" ... . A problem or issue is significant if it has potential for overcoming barriers to understanding other questions or issues that are themselves significant in the same sense. Operationally, a referee or other reader <u>has</u> treated a paper as significant in some sense if he cited it. His significance judgment should therefore correlate with the likelihood of citation by him. If a referee has cited many papers by an author of a manuscript, he is perhaps more likely to regard it as significant. Some might call such a referee biased, and it is an interesting question whether such referees should be selected often.

A referee's best possible judgment is assignment of a score on the five scales, combined with persuasive reasons for that score. It is the editor who must be persuaded by R's reasoning. R is not expected to provide objective data but to rely on his experience, intuition and willingness to take risks in the face of uncertainty.

A good editorial decision is one based on consensus. Our experience has shown that over the past year, of the 12 papers reviewed by 24 referees, there has been consensus in all but 3 cases. Science depends a great deal on consensual agreement about major paradigms, criteria of validity, etc. Consensus does not mean unanimity, but agreement among those most vitally concerned, without strong dissension by such people. Consensus may encourage inbreeding, in that adherents of a paradigm may reject anything that threatens it or does not fit into it, thus decreasing the chances for publication of genuinely new ideas. If all but one of two or three well-qualified referees, in rendering their best judgment, agree that a manuscript should be accepted (or rejected), then the editor can resolve the conflict, or better, try to get the referees to resolve the conflict among themselves. Often a new "super-referee" is called in to resolve such conflicts (13).

The essential core common to the various problems in creating a scientific basis for improved referee-selection and manuscript evaluation is that of measuring similarity among manuscripts with regard to language, and to representation of the key issue and background required for comprehension. One recently developed research technique with possible promise in this direction is the "mapping of science" (25). Two papers are similar according to the number of papers each of which cites them both. This measure works only on papers that are already published and which can, therefore, be cited. We could test if two papers that are very similar by this measure are also close along the three attributes of interest here. But we could not apply the co-citation similarity measure to judge similarity between manuscripts and papers in a standard test set, though we could do so for their authors.

It seems plausible that if a manuscript M is eventually published and cited, and if the referees were properly chosen, then there should be a high probability that the referees will be among the authors who cite M. Moreover, their papers have a good chance of co-citing M and one of the papers P in the standard test set which they presumably have already cited. Allowing the referees to self-select papers might therefore be a promising alternative procedure. Those readers of M as a publication who cite it may have profiles similar to those of the referees.

3. An Algorithm For Improving the Pool of Potential Referees

The procedure now in use for manuscripts submitted to JACM in the area of information retrieval adds referee-candidates by the following devices: (a) recommendations from referees in the pool when they respond to correspondence from the editor (they are asked to suggest one or more colleagues likely to be good referees for a manuscript every time they are sent a manuscript for refereeing); (b) authors on a topic of a manuscript who in the editor's opinion would be good referees; (c) authors of papers listed in the references of the manuscript; and (d) experts suggested by the authors of the manuscripts.

As can be seen in Table 1, the pool of referees for the information retrieval section of JACM has expanded from 36 in 1974, to 81 in March, 1977. It has more than doubled over the past three years.

TABLE 1

BREAKDOWN OF HOW POOL OF REFEREES HAS EXPANDED

OVER THE PAST THREE YEARS

Year	1971-1974	1975	1976	1977
Number of Referees	36	57	78	81

Of the 45 that have joined the pool since 1974, 21 (46%) have been recommended by other referees in the pool, 15 (33%) have been chosen by the editor because of their knowledge of a particular subject, 5 (11%) have been authors listed in the references of the manuscripts and 4 (10%) were experts suggested by the authors (see Table 2).

TABLE 2

METHOD WHEREBY REFEREES WERE ADDED TO POOL

(all are approved by the editor)

Year	1971-1974	1975	1976	1977
a) Recommendations from Referees in Pool	9	12	9	-
b) People suggested by Editor and not (a) (c) or (d)	4	2	3	-
c) Authors of Papers Cited in References	23	7	5	3
d) Experts Suggested by Authors of Manuscripts	-	-	4	-
TOTAL	36	21	21	3

The proposed improvement of the algorithm adjoins the following devices: (e) authors of papers that are co-cited a great deal with papers listed in the references of the manuscript; (f) authors who have cited earlier papers by the submitting author; and (g) authors of papers in a co-citation cluster (if there is one) dealing with the key question or issue of the manuscript (25).

A simple mathematical argument shows that even mechanisms (a) and (b) alone would, under plausible assumptions, cause the pool to grow faster than linearly over a certain period before converging. The method could be extended to include the effects of all seven mechanisms, though the analysis is quite complex; the other five mechanisms will accelerate the rate of growth of the referee pool. The mathematical formulation is of interest in suggesting how to proceed with the more complete, complex analysis.

Let:

n be the average size of the referee pool in any year t

R be the number of manuscripts submitted per year

m be the number of referees assigned to a manuscript;

p_R be the probability that one of these referees suggests a colleague as an addition to the referee pool; assume that the referees suggest independently, and that each does not suggest more than one at a time;

K_R be the number of colleagues in the area known to a referee.

Assume that the probability of a recommended colleague being in the referee pool already is K_R/n, $K_R < K_E \leq n$, where K_E is the number of colleagues in the area known to the editor.

The rate of change in the expected number of referees is, then, $n = Rmp_R(1 - K_R/n) + Rp_E(1 - K_E/n)$, where p_E is the probability that the editor suggests a referee by device (b). Interpreting this as a simple differential equation in n with the initial condition that $n = K_E$ at $t = 0$, leads, upon integration, to the transcendental equation:

$$n - K_E + (b/a)(\log(an-b)/(aK_E-b)) = at,$$

where

$$a = R(mp_R + p_E), \qquad b = R(mp_R K_R + p_E K_E).$$

If K_R and K_E were very small, so that b is small, then n would increase linearly with t, $n = K_E + at$, until the pool begins to get close to $nK_R + K_E$; then the editor and the referees in the pool will begin to recommend colleagues already in the pool. If K_R and K_E are large, which is likely, then the number added to the pool will grow nearly like a chain reaction at first, then to be limited by recommendations of colleagues already in the pool to which fewer and fewer new names get added.

Merely increasing the size of the referee pool does not necessarily improve it. Let R_T denote the set of all scientists interested in a topic, with R_E the subset of these known to the editor and considered as candidates for the referee pool. The competence of people in R_T to judge papers is distributed according to a distribution function F(x), interpreted as the fraction with competence $\leq$ x. We could characterize the editor's competence by

$$\int_{R_E} x dF(x).$$

If R_t denotes the referee pool at time t growing according to the above mentioned mechanisms, then its quality could be similarly measured by

$$\int_{R_t} x dF(x).$$

It is possible that with a good editor, a conventional system in which he relies on his own memory to select referees, is better than the proposed innovation. In formal terms, we could have $\int_{R_E} x dF(x) > \int_{R_t} x dF(x)$, even though $|R_E| = K_E < n = |R_t|$, where the vertical bars denote the number of elements in a set. That should be rare, however, because R_t is the result of suggestions not only from the editors, but from referees he chose who presumably suggest other referees of high competence. Almost certainly $|R_t| > |R_E|$ and $R_E \subset R_t$ so that $\int_{R_t} x dF(x) > \int_{R_E} x dF(x)$.

In some fields it is possible to obtain a list of all the active scientists in the area and use that as the pool from which to choose referees. Such a set--say of all anthropologists in the world--is not the same as R_T, because R_T are all scientists interested in a specific topic, issue, question and that may include specialists in various disciplines. In highly insular disciplines, authors would object to evaluations of their papers by people outside their disciplines. Often, however, they would welcome such criticism and realize how this might enrich their area.

4. An Algorithm For Assigning Referees from the Pool to a Manuscript

Every manuscript submitted to JACM contains index terms selected by the author. The editor may add additional terms to the manuscript or question those selected. These are generally quite specific.

For every referee in the pool there is a record of his profile listing the topics, methods, problems of special concern, together with R's degree of concern and critical competence (see Appendix B for a typical example). This data is generally supplied by the referee-candidate. A record is also kept of the referee's past performance, i.e., time taken to review, any bias shown, detailed or superficial review, etc. Each time a particular referee is asked to review a manuscript he is sent a copy of his profile and is asked if there are any additions or changes he wishes to make. Additional terms may be added to the profile by the editor based on his experience. These are not quite as specific as the index terms assigned to manuscripts.

The index terms of a newly arrived manuscript are matched against the profiles of referees in the pool using a thesaurus where terms are related by "broader than" or similar relations. If the referee pool is large enough in all topics, then a reasonable number of successful matches should be found. For each paper that we receive, there are usually somewhere between three and five possible successful matches. Two referees are selected and the remaining ones used if more referees are needed.

Further selection is based on data about the referee's past performance and the editor's preliminary evaluation of the paper. If a referee has shown a consistent bias for or against certain manuscripts, the editor will try to add another referee with a compensatory bias in the other direction. Our experience has shown that referees tend to show very little bias and usually seem to do a fairly objective assessment. If a referee has already judged quite a few manuscripts in the recent past, or if he has declined or delayed most of those previously sent to him, he may be omitted. That is also the case if his responses were not well reasoned and persuasive. Analysis of our data suggests that the promptness of referees' responses correlated highly with their competence and performance as referees.

The following features will be added to an improved version. If the set of references cited by, or citing a reference in any of his own publications has a large intersection with the set of references in the manuscript--or to broaden it, if just the sets of authors of these papers have many names in common--then the referee is more likely to be assigned. The method of identifying related papers by commonalities in the cited references is known as bibliographic coupling (14), and was found to be remarkably well correlated with methods of indexing.

One of the most important new features to be added is this: A brief description of each newly arrived manuscript is broadcast to all well qualified referees in the pool, and they can volunteer to review it. The description contains the author's representation of the key question, issue Q, his main claim, a sample of his language, and his list of references. A good title and abstract together with key words and bibliography might serve this purpose. The title alone might also do.

A fairly important and fairly simple alternative addition, which will improve the referee selection and therefore influence the quality of papers to be published, is to computerize all information about referees and set up a computer program to match the profiles of referees and the index terms supplied by the author of a newly arrived manuscript. Thus, instead of the editor being required to rely on his memory and spend time arriving at the best judgment as to who the most qualified, competent referees are for a particular manuscript, the computer could do the matching much more quickly, more efficiently and with less chance of error or leaving out some potentially good referees. The computer could be programmed to automatically search through the pool of referees and list those most qualified. The editor could then make a decision based on this list. Or, suppose ten possible names were matched by the computer, he could send out this list to the ten people, together with a brief description of the manuscript, and ask for volunteers.

Computer conferencing (17) could take us one step further toward improving referee selection and manuscript evaluation. It has already been used by the editor in obtaining names of referees to add to the pool, where the pool was deficient in a particular area. The editor sent a message over the "Stanford University Medical Experiment - Artificial Intelligence In Medicine" (18) network to two colleagues, one in New Jersey, the other in California, asking them to recommend potential referees of great competence for a particular paper. Responses came within a day. The named candidates were also on the network and were contacted with a message. Again useful responses arrived within a day.

5. An Algorithm for Eliciting Good Judgments from Referees

In our present system, the only means of communication about the paper between the referee and the editor is the review form with its seven questions, the referee's data sheet and associated cover letter and courtesy correspondence (see Appendix A). A referee is rarely asked to reevaluate a revised manuscript. There is seldom more than one pass of such substantive correspondence from the referee to the editor, and there is no communication among the referees. Referees' comments are usually forwarded to the authors, anonymously, by the editor-in-chief. Occasionally they choose to reveal their identity and communicate with the author directly.

At this point we wish to introduce the potential of computer conferencing, not only for improving the quality of the referee's judgments, but for improving the pool of referees and referee assignments as well. Here is how it would work. When the editor receives a manuscript he enters the author, title, index terms, abstract (including the author's representation of his key issue and claim and a sampling of his language), and his bibliography -- all from the manuscript -- as an item in the conference. All referees in the pool with terminals can, upon signing on the conference, read the new items. In addition, those whose profiles match the index terms of the manuscript, will receive a personal message notifying them about the new manuscript as one of potential special interest. Those who do not have terminals or do not sign on frequently enough will be notified by mail.

Immediately upon reading a message, the recipient is asked whether he wishes to reply, delete, forward or ignore. On reading an item he is also asked to "vote" or react immediately. The reply (or vote) might be an expression of interest in refereeing the manuscript (see Appendix C).

What is to motivate anyone to referee a manuscript and to render his best judgment? Aside from a sense of obligation to participate in this vital aspect of science, a sense of power and the opportunity to be among the first to read a potentially important paper, there are more reasons not to do it: It diverts the reader from his work; it forces him to make painful decisions; it may hurt others and possibly him; he has to take a risk in the face of uncertainty; etc.

A positive reply from a referee would be announced by the editor as an item, if the referee agrees, and positive replies are tallied for each referee for display to all other referees in the pool. These tallies may become the currency in the social system that could help to motivate referees, provided such tallies become a recognized and valuable part of a professional resume and reputation.

If a referee decides to evaluate a manuscript, that decision will immediately be known to every participant on the conference, except the author, though everyone will be instructed to observe the confidentiality of this information. A referee has the option of disclosing his identity to the author as well, or he may choose to remain anonymous to everyone. In that case, messages can still reach him at his computer mailbox. Any participant in the conference may challenge the qualifications of a referee for a particular manuscript. Special messages will, of course, have been sent to the best qualified referees, based on the data in the system. There are obvious pros and cons in allowing the author to challenge the qualifications of a referee, and the balance of the argument favors giving the author some say. Conflicts are resolved by reducing the arguments for and against the sufficiency of a referee's qualifications to a few basic value differences, with the editor making the final decision, but making explicit his reasoning and his values.

6. Reaching an Editorial Decision

A referee enters his report as a message or as an item with a restricted circulation. It is filed in the mailboxes of all the other referees assigned to that manuscript. The recipients are requested to comment, to state their agreement or reasons for disagreement. The discussion is allowed by the editor to proceed, possibly using a variant of the Delphi technique (19) until some kind of consensus emerges or until it becomes clear to him that there is a deadlock. He may forward selected, edited excerpts from this discussion to the author, without revealing the commentators' identities, giving the author a change to join the discussion. He decides when to terminate discussion and inform all concerned of his decision, together with his reasons.

The resulting judgments, after repeated review and revision by qualified peers should improve. So should the final editorial decision. The arguments follow.

Consider the following "conventional" editorial decision algorithm. Let p be the probability that a conventially selected referee votes to accept a manuscript given that it would be accepted by most of the readers, and q the probability that such a referee votes to reject a manuscript that would not be accepted by most of the readers. Suppose that the editor chooses m referees and accepts it if they all vote acceptance. The probability of accepting an "acceptable" paper is then p^m, assuming they all vote independently. Let ϕ be the fraction of "acceptable" papers. Then a plausible performance measure is:

$$r = (p^m \phi + (1-p)q^m) \quad .$$

The algorithm we claim to improve on the conventional one, differs from the above in that a manuscript is accepted if at least m out of n highly interested and well-qualified referee-candidates volunteer to review the manuscript, each with probability v, and independently vote to accept it, with none voting rejection. Let P be the probability of such a referee voting, after discussion, to accept a manuscript that is "acceptable" by the higher standards of this group. The probability of accepting such "acceptable" manuscripts is now $\sum_{k=m}^{n} \binom{k}{k} v^k (1-v)^{n-k} p^k$. If one of the k volunteers, after discussion, still rejects it, it is rejected; if there are not at least m volunteers, then the editor drafts needed referees, as above.

Let ψ be the fraction of manuscript that are "acceptable" by these higher standards. Presumably, $\psi < \phi$. Let Q be the probability of one of the more especially interested and qualified referees rejecting such an "unacceptable" manuscript. The corresponding performance measure is now:

$$R = \psi \sum_{k=m}^{n} \binom{n}{k} (Pv)^k (1-v)^{n-k} + (1-\psi) \sum_{k=m}^{n} \binom{n}{k} (1-Qv)^k (1-v)^{n-k}.$$

To simplify the analysis, let m = 2. By the binomial theorem,

$$\sum_{k=0}^{n} \binom{n}{k} (Pv)^k (1-v)^{n-k} = (1-v(1-P))^n \quad \text{and}$$

$$\sum_{k=2}^{n} \binom{n}{k} (Pv)^k (1-v)^{n-k} = (1-v(1-P))^n - (1-v)^n - nPv(1-v)^{n-1}.$$

If v is small and n large so that vn is small too, then this is approximately

$$1 - nv^{n-1}(1-P)^{n-1} - (1-nv^{n-1}) - nPv(1-(n-1)v^{n-2}) \text{ or}$$

$$nPv[v^{n-2}P^{n-2} - 1 + (n-1)v^{n-2}] \doteq Pnv.$$

Thus, R increases with n while r does not. In other words, an expanded referee pool in a good peer-review system increases performance.

De Solla Price (private communication, April 18, 1977) points out that the scientific merit and technical competence of work reported in a paper is not the sole nor perhaps even the major determinant of quality as a contribution to scientific output as a whole. Authors are often surprised when the papers of which they are most proud are ignored or regarded as insignificant, while papers they had written, with certainty that these would have little impact, make quite an impression. To be a contribution to scientific output of high quality, a paper must make some impact. It should be cited. Whether a work "catches on" depends in some measure on chance and also on how effectively it is presented, an art many scientist-authors have yet to master. Though most scientific papers published in refereed journals are eventually cited (three years after publication is the peak year, and if a paper is not cited in the first five years, the likelihood of being cited drops drastically), it is usually the most recent papers that get cited. Often older or highly original papers are not cited. The notion of a "citation profile," for example (15), which includes both the references of a paper and the growing set of citations to it, for purposes of coupling with other papers, was first introduced by the author in August, 1961 (personal communication to F. E. Heart, who worked with M. M. Kessler), but neither the literature on bibliographic coupling nor co-citations links to that early notion.

An improvement in the final editorial decisions-- i.e., an improved scientific literature-- would be observed if: (1) there were an increase in appropriate citations that are suggestive of scientific use of the literature; (2) and more authors of papers likely to be cited in that way submit manuscripts. The total number of citations can increase but only up to a point; and (3) certain errors of omission were avoided.[2] There are only so many references an author can profitably cite, and the literature, even in a narrowly specialized field, is already too voluminous for a single author to utilize fully. To be sure, if total publication volume increases, it is because the number of authors increases and hence the total number of citations should increase at the same rate. But, if some published articles are cited proportionately more, it must be because other published articles are cited proportionately less. This indicates greater inequality in the overall distribution. As authors become more selective, many good papers would be unused. That would not, however, keep authors from trying.

There must be limits, too, on the rate of publication. We may be at the point where the introduction or expansion of new journals can economically be justified only if little used journals are allowed to die gracefully and make way for the new ones. Contrary to cherished ideals, the scientific journal exists primarily to serve the author rather than hypothetical readers, a point that de Solla Price has forcefully made. Unlike popular journals that cater to, and to some extent shape, the appetites of readers in search of entertainment, diversion, and ways of meeting personal needs, scientific journals aim to help scientists to do science, the output of which is the scientific literature.

Because of the change elements in whether a particular article is used and has an impact, we must publish a certain average number of articles that will not be used a great deal for every publication that is. This is a safety buffer, a margin of error. Improvements in referee-selection and in manuscript evaluation may reduce to some extent this uncertainty about what papers will ever be used, but not entirely.

The editor takes a risk in accepting or rejecting a manuscript. As is the case for any risk-takers, for controversial manuscripts, the good editor may simply be lucky and appear wise in retrospect, but he seems to be luckier more often than the less mature and less experienced editor. The worst editor, however, is one who does not make timely decisions at all. He waits for data to reduce the risk of error and that impedes the growth of science, which is amply supplied with self-regulating and error-detecting/correcting mechanisms.

Science advances and maintains its health, vigor, and strength via processes in which: Authors take risks in their choice of problems, methods, and claims; referees take risks in their judgments; editors take risks in their decisions; funders take risks in their investments, etc. If we want great science, we must be willing to let the best among us take bold, imaginative risks. Identifying, engaging and helping those best among us in their roles as editors, referees, and authors is perhaps the best way we can improve science and its output. The procedures described here are a step in that direction.

7. What Improvements Are Needed and Why

In highly structured and rapidly progressing fields of science, scientists appear to work together almost as if they were a team, despite vastly diverse organizational and other affiliations. The scientific literature is not the immediate mechanism through which this remarkable self-organization and communication occurs. It serves rather as an ultimate incentive and record. Scientists in invisible colleges communicate with those they most need to communicate with by telephone, mutual visits and informal exchanges of preprints. Those they need to communicate with are remarkably the same ones who need to receive the communication, and the exchange is often reciprocal. Moreover, there is great economy in these exchanges: Communicators miss few of those they need to communicate with and communicate to few they do not need to communicate with. Our experience with JACM referees has shown that many of the referees chosen to review a particular paper have already been sent a copy of the paper by the author. In the questionnaire sent to referees, those who said they read or referred to papers they reviewed all used the original manuscript or a preprint sent to them by the author, and not the final, published version. On the average, it takes between a year and eighteen months after a paper has been submitted to be published, often even longer.

The published literature serves more of an archival function. Awareness of published findings is not sufficiently current to compete with fast-spreading news in an invisible college. Its value lies largely in its relative permanence as a record, available for retrospective searching, and, above all, for citing, even if that permanence has only a half-life of five years or so. It is more an outlet for the author than a service to the reader. It is the output of science.

Studies of how many of the papers in a leading journal are never cited shows very low percentages, less than 1% for J. Am. Chem. Soc. in a six year period (9). For the scientific literature as a whole, estimates of that fraction have been estimated to be about 10%[3]. If it is increasing as the total volume of publications increases, that may signal an increase in the number of journals consisting mainly of rejects or anticipated rejects from the more prestigious journals[4]. Perhaps there is a need for a number of such journals. Whether the market mechanisms and political and social forces that determine the survival of such journals are functioning adequately to maintain the vigor of science is an open question.

Insofar as much of science is cumulative -- the lastest findings transcend, encompass, build upon previous ones -- most very old journal articles have little value for contemporary science. They may be indispensible for the history of science. But they need not be made part of the scientific communication system. Referees should judge the usefulness of articles over their expected lifetimes. Should a manuscript judged to be "a century before its time" be published? Perhaps, and perhaps in a journal of such articles, slated for re-examination a century later. There should probably be a limited number of futuristic journals. Referees should ask about each article, who, if anyone, is likely to use it and when.

Let us raise again the question about who benefits from the publication of a manuscript--the author or the scientific community. In the former case, one would tend to publish any good paper, regardless of its current value to the scientific community, whereas, in the latter case a much tighter, stricter control of manuscripts accepted for publication would result. The question, however, assumes that there is a distinction between authors and the scientific community, when they are essentially the same. An author is a member of the scientific community in the temporary role of a contributor. But he necessarily builds on prior contributions. Hence he is also a citer. As an author his interest is to get his work published. As a reader/citer his interest is to save his time, to efficiently identify and use the best published literature that concerns him and, ideally, only that. Since he cannot satisfy both of his conflicting interests, he tends to give top priority to his role as author. It would follow that we should err on the side of publishing too much, and publish any good paper.

If the published literature is the output of science, then its faults are the faults of science. A science editor plays a vital role in science. He performs this duty at the expense of doing original science and there must be compensating motives. One is that he recognizes his obligation to contribute part of his time to this essential function. Another is that if he were to refuse, another possibly less competent editor would do it in his place, a circumstance he views more costly than that of contributing his time. A third is the recognition and prestige this earns him among his peers; it helps him maintain his leadership, it strengthens his resumé, and may help him earn raises and promotions. A fourth is that it is an easy way of keeping very current with work in his field, being among the first to learn of important discoveries; he remains a key member of his invisible college. Finally, it may help him exert intellectual influence and shape an area of investigation.

Quite possibly, science as well as its literature is not in dire need of improvement. It is, as it always has been, one of the most remarkable success stories in human history. That science, as well as its literature, is today probably overspecialized, excessively fragmented and under or mis-utilized, does warrant close watching as does the cost of journals. Most probably, there will emerge, in a natural way, mechanisms that provide for screening, evaluation and synthesis of specialized works as rapidly as such works are generated, and for bringing them into some coherent unity as fast as they are scattered. An example might be the idea of an integrative journal system in which the publication of what are now piecemeal articles would be delayed until a coherent series of research works can be synthesized into a simple major article, as proposed by Garvey and Gottfredson (8). This idea might be modified by ensuring such a synthesis article in addition to timely publication of its component fragments.

It is likely that the literature will continue to grow exponentially as long as population does. Even though the countries contributing most to world scientific output are decreasing their rate of production, newly developing countries are compensating for that decline by extraordinary growth. Such total growth places increasing burdens on libraries and information services that attempt to maintain current awareness, help with retrospective searches, state-of-the-art reviews and exert bibliographic control. Rising cost of scarce materials and human labor also make scientific publications much more costly to produce, distribute and acquire.

If these pressures provide a push for innovations to improve the quality of what is published by improved referee-selection and manuscript evaluation, then the trend toward decreasing costs of information technologies and their imaginative use in revising procedures for referee selection and manuscript evaluation provides an additional pull. This push-pull combination should propel these innovations forward were it not for major barriers, costs and resistances.

8. Barriers to and Cost of Proposed Innovations

An experienced, competent editor can, after scanning a manuscript in less than half an hour, select referees from his memory in a few seconds. The efficiency, simplicity and effectiveness of this conventional and time-honored procedure is very hard to rival. He may, however, not remember off-hand the most qualified and he may err. We have shown that effectiveness could be improved by extending and amplifying the editor's memory and systematizing his procedures. That requires setting up, maintaining and using filing systems of varying degrees of sophistication. Setting up a new filing system, especially if it is computerized, incurs a considerable start-up cost and that may be a major deterrent. Maintaining a file is a tangible overhead cost that cannot be recovered, and using a file is an operating cost, the justification of which is felt only very indirectly, months later, if at all.

To use a computer terminal, an editor must enjoy using a computer system for reasons other than improving his editing work. He would most probably regard the trouble of turning aside from his desk -- even if the terminal were as close as his telephone -- signing on, etc., as a diversion from the task of reading a manuscript and assigning referees, one that impedes rather than helps him get on with the task. It is not merely the cost of the extra time it takes the editor to sign on that might deter him. Signing on a computer system with which he is not fluent may cause him to lose his train of thought about who might be a good reviewer while he reads the paper, because he has to concentrate on what keys to press and about the mechanics, as well as the different intellectual mode of using the computer system.

If he uses the computer system so frequently that he is fluent, then the mechanics become second nature, and switching to another mode may not be a burden. It is difficult, however, to maintain the frame of mind--best described as "leisure" when an editor reads a manuscript in a conventional way-- when reading through a manuscript, either on a hard copy while a user is signed on, or on-line on a cathode ray tube, while conscious that the on-line system is costing him money each minute it is on. That cost is totally absent (or forgotten as the past cost of supporting typing and retyping) if the manuscript were just being read without simultaneously being signed on to a computer system. As computer costs continue to go down, perhaps this "leisure" frame of mind can be maintained. Perhaps a more valuable new frame of mind -- "structured leisure" -- could emerge out of the present "leisure" + "pressured leisure". Eventually, editors may find the benefits sufficiently attractive to overcome these barriers, but that can only happen after they have used the system enough to experience the benefits. That, however, is a feature common to all technological innovations.

The cost of an on-line system is now about \$6 per hour including the use of long-distance telephone charges over a network in the U.S. and Canada; this would be about \$2 for a typical 20-minute on-line session or about \$6 for a typical session involving transatlantic lines. As volume increases and technology improves, these costs may well continue to decrease for a decade or two at the dramatic rates that characterized cost reductions in computer processors (due to large scale integration of the solid state electronics). Setting up, maintaining and using the kinds of files we have discussed in a computer system is well within the state of the art, but it is an added cost-- both for the use of hardware, software and human labor (both clerical and managerial). Processing the incoming manuscript to determine how similar it is in language, references and problem-representation to each paper in a standard test set could also be partially automated if the manuscript arrives in machine readable form and in a code and format acceptable to the system, though the scientific problems underlying these procedures should not be underestimated.

The computer art is still one of the least standardized with regard to just those aspects. Some authors might ship a cassette tape to the editor, produced as a by-product of typing the manuscript, as by WANG, MTST or TI-700 dual cassette machine, for example. (No two of these three tapes or their codes are compatible. At a cost, however, each can be converted to the other.) Other authors might store their manuscripts as a series of connected "items" in a file on a computer conference and then send a message to the editor, who is also on the conference, asking him to read the items and assign referees to read them. Other authors might send the manuscript in conventional typed form to the editor who would then have to get it entered into a computer system.

Editorial correspondence, of course, could be made an easy automatic by-product, including alerting and reminder messages. Doing this by computer might become important as secretarial services become scarcer and more costly. That, however, is an insignificant aspect of referee-selection and manuscript evaluation, and is easily subsumed among the many functions of an editorial processing center (1).

The costs in the above operations may be sufficiently substantial to dampen the enthusiasm of even the most imaginative editor who can visualize how he, the referees, the author and special editorial assistants might improve the content and style of manuscripts by making changes via editing programs at the instant they think of it. An even greater barrier than the cost is the uncertainty about how these costs and benefits will be distributed. A scientist-editor generally receives no pay for his labor and for the supporting infrastructure except what his employer contributes to help him discharge his professional duties. Editorial work by a scientist rarely contributes to his employer's revenues, and the employer will naturally try to minimize that overhead. Improvements in quality of published manuscripts are of benefit to the scientist-editor and even more indirectly to all the scientists in the area. It is not likely to have much direct effect on the journal's circulation. Hence, the publisher is not likely to regard his benefits great enough to underwrite most of the costs. The benefits are, indeed, very intangible and widely diffused, while the costs are concentrated where the benefits are least. Insofar as the ultimate beneficiaries are among the general tax-payers, the cost of such innovations should probably be underwritten by national governments and, possibly, multinational consortia.

The notion of an editorial processing center (3), regarded as a computer facility for use by authors, referees, editors and publishers, may have some potential if it would be shared by a number of different user groups. While publishers may be attracted by the promise of cost and time savings, they may not be willing to relinquish the degree of autonomy and control they now enjoy. If a plurality of smaller "centers," based on mini or smaller computers would be economically viable for individual publishers or user groups, then cost savings might be realized without the necessity of pooling efforts and resources into a single center. Standardization would become an even more important issue, especially standardization of references.

The idea of editorial processing centers has been presented as if it were a solution in search of a problem, an attempt to fund important new uses for computers. The use of computers is likely to be important but not all-important. It should not dominate the basic research that is needed to explain the basic phenomena of science and its outputs and to provide a sound foundation underlying systems design.

Probably the greatest barrier to undertaking the kind of improvements that are suggested and which are technically and perhaps even economically feasible stems from psychosocial, organizational aspects. Professionals in almost all fields measure their power and prestige by the people over whom they exert control. Uncertainty about the redistribution of control may be the major obstacle to change. There is hardly a scientist-editor who would prefer an attractive computerized system to do all the above to an attractive secretary or editorial assistant using present methods, if he had to control the computer himself.

Professionals and managers prefer to interact with other people, and to do so verbally rather than in writing, to an extent that increases with their rank in the organization. Having to communicate, instead, with a computer terminal, and via keyboard, is often interpreted as dehumanizing, demeaning and as a demotion. People, especially privileged people, who are forced into communicating with equipment, rather than other people, perceive this as a status-lowering and they may subconsciously sabotage its use. Previously disadvantaged people, on the contrary, perceive the use of new technologies as status-raising, use the technology eagerly and imaginatively. Editors, authors, and referees of scientific manuscripts do not yet, however, count many such previously disadvantaged people among their numbers.

In a computer conference, the influence of the organizer or chairman differs markedly from what is the case in a conventional face-to-face or social situation. If each of a group of referees can select manuscripts they want to evaluate, and if referees can communicate with one another and even the author more freely and directly, then the editor's control is decreased. The author's options for appeal are increased. Social and political forces may play a greater part in the acceptance/rejection decision along with more thorough technical and intellectual criticism. Uncertainty about whether the end result, on balance, will be better may be overshadowed by the fear on the part of an editor that he is relinquishing control over an efficient and effective procedure for one that has an element of risk and over which he has less control. Depending on whether he has faith in the efficacy of democratic and self-correcting mechanisms in science or whether he adheres to a more elitist view that most and the best of science is controlled by just a few, he may regard steps toward the proposed innovations as a boon or a threat.

9. How and Why, On Balance, The Proposed Innovations Leads to Improvements

The promise of new methods of selecting referees and evaluating manuscripts, stimulated by the potentialities of new information technologies is alluring. But the barriers and constraints to its use are severe. Are secular trends such that in time the barriers will be overcome? Are the potentialities of newer procedures likely to be realized as sufficiently attractive to outweigh, on balance, the barriers? If so, what steps seem worthwhile taking now, with what likely consequences?

In the long run, it is almost certain that human labor costs will be much higher while the costs of using information technologies will be lower. There are well over a million computer terminals in the U.S. now. Over two-thirds of the respondents to our questionnaire said they had terminals and would be willing to participate in a computer conference to referee manuscripts. Within three years at most--by 1980--most scientists in their roles as authors, referees, editors will be able to easily communicate over international networks with some computer capabilities--at least message storage and forwarding, file management, and some editing. They may even have output terminals in their homes (in attachment to a home TV set) and some audio response facilities over their telephone (Votrax devices are available at costs less than $1000). Data entry devices, e.g., electronic pens that accept handwriting input, will soon decrease in cost. Secretarial duties and procedures are likely to change drastically as office automation takes hold.

The division of labor between man and machine is likely to shift as machines can outperform people in most of the routine clerical chores and even some tasks requiring intelligence now performed by most white-collar workers. People will want and be required to devote more of their energies to making subjective judgments, coping with risk and uncertainty.

Those with expertise and direct control over, and contact with, the information technologies--including many of the previously disadvantaged--are likely to exert an increasing degree of control wherever these technologies are making an impact. It may be the fear of this trend that threatens professionals in the knowledge industries (e.g., publishing, mass media, health care, education) to resist the widespread introduction of the technologies. The technologies, however, may well become much more humanized, evidenced, for example, by the Stanley House criteria (26). If that is so, the presently privileged professionals need not feel as threatened: They are likely to continue interacting with other people, even if their control is lessened; their judgments are likely to be needed even more sorely than before. In sum, there is likely to be a greater degree of equity in the distribution of control. Thus, sooner or later, information technologies will impact this area as well.

Whether that will result in a net improvement in the quality of published manuscripts depends on whether we can learn enough from experience with pilot experiments in the few years before the technology is in widespread use. What we can learn from experience with a very modest volume in one area of one journal is illustrated in this paper. We cannot extend our conclusions to larger-scale journals or groups of journals. That should be done next.

We recommend three stages of activity. In the first stage, we suggest immediate steps toward an agreement among editors for maintaining excellence in the current practice of referee-selection and manuscript evaluation. The issues on which agreement could be reached are rather evident (13) and include: Having at least two referees for each manuscript; giving referees detailed guidelines, perhaps standardized as the form in Appendix A; requiring referee reports within 4 weeks; acknowledging and rewarding referees; informing referees of the fate of manuscripts they refereed; possibly offer them some kind of selective dissemination about articles in their field in the journal as compensation. There should be more two-way communication between authors and referees, possibly among referees and with the editor. Unrefereed journals should be identified as such. Possibly there might be an editors' clearinghouse to identify previously refereed manuscripts that were rejected by another journal or that were previously published or simultaneously submitted to several journals. Exemplary practices, such as the procedures used by Science (10), should be disseminated. Attempts should be made to monitor if the selected referees are representative of the journal's users. A log of each referee's performance should be maintained to evaluate his competence, impartiality, promptness, sense of responsibility, incisiveness, thoroughness, and interest, and this should be used in assigning referees and recruiting new ones.

The second stage, starting now and extending over a decade or so, is one of both basic and applied research, in which some of the present practices are modified and automated under controlled condition. Priorities must be kept in mind. Such experimental systems exist primarily to establish conditions that improve literature, secondarily to improve it. Only in the third stage, which might actually be thrust upon us in a year or two -- before we are ready-- should we aim to improve the processes in practice. That is dangerous because in the absence of understanding, our interventions might do more harm than good. Computer conferencing has great potential for reducing the publication log; it also has constraints.

For the second stage, research, we recommend that a multinational consortium of journals solicit research support over the next three years from science-supporting foundations such as Volkswagen, Club of Rome, National Science Foundations, etc., to implement and test the claims made for these and better procedures for selecting referees and evaluating manuscripts as reported in this paper. A more thorough mathematical analysis replacing the probabilitistic approach presented by the calculus of possiblities (e.g., 28) is needed. So is an analysis of the complexity of the algorithms that are proposed. It would be premature to attach such an experiment directly to the more ambitious proposals for computerized editorial processing centers (1). Possibly an experimental system that deals with a number of interdisciplinary journals focussing on articles that rely on computer-stored data bases and computer analyses would be a good starting point. The chances are excellent that much will be learned and this may well lead to significant improvements in the quality of what is published.

FOOTNOTES

1. If N denotes the total number of articles, then the ratio of those that are good is $\sqrt{N}/N$, which is $1/\sqrt{N}$, and that tends to 0 as N tends to infinity. The worldwide output of chemical and chemical enginering and articles and patents, for example, doubled every 8.2 years in 1966 compared with a doubling period of 9.5 years in 1960 (2). In several fields of science, such as chemistry and physics, output has leveled off in the 70's (22, p. 50).

2. A serious error of omission in the published literature is illustrated by the tendency, in fields where statistical tests of significance are used, to publish only results that are statistically significant, and not even submit for publication results which fail to reject a null hypothesis (e.g., only 1 in 100 articles in Experimental Psychology seems to report it) (27). If such results cannot be cited, then investigators would unknowingly duplicate such experiments until, by chance, one of them obtains a rejection of the null hypothesis, which he then publishes. The published literature then gives a false picture of science.

3. The actual figure is not accurately known. One guess of 50% was printed in Science, for all scientific journals, but it seems to have no empirical basis. E. Garfield, judging from his experience with Science Citation Index and the scientific journal literature estimates about 30% (personal communication, 4/25/77).

4. In sociology there appears to be a rank order of journals, with the top-ranking journal rejecting over 80% of the manuscripts submitted, but with 60% of all such manuscripts being published in some journal within 5 years (21).

ACKNOWLEDGMENT

We thank Flora Wallace and Robin Crickman who in the past have helped to carry out some of the procedures described here for evaluating manuscripts submitted to the JACM in the area of information retrieval, for which M. Kochen is the area editor.

REFERENCES

1. Aspen Systems Corporation, Editorial Processing Centers: Feasibility and Promise. Aspen Syst. Corp., Rockville, Maryland, undated, post-1972.

2. Baker, D. B. "Chemical Literature Expands," Chem. & Engin. News, 44, No. 23, June 6, 1966, p. 84.

3. Bamford, H. E. "A Concept for Applying Computer Technology to the Publication of Scientific Journals," J. Wash. Acad. of Sci., 62, 1972, 306-314.

4. Branscomb, L. M. "Is the Literature Worth Reviewing," Sci. Res., 3 No. 11, 1968, p. 49.

5. Crane, D. "The Gatekeepers of Science: Some Factors Affecting the Selection of Articles for Scientific Journals," The Amer. Sociologist, 1967, 195-201.

6. Etzioni, A. "The Need for Quality Filters in Information Systems," Science, 171, 1971, p. 133.

7. Faegri, K. "Report to the First General Assembly on Refereeing in Scientific Periodicals," ELSE (Association of European Editors of Biological Journals), unpublished, 1970.

8. Garvey, W. D., & Gottfredson, S. D. "Changing the System: Innovations in the Interactive Social system of Scientific Communication, Information Processing and Management, 12, No. 3, 1976, p. 165.

9. Ghosh, J. S., & Neufeld, M. L. "Uncitedness of Articles in the Journal of the American Chemical Society," Information Storage and Retrieval, 10, 1974, 365-369.

10. Goudsmit, S. A. "What Happened to My Paper," Physics Today, 22, 1969, 23-25.

11. Herring, C. "Distill or Drown: The Need for Reviews," Physics Today, 21, 1968, 27 33.

12. Ingelfinger, I. "Peer Review in Biomedical Publications," Amer. J. Medicine, 56, 1974, 686-692.

13. Juhasz, S. "Acceptance and Rejection of Manuscripts," IEEE Trans. Prof. Communication PC-18, No. 3, Sept. 1975, 177-184.

14. Kessler, M. M. "Bibliographic Coupling Between Scientific Papers," Amer. Documentation, 14, No. 1, Jan/Feb. 1963, 10-25.

15. Kochen, M. Principles of Information Retrieval, Los Angeles: Melville/-Wiley, 1974, p. 25.

16. Kochen, M. "Quality Control in the Publishing Process and Theoretical Foundations for Information Retrieval," in Software Engineering, J. Tou (Ed.), New York: Academic Press, 1971, 19-53.

17. Kochen, M., chairman. "On-Line Intellectual Communities," introduction to papers by J. Vallee, K. R. Hammond, R. Crickman, K. Samuelson, & L. Uhr, panelists on computer conferencing at Annual Meeting of the Society for General Systems Research, 1977; published in Proceedings.

18. Levinthal, E. C., Carhart, R. E., Johnson, S. M., & Lederberg, J. "When Computers Talk To Computers," Industrial Research, Nov. 15, 1975, 35-42.

19. Linstone, H. A., & Turoff, M. The Delphi Method: Techniques and Applications, Reading: Addison Wesley, 1975.

20. Manheim, F. T. "Referees and the Publication Crisis," EOS, 54, 1973, 532-537.

21. McCartney, J. Proceedings of the First International Conference of Scientific Editors. April 24-29, 1977, Jerusalem. Dordrecht: Reidel, 1977.

22. National Science Board, Science Indicators 1974, NSF, U. S. Govt. Printing Office, Washington, D. C., 1975, p. 50.

23. Noble, J. H., Jr. "Peer Review: Quality Control of Applied Social Research," Science, 185, 1974, 916-921.

24. Price, D. de Solla. "Some Remarks on Elitism in Information and the Invisible College Phenomenon in Science," JASIS, 22, 1971, 74-75.

25. Small, H., & Griffith, B. C. "The Structure of Scientific Literatures I: Identifying and Graphing Specialties," Science Studies, 4, 1974, 17-40.

26. Sterling, T. D. "Humanizing Computerized Information Systems," Science, 190, No. 4220, Dec. 19, 1975, p. 1169.

27. Sterling, T. D. "Publication Decisions and Their Possible Effects on Inferences Drawn from Tests of Significance -- Or Vice Versa," J. Amer. Statistical Association, 54, March 1959, 30-34.

28. Tahani, V. "A Fuzzy Model of Document Retrieval Systems," Info. Processes and Management, 12, No. 3, 1976, p. 177.

29. Ziman, J. "Information, Communication, Knowledge," American Psychologist, 26, 1971, 338-345; also in Public Knowledge. Cambridge: Cambridge University Press, 1968.

30. Zuckerman, H., & Merton, R. K. "Sociology of Refereeing," Physics Today, 2, 1971a, 28-33.

31. Zuckerman, H., & Merton, R. K. "Patterns of Evaluation in Science: Institutionalism, Structure, and Functions of the Referee System," Minerva, 9, 1971b, 60-100.

APPENDIX A

JOURNAL OF THE ASSOCIATION FOR COMPUTING MACHINERY

1133 AVENUE OF THE AMERICAS NEW YORK, NEW YORK 10036 212 265-6300

acm

EDWARD G. COFFMAN JR., Editor-in-Chief
MYRTLE R. KELLINGTON, Executive Editor

Reply to: Manfred Kochen
Area Editor for Info. Retrieval
University of Michigan
Mental Health Research Institute
205 Washtenaw Place
Ann Arbor, Michigan 48109

REFEREE EVALUATION

TITLE: ____________________

____

AUTHOR: ____________________

REVIEWER: ____________________

Please return this form with the manuscript after you have evaluated it. Feel encouraged to mark your comments, criticisms, and questions to the author on the manuscript as extensively as you like. In addition, formulate comments for anonymous transmittal to the author on a separate sheet. Please answer as many of the following questions as you can and wish by placing cross-marks (X) at an appropriate on the vertical/horizontal "scales" that best express your judgment; please state reasons for it in your own words or substitute your words for the suggestive phrases on the "scales."

(1) How clear, organized and well-written or readable do you find this paper?

— Absorbing; pellucid; masterpiece of clear thought and composition, and made me realize that I didn't know something I need to know and didn't even think of looking for.

— Interesting, clear; my attention didn't wander; helped me find some knowledge I was looking for.

— Required effort to maintain attention; I didn't need to learn what I was being taught.

— My attention wandered; I couldn't read it in one sitting; I would have preferred not to have to read it.

— Too boring; irritating, hard to read through, except I promised to; would love to get out of reading it.

(2) How do you rate the importance of this paper with regard to:

(a) The significance of the key questions, issues, concerns it raises, if any?

Extremely important in any discussion of this topic	Of average significance	The issues raised are of little importance or not pertinent in discussing this topic

(b) The depth of insight or light it brings to bear on these issues; power of ideas and methods?

Penetrating analysis; profoundly enlightening, inspiring	Reasonable analysis or a modest technical contribution	Very shallow analysis or lack of appreciation of the nature of the problems in this area; no stress on ideas; intellectually sterile.

(c) Would you use and cite this paper in one of your own papers?

Definitely, in the next paper I write	Almost certainly, if not I, then one of my colleagues	Probably; I wish I had seen this paper before	Probably not	No

APPENDIX A (continued)

(3) How persuasive to you find the main claims and the arguments or evidence the author presents to support them?

(a) Empirical papers:

Claims valid and supported — Claims false and unsupported

(b) Mathematical papers:

All theorems and proofs are correct — All theorems and/or proofs are incorrect or defective

(c) Other papers:

Compelling — Persuasive — Weak

(4) How original, novel, substantive do you find this paper?

Contains very much of solid substance which is totally new to me — Nothing new to me; little substantive content

(5) To what extent does this paper fit the journal's objectives in this area of information retrieval of:

(a) Advancing the sciences and arts of information processing by reporting results of original, analytical research on fundamental questions about the representation, storage, organization, retrieval, and communication of non-arithmetic information.

Perfectly; ideal for journal — Like most articles in journal — Definitely not

(b) Promoting free interchange both among specialists and among the public.

In the best scientific and professional tradition — Only a few fellow specialists are likely to be reached

(6) If you would like to express your overall opinion about the suitability for publication of this paper, circle your recommendation:

A - Publish; a must

B - Publish, if and when there is space

C - Publish, after some editing

D - Reconsider again, if resubmitted with changes (as indicated)

E - Reconsider if resubmitted after a complete rewrite into a different paper

F - Reject or submit to another journal

(7) If you wish, express your confidential evaluation of this paper and your reasons for the above judgments. This is for the editors' use only and will not be forwarded to the author.

APPENDIX B

AN EXAMPLE OF ALL INFORMATION RECORDED ABOUT A REFEREE IN THE POOL

Name: ____________________ Date Entered Pool: ____________________

Address: ____________________ Source: (a) Recommended by referee in pool: ____________

____________________ (b) Recommended by Editor ______

____________________ (c) Cited in manuscript ______

____________________ (d) Recommended by Author ______

TEL. NO.: ____________________

EXAMPLE OF REFEREE'S PROFILE	Degree of Competence as Judged by Referee (date)	Subsequent Update by Referee (date)
Structural Linguistics	A	A
Computational Linguistics	A	A
Computer Languages	B	B
Information Retrieval	C	B
Text Processing	C	B
Artifical Intell.		C

Additional comments by Editor ____

REFEREE'S PERFORMANCE LOG (Placed on a separate card, one for each manuscript sent to a referee)

Title of m/s: ____

Date m/s sent to referee: ____________ Date m/s review received from referee: ____________

Willingness to Referee: Yes/No

Speed of Response: ________ Date promised: ________ Date received: ________

Quality of Review:

Bias: Yes/No

Thoroughness: (additional pages of comments) ____________

Degree to which judgments justified: ____________________

Superficial: Yes/No

Additional Comments by Editor: ____

____

____

APPENDIX C

SAMPLE COMPUTER CONFERENCE

(Underlined items are typed in by the user, here called REF for referee, at a terminal, after he dialed a tel. nr. on a MODEM and received acknowledgement that he was connected, by automatic appearance of first line; each message sen by the user is terminated by pressing the key called "Carriage Return").

```
ENTER TERMINAL TYPE:     7

      MTS: ANN ARBOR (TI 700,LF10-CA06,02288)

# SIG CCID PW=XXXX

# TERM, NORMAL, UNIV
# **LAST SIGNON WAS 13:21:08
# USER "REF" SIGNED ON AT 15:59:41 ON WED APR 20/77

# SOURCE JACM

CONFERENCE FOR CCID:JACM
LAST ITEM IS 51
LAST YOU HAVE SEEN IS 50
LAST MESSAGE IS 2
DO NEXT?  M

     1_ MESSAGE FROM: EDITOR, IR, JACM 18:09 03-25-77 XXID:ED

     2_ MESSAGE FROM: ANON

WHICH MESSAGE?  1

1 LINES=10           16:01:15            04-20-77
MESSAGE FROM: EDITOR, IR, JACM          18:09  03-25-77 XXID:ED
```

"On local Properties of Binary Trees" by R. Sprugnoli, from Italy, has been received by JACM. It is about 21 pages, many with solid mathematics, mostly combinatorial analysis and graph theory, applied to retrieving information from a data base with weighted binary search trees. Offers a simple new technique to study such trees for large data bases.

INDEX TERMS: Binary Trees; Searching; Information Storage and Retrieval; Data Organization; Paging Binary Trees; Combinatorial Properties of Trees.

CITES: 10 Refs, including Knuth, Severance, Wilson.

Will you review in 3 weeks?

```
ENTER FORWARD, REPLY, DELETE OR IGNORE:    R

ENTER YOUR MESSAGE.  TERMINATE WITH $ENDFILE.

> OKSsend it over
> $ENDFILE

TO EDIT COPY OF MESSAGE ENTER E:  E
:ALTER 13 'Ss',S'

13 OK, SEND IT OVER
```

APPENDIX C (Continued)

```
:STOP
OK? TO SEND TO EDITOR, IR, JACM:  OK
MESSAGE SENT
ENTER RECIPIENT (OR PRESS RETURN)
ENTER FORWARD, REPLY, DELETE OR IGNORE:  D
DELETED

2  LINES = 6            16:02:30            04-20-77
MESSAGE FROM: ANON          15:03           05-01-77

    "I agree with your comments on Rice's paper, but think he should be given
a chance to correct the error on p.15 and get it accepted.  Would you concur?"

ENTER FORWARD, REPLY, DELTE OR IGNORE:    R
ENTER YOUR MESSAGE, TERMINATE WITH $ENDFILE
>OK.  CHANGE MY RECOMMENDATION FROM 'REJECT' to 'ACCEPT CONDITIONALLY'.
>$ENDFILE
 OK? TO SEND TO ANON:  OK
 MESSAGE SENT
ENTER RECIPIENT (OR PRESS RETURN):      ED
 OK? TO SEND TO EDITOR, IR, JACM:       OK
MESSAGE SENT
ENTER FORWARD, REPLY, DELETE OR IGNORE:  D
DELETED
DO NEXT?   ITEM 51
ITEM = I51                  16:13:30     04-20-77   LINES = 9 PRIME = NEXT =
        (same item as message 1 above)
TOO FEW VOTES ON THIS ITEM AS YET
GIVE YOUR FEELINGS AND/OR CURRENT VOTE ON ONE LINE
I ALREADY REPLIED TO EDITOR'S MESSAGE ON THIS ONE
DO NEXT:  STOP
YOU ARE LEAVING CCID:REF.  ENTER SIGNOFF TO LEAVE COMPUTER.
#EXECUTION TERMINATED
# SIG $
K1A7 15;59:41-16:32:36 WED APR 20/77
#      $1.41        (cost of this session)
#    $320.06        (amount left in account CCID)
```

REFEREEING RECONSIDERED: AN EXAMINATION OF UNWITTING BIAS IN SCIENTIFIC EVALUATION

Michael Gordon
Primary Communications Research Centre
University of Leicester, Leicester LE1 7RH, UK

ABSTRACT

This paper argues that while the majority of referees may attempt to give unbiased evaluations, they must inevitably bring their own particular sets of scientific predispositions to bear upon the recommendations they offer to journal editors. Qualitative and quantitative evidence is presented which demonstrates that resulting biases are systematically observable in physical as well as social and behavioural science.

It is concluded that editors must be sensitive to the intellectual affiliations and preferences of potential referees, and bear these in mind when selecting evaluators and interpreting their reports.

Publication of papers in research journals is widely accepted as having a vital role to play in the continuance of science as a social and intellectual activity. In particular, it is recognized as being both the medium through which scientists are able to establish and advance themselves professionally, and the means through which contributions are made to the archives of scientific knowledge.

This being so, the systems used by research journals to evaluate and select manuscripts for publication are seen to be of vital importance to science. And, not surprisingly, one finds the behaviour of individuals taking part in these systems, especially the behaviour of the anonymous referees consulted by journal editors, to be a matter of great sensitivity and heated debate.

Within this debate, accusations of deliberate referee bias are sometimes presented, usually by aggrieved authors. Such deliberate bias is not, however, thought to occur very often, and as it can usually be identified by the editor, with or without a second referee's report or author's letter of appeal, its influence upon the overall pattern of growth of a discipline's literature can be considered minimal.

The more important area of possible bias might therefore be that of subconcious bias, as exercised by the majority of referees who, one assumes, try to give sound, reasoned and "as objective as possible" judgements on the quality of scientific work. While these referees are not deliberately biased in their evaluations, it remains legitimate to enquire whether they might not have specific predispositions which affect the outcome of their refereeing activity.

One would anticipate that if this type of subconcious bias does exist, it will most clearly show itself in the performance of referees evaluating

Miriam Balaban (ed.), Scientific Information Transfer: The Editor's Role, 231-235. All Rights Reserved.

the work of authors who are identified with rival schools of thought and who, consequently, hold different beliefs in relation to their areas of research.

Such schools are more likely to be found in the behavioural and social sciences, and we may therefore quote some research that has been carried out on a community of American psychologists.[1] This set out to assess the nature and extent of biases involved in the making of referee evaluations, when referees were affiliated to identifiable schools of thought.

Seventy-five anonymously authored manuscripts were distributed to the referees of a U.S. psychology journal; all papers having the same bibliography, introduction and methods sections. The data sections were, however, adjusted to give results which were either congruent with, or contradictory to, the referees' presumed theoretical standpoints. When the referees' reports were returned, evaluations were seen to be dramatically affected by the particular slant of the data. Manuscripts which were identical in topic and procedure were recommended for publication when their results were congruent with referees' theoretical positions, but were consistently rejected when the results were opposed. Referees were also found to rate the same experimental procedures as significantly more adequate when they were alleged to yield the preferred type of data. Finally, referees were seen to scrutinize much more closely those manuscripts that bore results conflicting with their preconceptions.

So it is apparent that when non-complementary theoretical positions co-exist within a given research area, referees sometimes quite discernibly return evaluations which are biased by their own predispositions. In consequence the selection of referees can be critical in dictating which papers are recommended for publication.

This has been shown in a study of three American social science journals: *American Sociological Review*, *American Economic Review* and *Sociometry*.[2] Analyses of the past and present institutional affiliations of authors and evaluators were made for each of these journals over a number of periods. For each journal it was then found that as the proportion of evaluators chosen from certain groups of institutions increased, so the proportion of successful authors from those groups of institutions similarly increased.

It was concluded by the authors that the evaluation of papers is 'affected to some degree by non-scientific factors', and two possible interpretations were offered.

(1) As a result of academic training, editorial readers tend to respond to certain aspects of methodology, theoretical orientation and mode of expression in the writings of those who have received similar training.
(2) Doctoral training and academic affiliations influence personal ties between scientists which in turn influence their evaluation of scientific work. Since most scientific writing is terse, knowledge of details not usually included in journal presentations may influence the reader's response to an article'.

Their data were found to support the first of these interpretations more strongly than the second. In all three journals, the majority of authors and evaluators had degrees from U.S. major universities. This suggested that evaluators and contributors shared common viewpoints based on training, rather than personal ties. However, as the academic affiliations of evaluators became more diverse, the academic affiliations of successful authors became more diverse also. This suggests that personal ties may play a secondary role.

One might, of course, look at the findings of the two studies above and suggest that the systematic biases which have been observed are only characteristic of social and behavioural sciences, where a plurality of non-complementary methodological and theoretical positions can co-exist. Some might argue that such findings as have just been presented could not have emerged from studies of evaluation in the natural sciences, where such fragmentation is assumed not to occur.

Data comparing the frequency of agreement between pairs of referees do not, however, reflect totally uniform objective criteria being applied in the evaluation of research papers in natural sciences.[3]

AGREEMENT BETWEEN PAIRS OF REFEREES

	No. of pairs	Frequency of agreement between pairs
Physical Science	172	93%
Social Science	193	73%
Biomedical Science	1572	65-75%

Rather, they show that varying degrees of subjectivity enter all evaluations of the quality of scientific work. As such, one might question the biases displayed in the refereeing judgements of what appears to be the area of science in which the highest levels of consensus are displayed - physical science.

It would seem that even in this area, where clearly identifiable schools of thought are rarely in evidence, different sets of evaluators display slightly differing sets of preferences, predispositions and preconceptions of what constitutes 'good science'. While these non-concurring attitudes may reflect only small departures from the consensus of the discipline in question, they are, it seems, still detectable. During a series of interviews which we conducted, the editor of a prestigious journal covering the physical sciences offered the opinion that 'Scientists cluster in schools of varying degrees of visibility. The problem resulting from this clustering comes when an evaluation is required on a paper's interest value, degree of novelty and breadth of implication. When it is a matter of straight scientific soundness, referees may still have their objectivity impaired, but to a lesser extent'.

Another very experienced editor of the same journal commented: 'If a referee is a member of the same school as the author, he will likely as not be under-critical of methodology and over-emphasise the importance and relevance of the problem area to those working in adjacent areas. If, on the other hand, the referee is a member of a rival school, he might then assume an opposite stance'.

In the light of these informed opinions, it is of interest to investigate whether the biases referred to are reflected in the statistical analysis of refereeing judgements.

We have looked into this question by analysing data obtained from the records of a society which publishes research papers in prestigious areas of physical science. The analysis set out to identify the relationship between referees' evaluations and the nationality and institutional affiliation of referees and authors. The hypotheses under test were thus, firstly, that particular sets of research styles, values and beliefs would command higher levels of consensus within national and institutional groups, than across them. And, secondly, that this situation, compounded perhaps by the influence of personal ties, would result in authors receiving less critical evaluations from referees with similar national and institutional identities.

NATIONALITY. Looking firstly at the relationship between authors' and referees' nationalities, and resulting evaluations, the following data emerged.

PERCENTAGE OF PAPERS RATED 'GOOD' BY REFEREES

	U.K. Authors	N.American Authors	Totals
U. K. referees	70% of 600 papers	65% of 307 papers	907
N. American referees	60% of 35 papers	75% of 20 papers	55
Totals	635	327	962

Significance levels are limited, with the number of North American referees being so small, but it is apparent from the above table that U.K. authors are, on average, less critically evaluated by U.K. referees than they are by North American referees ($\chi^2 = 3.365$)*, while North American authors are less critically evaluated by North American referees ($\chi^2 = 0.8603$)*. Or, U.K. referees evaluate U.K. authors' papers less critically than they do those of North American authors ($\chi^2 = 2.8584$)* and North American referees evaluate North American authors' papers less critically than they do those of U.K. authors ($\chi^2 = 0.7379$)*. So it is seen that while levels of significance are limited, a higher frequency of favourable evaluation of papers by co-nationals is indicated in each of the four possible cases.

INSTITUTION. U.K. Universities were split into two groups. The first group, which can for convenience be called 'major universities', are those which stand apart from the rest in terms of the outstandingly high levels of funding they receive from the Science Research Council and their perception by researchers in the field as elite institutions. The second group, 'minor universities', consists of all the other U.K. universities.

Having established such a dichotomy, the table of authors, referees and evaluations appears:

PERCENTAGE OF PAPERS RATED 'GOOD' BY REFEREES

	Minor univ. authors	Major univ. authors	Totals
Minor univ. referees	65% of 120 papers	67.5% of 80 papers	200
Major univ. referees	55% of 110 papers	82.5% of 309 papers	419
Totals	230	389	619

It is seen that there is little difference between the frequencies with which minor university referees evaluate minor and major university authors' papers as 'good' ($\chi^2 = 0.13369$)* but a massive difference between the frequencies with which major university referees favourably evaluate major and minor university papers ($\chi^2 = 44.585$)*.

From the perspective of the author, minor university authors are more frequently evaluated favourably (i.e. less critically) by minor university referees ($\chi^2 = 5.295$)*, while major university authors are more often evaluated favourably by major university referees than they are by those affiliated to minor universities ($\chi^2 = 8.78$)*.

So it appears that when referees and authors share membership of national or institutional groups, the chances that the referee will be less critical are increased, and this can be put down to there being higher levels of consensus on research beliefs within these groups than there are across them. Associated with this situation are the preferences for particular areas or types of research which may exist within particular national or institutional groups. These problem areas or types of research (e.g. experimental/theoretical) may, by their nature, lead to the possibility of higher levels of consensus being shared on standards of good work.

*All χ^2 values are associated with 1 degree of freedom: 5% significance (extra-chance probability ≡ 20.1) $\chi^2 = 3.8$, 1% significance (extra-chance probability ≡ 100:1) $\chi^2 \simeq 6.5$, 0.1% significance (extra-chance probability ≡ 1000:1) $\chi^2 \simeq 11.5$.

Personal ties and extra-scientific preferences and prejudices might, of course, be playing a part as well. But it appears that even in the absence of these personal factors, the scientific predispositions of referees still bias them towards less critical evaluation of colleagues who, coming from similar institutional or national groups, share to a greater extent sets of beliefs on what constitutes good research.

A quote from a refereeing form completed by a referee at a 'major' U.K. university 'A', reporting on a paper by an author from another U.K. 'major' university 'B' further illustrates the point. "I was recently talking to (a non-U.K. Professor) about refereeing, and he volunteered the suggestion that some 'B' papers (and other U.K. papers) should be sent to 'C' for refereeing. He thought that there were some deficiencies which tended to recur, and which he and others at 'C' (and elsewhere) might be able to point out. There is a natural exchange of washing between 'A' and 'B', and I am concerned that I may too easily be giving papers a completely clean bill of health".

This illustrates how even the most conscientious referees can be limited in their critical perspective when evaluating the papers of colleagues who work in similar institutional environments and who, in consequence, often share similar sets of research beliefs. Concomitantly, it shows how criticism may be more forthcoming from referees who do not share these institutional and intellectual affiliations.

It can therefore be speculated that biases may be systematically operating within refereeing systems in such a way as to give advantage to those elements of the research community which supply the largest proportion of the referees used by the prestigious journals of that community. The papers of authors within these elements may, on average, be less demandingly evaluated than are those of authors outside these privileged groups, and their access to publication may, in consequence, be that much easier.

This situation appears dysfunctional to science, and as such, one would hope that the extent to which it may be apparent can be lessened.

The first step towards such an amelioration would appear to be the identification by journal editors of elites, groupings and schools of thought within their disciplines. These may be associated with a variety of professional characteristics beyond those investigated above and, in consequence, such acts of identification may prove difficult to make. Once they have been made, however, editors using extensive refereeing systems might profitably consider introducing greater heterogeneity into the professional characteristics of the evaluators they choose to consult.

When the institutional and intellectual affiliations of referees are made more diverse, in this way, refereeing systems may then be seen to operate not only less prejudicially, but also in such a way as will bring to bear a wider range of constructive criticism.

REFERENCES

1. MAHONEY, M.J. *The Scientist as Subject: The Psychological Imperative*. Camb. Mass. Ballinger. 1976 ch.5.

2. CRANE, D. The Gatekeepers of Science: some factors affecting the selection of articles for scientific journals. *American Sociologist* vol 2, 195-201.

3. ZUCKERMAN, H. and MERTON, R.K. Patterns of evaluation in science: Institutionalisation, Structure and Function of the Refereeing System. *Minerva* vol 9, No. 1, 66-100.

DISCUSSION on "Editorial Judgement in Scientific Periodicals" by Mark Keller
"Improving Referee-Selection and Manuscript Evaluation" by Manfred Kochen and Barbara Perkel
"Refereeing Reconsidered: An Examination of Unwitting Bias in Scientific Evaluation" by Michael Gordon

JOSEPH GILLIS - I'd like to make one or two points myself. One is that all this discussion of refereeing completely ignored the question of the referee who was not conscientious. And I've known some terrible cases of that. I know at least one case for example of some people who wrote an excellent paper and it was held up by a referee for about six months for no better reason than that he had a student who was working along the same lines and he wanted his student's paper published first. I know of at least two cases of such behavior. A different example came up when I was spending a summer in a large industrial research lab in the United States. I was shown some work that had recently been done and asked to comment on it and it looked excellent. In fact I was surprised, as it was a very difficult problem, one of the outstanding problems in applied mathematics which no one had managed to tackle before. And this had been tackled so beautifully. But when I examined it more closely I found that there was a very silly mistake in the elementary differential calculus at the beginning and that the solution which they got, though it looked elegant, just did not satisfy the differential equation that it was supposed to solve. It was way off. So I called the group together and mentioned this and their faces went very red. I pointed out that anybody could make a slip of that kind and it was better that I should have found it than that a referee should point it out later, or even worse that it should be pointed out by a reviewer after publication! Well then they went even redder. It appeared that the paper had already been submitted to a very reputable physics journal and had been refereed and accepted for publication. Up to that time, when refereeing papers, it had never occurred to me to check the elementary calculus and algebra. I had assumed they could be taken for granted. Since then I have been very careful.

JOHN ZIMAN - I would just like to say that when you come to refereeing you come to the heart of the scientific enterprise. These anecdotes which we have heard are extremely significant. They are part of the whole scientific activity. What Mr. Gordon has said puts us very clearly into that human activity side of things; we see the biases and prejudices. I don't believe that the machine will do all the things that it might do. We have to ask ourselves whether it will, for example, whether it would have done, the one thing that the referees need to do which is not to reject Waterston's paper on kinetic theory of gases and say that it is nonsense, five years before or ten years before Drew, and not to have rejected a paper by a friend of mine on entomology before the war because it said continental drift; and continental drift was out at that time. Now what I want to say very specifically is this: When you get into the referee thing you are in the human side of science and there isn't a machine answer to it and the thing to do is not to do a technological fix of it but rather to let all of us concerned with it, students, editors, referees and everybody else and the public at large - realize that here we are dealing with a very subtle and expert art which has to be learned by experience, often the hard way. I would make an observation: The moment that a scientist becomes really mature and really begins to see the meaning of his life is when he himself has to referee somebody else's paper. The role reversal situation there (up to that moment he has been producing his stuff and thinking how wonderful he is and how stupid the referees are not have understood his papers and so on, and the role reversal of being the critic of somebody else's work and having to decide whether it ought to go forward or not) is the most educative moment in his whole scientific career. And if he passes that properly, conscientiously and understands the meaning of it then he has become that responsible member of the scientific community which we are appealling to. And I don't think any technological fix will solve that problem. The job of editors is to make sure that those occasions are presented to people appropriately and that they are respected when they take their decisions.

JO AGASSI - What we have heard described here is a double standard. When a paper describes routine facts in a routine manner the editors and the referees throw a glance at it and see that it is more or less okay and get it published. Whereas if a paper is daring, unusual, and hence looks wrong at first sight, they don't find it really in their heart to say oh it looks wrong therefore it is wrong. They scrutinize it and find misspellings, wrong punctuations,faulty composition and so on. But all I wish to say is that I think we should defend this double standard because after all these are the papers that are read and therefore they should be written properly. The others are not read so it really doesn't matter that they are written, not only as poor science, but also in poor English.

CYRIL BELSHAW - I just want to say that my subjective experience is fully in accord with what Mr. Gordon has said and I'm coming more and more to the opinion that what is needed particularily with respect to review articles, controversial articles and things of this kind, and even those that don't look quite so controversial, is a balance in the referees. I must say that one of the most difficult things that I find is to break away from the citation group with respect to articles and also to break away from the noncited group who can also be emotionally involved and can be involved in biases. So I am thankful at the moment that the previous editor of my journal introduced this notion of twenty referees, so that by doing that we can choose people out of the citation group or avoid them or balance them depending upon the nature of the article. I think it's extremely important to do this in relation to the nature of the arguments that are contained within the article.

DIMITRI VIZA - I think we all agree that there is a referee problem and personally I'd really like to see the referee system abolished, but unfortunately there is no alternative and we have to accept it as the lesser of two evils. However, what I would really like to draw attention to is that by accepting the refereeing system, we take the risk of crystallizing the whole communication system of scientific reports - we create indeed a self perpetuating and closed system being a priori repellant to new and, by definition, disturbing ideas. From what I have heard so far, the only suggestion to get round this problem is to use young scientists as referees, and I agree with this. Now I have two more suggestions: to keep the referees anonymous has been a controversial issue; then one could propose to keep the authors anonymous when you send the report to the referees and, furthermore, when it is appropriate, to publish the referees' objections with the author's report and with the author's reply to the referee. This sort of mild public controversial discussion is being introduced timidly in the correspondence columns of *Nature*. We could probably consider it for all Journals.

STEPHEN KERTES - My question goes to Drs. Gordon and Kochen. Would you have any statistical information on the relevance, quality and depth of reviews submitted by referees if they are identified by name. Most of the American Chemical Society's periodicals have the question on their reviewer's form asking the referee whether he wants to be identified or not.

The background to my question is that the experience accumulated within most of the scientific commissions of the International Union of Pure and Applied Chemistry regarding the "by name refereeing" procedure is absolutely positive. Would you care to comment please.

MYRA KAYE - I think we have all seen the misery of a bad marriage where two people who are quite unfitted to live together really try, but are inevitably at cross purposes. One example perhaps is information requirements and computers, where, say, the male partner - the scientist - wants new information, mainly *ideas*, and the lady - the computer - can only give him year or even years' old data. The scientist wants ideas which will stimulate him and that might originate from regions far out of his field and the well-programmed computer gives him data from as exactly a matching area as possible to his own work! The scientist is working with ideas which are changing very fast and the computer is working on a standradized programmed scheme which can change only very slowly. I see now with great distress that this kind of bad marriage is apparently being considered with regard to referees. Referees in a journal belong to some kind of community where you have editors, referees, authors and so on and these people are linked by a need for communication. We have excellent telephone communication, magnificent communication technology and frankly I didn't see in the computer programs which were described anything that could not be done by telephone rather more easily in any country (except perhaps Israel!), and I would really like to know, what is the idea behind introducing a retrieval technology into what seems to me a purely communication problem when we have a very fast advancing communication technology. I might say that this is my gripe about information science in general.

NORMAN CHIGIER - I think the speakers here have shown us that our selection system is not perfect and all of us who are involved in the selection system realize that being human we are vulnerable. But we should be careful about throwing away this selection system because of the fact that we have all these new techniques and I include in this the computer, the telephone and everything else which, by doing things quicker, is expected to be able to do them better. I think there is a demand from our readers that we do something about it. They are saying that the present system of just throwing huge quantities of papers at them is unacceptable to them and they cannot find their way through this. So they are asking for a selection. I do not believe that the machine system is a more effective way of selecting than what we have at the present time. It is true of both referees and editors that we have both good ones and bad ones. And perhaps there again we have to get another selection system to try and improve the quality of the referees and the quality of our editors. I think that most of us are trying to do this, but you will always be able to point out our deficiencies. And indeed we are not doing it perfectly. I feel that we should strongly avoid the alternative, i.e. throwing out the refereeing system and allowing a computer to try and do it instead of us.

KNUT FAEGRI - It is a common experience that if you have a complicated machinery you are trying to make things in a complicated way. With all this beautiful technology the usefulness or not of which I can't go into, I was missing a very simple trick which the Nordic publication board has been using for years now. It is simply a workshop to train your referees - call it a training camp for referees - and in this way you get again ideas to a large group. Why not try that? But my main reason for asking for the floor is with one point I didn't quite understand and that was the compatibility of language between author and referee. It so happens that usually papers from this country fall into my lap. I'll not mention it, but the level of English is a little below par and you can't demand that we have referees with a similar knowledge of English. You must mean something definite by language there, and I'd like to know what you meant by it.

FRANCIS DE KÖRÖSY - I think the presidential republic works quite well in the United States and we ought to try something similar. This is a very very difficult situation we are in with scientific publications. I think an editor has to be responsible. The author is vis a vis the editor and he has to know him and he has to sign. Now of course an editor can't know everything but he ought to be a scientist of high reputation, of very good character and intelligent. Therefore editors have to be from among the best. And then still they need referees. I think it is very wrong that referees can remain anonymous. You sometimes get responses from unsigned referees which are simply fantastic. I'm quite awed when reading them. I am nearing my hundredth publication and everything I wanted to publish went actually through in one of the journals or another. But sometimes I got rejections in my first or second journal of choice and I'm sorry that I haven't them here. The same manuscript was eventually accepted in another journal, and very many people wrote for reprints, so I don't think it was bad. I think a referee has to sign, I always do it. I use a phone if the author is in the same country and I use the mails for other countries, and work something out with the authors. Generally something comes out of it. Not always an interesting paper; very often the paper is mediocre but it might contain facts which are important. Perhaps it would be worthwhile to think of a system like that proposed by Wiehelm Ostwald, where important facts are being recorded and only short abstracts are given to the general public. Photocopies can be obtained from the publisher of the full paper on request. Not everything has to be printed in extenso but good things have to be made available to the man who is interested in it. The present system is bad. The system of anonymous referees is certainly very bad and giving over the work to the referee is not worthwhile for a good editor.

LORE HENLEIN - I should like to ask the panelists and the chair whether any consideration has been given to the system of abandoning referees, but putting to work the members of editorial boards in such a way that the paper when submitted is submitted to two or perhaps more of the members of the board, who would then sponsor it and disclose their identity. You will have on the headline of the paper the name of the author and his affiliation and the fact that it was recommended or submitted through Professor X and Dr. Y. I think you might find that that system does work. It has been used and it's reasonably speedy. I'd appreciate comments if there were time.

JOSEPH GILLIS - Well the last question I want to ask is addressed to Prof. Kochen. Have you ever in fact used this mechanized outfit for selecting referees, or having put the program through the computer you then put it aside and looked at the list to decide that this is the man we want. I am reminded of a relative of mine who after qualifying through seven years of hard study became a doctor. The first thing he did when he had a sick baby was to go and ask his mother what she would do. Is this the sort of situation? But I now ask the members of the panel to take up the answers to the question.

MICHAEL GORDON - Well I don't see that the questions are particularly directed at me, but I'll try to make some comments.

Firstly: on whether referees should sign their reports. My personal opinion is 'yes'. I would prefer that referees sign their report; it would lead to more discussion on the points of intellectual difference. But, as you as editors must realize, if your journal is going to have a policy of publishing the names, or at least telling the author the names, of his referees, then a lot of these referees will be very reluctant to be as critical as they might otherwise have been, or to offer any reports at all.

On the question of sponsorship of papers, were you suggesting that if papers are endorsed by a recognized sponsor then they need not be refereed? (Affirmative nod from the floor).

Well this would presumably lead to a faster flow through of manuscripts but at the same time it would be leaving editorial judgement in the hands of a relatively small nucleus of evaluators who might not have sufficient specialized expertise and who might not improve papers as much as they could be improved if they were to go to experts in their field. Also, of course, the panel of sponsors might well represent one standpoint within the field and this could preclude publication of papers written from other standpoints, at least in that journal.

A further question asked if anyone had tried to send manuscripts to referees with authors' names removed and whether we knew of any studies or attempts to do this. I know of one instance in which the introduction of such an innovation was subsequently studied, and it was found that it had had little consequence. It was assumed that this was because referees often guessed authors names from their style of work, and self citation.

MANFRED KOCHEN - I will try to respond to some of the questions and comments about the use of information technology in refereeing. Professor Gillis asks whether I have ever used the mechanized referee-selection algorithms that I proposed. My only use of computers to select referees involved sending messages over a computer network to which I belong to several well-connected computer scientists, asking them to recommend potential referees for a manuscript on data security. I was surprised to see two suggested names and addresses in my mailbox the next day. I contacted them over the network and they agreed to review the manuscript. Use of the network really helped in this instance by saving time and raising quality.

Most of the algorithms have not been implemented by computer. I have, however, implemented some of them manually and the data reported in the paper is based on these procedures. Even in my manual execution of the referee-selection procedures based on matching profiles, I often chose to let my intuitive judgment override the output of the algorithm in finalizing the choice of referees. How often that occurred is also given in the paper. But I cannot say, from my experience, whether a matured, tested and systematic procedure would out-perform a good editor's intuitive judgment. The arguments presented in the paper persuade me that it would.

I agree entirely with Dr. de Körösy that the editor has to make the final decision and is ultimately responsible to both the author and the scientific community. The editor should be a key scientist, an eminent leader in his invisible college. Referees assist him by providing a consensus of peer evaluation.

I am not as convinced as he is that the pros of depriving referees of anonymity outweigh the cons. It is still controversial. I agree that a referee should accept responsibility for his evaluation and should be prepared to defend it. Under no circumstances should anonymity become a cloak to hide incompetence or malice. But I would not want lack of anonymity to discourage a referee from being entirely frank and critical or concerned about the possible acrimony that might result from public evaluations. In any case, I would hesitate to deprive every referee of his right to anonymity.

Dr. Kertes asked for my experience on the questions of referees identifying themselves. In one instance, a referee called the author directly. As a result the author withdrew his manuscript and submitted an improved version. I have not noticed any difference in the quality of the reviews by referees who have identified themselves and those who preferred to remain anonymous. Often younger, little known but very competent referees produce excellent reviews, but prefer anonymity because their youth-related lack of eminence may embarass the editor and the author.

Professor Ziman expressed my own view perfectly when he called peer review the very heart of science, and that refereeing is that aspect that requires the most human judgment. I would add that controversies in science are settled by attempts to reach consensual agreement among referees, authors and editors on both objective criteria, such as validity, and subjective ones such as significance, elegance rather than by negotiation or adversary proceedings. Rather than suggesting a technological fix that would in any way displace human judgments on striving for consensus based on a common search for truth and faith in the methods of science to attain it, I suggest the use of information technology to free people from constraints on their human judgments. Thus, by asking colleagues to recommend competent referees over a computer network rather than by telephone, I give them more time to think of a good response, to look up names and addresses; I also gain nearly the speed of the telephone.

Manfred Kochen

Mrs. Kaye's point about the sufficiency of the voice communication by telephone for referee-selection: It seems to me that this medium may be an improvement over the present system that relies primarily on mail, especially if the two are combined. I frequently call tardy referees. But a telephone call may interrupt a recipient at a time at which he is not well prepared to produce his best response, it pressures him to produce it "on-line" without adequate time for reflection, and it does not produce a reliable record for forwarding. There are many advantages and few disadvantages to an appropriate mix of computer conferencing, the mails and the telephone in improving communication between the authors, editors and referees. Computer conferencing is, after all, the use of the telephone--and other communications technologies--to transmit digitally encoded information that can also be stored, forwarded, processed, retrieved, etc. Indeed the newer technologies have been called compunications.

Prof. Knut Faegri asked what I meant by matching the language of the author with that of the referees. The referee's fluency in the language of the speciality to which an author is claiming to contribute should be no less than that of the author. That applies to both the technical and the non-technical language used in the manuscript. A referee must have a good reading knowledge of English to understand both papers written in good as well as in bad English. To his excellent suggestion of a workshop for referees, I would add why not have it on-line--as a computer conference--and why not start it in school.

Dr. Gillis' first comment about the non-conscientious or dishonest or careless referee stimulates me to suggest that: (1) all manuscripts should be routinely checked for technical errors--in arithmetic, syntax, conformity of rules of style, etc., by someone on the editor's staff; (2) the most deviant acts of referees could be publicized, subjected to peer judgment and the consequent social ostracism from the scientific community; (3) to protect the author from referees' excessive tardiness, the author's first date of submission is often printed; also who gets into print first, who wins a priority dispute is really not so significant in paradigmatic fields where multiple, independen discoveries are the rule rather than the exception.

Finally, I cannot share the doubts about the effectiveness of the refereeing system that have been expressed. Esteemed (and busy) members of editorial boards seldom invest as much care and effort in criticizing manuscripts as do selected referees, especially very competent younger ones. It seems to me tha the scientific enterprise is still one of man's most monumental and successful achievements, and it owes much to the free and open peer review system. If we find that less than perfect, then we should look the unique developments of our age - e.g. the information technologies -- to improve the process so that its full potential can be more nearly realized.

MARK KELLER - An anecdote on the subject of referees and anonymity: I have always preferred to keep my referees anonymous, because the most important thing is for the referee to tell me what he really thinks. And in order to ensure that, I am not going to let anybody know who my referee was on any paper. However, many times the referee's comments ought to be made known; and what I have done is subsequently to publish them, as comment on the paper, without revealing that the commentator was a referee. And I have had, now let's take this for the anecdote, I've had the experience that two referees disagree thoroughly with a paper. Both recommended, however, that it should be publishe I published the paper. I also published their disagreements, and got thirteen additional comments on one paper - because of starting that discussion. So this is a good scheme to use: Publish referees' comments without revealing tha they were referees and get additional comments. And, of course, one of the techniques that I have used, since I know a great deal about the referees - inevitably I have used referees a long time - I select two referees on a paper w represent possibly different viewpoints, and very often get conflicting advice from those referees. I like that very much because, then, you can guess who casts the deciding vote.

PUBLICATION, INNOVATION AND CITATION

Naomi Shanin-Cohen, *The National Council for Research and Development, Prime Minister's Office, Jerusalem* and *Department of Sociology, Hebrew University, Jerusalem, Israel*

ABSTRACT

A case study in the field of plant physiology (kinetin research) shows a strong correlation between the innovation quality of scientific papers and their citation rate - suggesting that factors extraneous to the content matter of a paper have little effect on the paper's citation rate. Methods are suggested for determination of scientific papers' "innovation quality", through systematic content analysis or citation counts.

The need for a yard-stick to measure the quality of scientific output has long been felt among sociologists of science, science administrators, and science policy-makers (1). This need on the one hand, and the relative ease with which a citation count can be obtained through the "Science Citation Index" (SCI) (2) on the other, has tempted many to use citation counts as a measure of scientific quality. In the last decade, the frequency of citation has increasingly served as a criterion for judging individual scientific papers, the life work of researchers, the scientific standing of research groups, and even the quality of national research output (3).

The correlation between the quality of the scientific paper and its citation rate has not been thoroughly examined and we do not yet know what is the meaning of "the quality of the paper" when it is evaluated by means of citation counts. The present study is an attempt to fill this gap by analyzing the correlation between the "innovation quality" of papers in a selected research area and their citation rates, and by systematically examining the reliability of citation counts as a means of evaluating their "innovation quality".

In so doing, it was necessary not only to find ways to determine the scientific quality of any paper, but also to devise a procedure by which the citation rate of each and every paper can be determined in a consistent way.

THE PROCEDURE FOR "INNOVATION QUALITY" DETERMINATION THROUGH CONTENT ANALYSIS

In the first stage, evaluators tagged and flagged the content of 187 papers for phenomena described, models and links suggested, and other data presented. The newness of the information presented in the paper was then checked by relating it to the material already published on the subject, using as a source of information a collection of abstracts drawn from Biological Abstracts (4) by relevant keywords. On the basis of these analyses each paper was classified under one of the four following categories:
1. Primary Finding (PF) Papers; 2. Primary Empirical Evidence (PEE) Papers; 3. Empirical Reinforcement (ER) Papers; 4. Non-Contributing (NC) Papers.

Primary Finding (PF) classified the papers that presented new observations of phenomena, providing that new physiological models or links between phenomena and models were suggested, on the basis of the observations.

Miriam Balaban (ed.), Scientific Information Transfer: The Editor's Role, 243-247. *All Rights Reserved.*

As we found that these PF papers generally did not present strong empirical evidence, we set the category of Primary Empirical Evidence (PEE) for papers that provided the first published empirical validation of the findings presented in the PF papers.

Empirical Reinforcement (ER) classified the papers that in the main presented findings already published in the PE and PEE papers and yet contributed to the validation of the innovation. This category allows for those cases where the biological object or the physiological state in the experiment or observation differed from those previously published. The contribution of such studies to the reinforcement of the innovation is linked with the varied nature of the biological object, which is diversified and dynamic.

Non-Contributing (NC) papers classified the papers presenting more or less a repetition of findings already published (more than twice), without providing results of higher statistical or other significance. We would like to stress that by considering these papers as not contributing to the validation of an innovation we do not imply that they cannot contribute in another way to the growth and development of science.

The 187 papers were analized and classified in the above mentioned procedures by two independent assessors. This population of papers includes all the research papers published between 1959-1966 on the role of kinetin (a plant hormone) in senescence, growth and stress, that we were able to compile from Biological Abstracts. They cover most of the papers that were published during the early period of kinetin research. Kinetin was mentioned the first time in the literature in 1956.

"PEAK CITATION COUNT" PROCEDURE FOR CITATION RATE DETERMINATION

We tried to establish this procedure in such a way that the citation rates determined will be affected as little as possible by factors such as the citation delay, the rates of self-citation, and the different citation trends of different types of papers, and thus to make the determinations as consistent as possible for any paper analyzed.

In order to do so, the general citation trends in the research area had to be examined first. In this examination a citation curve was plotted for each paper individually according to the number of times it was cited each year, beginning with the year of publication.

Observing these curves, we detected the following tendencies: When dealing with conceptual papers, in this research area as in many others, the annual citation rate of the papers increases and decreases following a parabolic curve pattern. However, the shape of the curve differs from one article to the other (5).

The later in the development of the specific line of research the paper is published, the shorter is the time-lag between its publication and the time its citation reaches a peak. For the papers that were the first to be published in the area, the citation peaked about 10 years after publication. The time-lag between publication date and citation peak period gradually diminished for later papers down to a 3 year-lag for papers published 8 years after the first group.

In order to diminish the effect of the citation delay, and taking into account the tendencies mentioned above, we determined the paper's citation rate by summing up the total number of times the paper was cited during the 5 years - Citation Peak Period (CPP).

The citation counts were carried out manually. To obviate errors in the SCI source, self-citations were excluded from the count. Review papers and

Papers defined as Empirical Reinforcement (ER) were less cited. The majority (72%) were cited between 5 and 15 times, and non-innovative papers classified under the Non-Contributing (NC) category were the least cited, most of them (84%) less than 5 times during the CPP.

It seems that a rate of 15 citations during the CPP is the minimum for innovative papers (PF and PEE) in this research area and also the maximum for other papers (ER and NC), and that a rate of 5 citations is the borderline between papers contributing to the innovation (PF, PEE and ER) and Non-Contributing (NC) papers.

TABLE 2
DISTRIBUTION OF PAPERS ACCORDING TO THEIR "INNOVATION QUALITY" IN A SAMPLE OF JOURNALS

CLASSIFICATION ACCORDING TO RATE OF CITATION	NATURE	SCIENCE	PLANTA	PLANT PHYS.	PHYS. PLANT	P. NAS US	J. EXP. BOT.	BERLIN DEUTSCHE BOT.	AMER. JOUR. BOT.
PRIMARY PAPERS	2	1	6	6	7	3	2	2	1
EMPIRICAL REINFORCEMENT	2	3	8	5	4	-	3	3	2
NON-CONTRIBUTING	2	-	11	8	6	2	-	1	1

CONCLUSIONS

For the research area we are dealing with, the findings of this study indicated that a paper's citation rate is determined in the main by the content matter of the paper, its innovative contribution, and that elements extraneous to the content matter, like the authors' prestige, the standing of the journal in which the paper is published, etc., can not have a considerable effect on the citation rate (from Table 2 one can see that many papers even when published in journals of high standing, are cited very poorly).

When research fields and areas are classified according to their citation patterns, kinetin research and physiology in general, falls into the group variously defined as "less descriptive", "highly codified" or "hard science" (6). It follows that there is a higher probability of finding correlations similar to those found in our study in other "highly codified" areas, such as physics, biophysics, chemistry, physiology, etc., rather than in the more "descriptive" research areas such as botany, zoology, palaentology, etc., in which citation assumes a different pattern (7).

We suggest that, once one knows the citation rates which demarcate between innovative papers and not-so-innovative papers, or between papers contributing to innovation and those which are not in a research area, one can use citation rates for the evaluation of the "Innovation Quality" of the papers in the area.

methodological papers (papers presenting in the main a technical method) were not included in the comparative analysis because of their different citation patterns.

THE CORRELATION BETWEEN THE RATE OF CITATION AND THE "INNOVATION QUALITY" OF SCIENTIFIC PAPERS

A surprisingly high correlation was found between the rate of citation and the "innovation quality" of the paper. The papers contributing the most to the innovation, and thus classified under Primary Findings (PF), were most highly cited. The papers in the category of Primary Empirical Evidence (PEE), were cited slightly less and papers in the Empirical Reinforcement (ER) group less still. The least cited belonged to the last group, the Non-Contributing (NC) papers.

TABLE 1a
"INNOVATION QUALITY" OF SCIENTIFIC PAPERS VERSUS RATE OF CITATION

CLASSIFICATION		NO. OF PAPERS CITED AT FOLLOWING RATES IN THE 5 YEARS CITATION PEAK PERIOD									
		0	1-4	5-9	10-14	15-19	20-24	25-29	30-34	35-50	50-100
PRIMARY FINDINGS	N 30	0	0	0	2	11	5	4	2	3	3
PRIMARY EMPIRICAL EVIDENCE	N 15	0	0	2	2	3	3	2	0	2	1
EMPIRICAL REINFORCEMENTS	N 47	5	6	28	6	0	2	0	0	0	0
NON-CONTRIBUTING PAPERS	N 95	32	48	13	2	0	0	0	0	0	0
TOTAL	N187										

Most of the papers (87%) classified under Primary Findings (PF) and Primary Empirical Evidence (PEE) were cited 15 times or more during the citation peak period (see Table 1b).

TABLE 1b
"INNOVATION QUALITY" OF SCIENTIFIC PAPERS VERSUS RATE OF CITATION

CLASSIFICATION	TOTAL PAPERS IN CLASSIFICATION	NO.OF PAPERS CITED 15 TIMES OR MORE*	NO.OF PAPERS CITED 5-14 TIMES*	NO.OF PAPERS CITED 0-4 TIMES*
PRIMARY PAPERS** (PF+PEE)	45	39 (86.7%)	6 (13.3%)	0
EMPIRICAL REINFORCEMENT	47	2 (4.3%)	34 (72.3%)	11 (23.4%)
NON-CONTRIBUTING PAPERS	95	0	15 (15.8%)	80 (84.2%)
TOTAL	187			

* During the 5-year citation peak period.

** As no significant difference was found between the citation rates of papers classified under Primary Finding and those classified under Primary Empirical Evidence the 2 categories are united as one category of Primary Papers.

REFERENCES

1. J. Margolis. Science, v. 155, 1967, p. 1214.
S. Cole and J.R. Cole. American Sociological Review, v. 32, 1967, p. 378-379.
J.R. Cole and S. Cole. American Sociologist, v. 6, 1971, p. 23.

2. Science Citation Index (SCI). Philadelphia, Pa., Institute for Scientific Information.

3. G. Lemaine, B.P. Lecuyer, A. Gomis and C. Barthelemy. Les Vois de Succès. Paris, Group d'Études et de Recherches sur la Science, 1972, p. 245-262.
J.T. Wallmark and B. Sellerberg. IEEE Transactions on Engineering Management, v. 3, 1966, p. 137-142.
H.W. Menard. Science: Growth and Change. Cambridge, Mass., Harvard University Press, 1971, p. 94-95.
J.R. Cole and S. Cole. Science, v. 178, 1972, p. 368-375.
S.Y. Crawford. Informal Communication among Scientists in Sleep and Dream Research. Chicago, Ill., Chicago University, Thesis, 1970.
Science Indicators 1972. Washington, D.C., National Science Board, 1973.

4. The Biological Abstract BA. Philadelphia, Pa., Biosciences Information Service of Biological Abstracts.

5. G. Lemaine, B.P. Lecuyer, A. Gomis and C. Barthelemy, op. cit. note 3, p. 256.
S.Y. Crawford, op. cit. note 3.
H.W. Menard, op. cit. note 3, p. 125-128.
H. Zuckerman and R.K. Merton. Age, Aging, and Age Structure in Science: a Sociology of Age Stratification. In: Age and Society, v. 3, 1972, p. 302-305, M.W. Riley, M. Johnson and A. Forner, eds. New York, Russell Sage Foundation, 1972.
P. Wiess. Science, v. 131, 1960, p. 1716-1719.
D. de Solla Price. Citation Measures of Hard Science and Soft Science Technology and Non-Science. In: Communication among Scientists and Engineers, 1970, p. 3-22, C.E. Nelson and D.K. Pollack, eds. Lexington, Mass., Heath Lexington Books, 1970.

6. P. Wiess, op. cit. note 4, p. 1716-1719.
D. de Solla Price, op. cit. note 4, p. 3-22.
H. Zuckerman and R.K. Merton, op. cit. note 4, p. 302-305.

7. P. Wiess, op. cit. note 4, p. 1716-1719.
H. Zuckerman and R.K. Merton, op. cit. note 4, p. 302-305.
D. de Solla Price, op. cit. note 4, p. 3-22.
N. Kaplan. American Documentation, v. 16, 1965, p. 179-184.

SEX AND MONEY

Alexander Kohn, *Editor, Journal of Irreproducible Results*
Department of Human Microbiology, Faculty of Medicine,
Tel Aviv University, Ramat Aviv
Israel Institute for Biological Research, Ness-Ziona, Israel

ABSTRACT

The title is misleading and is used to show how to obtain good indexing coverage. The paper satirizes the obscurantist excess in published scientific literature by providing many examples of what should not be done.

Motto:

$$\boxed{\int e^x = f(u)^n}$$

There are two points about writing papers which I should like to analyse: whofor and whatfor. Whofor means evaluation of the potential audience which might be either your betters, your peers, your subordinates, or the public at large.

If you write for your betters, include as many references to the ideas of your superiors as possible. It is not relevant or important whether these ideas were experimentally sound. Saunders (1) calls this technique the "White headed boy approach".

When you write for your peers- and this usually involves submitting the paper to a professional journal - you have to appear erudite (see later), but in any case the writing must not appear to be too simple.

For subordinates or public use some such well known phrases as: "Everyone knows that...;""It is obvious that"..." followed by some esoteric remark. These statements help to build up your image as an expert, and that is what the public expects of you.

Parenthetically, the word "clearly" is a swindle: it means that the author is unable to provide any good evidence in support of what he is about to say and hopes to get away with it by persuading his readers that it is obvious or so universally accepted that it is a waste of time to go over the arguments (2).

Fraenkel-Conrat, the famous virologist, gives you this advice (3): "Publishing a lot may lead any scientist into publishing some errors, be they due to faulty experimentation or to premature interpretation ... Such publications will be rebutted and criticized, which means that the paper is referred to, and a counter rebuttal supplies an additional publication without the

Miriam Balaban (ed.), Scientific Information Transfer: The Editor's Role, 249-255.

need of additional data or thoughts. In sort, if publishing a lot has been selected as the path to success, the quality and veracity hardly matter. A man endowed with some political astuteness and a sense of salesmanship can count on this path to make him a man of note, if not of notoriety. He may soon play his role on the wider stage of scientific power policy. This may fulfill his heart's desire, but it may also be of advantage for science since it means his removal from the scene of publishing investigators".

Whoever the audience - the paper should have a traditionally established frame. It should start with a title. A paper without a title leads you nowhere (except perhaps to the House of Lords). The practice to indicate in the title what the article is about is now obsolete. Its main function nowadays is to keep librarians and indexers busy and to get the maximum number of citations and cross references possible. As an example, see the title of this paper. You may have gathered by now that there is no connection whatsoever between the title and the content of this presentation. Dr. Emile van Handel said in this connection: "There has always been a much greater demand for my method papers than for my biological results except when I started the title of an article with the word "sex" (4).

The technique of proper choice of titles is known as the Citation Index Game (5). According to its author, Leopold, the purpose of this game is to increase the extent to which your papers are quoted. The first step towards success in this game is of course to cite your own papers as often as possible.

Whitebottom for instance publishes an original paper in some important field; Papierfresser, who is competing in this - publishes a paper on the same subject and is big enough to quote Whitebottom. In all his later papers, however, Whitebottom no longer appears - all that remains is the reference to Papierfresser's first paper where he gave credit to Whitebottom.

This ploy was used with great effect when Nirenberg published his revolutionary finding that polyphenylalanine was coded by poly-U. He was quoted by Ochoa, his competitor, only twice and never again (6).

The following are some good examples of intriguing titles:

"An instant of pitfalls prevalent in graveyard research" (Biometrica 19:683, 1963), and

"A sociosexual behavioral abnormality in the dog resembling Oedipus complex in man" (J. Am. Vet. Med. Assoc. 144:868, 1964) by Fox (Fig. 1).

Fig. 1.

The next part in a scientific paper is an Abstract. Its purpose seems to be to contain enough information to make the reading of the publication itself unnecessary and actually undesirable. The opposite view is: "the abstract must present information in such a way as to make the reading of the article absolutely essential, so as to understand what the article is all about". (1)

Two examples of the second approach are shown:

"Simulation by Monte Carlo methods of the effects of selection can produce a degree of genetic canalization which is more restrictive than that indicated by the limits of selection showing the canalization of a rigid degree can be caused by loose selection" (7).

"Communication is an important daily aspect of a librarian's profession. The horizontal relationship is described and a distinction is drawn between it and the vertical relationship. The horizontal relationship is an important communication concept which when mastered by the librarian accrues many personal satisfactions" (8).

Introduction. As in the case of the title, the last decades have witnessed a change in concept of the introduction. The old custom of reviewing the field and outlining the scope of the work has been discarded. The modern approach is that the initial revelation of the scope of the paper should not appear till the discussion. The "sudden death" approach calls for a technique that forces the reader to figure it all out for himself, the author presenting no clues whatsoever.

Results. In this section one presents all experimental details except one, like a woman giving a recipe for a cake to her neighbour. This actually means that no one would be able to repeat the experiment. When final results are given and the author does not intend to continue research on the subject, full details are given so that the paper may serve eventually as a textbook method (e.g. ref. 9).

Fig. 2 is a facsimile of the most concise presentation of results.

JOURNAL OF APPLIED BEHAVIOR ANALYSIS 1974, 7, 497 NUMBER 3 (FALL 1974)

THE UNSUCCESSFUL SELF-TREATMENT OF A CASE OF "WRITER'S BLOCK"[1]

DENNIS UPPER

VETERANS ADMINISTRATION HOSPITAL, BROCKTON, MASSACHUSETTS

REFERENCES

[1]Portions of this paper were not presented at the 81st Annual American Psychological Association Convention, Montreal, Canada, August 30, 1973. Reprints may be obtained from Dennis Upper, Behavior Therapy Unit, Veterans Administration Hospital, Brockton, Massachusetts 02401.

Received 25 October 1973.
(Published without revision.)

Fig.2.

Discussion. We have already stated (10) in the past that in the discussion one simply repeats the predictions made in the introduction (if there were any), and rehashes the results to weave them together into plausible theory, even if it means that the original cause-effect assumption turns out to be effect-cause relationship. Any results that do not fit the explanation are simply omitted.

References. For those beginning their publishing careers it is best to include as many references to their superior's ideas or publications as possible; ditto your referee's, if you happen to know who he might be. There only remains the problem of a head of a department who automatically puts his name on any publication coming out of his lab (this is the so-called droit de signer). The important point to grasp is that as long as you don't use references that contradict your results or conclusions, you may put in as many supporting or neutral references as decently acceptable (the length of the paper).

Omission of your name in bibliography "is essentially irreversible, since bibliographies usually are wafted in their entirety from one paper to the next, except for the insertion of the respective author's own contributions, which, if luck has it, may then accompany, plasmidlike, the standard package in its subsequent passages" (11).

Acknowledgements. The following example will suffice: "We are indebted to Dr. James Dizikes and Dr. George Sherman, Grading Branch Dairy Division, Consumer and Marketing Service, U.S.D.A,, Chicago, for the microbiological analysis, and to Mr. G.R. Anderson, Acting Area Supervisor, Grading Branch, Poultry Division, Consumer and Marketing Service, U.S.D.A., Philadelphia, for supervising the sampling of products, to Dr. D.H. Berquist and Mr. P.W. Klusmeyer, Henningsen Food, Inc., and to Mr. G.W. Patterson and Dr. A. Brant, University of California, Davis, for allowing us to use their data on the number of Salmonellae in commercial albumen and whole egg samples respectively' (12).

As to the whatfor of publications, obviously the most common reason is to improve one's standing in the lab., department, University, research institution, or society at large. The idea that those who decide your promotion are weighing your papers is now almost completely erroneous, since most of the deans or ad hoc committee members simply count the papers. Some more sophisticated judges use the so-called Parse analysis method (13).

This method meets the need to evaluate the relative contributions of authors to papers with more than one author, and also takes into account the degree of difficulty of the problem attacked and the quality of execution. Each author is assigned a decimal fraction according to his relative contribution. To take an actual example (Fig. 3):

[Reprinted from the Journal of Pharmaceutical Sciences,
Vol. 59, No. 4, April, 1970. © 1970 by the American Pharmaceutical Association]

Odor Threshold and Gas-Chromatographic Assays of Vaginal Odors: Changes with Nitrofurazone Treatment

0.24 0.08 0.36 0.27
A. DRAVNIEKS, B. K. KROTOSZYNSKI, L. KEITH* and I. M. BUSH*

Received August 27, 1969, from *IIT Research Institute, Chicago,*

* Present address: Department of Urology, Cook County Hospital, and the Chicago Medical School, Chicago, Ill.

Fig. 3.

We have assigned arbitrarily a fractional number to each author. We assume that the first name is of the investigator who did the actual work, thus we assign to him the major fraction. We also assign the lowest value to one of the middle authors by assuming that he has the droit de signer (of course, we might be entirely wrong). In any case the total, which should add up to 1.0 is only 0.95. This is not because the work was significant at the 5% level, but we assume that there participated in this study a visiting investigator who left, or because we did not know how to evaluate the work of the technician who did most of the work.

The Parse analysis involves two additional scoring devices by taking into account the difficulties of the task and the quality of execution, each with a potential maximum value of 3. In the case shown here, though the subject matter was rather easy to conceive, the design of the experiments was complex, ingenious and novel, so one takes 1.5 x 3 = 4.5, i.e. 50% of the maximal 9. This factor, 0.5, is then multiplied for an author. This is a very nice method, but which dean has the time and patience to apply it? Nevertheless, the Parse analysis has immense potential when it comes to defining academic success in easy mathematical terms, which can be computerized.

As for the evaluation of a paper by the Editors, there exists a method called TISC-TISC, meaning The International Scale for Critical Evaluation, or Terse Index of Scientific Censure (14). The TISC-TISC coefficient is the natural logarithm of astonishment experienced by the reviewer upon confrontation with a particular statement from a prospective author. The integral values of this coefficient are shown in Table 1.

TABLE 1

TISC-TISC Coefficient of Astonishment	Mean Observation
--7	Yes, I observed this myself in 1932
--6	o
--5	o
--4	o
--3	o
--2	o
--1	o
0.	So What?
1.	This seems to need clarification.
2.	I don't quite follow this argument.
3.	I doubt this.
4.	Heavens no.
5.	Birds, for the.
6.	Perhaps another journal would be more appropriate for this interesting work.
7.	Is he some kind of a nut or something?

Another important condition for advancement in the field is the image. You create a proper image by improving the length and the style of the publication.

One way to increase the length of a paper is to use graphs, tables, and diagrams. Accuracy of the results may be disguised by the use of log-log paper. The second method for elongation is the use of scientific gobbledygook, very aptly named NONSCIENCE by Ford (15). By using the right phrases and jargon you can extend the paper at least twofold. A few examples are taken from ETC (16):

"At this point in time it is possible to arrive at an approximation as to the optimum magnitude and geometric profile of the experimental systems module and the construction of such a systems module will be proceeded at some not too distant future point in time".

meaning: We can now estimate the best size and shape of the apparatus and it will be built soon.

"The absence of fully programming to these enlisted manpower requirements or endurance parameters forms a dichotomy between our stated mission/objectives and the individual unit's ability to carry out the stated mission in a peacetime environment" (Navy document).

meaning: Because it's peacetime, the Navy won't assign us enough men so that we do our job. It would be only fair therefore, if they would cut down the number of tasks we have to do.

From a document of the Department of Interior: "Our office has become cognizant of the necessity for eliminating the unnecessary vegetation surrounding the periphery of this facility"

meaning: Get rid of those weeds.

Two examples which Gower in his Plain Words has quoted from the New Scientist, and erroneously attributed to me, are (17):

"The allocation of a very limited portion of a person's resources, abilities and energies to the ownership, maintenance and adornment of residential structures".

means: they live in slums.

and: "A set of arrangement for producing and rearing children the viability of which is not predicated on the consistent presence in the household of an adult male acting in the role of husband and father".

means: Dad is not home much of the time.

(Both examples appeared in a report of a seminar on Poverty by the American Academy of Arts and Sciences, Chapter 9).

The meaning of the following really escapes me:

From the National Aeronautics and Space Administration (NASA) letter on procurement policy.

Page 2 NASAHQ 285 UNCLAS:

"The purpose of this message is to provide clarification, on an interim basis, to be followed until such time as PRD 7015 is formally revised; note that this interim guidance on the conduct of discussions is not repeat not a change in policy or concept, it is solely clarification, and should be so construed. While the clarification is presented in the form of a partial revision of PRD 7015, this does not necessarily mean that formal revision will follow the same format or composition; but rather it is so presented so that it may be read in context with 7015 which, of course, remains effective. The affected part of 7015 is paragraph III D(2) covering conduct of discussions in cost-reimbursement type contracts and all R&D contracts".

One may note here that the Orwellian Newspeak and Doublespeak (18) has been transformed by the scientists to Gobbledygook (16).

While Doublespeak has intentionally limited grammar and vocabulary, our Gobbledygook is very complicated in structure and in vocabulary, and although it is supposed to be a clear way of transmitting messages, in fact it often does mislead. Some scientists actually believe that elegant latinate vocabulary, complex sentence structure and sophisticated writing style are supposed to impress the readers. There may however also be some ulterior motivation in the use of gobbledygook. "The gobbledygookers assume that they protect their tails by leaving plenty of escape routes in case their message is understood better than they wanted it to be; ambiguity rather than clarity is therefore the goal" (16).

According to Chernin (19) some scientists tolerate jet-lag better than publication lag, and value speed over sound scholarship. It is hard to imagine that scientists who are deliberately exposing themselves to motion sickness, writing grant applications, teaching students, supervising theses, serving on committees and boards - can really do good science and write decent papers when they touch down in their lab.

The net result of the "publish or perish" axiom is that some of the stuff that flows through journals represents unprocessed sewage (20). Is it really true that every thought is better published than perished?

References

1. Saunders, R.F. How to become a successful author. Technology, Ireland, Jan. 1976 p. 28.
2. Pollack, M.R. From Pangens to polynucleotides. The evolution of ideas on the mechanism of biological evolution. Persp. Biol. Med. 19: 455-472, 1976.
3. Frankel-Conrat, H. Design and function of the threshold of life: the viruses. Academic Press, N.Y. 1962.
4. Van Handel, E. Citation Classics, Current Contents. 1977. April 18, p. 13.
5. Leopold, A.C. Games scientists play. Bioscience 29: 590, 1973.
6. Lengyel, P., Speyer, J.F., Ochoa, S. Proc. Nat. Acad. Sci. 47: 1936, 1971.
7. Fraser, A.S. Simulation of genetic systems by automatic digital computer. Austral. J. Biol. Sci. 13: 344, 1960.
8. Smith, N.M. and Fitt, S.D. Vertical-Horizontal Relationships. Their applications for librarians. Special Libraries 66: 528, 1975.
9. Lowry, O.H., Rosenbrough, N.J., Farr, A.L., and Randall, R.J. Protein measurement with the folin phenol reagent. J. Biol. Chem. 193: 265, 1951.
10. Kohn, A. Principles and methods of obsurantism. New Scientist, Jan. 29, 1969.
11. Chargaff, E. Triviality in Science. Persp. Biol. Med. 19: 324, 1976.
12. Garibaldi, J.A., Lineweaver, H. and Ijichi, K. Number of Salmonellae in Commercially Broken Eggs before Pasteurization. Poultry Sci. 48: 1096, 1969.
13. Davis, P.J. and Gregerman, R.I. The parse analysis. New Engl. J. Med. 281: 989, 1969.
14. Sorry, I cannot find the original paper. My apology to the authors.
15. Ford, B.J. Nonscience. Wolf Publishing Ltd., London WC2, 1971.
16. Ralph, R.E. Illuminating effectively. ETC 33: 445, 1976.
17. Gower, E. The complete plain words, HMSO, 1973, p. 204.
18. Orwell, G. Politics and the English language. Collection of essays by George Orwell. Garden City, Doubleday, 1954.
19. Chernin, E. The worm's eye view of biomedical journals. Fed. Proc. 34: 124, 1975.
20. Kaye, M. Preparation of pure crap. Proposal for research project, in: (Ed)J. Ertel: Selected papers. The Journal of Irreproducible Results JIR publishers, Inc., Chicago Heights, Ill. p. 249.

DISCUSSION on "Publication, Innovation and Citation"
by Naomi Shanin-Cohen
"Sex and Money"
by Alexander Kohn

YEHUDAH WERNER- Referring only to the first of these two capers, I think that the game of counting citations, in order to be scientific, should be played a bit differently. It is not logical to count them in numbers. They should be expressed as a percentage of the papers in which that original paper should have been quoted. Since this apparently is impossible to define the whole game should be abandoned.

ELEAZAR SHAFRIR - I also want to point out some of the pitfalls of using citations as an index of scientific quality. I think that they reflect scientific popularity rather than quality. For instance a very good example of how to get a large number of citations is to publish a good method. The most often cited scientific paper in biological literature is that of Lowry et al. on how to determine protein. Because most researchers determine biological variables per milligram of tissue protein, they have to quote the modification of Lowry which has become an accepted method for determining protein. Several other examples may be mentioned such as determining glucose or DNA content in biological fluids. And this doesn't need to be a methodological paper or a paper published in an Analytical Journal. Sometimes you publish a good scientific paper with an adapted method built into it. Then many cite the paper for the sake of the method, not for the science in the paper.

Another aspect related to citations is that there are fashions in science. For instance today we are living in the period of the big controversy on sweetener use in the United States. Any paper which would be published on saccharine this year will be very extensively quoted, far in excess of its scientific value, but a few years hence nobody will talk about it anymore. So if you would judge according to the number of citations of a paper on saccharine today, you would arrive at a disproportionate importance. I could cite many other examples of this type.

One should also remember that there are some journals which are more widely read than others, without relation to the quality of papers they contain. People who often read only a small number of popular journals consequently quote from a selected pool and this also produces a skewing in appraisal by citation. I'd like those who really want to use citations as an index to be very well aware of all these pitfalls.

EUGENE GARFIELD - I think that on the basis of the previous comments I could state that, based on statistical evidence, the average temperature in Jerusalem is fifty degrees because I could cite the weather of the one day I have been in the city. Similarly, we can all cite the kinds of examples that were just given to "prove" many myths about citation practices. Instead of citing these anomalies, people should look at the hard statistical evidence based on thousands of cases. But they prefer to cite as proof of something one paper out of the entire literature of more than ten million papers which have been published. Using the Lowry paper, which has already been cited over fifty thousand times, to "prove" anything is absurd.

I would also take issue with the assertion that method papers are necessarily highly cited. I can show you thousands that have never been cited. Moreover, ISI has done several studies of this question. Henry Small of our research staff examined the Chemistry literature and found that method papers are not more highly cited than theoretical papers. We just remember them a little better. Granted, there are examples in certain fields of the life sciences of method papers with extremely high citation rates, but there are many method papers that have very low citation rates. Contrary to what we would like to believe - that friends cite each other heavily or papers get citations for negative reasons - the correlations between quality of papers and their citation frequency is very high.

Now the purpose of citation analysis is not to eliminate the human judgment that is necessary in enaluation processes. As a matter of fact the reason why I have devoted so much time to this is to prevent a disservice often done to people whose work is in fact not necessarily exceptionally highly cited but cited quite well and who somehow are outside what I call the political mainstream of science.

The fact is that through closer analysis--or the kinds of things Kochen was referring to--we can get very good correlations between citations and peer review. After all, we have sent clusters identified by citation analysis to the experts in the field and asked them to name the ten leading people in their own field. The correlation is remarkably high, and it is not surprising when you think about it because the people who are going to pass on peer reviews are the very same people who are in the citation circle.

Now what we have found however is that in various political sides of science, certain types of personalities will be excluded from consideration for awards, elections to academies, you name it, simply because they are not political. If we can show the scientific community that here is a list of people who have done very respectable work in terms of the criteria that we are measuring, all that we are saying is that they sould be given equal consideration with those who are active politically.

ANDERS KYLIN - As editor of one of the Journals cited in this investigation of citations, I think that maybe I should say a few words. I was very interested and happy about one thing in the investigation. And this is that it solved the problem of the time lag. It is quite evident that people asked several years ago whether Science Citation Index can be used to evaluate the scientific merits of a person. My first reaction was that this was impossible due to the fact that the really new ideas will not be cited in reasonable time. This has evidently been taken care of in the investigation. So, I am happy to accept that point.

On the other hand, it is also interesting to see and recognize the duties of a journal. It is not only to get the scientific Rolls Royce out of the garage, as it has been expressed here. It is also a duty to serve as a data bank - to provide papers, which do not necessarily provide new ideas to science but only provide data around a problem. Just so that people know that the data are available. These papers can be published in different forms; but I would say that the pool of "lost" papers in the evaluation, the papers not much cited they can still be rather valuable as data banks. I think that this is one of the functions that we cannot yet escape as editors of journals in normal fields under normal standards. We have to let our journals serve as data banks. We should see to it that papers of this type are reasonably easy to read and then published, until better methods are found for the storage of primary data. The people behind these papers are often "the hewers of wood and the fetchers of water", and we cannot manage in our scientific society nor in the society as a whole without them; and we have to respect them and to publish their work in some form.

NAOMI SHANIN-COHEN - Citation counting, even turned into a defined evaluation method, is only a tool - a measure of the "innovation quality" of scientific papers. This measure, like any other, can be used only for what it is made for. Therefore, even if non cited papers may contribute to the growth of knowledge by providing new data, as Prof. Kylin has suggested, this is not the tool to detect it with. What Prof. Kylin is referring to is not the method, but wrong interpretations of the measurements that might and can occur. Any tool has to be "equilibrated" before it is used, and that is what we tried to do in this work, by equilibrating the citation tool according to the "innovation quality" of the scientific paper. I do agree with Dr. Werner and Dr. Shafrir that this tool has to be applied cautiously; for example, I think that it should not be used for comparing the "innovation quality" of papers from different research areas, unless the validity of such a comparison is proven. Personally, I found this tool very valuable for a comparative evaluation of research groups working in the same field of research in a certain period. As to Dr. Shafrir's comment that the visibility of a paper is related to the rank of the journal it is published in, I would like to say that we could not find this to be true. For example, we found that many of the papers defined as "non-contributing" papers through content analysis were published in journalsof high standing in which the PF, PEE and ER papers were published; yet they were not cited more than the other "non-contributing" papers which appeared in publications of lesser standing.

THE CREATIVE ROLE AND FUNCTION OF EDITORS

Mary L. Borysewicz, American Journal of Ophthalmology
Chicago, Illinois

Abstract

The primary role of an editor, specifically as the proprietor of a scientific journal, is not to be creative but to be self-effacing. An editor must function as an intermediary between author and audience. The editor must provide a network of review, scientific and editorial, for the author; clarification of the message for the audience; and massage for both by sensitive handling of the peer review process and by maintaining a sense of common purpose and balance through the various crucial aspects of publication.

The primary role we as editors have as the proprietor of a particular publication is act as the intermediary for the creative efforts of those seeking publication in our journals and as the interlocutor between author and readers.

One might argue that, in the case of scientific writing, creativity is not a function of author or editor; it belongs to the realm of the humanities and possibly to the soft sciences. Unless we consider only those publications which may be, for whatever reasons, pure archives of accumulated data, then creativity must be a consideration as the force by which the new, true, and important are presented to a particular scientific community. As such, then, it is the function of the author.

As good editors we must function in many other roles; as colleague, counsellor, advisor. More than likely, we find ourselves acting as a stimulus, convincing authors that creativity is a complement to preciseness. We in fact use our own particular creative energies to cultivate and encourage what we see as a potentially valuable contribution in a paper judged scientifically accurate and worthy of publication, but also needing extensive editorial revision. Where we can, we must offer guidance and direction. After numerous exchanges with an author, if a paper still does not seem suitable for publication, then we are faced with the dilemma of weighing the merits of publication of a poorly presented piece of work, and perhaps raising doubts as to the validity and worth of the study itself, against the responsibility we have to our readers and their need to be comprehensibly provided with the new, the true, and the important. A delicate balance. Rarely do any of us find ourselves in the happy circumstance of achieving such a balance at the outset. Nor do I believe we really expect it.

As editors we must act as a stimulus to authors and as a catalyst between authors and readers. While no negative connotation is implied or intended, I think in the biological sciences the role of journal editor is complicated by the peer review system. It is not enough to have encouraged an author through several revisions of his paper only to have it returned as wanting by the scientific referee. We must then draw on our powers of tact, ingenuity, and diplomacy. We must use ingenuity after evaluating the critique of the scientific reviewer in order to interpret the strengths and weaknesses of a paper in such a way that the author will have the constructs he needs to bring his writing to completion. Tact and diplomacy are imperatives. They are most needed when an editor, perhaps in desperation and frustration, loses his temper.

The increasing emphasis in the last few years on continuing education and self-assessment courses, particularly in the health

Miriam Balaban (ed.), Scientific Information Transfer: The Editor's Role, 261-264.

sciences, has resulted in a kind of course population explosion. The problem for the journal editor is that often the author-participants of these courses, in order to get the maximum mileage out of their material, will submit a manuscript for publication that was originally prepared for presentation before an audience. Anyone who has been involved in revising such material for publication knows what a task it is. Because courses generally are geared to a show and tell format--that is, numerous slides, flip charts, models and other visual examples--the text is meagre. When subjected to editorial scrutiny the weaknesses of results obscured by generalities or of arguments masked by series of visually presented case reports is apparent. Even an elegant study suffers in the translation. This particular problem was emphasized for me because I had occasion to hear a major lecture and subsequently edit it for publication. The author, unknowingly of course, obliged my example here by submitting a verbatim transcript for publication. Remembering that the presentation had been some 30 minutes in length I was surprised to find the body of the typescript was only 20 pages long. Add to this the 40 references, which had not been cited in the presentation and still it seemed the paper had no where near the impact of the original lecture. Not until I had merged text, tables (20), figures (75) and their legends did I really see the problem. In speaking, the lecturer had moved easily through his commentary from slide to slide, building to his conclusions. In print that fluid movement was not possible. His statements were in the text, his proofs either in the tables or the figure legends, and his conclusions split between. Thus the impact was diffused to the point of ambiguity. As I said at the outset the basic study was sound. What was needed was a form that would reflect the study as clearly as the lecture had. Fortunately, the author-lecturer recognized the editorial problems and he was most cooperative throughout the cut and paste necessary to tighten the focus and strengthen the published article. More often of course papers originally intended for course presentation defy the editorial process.

The review process requires our full attention and diplomatic talents. The approach to this aspect of publication is equal in importance to the exchange between author and editor. Once the review of a paper is complete we must evaluate the impact, pertinence, and over-all merit of the review itself, before referring it to the author. Frequently we find that we need several opinions and views to assess a study fairly and objectively. Even so, having had to rely heavily on the judgments of referees we may still find ourselves having saved the chaff and discarded the wheat. We may then redeem ourselves with a gracious letter of congratulations to the author and one to the sagacious editor who published the paper.

I often wonder, in the case of the lecture referred to earlier, if the manuscript had been sent out for review whether the scientific referees, most of whom have heard the original presentation, would have recognized any of its editorial problems.

I don't believe a paper should ever be returned without a transmittal letter, however brief, to the author, explaining the editorial board's decision. Unless the remarks of a reviewer are critically destructive, in which case we will have sought other opinions, they should be sent to the author for reevaluation of his work. We may exercise our editorial prerogative and edit or paraphrase a review, without changing its meaning, to reinforce our decision or to preserve the integrity of a reviewer through anonymity. I believe the anonymity of the reviewer must of course be preserved. The remarks of a reviewer to the editor can, because of their candor, be valuable in helping an editor make a decision with regard to marginal studies. Such confidential statements should be encouraged for the insights they may contain.

Depending on the procedural policy of the journal, a thorough editorial review of the paper should be sent to the author as well. For the author's sake, particularly where revision is encouraged, sending the editorial review together with the scientific review is expeditious. Thus, the author can integrate the two and incorporate the necessary changes in his paper before resubmitting it for further consideration. I think we defeat the purpose of the editorial process when we separate the two and ask the author first to tighten up his data or sharpen the focus in his photographs, and then to concern himself with agreement, voice, and clarity in writing. One must serve the other, if an author is to be made aware that the importance of his message can best be appreciated by direct writing. Recently an author wrote that he believed the material in his paper was so exciting and so interesting "that you should not be concerned with style." What he failed to comprehend was that neither the scientific reviewer nor the editorial board could understand his message because of the manner in which the paper was written.

It seems to me that we are more than surrogate readers. We are rather the mediators between authors and the particular audience that concerns us. Because authors and editors must relate on a one-to-one basis this often appears to be the first priority. However, we must always reflect the needs of the broader relationship we have with our readers and our responsibility is to communicate the scientific message as clearly and directly as possible.

Ideally, readers assist the editor by making their views known. Unfortunately, I think the distance between the norm and the ideal is vast in many scientific publications. One has only to ask a variety of journal editors to find this distressingly true. With the possible exception of those journals, New England Journal of Medicine, Lancet, British Medical Journal, and the like, whose correspondence section is a vital part of the publication, reader response is often nonexistent or it is politely apathetic. And yet, response or no, we must determine the needs of those readers and function accordingly. Expectations for an exchange of correspondence may be aroused by the confidential statement of a reviewer that if X's paper is published it will draw the sharp pens of correspondence to the editor. Scepticism is reinforced when no such exchange occurs. The resilient editor copes equally well with apathy and with adversity.

In a journal where correspondence is encouraged and reader opinion is high, the editor's role as intermediary is easier, inasmuch as he has some tangible basis for his editorial decisions. But what of a journal in which there is little or no communication. The readers' needs still must be met and their views reflected. The editor of such a journal, in any given discipline, must then seek out the readers. Although it is more realistic to note that more often readers and authors seek out editors for vis-a-vis confrontation, we must be available at conferences, meetings, and the like. Moreover, we must make our availability, however restricted by other commitments, known to readers as well as authors. That we will not commit ourselves to an editorial decision by listening and discussing must be clear at the outset. We may not maintain attitudes of isolation or anonymity at least not if we wish to see our journals expand and broaden in thought and thus reflect the evolution of a particular scientific discipline.

Finally, the mark of a successful collaboration between author and editor that corresponds to what readers expect of a publication is that accolade from the author, which every editor cherishes, "Thank you for making my writing say exactly what I wanted to say."

DISCUSSION on "The Creative Role and Function of Editors"
by Mary L. Borysewicz

ANDERS MARTINSSON - I would like to kindly take issue with the speaker on the point of the non-creative role of the editor as idealized by her. Actually, the speaker did make reference to situations in which the editor is creative. The editor should certainly be self-effacing whenever possible and try not to dominate the contents of his journal. He should not be a censor of intellectual contents and not rewrite articles, adding new contents. But he should certainly be creative whenever possible in the promotion of his journal, its programme or profile (every journal should have one) and his scientific discipline. He should be creative in developing his serial as a mechanism integrated with research and education, and he can do this for instance by means of editorial articles and comments. I have repeatedly described the creatively managed journal as an international or national seminar. Norman Chigier has today drawn a close parallel between the role of the editor and that of the examiner in academic education. Many of us are trying to get away from the letter-box type of scientific editor. The creative role that he should take is the one of a *promotor of science*.

This is relevant even in the apparently very rare case referred to when the editor is the proprietor of the journal or associated with the ownership in some way. It is likely that he has come into this position by a highly creative act. For example, I find it both legitimate and praiseworthy when the editorial director of a once innovative secondary service appears in the pages explaining what he has done and how he continues to develop his system.

CAN EDITING BE QUANTITATED?

Franklin H. Portugal and Bernard K. Forscher
Proceedings of the National Academy of Sciences
Washington, D.C. 20418, U. S. A.

ABSTRACT In order to determine how uniformly editors handled their editing assignments and to determine which areas of manuscript preparation caused authors the most difficulty, manuscripts from the PROCEEDINGS OF THE NATIONAL ACADEMY OF SCIENCES were analyzed for the number and type of corrections made by the editors. Four broad defect areas were established: nomenclature, spelling, punctuation, and construction. The results suggest that there is a remarkable degree of uniformity among the editors who have similar training but different levels of editing experience.

The first great editor was George Alterer who lived in Edere and edited during the first century B.C. Editors would enter the publishing trade as apprentices usually around the age of seventeen. In that era, incorrect use of scientific nomenclature, improper sentence constructions, and failure to punctuate sentences adequately were considered capital offenses because officials realized that without adequate documentation their claims to priority of publication in the scientific literature would go unrecognized in the twentieth century. George Alterer lasted longer than most editors in his time, but he was put to death within two years for failure to enclose a parenthetical expression between commas.

Editors have come a long way since those early, dark ages Social attitudes toward editors have become more restrained and civilized, but scattered exceptions here and there have been noted. These calmer attitudes have nurtured the explosive increase in scientific journals and scientific editors, as editors realized that their life was no longer inseparable from their work.

As a scientific editor at and representative from the PROCEEDINGS OF THE NATIONAL ACADEMY OF SCIENCES OF THE UNITED STATES OF AMERICA, I share your thankfulness for increased longevity, and now would like to share with you the results of a study to determine whether editing can be quantitated.

We had several reasons for undertaking this analysis. First, we wanted to determine how uniformly the editors handled their editing assignments at the PROCEEDINGS OF THE NATIONAL ACADEMY OF SCIENCES. Second, we wanted to determine which areas of manuscript preparation caused authors the most difficulty and were most likely to be the source of errors. And finally, we hope to be able to use this information for the training of new staff members and for the development of guides to assist authors in preparing new manuscripts.

Despite the relatively high quality of manuscripts received by the editorial office of the PROCEEDINGS OF THE NATIONAL ACADEMY OF SCIENCES, considerable time is spent on editing them. The editorial staff consists of

Miriam Balaban (ed.), Scientific Information Transfer: The Editor's Role, 265-270. *All Rights Reserved.*

four members with editing experience ranging from fifteen years to one year. Each editor has a doctoral degree in biochemistry. The editors have various amounts of actual laboratory experience, but only one has worked as an author's editor. The manuscripts are given to the editors in a random fashion based on the manuscript number assigned as the mail is opened.

The manuscripts used for the quantitation studies were selected from the December 1976 issue at random, plus seven chosen from the January 1977 issue to increase the total number of manuscripts used for one editor. The 83 manuscripts studied represented 384 printed pages (mean, 4.6 printed pages per paper). Sixty-six papers came from authors who reside in the United States; the seventeen from foreign authors included 7 from Israel. Because the PROCEEDINGS has a strictly enforced maximum of 5 printed pages per paper, our articles tend to be of relatively uniform length except that mathematics papers run a little shorter.

Manuscripts published in the PROCEEDINGS are divided first into two broad categories, physical sciences and biological sciences, and then into classes reflecting the sections into which the membership of the NATIONAL ACADEMY OF SCIENCES is divided. The author selects the classification from a standard list. For the purposes of the present study, biochemistry, which represented 33% of the total manuscripts, and cell biology, which represented 24%, were considered separately. Chemistry, mathematics, astronomy, and physics were grouped under physical sciences; immunology, biophysics, physiological sciences, botany, genetics, microbiology, medical sciences, and neurobiology were grouped as biological sciences. In the sample studied, the distribution of the classes among the editors was quite heterogenous. For example, papers in botany, genetics, neurobiology, astronomy, and physics were each edited by only one editor. Botany, neurobiology, astronomy, and physics were each represented by only one manuscript (see Table I).

A "defect" was defined as an entity that was corrected by the manuscript editors Changes made by an editor to make a manuscript conform to house style were not counted if, by general standards, a correction was not needed. Four broad defect areas were established: nomenclature, spelling, punctuation, and construction. A nomenclature defect was the use of a scientific term that was either undefined or defined incorrectly, incorrect use of isotope symbols, incorrect abbreviation of a biological or chemical term, or incorrect spelling of a chemical or biological term. If the same error was made repeatedly in a manuscript, it was counted only once.

Spelling defects included the incorrect use of letters to form a word (other than a word naming a chemical or biological material) and the incorrect use of a prefix or suffix. No attempt was made to distinguish spelling errors from typographical errors.

Punctuation defects included the incorrect use of a semicolon, colon, comma, period, hyphen, or one-em dash. Every punctuation defect was counted. Most entries in this category refer to the use of commas. Corrections for punctuation included both additions and deletions, although almost all changes were additions to the punctuation already present. Words that were hyphenated incorrectly and used repeatedly, such as cross (hyphen) section when used as a noun, were counted only once in each manuscript.

The fourth area, construction, included agreement between subject and verb, use of the correct verb tense, and sentence rearrangements by repositioning, adding, substituting, or deleting individual words or groups of words. Again, an error of a single type was counted only once in each paper even though it may have been made repeatedly.

Table I. Distribution of Manuscripts by Discipline for Each Editor

Discipline	Editors K	H	B	T
Biochemistry	7	37	37	19
Cell biology	10	25	35	30
Immunology	25	75	0	0
Biophysics	33	33	33	0
Physiological sciences	0	50	25	25
Botany	0	100	0	0
Genetics	0	0	100	0
Microbiology	0	0	50	50
Medical sciences	0	0	83	17
Neurobiology	0	0	0	100
Chemistry	13	13	37	37
Mathematics	0	0	50	50
Astronomy	0	0	0	100
Physics	0	100	0	0

With the exception of two disciplines, biochemistry and cell biology, the disciplines were grouped together as either biological sciences or physical sciences. For each category or discipline, the mean number of editorial changes per manuscript by the four editors, as a single group, was calculated for each of the four defect areas. The results are shown in Table II.

Table II. Editorial Changes for Each Defect Area in Selected Discipline or Category

Discipline	Defect area	Editorial changes, mean no.
Biochemistry	Nomenclature	7
	Punctuation	21
	Spelling	2
	Construction	15
Cell biology	Nomenclature	3
	Punctuation	19
	Spelling	4
	Construction	17
Biological sciences (other)	Nomenclature	2
	Punctuation	15
	Spelling	3
	Construction	15
Physical sciences	Nomenclature	2
	Punctuation	15
	Spelling	1
	Construction	8

It is apparent that two defect areas, punctuation and construction, required the most changes. For example, 24 of 27 biochemistry papers required punctuation corrections that clearly exceeded the nomenclature corrections by 1.5 times or more. These values are the minimal number of changes made by the editors because, as mentioned above, the same error repeated throughout a manuscript was only counted once. The nomenclature defect area had an apparently higher value for biochemistry than for the other three categories, but this value was within two standard deviations of the other nomenclature values and probably is not significant. The construction defects showed an apparently lower value for the physical sciences than for the other categories. However, biochemistry construction values were within one standard deviation of those in the physical sciences construction category, and so these differences also may not be significant. Otherwise, the values for each defect area showed considerable uniformity among disciplines. Incidentally, the mean lengths of manuscripts for biochemistry and cell biology disciplines were identical; any apparent differences in editorial changes, therefore, cannot be due to differences in manuscript length.

Next, we considered the question of uniformity of editing among the four individual editors. The mean number of changes made by an individual editor for each defect area, regardless of discipline, was calculated, and the results are shown in Table III.

Table III. Mean Number of Changes by Each Editor

	Editor			
Defect area	K	H	B	T
Nomenclature	3	4	4	4
Punctuation	21	19	18	16
Spelling	4	2	3	2
Construction	41	12	4	9

The number of changes made by each editor for a specific defect area appears to be quite uniform with one exception. Each editor made changes more frequently in the punctuation and construction defect areas than in the nomenclature and spelling ones. Editor K with the most editing experience in general but the least specifically with the PROCEEDINGS, however, made changes more frequently in the construction defect area than any of the other editors. These construction changes occurred at a rate per manuscript that was 3 to 6 times greater than for the other editors. This editor, only, had experience specifically as an author's editor.

Construction changes involved a broad number of alterations. As mentioned earlier, these alterations included using the same tense for two different verbs in one sentence, making the verb agree with the subject rather than the subject modifiers, distinguishing between frequently misused words such as "that" and "which" or "following" and "after", deleting extensive but unnecessary phrases such as "in the case of," substituting one word for another such as changing "the dishes were renewed" to "the medium was replaced," etc.

All of the data are summarized in the last table in which editorial changes by individual editors are shown distributed both by defect area and by discipline (see Table IV). These data confirm our earlier impressions from the previously shown condensed data. Punctuation and construction defects were changed more frequently than nomenclature and spelling ones, regardless of which editor was editing that category. With the exception of construction defects for Editor K, editing appeared quite uniform for the four defect areas in each category or discipline. For the defect area of nomenclature, the mean number of changes made by individual editors within each discipline or category was similar. Whereas some variation was noted for individual editors for nomenclature changes when comparing one discipline or category with another, these differences do not appear to be statistically significant. Editor B, for example, had a mean of seven nomenclature changes for biochemistry but only two for cell biology. Yet these values fall within one standard deviation of another. Similarly, values for spelling changes made by individual editors were quite uniform regardless of whether values for individual editors were compared within a specific discipline or values for one individual editor were compared among each of the four disciplines or categories.

Table IV. Editorial Changes by Individual Editors

Defect area	Editor				Number of Manuscripts
	K	H	B	T	
Biochemistry					
Nomenclature	4	7	7	6	2/10/10/5
Punctuation	21	24	19	17	
Spelling	3	2	3	2	
Construction	49	12	15	8	
Cell biology					
Nomenclature	2	4	2	5	2/5/7/6
Punctuation	23	15	19	20	
Spelling	7	2	4	3	
Construction	49	12	13	16	
Biological sciences (other)					
Nomenclature	4	1	3	1	2/7/11/4
Punctuation	19	10	20	9	
Spelling	2	1	4	2	
Construction	33	12	15	8	
Physical sciences					
Nomenclature	1	2	2	2	1/2/4/5
Punctuation	20	30	7	14	
Spelling	2	0	2	1	
Construction	27	14	4	4	

Punctuation values generally appeared uniform between editors within a given discipline or category as well as between disciplines for any one specific editor. Statistical analysis for some apparent variation as between Editor H (punctuation value of 30 for the physical sciences) and Editor B (punctuation value of 7 for the physical sciences) suggested that the differences were not significant.

Finally, values for construction changes were uniform for individual editors in different disciplines. Comparison of values for Editor T between cell biology and physical sciences revealed an agreement within one standard deviation.

Finally, values for construction changes were uniform for individual editors in different disciplines. Comparison of values for Editor T between cell biology and physical sciences revealed an agreement within one standard deviation.

However, the consistently larger values for construction changes for Editor K compared to the values for the other three editors in each of the disciplines generally does appear to be significant. Although Editor K edited only two manuscripts in three of the categories and only one in the other category, the larger value occurred in all four disciplines or categories studied.

Several conclusions can be drawn from these studies. There is a remarkable degree of uniformity among the editors who have similar training but different levels of editing experience. The editing changes frequently made to correct punctuation reflect an author's lack of familiarity with the uses of the comma. The small but apparent spelling errors generally involve the incorrect use of a prefix such as non(hyphen) malignant, pre (hyphen) heat, or semi (hyphen) liquid. Examples of the elements of construction that frequently merit attention have already been mentioned. The need for a small but constant amount of editing of nomenclature suggests that authors do not always pay careful attention to this element of the manuscript, perhaps expecting the editor to detect and make the necessary corrections. Authors may also feel that it is impossible to keep up with changes in nomenclature rules and that simply being "understood" is all that is important. The greater number of construction changes made by Editor K may reflect his background training as an author's editor, his considerably greater editing experience, his personal style of manuscript editing, or a combination of these factors.

Writing, even for scientific journals, is a highly individualistic activity nurtured through creativity and marked or marred by "style." In their efforts to introduce uniformity and clarity of expression, editors must be careful not to interfere with the delicate forces generated by the manner in which words are put together to form sentences or with the fine stitching that sews one sentence to another to form the fabric of the completed manuscript. That is not our intention. So-help-me George (Alterer).

STANDARDS AND STYLE

Knut Faegri, Chairman

A COHERENT PHILOSOPHY OF TITLES, ABSTRACTS AND KEYWORDS IN SCIENTIFIC ARTICLES

Anders Martinsson, *Editor, Lethaia*
Department of Palaeobiology, University of Uppsala
Box 564, S-751 22 Uppsala, Sweden

ABSTRACT
Title, abstract and keywords should be regarded and developed as parts of an integrated system designed to cover the needs of selective readers, librarians and secondary services. (1) If combined with an abstract, the title can be relieved of its burden of excessive information of keyword type and so better suited for its function of catching the attention of the selective reader. (2) The abstract, then, should be regarded as an extended title in the form of connected sentences. (3) If the abstract is supplemented by a line of keywords (which are indicative by definition), the abstract can be relieved of its indicative function (serving the documentalists in the first place) and concentrate upon its informative function (which is what the scientific reader wants). — Consequently, abstract and keywords should together with the title form part of the article-head, and in standardization the three components should be treated together. — Although remaining a self-contained item of information, the abstract in an international language that is added to an article in a national language can be coordinated with other components of the article (summaries and captions) in the service of the selective reader.

The title, the abstract and the keywords are all derivatives of the main text of a scientific article. Only the abstract has been the subject of an international standard, namely ISO 214 (Documentation — Abstracts), and there are corresponding national standards. The title and its style are mentioned in different draft or valid standards and in codes and guides to good practice. Keywords do not lend themselves to standardization in the same sense. Without reviewing all possible documents which would swamp my message, it should be acknowledged here that it is frequently urged that the abstract should be placed at the beginning of the article, that the article title should be placed in close context with the abstract or repeated within the same block of text, and that the abstract may be followed by a line of keywords (descriptors).

What I am going to show here, however, is that there is a functional relationship between title abstract and keywords that does not only affect their physical position in the document but also their intellectual contents and style. They have to be regarded and treated as components of an integrated system and should be standardized as such.

Miriam Balaban (ed.), Scientific Information Transfer: The Editor's Role, 273-275.

The title

The system still prevails where the title is as crowded with keywords as possible. When the title proper is not sufficient to accommodate enough information of keyword character, a subtitle is added. When the title reflects a multidimensional classification it can be very long. In my own field (palaeontology), for example, the title of this kind mostly contains information about the place of the material in the botanical or zoological system, its stratigraphical position (i.e. age), its geographical provenance, methods used for the investigation and often some noise-words to tell the reader that the author has not treated the subject exhaustively but merely made a contribution (cf. Martinsson 1972). This usage leads to the introduction of abbreviations and parentheses in order to shorten the title in general and particularly to avoid connecting words. Documentalists do not like abbreviations, formulae and code expressions in the title, and from the editors' point of view titles constructed in that way are simply ugly and cumbersome.

However, this usage developed when the article contained no other conventional component than the title in which information of keyword type could be placed. It also developed at a time when scientific readers had not been forced by the quantity of publications to be as selective as they are today and when even the headlines of the mass-media were exhaustive.

The existence today of an abstract placed in close conjunction to the title should relieve the latter of its excessive burden of keywords and make most subtitles superfluous. The title can concentrate on the function of turning the readers' attention to the actual results of the investigation, and the derivation of the material, the methods applied and the main details of the results can be left for the abstract.

This might disturb those indexing services which are based on keywords in title context alone. However, we do not write our publications for the secondary services but for the user — the selective reader — even if we try to do our best to prepare the publications so as to be as easily and efficiently abstracted and indexed as possible. However, greater selectivity with regard to keywords in titles will lead to less cheating of the user of title-based indexing services where unnecessary attention has been drawn to excessive keywords that refer to the categorization of the material investigated rather than to the results.

The abstract

At this point it helps if we look at the abstract not as a direct derivative of the text (which it certainly remains with regard to its intellectual contents) but a *self-contained extension of the title, consisting of connected sentences* (which settles its function both for the selective reader and in the secondary systems).

Looking at the abstract in this functional perspective immediately explains the difference between the abstract on one side and the summary (annotation, extract) on the other, as ISO 214 endeavours to point out. At the same time, however, ISO 214 defines the abstract as "an abbreviated representation of a document", which is a good definition of a *summary*, except that it is permitted for a summary to contain "interpretation or criticism". The definition is the weakest point in ISO 214. What this document says about the purpose, use, contents and style of the abstract is, however, in full harmony with our present philosophy (my reservations in other contexts against the all-capitalized, centred abstract titles are only of technical importance).

It is difficult in most languages to find a vernacular or more easily assimilated term in translation of the Latin-derived, English word "abstract". "Summary" (Zusammenfassung, résumé, резюмé, sammanfattning, etc.) refers to a piece of text containing any of the elements but serving a different purpose. The term "synopsis" has fortunately become obsolete for designating the author-generated abstract and has now been taken over to designate a specific type of condensed scientific article. The French term "analyse" is even more in conflict with our philosophy than terms implying that the text is a condensate or a review (реферат, referat, etc.). The German term "Kurzfassung" seems specific enough to be acceptable, but otherwise, till we have become adapted to something, many of us will in different languages use "abstract" as an unassimilated term with English spelling and pronunciation.

The place of the abstract, in harmony with many recommendations, should according to our philosophy be immediately after the title of the article.

The keywords

There has been a protracted discussion, still not concluded but very diplomatically reflected in ISO 214, whether abstracts should be indicative or informative. It is natural that the documentalists should want subject-oriented, indicative abstracts but that the selective reader (the primary user for whom we write our publications) wants a findings-oriented, informative piece of text.

Keywords are by definition indicative. If we break out the keywords as a separate line or sequence, we relieve the abstract text of its indicative function and can style it for informative purposes without having to watch out so that all keywords are crowded into the text and none lost.

This solves the problem and should conclude the discussion. To implement the system will not be entirely easy since in several branches of science it seems meaningless to operate at all with keywords until we have controlled vocabularies. The place of the keywords is evident, however — in close connection with the title and abstract and in a form that makes them stand out clearly and allows them to be easily retrieved.

Conclusions

The titles, abstracts and keywords belong to one and the same functional system that should be standardized as such. The obstacles to implementing the system are none as far as the primary publications are concerned. For the indexing services that exclusively use keywords in title context it implies a new situation that might lead to new strategies. In order to further acceptance of the system and to make it more efficient, controlled vocabularies for the keywords should be worked out.

For practical purposes I would like to recommend a block of text following the article title and author's name consisting of (1) an abstract biblid according to ISO DIS 30 and (2) the abstract text, followed by (3) keywords in italics, all to be accommodated within the size of a standard library catalogue card. This means an arrangement in close accordance with Example 6A in ISO 214, except that keywords are added and the article title is given in lower-case text that can be used with correct capitalization and accents in reference lists.

The arrangement might modify the concept of "self-contained" as so often mentioned in connection with standardization of abstracts. The block of title, abstract and keywords is the unit that should now be self-contained.

Often the abstract in an international language that goes with an article in a national language can be coordinated with international versions of the figure and table captions to give a generous synthesis of the contents of the article. Moreover, the presence of an abstract does not exclude the addition of a summary.

Reference

Martinsson, A. 1972: Editor's column. Titles of palaeontological papers. *Lethaia*, Vol. 5, pp. 126-128. Oslo.

THE CSIRO STYLE MANUAL

Basil J. Walby, *CSIRO Editorial and Publications Service, 372 Albert Street, East Melbourne, Victoria 3002, Australia*

CSIRO publishes seventeen journals, many irregular serials and books totalling some 24000 pages each year. Because most of these are printed in-house in Melbourne it is necessary to have standard routines that can be followed by the forty or so Research Divisions scattered around Australia. Apart from the 6000 people working for CSIRO another 1000 scientists submit manuscripts for publication in one of the nine national journals of scientific research that are managed by CSIRO.

The Foreword to the *CSIRO Style Guide* states that "it is designed to help all members of the staff of CSIRO to prepare typescripts for publication within the Organization. Many of the suggestions included are equally applicable to submissions to other publishers."

The Guide has been designed to be as short and as readable as possible. There are only 12 pages apart from 4 appendices. It does not attempt to deal in a comprehensive way with any particular publication. Further information is available from the Notice to Authors published in each of the main journals. Authors are referred also to international standards and other authorities that have issued codes.

The following items are covered as succinctly as possible:

Miriam Balaban (ed.), Scientific Information Transfer: The Editor's Role, 277-278. *All Rights Reserved.*

Materials and format	Pagination
Length of papers	Spelling and punctuation
Units, abbreviations and symbols	Footnotes
Numbers	Dates
Trade name and trade marks	Title
Authors	Abstract
Headings	Tables
Illustrations	References
Appendices	Accessory Publication
Mathematical papers	Chemical papers
Biological papers	Proof correction
Selected bibliography	

Appendix 1. Standard and preferred spellings

Appendix 2. International system of units

Appendix 3. Abbreviations, symbols and conventions

Appendix 4. Proof readers' marks

The most recent edition was published in 1975. Over 3500 copies have been distributed. Requests come from a wide variety of people including commercial publishers, university staff and societies that wish to adopt the Guide's recommendations for their own publications. Experience shows that it is rarely necessary to send submissions back to authors for stylistic revision.

PROGRESS WITH EDITERRA EDITORS' HANDBOOK

J.W. Glen, *Editor Editor's Handbook*
Department of Physics, University of Birmingham, P.O.B. 363, Birmingham B15 2TT, U.K.

and

K. Helbig, *President of Editerra*
Rijksuniversiteit Utrecht, Vening Meinesz Laboratory, Lucas Bolwerk 7 Utrecht, The Netherlands

ABSTRACT

Since its foundation in 1968, the European Association of Earth Science Editors (EDITERRA) has regarded the preparation of a Handbook as a natural means by which the discussion of problems which editors encounter might be codified for the benefit of all editors. At early meetings the general pattern of the proposed Handbook was decided, and a number of draft sections were prepared and discussed in Assemblies of Editerra.

Later it was decided that the individual sections might be discussed more rapidly if small committees were set up, one for each section, and these committees sent their proposals to a Handbook Committee which would correlate their work and approve it for publication. Several of these committees produced drafts, but this mechanism also proved too slow, and so it was decided to commission specific persons to prepare drafts which were then circulated for comment and approval to the Handbook Committee. A number of sections have been approved in this way, and these are now being published as separate pamphlets which can be assembled to form the growing Handbook. The first five sections are being issued at the same time and a further two have been approved by the Handbook Committee.

Editerra Council recently reviewed the structure and dissolved the Handbook Committee, coopting the Handbook Editor onto Council. The future development of the Handbook will be under the direct control of the Council and will be reviewed by General Assemblies of Editerra.

Editerra, the European Association of Earth Science Editors, was founded in 1969 when UNESCO encouraged the formation of several regional editors unions each covering a closely related group of disciplines. The purpose of these associations was to stimulate cooperation and exchange of experience

Miriam Balaban (ed.), Scientific Transfer Information: The Editor's Role, 279-281. All Rights Reserved.

Within Editerra it was felt that the main goal should be standardization and what can be succinctly described as "professionalization" - to bring the criteria and standards used by the part-time editor to a professional level. As a means to achieve this end, the first Editerra Circular Letter proposed an "Editerra Editors' Handbook". This is not a style guide, but rather was intended to be a book which every editor would keep at his desk to check on questions of procedure and on details which were outside the range of experience.

The format and scope of this Handbook and the mode of operation were discussed during the First General Assembly in Gent (December 1969) and has been the current theme throughout the subsequent General and Extraordinary Assemblies of Editerra. Scope and purpose of the Handbook have remained essentially unchanged in the 7½ years, but the format and the mode of operation have been under constant revision. This is not surprising in view of the fact that the experience on which an Editors' Association can draw is in editing, while writing, production, and marketing are areas where no common pool of experience exists.

The first working groups were established on the following subjects: Standardization of references and key-words, Earth-scientific nomenclature and terminology, Publication of computer-programmes, Refereeing prodecures, Criteria for publication of scientific articles, Review articles and solicitation of manuscripts, Cooperation between editors, Inventory of primary and secondary publications, Dissemination of geo-scientific intercommunication.

These working groups were to report to a Joint Editors' Handbook Committee (jointly with the AESE). Drafts were to be circulated and, if and when approved at an Editerra Assembly, distributed in final form to members with the possibility of distribution beyond the membership. In a sense, the working group was to act as author, the membership as referees, the assembly as editor, and the association as publisher. In retrospect, it is perhaps not surprising that - though many important sections were circulated as drafts- few reached a final typed form, and the goal of an easy-to-use handbook that every editor would keep at his desk for constant reference - thus encouraging standardization and improved editorial skills by osmosis - was still far off. Sections which have reached the members at different levels of acceptance until 1974, when an Extraordinary meeting of Editerra took place, were:

Tables (A.A. Manten)
Standardization of references (B.C.M. Butler)
Key-words (J. Gravestijn)
Copyright (M8)
Policy regarding advertisements (M9)
Information on the cover of the publication (M10)
Stratigraphic and geochronologic terminology and nomenclature (M30)
Review articles M(14)
Geographical names and indications (M15)
Numerals (M16)
Brief analysis of scientific communication (M17)
Publication of computer programmes (M20)
Mineralogical nomenclature (M23)
Book reviews (M25)

During the meeting in Arhus (1974) and the Third General Assembly (Lammi, 1975) the format and the distribution of responsibilities was changed. Editerra now asks some members to write sections under their name and submit them to the Council of Editerra. The Chairman of the dissolved Handbook Committee acts as editor with the prerogative to call on referees, and the Council on behalf of the Association acts as publisher. This stream-lining worked: we are now on the threshold of issuing the first five sections as individual (color-coded) pamphlets, which is not only cheaper, it also allows the user to keep only those on his desk which he regularly needs.

The first five sections in preparation are:

Indexing (G.H. Brown)

Editorial handling of manuscripts and proofs (J.W. Glen)

Mineralogical nomenclature (M.H. Hey)

Rock nomenclature (P.A. Sabine)

Stratigraphic nomenclature (A. Martinsson)

Two further sections have been submitted for acceptance and will be published in the near future:

Book reviews (A.A. Manten)

Numerals (A.A. Manten and J.W. Glen)

Though we are still not in the possession of the first tangible results which would stand circulation outside the membership of Editerra, the process through which we have gone to this date is not entirely futile. Practically all members were involved in some aspect of the preparation of the Handbook and many idea has found its way through circular letters, through discussions at meetings, and through working groups to the membership at large. Our handbook is still very much alive and we hope that we have now passed the time of developing a method to produce this handbook and have arrived at producing it.

Since practically every member of Editerra participated in the early preparation of the sections mentioned above, it is not possible to acknowledge their work individually. However, special thanks are due to Dr. A.A. Manten, Editerra's First Secretary, who stimulated many of the early discussions and has produced a number of first drafts single-handedly.

THE FOURTH EDITION OF THE CBE STYLE MANUAL

Philip L. Altman, *Director*
Office of Biological Handbooks
Federation of American Societies for Experimental Biology
9650 Rockville Pike, Bethesda, Maryland 20014 U.S.A.

A fourth edition of the CBE (Council of Biology Editors) *Style Manual* is nearing completion and should be available later this year. It is being prepared for authors, editors, redactors, printers, and publishers of the literature of the biological and biomedical sciences. The chapters in this new edition have been rearranged with the first three chapters providing information on preparation of research articles for publication in journals; the next chapter presenting the criteria for editorial review; and seven chapters specifying the details of proof correction, indexing, general style conventions, style in special fields, abbreviations and symbols, and word usage. An annotated bibliography of books that might be useful to writers and editors of biology papers constitutes the last chapter of the fourth edition. The chapter on "style in special fields" has been expanded, and for utility, has been subdivided into animal sciences, plant sciences, and microbiology. The "abbreviations and symbols" chapter has been simplified for easier use, and derived primarily from internationally-recognized authorities, such as the General Conference of Weights and Measures, the International Organization for Standardization, and the IUPAC-IUB Commission on Biochemical Nomenclature. The new edition of the CBE *Style Manual* should prove even more useful than the third edition.

The Council of Biology Editors was established in 1957 for the purpose of fostering education and improving communications in the biological sciences. It does this by providing efficient means for cooperation among editors and by promoting effective preparation of serial publications such as journals, reviews, monograph series, abstracting journals and services, indexes, microforms, and similar media. Most of the work of the organization is performed by its standing committees. One of these is the Style Manual Committee which is responsible for preparation and constant revision of the *Council of Biology Editors Style Manual*. The aim of this widely accepted manual is to present the forms of editorial style in manuscript preparation and to encourage the general technical uniformity that is required by most journals in the field of biology. When possible, recommendations are based on internationally and nationally agreed-upon-systems of procedure, terminology and nomenclature, units of measurement, abbreviations, symbols, and constants. In aspects of usage pertinent to all of the life sciences but for which standards do not exist, recommendations are avoided or are based on widely accepted practices.

The first edition of the CBE manual was entitled *Style Manual for Biological Journals* and was published in 1960. The second edition, retaining the same

Miriam Balaban (ed.), Scientific Information Transfer: The Editor's Role, 283-285.

name, was published four years later. By 1972, when the third edition of the *Style Manual* was prepared, its coverage was broadened beyond biological journals to include other forms of publication, such as technical reports, dissertations, and books. Therefore, the title was changed to *Council of Biology Editors Style Manual.* This title is to be retained for the fourth edition which is in preparation and should be available some time in 1977.

The fourth edition of the *CBE Style Manual* is being prepared for authors, editors, redactors, printers, and publishers of the literature of the biological and biomedical sciences. Like the previous editions, the manual addresses itself primarily to journals within which there are many types of articles. There are research articles concerned with experiments or observations, reports of new methods or apparatus, reviews, historical essays, editorials, and book reviews. Chapters I to III of this new edition of the *Style Manual* provide information on the preparation of research articles for publication in journals, and cover "Planning the Article," "Writing the Article," and "The Final Draft." Although they have been prepared mainly for authors, inexperienced editors may find them helpful as well.

Chapter IV is entitled "Editorial Review" and is directed principally to manuscript referees or consultants, who are asked by journal editors to judge the suitability of manuscripts for publication. The criteria listed in Chapter IV also may be used by editors for decisions on acceptance or rejection of papers, and should be of interest to authors whose papers, submitted for publication, are likely to be judged by these or similar criteria.

Chapters V through XI--those on typographic style, proof correction, indexing, and style conventions, both general and specialized--are likely to be used principally by editors and redactors. Some of these chapters indicate the authorities for style and usage, and the sources to be consulted for further details. Authors should use these chapters, while preparing manuscripts, for guidance in punctuation, in acceptable abbreviations and symbols, in word usage, and in other details not discussed in the opening chapters on how to write an article. Although the primary concern in all of these chapters is with writing research articles for publication in journals, much of the content of this fourth edition of the *CBE Style Manual* is relevant to the writing of other kinds of articles, as well as books, technical reports, and dissertations.

The major changes in the fourth edition are worth noting. The rearrangement in the sequence of chapters enhances the utility of the entire manual as a reference source. Some of the chapters have been completely rewritten in order to update and improve their content.

The chapter on "Style in Special Fields" has been expanded considerably over that in the third edition. In recognition of the trend toward specialization, biology has been separated into its discrete major areas in this chapter. It has been subdivided into three sections--Animal Sciences, Plant Sciences, and Microbiology. Coverage for the Animal Sciences includes nomenclature and systematics, anatomy, genetics and cytogenetics, immunology, hematology, physiology, and experimental materials; and for the Plant Sciences, nomenclature and systematics, genetics and cytogenetics, and physiology. The section on Microbiology covers the nomenclature of bacteria, fungi, viruses, and their genetics, plus a classification and keys to the identification of microorganisms. In this same chapter the styles of Chemistry and Biochemistry, and of Geography and Geology, are retained from the third edition, but in revised form.

For each disciplinary subdivision in the "Style in Special Fields" chapter, international conventions, standards, abbreviations, and symbols have been used whenever they have been available. Where such internationally accepted rules do not exist, national standards have been used, and the differences between American and European usage have been indicated. Some fields, such as neurophysiology, have not been included because their conventions have not been established by appropriate agreement within their professional societies, either national or international. In the hope that all fields of science may some day agree to use the same conventions in style to facilitate communication among all scientists, the CBE Style Manual Committee has adopted those recommendations of the American Chemical Society, the American Geological Institute, and the American Institute of Physics that are applicable to the needs of authors in biology.

The chapter on "Abbreviations and Symbols" has been simplified and refined from the two lists and three types of abbreviations and symbols in the third edition to a single list arranged in alphabetical order by term, with prefixed Arabic numerals, Greek letters, and italicized expressions disregarded in the alphabetization. When multiple abbreviations or symbols are given for a term, they are listed in decreasing order of preference. Abbreviations and symbols for chemical elements, genetic and taxonomic names, and structured coherent systems (made up of interdependent symbols in a logically consistent relationship) are excluded from the list; the only exception to the latter is the International System of Units (SI). The abbreviations and symbols for this chapter have been derived from two groups of sources. The primary sources are the internationally recognized authorities--the General Conference of Weights and Measures, and the IUPAC-IUB Commission on Biochemical Nomenclature--which are referenced in the list by letters. The secondary sources are the international and U.S. standards organizations, international scientific bodies, and major U.S. scientific publishing groups. These include the International Organization for Standardization, the American Society for Testing and Materials, the World Health Organization, the American Chemical Society, the American Institute of Physics, and the BioSciences Information Service, all of which are referenced in the list by numbers.

The word list of the third edition has been changed to a chapter on "Word Usage" for the fourth edition. It has been streamlined to eliminate most of the words that were not annotated, and those that were included only because they often tended to be incorrectly spelled. It is easy enough to check the correct spelling of a word in any good dictionary. The chapter now becomes an uncluttered guide to the proper use of words.

As in the third edition of the *Style Manual*, there is an "Annotated Bibliography" which, though not exhaustive, presents a list of other guides, dictionaries, and source books which should be of use to authors, editors, and copyeditors.

Plans are underway for a greatly expanded, comprehensive index for the new edition. The earlier versions of the *Style Manual* were justifiably criticized for the inadequate number of entries in their indexes. There is no excuse for a poor index in a reference book; it dilutes the value of the publication no matter how excellent the content. The CBE Style Manual Committee is acutely aware of the need for improvement, and is determined that this edition shall be indexed in depth to provide easy access to any item of information.

Those associated with the preparation of the fourth edition of the *CBE Style Manual* are convinced that its usefulness has been enhanced considerably by improving its content and format; by promoting the International System of Units, and the international and national standards and conventions developed since publication of the previous edition; and by including an index that does justice to the contents of the manual.

STYLE MANUALS AND GUIDES FOR AUTHORS AND EDITORS: PRESCRIPTIVE OR DESCRIPTIVE?

Maeve O'Connor, *Senior Editor*
Ciba Foundation,
41 Portland Place, London W1N 4BN, UK

ABSTRACT

Style manuals began as books of rules for copy-makers in printing plants and later in publishing houses; their use soon spread to editors and authors. Manuals, or guides, addressed to authors should be as concise and as readable as possible. Manuals are more likely to be prescriptive, while guides are descriptive, though the distinction cannot be pushed too far. *Writing Scientific Papers in English,* the ELSE-Ciba Foundation guide for authors, is a guide in spite of its use of the imperative mood. A forthcoming book, *Editing Scientific Books and Journals,* will be even more descriptive than the book on writing.

In his essay 'Of Studies' Francis Bacon said there were three kinds of books: 'Some books are to be tasted, others to be swallowed, and some few to be chewed and digested; that is, some books are to be read only in parts; others to be read but not curiously; and some few to be read wholly, and with diligence and attention.'

I'll link this quotation to my subject later but first I want to say something about the development of style manuals and guides for authors and editors. The earliest manuals of interest to this meeting are *Hart's Rules for Compositors and Readers* and the Chicago *Manual of Style*, first published in 1893 and 1906, respectively. Basically these were expanded versions of the accumulated house rules of the Oxford University Press and Chicago University Press. They soon became bibles, carrying perhaps too much authority with editors and copy editors in many fields of knowledge. Their prescriptions were often imposed on authors, some of whom began to follow slavishly the rules laid down, not always with good results.

The earliest manual addressed directly to scientists was probably Sir Clifford Allbutt's *Notes on the Composition of Scientific Papers,* published in London in 1904; in the USA Trelease's first book on the *Preparation of Scientific and Technical Papers* was published in 1925.

Instructions or guidelines also began to be printed regularly in scientific journals in the first part of this century. For example, the American Psychological Association's *Psychological Bulletin* published its first set of instructions for authors in 1929, though at that time it was called a guide. The APA's guides then grew in line with the expansion or explosion of journals, and their history is perhaps typical of what happened in other fields too. The 7-page guide of 1929 mushroomed into a 32-page article in the *Psychological Bulletin* in 1944 and then into a 60-page supplement to the journal in 1952 (when it was first called a *Publication Manual)*. This manual became a separate publication in 1957, followed by

Miriam Balaban (ed.), Scientific Information Transfer: The Editor's Role, 287-290.

another separate publication in 1967, with a 2nd edition in 1974, as described by Anita DeVivo (1).

Another manual for scientists is the one whose 4th edition has been described at this meeting by Philip Altman. The Conference of Biology Editors' *Style Manual for Biological Journals*, first published in 1960, was originally designed for authors, but since then it has changed its original direction as well as its title.

In Europe, CBE's younger counterpart, ELSE, had been trying since 1967 to produce a manual based on the 2nd edition of the *CBE Style Manual*. This task presented enormous problems and little progress was made until 1973, when Knut Faegri proposed that there should be a central manual plus separately produced booklets for the different language groups and disciplines. That was when Peter Woodford and I began to write our first book together. This was *Writing Scientific Papers in English*, subtitled *An ELSE-Ciba Foundation Guide for Authors*, and it was published in 1975 (2). Table 1 lists its contents.

Table 1

Contents list: *Writing Scientific Papers in English* (2)

Introduction

Planning

Preparing

Writing the First Draft

Revising

Refining

Typing

Submitting the Final Version

Responding to the Editor

Correcting proofs

Appendices:

- Steps in Writing a Paper
- Units of Measure and their Abbreviations
- General Abbreviations and Symbols
- Abbreviations in Biochemistry and Taxonomy
- Expressions to avoid

Bibliography

Index

To judge from its sales this book has in fact found favour with its intended audience of authors, and not solely with editors and publishers, as sometimes happens with manuals and guides prepared for authors by editors (3). We believe it succeeded because it was addressed to a single audience, whose character we kept firmly in mind during the preparation of the book. When we began drafting it, we decided that the book should be as short and as readable as possible, while covering all the necessary ground: scientists want to get on with their next piece of work, not study a hefty tome telling them in minute detail how to write up the piece of work they have just finished. As our book was for people writing for journals in a very wide range of disciplines, we couldn't in any case lay down rules for terminology or details of presentation. That is how the book became a guide, a description of generally accepted requirements for the structure of a paper and its various parts. It is true that the book often sounds prescriptive, because for brevity and directness we use the imperative mood, but we always return to the basic piece of advice to authors: 'ignore our prescriptions if they conflict with the instructions of your target journal'.

In writing our second book, a guide for editors, to be called *Editing Scientific Books and Journals*, we couldn't refer to anyone else's instructions. There aren't any, for one thing; and editors wouldn't stand for it, for another. There are as many different ways of editing as there are editors. So in this new book, which we expect will be published in 1978 (4), we have been even more descriptive than in our book on writing. Where the imperative mood creeps in, we have tried to make it clear that we are making suggestions, not giving instructions (see Table 2 for the contents list).

Table 2

Provisional contents list: *Editing Scientific Books and Journals* (4)

Editing Scientific Books and Journals is addressed to scientists who are part-time editors as well as to experienced or inexperienced full-time editors and copy editors. Three types of scientific publication are considered: multi-author monographs, conference proceedings, and journals.

Some of the problems of editing, and of copy editors, are common to the three types of publication, while others are characteristic of only one of them. By exploring policies and practices in the editing of journals and both kinds of book, we attempt to illuminate the subject in a totally new way.

As I indicated, our first book became a guide, not a manual. But what are the differences between style manuals and guides? Style manuals I think tend to enunciate rules, to go into great detail, and to include many lists. They are really reference books or textbooks, and are not for reading straight through. Guides, on the other hand, give a general picture rather than a fully detailed one, they describe good practice and principles based on the rules found in style manuals and elsewhere, and they are, or ought to be, organized to tell a continuously readable story.

On Francis Bacon's description, then, one could say that style manuals fall into the category of books that are to be tasted or read only in parts - that is, when needed for reference purposes. Guides are also used for reference purposes, and must be designed so that busy readers can easily find what they want; but to attract the readers it wants to influence, any guide should first and foremost be wholly digestible and worth reading with all diligence and attention.

Of course, this arbitrary distinction between style manuals and guides cannot be pushed too far. For one thing, some manuals should have been called guides, and for another some guides are more like manuals. My point is that since few scientists are willing or have time to tackle reams of instructions about writing, or advice about editing, a guide that is easy to read, short and to the point is more likely to promote good practice than a full-scale manual. I hope that both of our books, the published one on writing and the forthcoming one on editing, can be classified as true guides, and that they can and will be 'chewed and digested; ... read wholly, and with diligence and attention'.

Acknowledgement

I wish to thank Peter Woodford for helpful discussions during the preparation of this paper.

References

1. DeVivo A (1975) A new publication manual for psychologists. *Scholarly Publ* 7:37-47

2. O'Connor M & Woodford F P (1975) *Writing Scientific Papers in English: An ELSE-Ciba Foundation Guide for Authors*. Elsevier/Excerpta Medica/North Holland, Amsterdam

3. Harman E (1976) A reconsideration of manuscript editing. *Scholarly Publ* 7;146-156

4. Woodford F P & O'Connor M (1978) *Editing Scientific Books and Journals: An ELSE-Ciba Foundation Guide for Editors*. In press.

ON THE PREPARATION OF A REFERENCE MANUAL FOR AUTHORS AND TYPISTS OF AN INTERNATIONAL RESEARCH CENTRE

Andreas Jesse
Avenue Wendt 22, CH-1203, Geneva, Switzerland

Abstract

The specific requirements for preparing the reference manual are briefly presented. Some of its features are discussed: the manual is fairly detailed (70 pages) and is intended "for reference"; for "quick reference", there are an "Author's Short Guide" and a "Typist's Short Guide" (4 pages each); the chapters are self-contained treating specific tasks (writing, typing, correcting, printing, distributing); entries in the detailed subject index refer directly to (numbered) paragraphs to facilitate quick finding; specific instructions are presented clearly and concisely (e.g. sample references to all kind of literature, sample pages for the typist, checklists for authors and typists).

1. The need for a reference manual

CERN, the European Organization for Nuclear Research, is an international centre with the aim of advancing research in the field of elementary-particle physics and its related technology, such as accelerators, detectors and computers. It employs some 5 000 persons, of which rougly one third are "visiting" scientists and engineers who stay only for a limited spell of time.

In addition to such important turn-over of personnel, which poses certain problems as to the production of scientific papers, there is a great diversity of document types: several kinds of reports and an important number of preprints (in the field of elementary-particle physics, such advance copies of papers submitted for publication or presented at a conference are legion; they are widely distributed in order to provide for rapid communication). CERN produces some 1 000 preprints and reports per year, of which a half will eventually be published in the "open" literature.

Because of the staff limitations of CERN's editorial, typing and printing facilities, the preparation of scientific papers is de-centralized and left, more or less, to the individual research Divisions. However, in order to obtain a fairly uniform presentation of all CERN scientific documents and in order to produce the in-house printed documents by the most appropriate means in respect of time and cost, some instructions were required to guide authors and typists.

Miriam Balaban (ed.), Scientific Information Transfer: The Editor's Role, 291-298.

2. Short-comings of style manuals elsewhere

Some 50 author's guides, style manuals and similar handbooks were consulted to see if they could not be used or adapted for the purpose just mentioned. Clearly, the specific needs of this research organization still would have to be taken into consideration.

Although much useful material (especially references 1-6) could be found, the following short-comings became apparent:

- author's guides were either too short, say 1-10 pages, which causes problems when detailed instructions are looked for; <u>or</u>
- they were too long and too detailed, with more than 100 pages; being aimed at the professional writer/editor/printer, the "average" author or typist would not readily consult them (this also holds true for other specialists' literature, such as handbooks for illustrators, abstractors, indexers, documentalists, etc.);
- many of the guides were lacking good subject indexes and clear instructions (e.g. style of references covering all types of literature; simple rules concerning the preparation of slides and illustrations);
- finally, many of the guides, however useful, serve only a specific purpose as they are meant for authors and typists who prepare a contribution for a particular journal or publisher; what was required, was a more generalized guide, treating in particular scientific und technical reports as well as preprints.

All these aspects were taken into account, along with many others, when elaborating the "Reference Manual for the Preparation of CERN Scientific Papers".[7] Some of its features are presented below.

3. Features of the CERN reference manual

3.1 Purpose

The manual [7] is addressed to authors as well as typists. It is fairly detailed (72 pages, 6 Figures, 18 tables, 2 flowcharts, 1 checklist, 58 references; 8-page subject index) and therefore is intended "for reference" rather than being read. However, for "quick reference" and in order to satisfy specific requirements, important extracts are presented in an "Author's Short Guide" and "Typist's Short Guide" (4 pages each; including a checklist, style of references, proof correcting signs).

3.2 Contents

Utilization of the reference manual is facilitated by having self-contained chapters on the different phases of a paper: writing, typing, correcting, reproducing, distributing. There is an introductory chapter on scientific communication, and nine appendices contain detailed information on abbreviations, units, copyright, typing of mathematical formulae, etc. (see Fig. 1).

Fig. 1 Contents of the reference manual

3.3 Subject index

Being a manual for reference, the subject index is placed in the beginning. The subject index includes a listing of the applied terminology in order to help users as to which term to look up (this listing also served to control the terminology when elaborating the index). As a quick finding-aid, the entries refer to the particular paragraph on a certain page (e.g. 10:3 = page 10, paragraph 3), where often the wanted term is printed in italics).

3.4 Typing of papers

Instructions for the typist are presented in form of several sample pages. As shown in Fig. 2, the salient data can be seen at a glance.

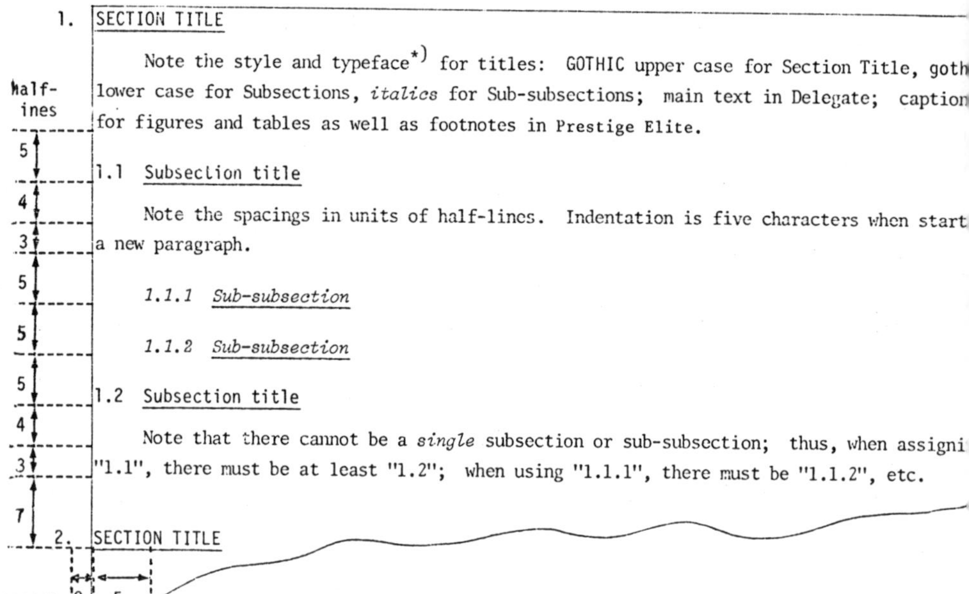

Fig. 2 Instructions for the typist

3.5 Preparation of figures and slides

It has been tried to give authors straight-forward, simple guidance. For example, Fig. 3 summarizes the line thicknesses and letter heights which are recommended for graphic representations, together with a warning that in case of high reductions (for camera-ready copies of papers with page limitations) these values are no longer valid.

Concerning slides (5 cm x 5 cm) it is said that projection in a medium-size auditorium will be satisfactory if the slide, i.e. its lettering, is legible at 0.25 cm distance from the naked eye; or, alternatively, if its 30 cm x 21 cm original is legible at 2 m distance, a postcard-size original (15 cm x 10.5 cm) at 1 m distance.

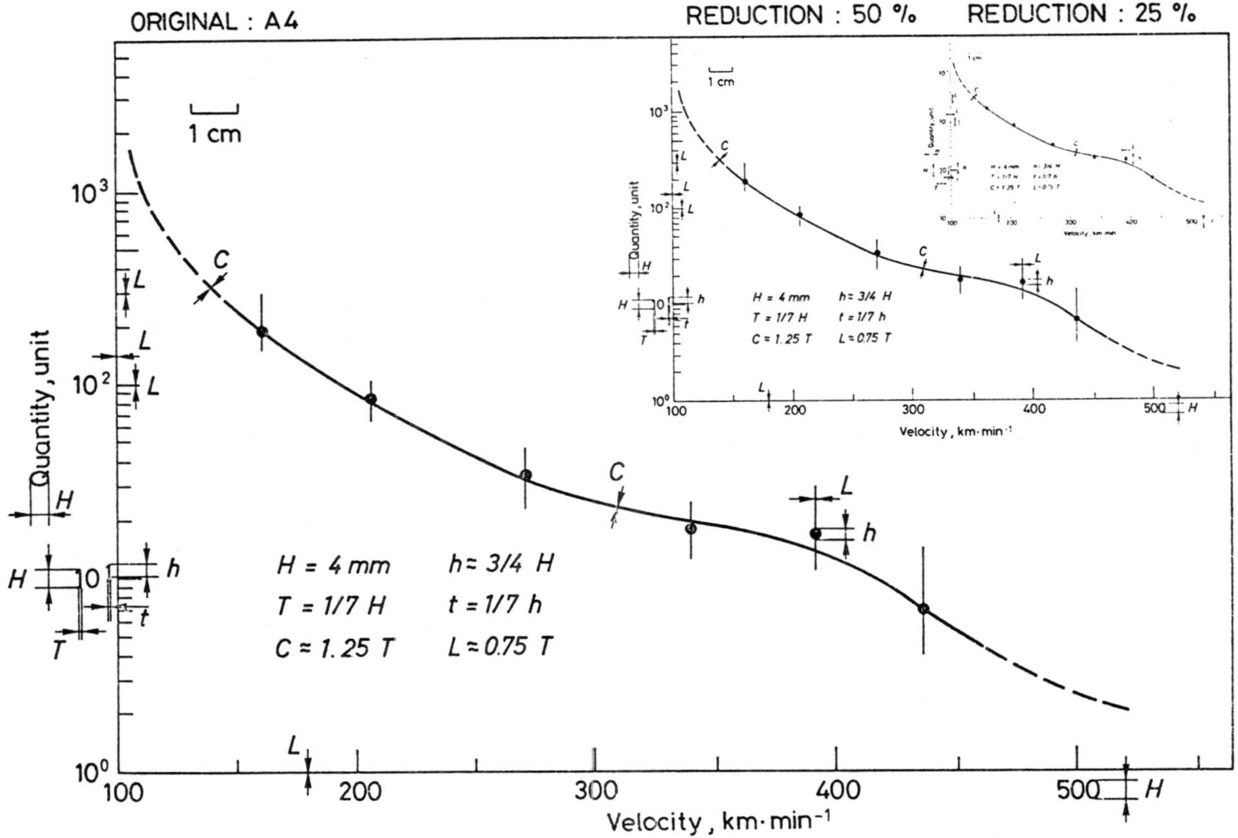

Fig. 3 Comparison of figures after various reductions (Note that the heights and thicknesses recommended for originals in A4 format are adequate for reductions to 50%; but not, however, to 25%)

3.6 Style of references

Great difficulties were encountered when trying to present rules how to cite references to the literature: 1. there are different standards among the different professions (librarians use other rules for catalogue cards than documentalists, which again differ from those applied by editors); 2. even within one and the same profession, there are different standards; 3. standards are not followed consistently (e.g. British publishers tend to apply a slightly different style of references than their American or continental-European colleagues); 4. most author's guides have the great difficiency in that they give only sample references for a journal article and a book, neglecting the many other type of documents. Fig. 4 shows the style of references as recommended by the CERN reference manual.

Kind of document		Examples	
		Detailed version	Short version
Periodicals	Article[a)]	F. Miller and L. Smith, Electron scattering, Nuclear Phys. B10 (1970) 99-101.	F. Miller and L. Smith[b)], Nuclear Phys. B10 (1970) 99.
	Preprint	J. Smith, Scattering amplitudes, Helsinki preprint TFT 71-22 (1973) (submitted to Phys. Rev.).	J. Smith, Helsinki preprint TFT 71-22 (1973) (subm. to Phys. Rev.).
Books	Whole book	D. Drijard, W.T. Eadie, F.E. James, M.G.W. Roos and B. Sadoulet, Statistical methods in experimental physics (North-Holland, Amsterdam, 1971).	D. Drijard et al.[b)], Statistical methods in experimental physics (North-Holland, Amsterdam, 1971).
	Contribution	C. Piney, Polynomial manipulation, *in:* Computer-aided design techniques (ed. E. Wolfendale) (Butterworths, London, 1970), 2nd ed., Vol. 1, pp. 283-314.	C. Piney, *in:* Computer-aided design techniques (ed. E. Wolfendale) (Butterworth, London, 1970), p. 283.
Conference Proceedings	Whole Conference proceedings	Proc. 7th International Conference on Cyclotrons and their Applications, Zürich, 19-22 August, 1975 (ed. W. Joho) (Birkhäuser, Basel, 1975).	Proc. 7th Internat. Conf. on Cyclotrons and their Applications, Zürich, 19 August 1975 (Birkhäuser, Basel, 1975).
	Contribution	F.J. Sacherer and T.R. Sherwood, The effect of space charge in beam transport lines, Proc. 4th National Particle Accelerator Conference, Chicago, 1971 [IEEE Trans. Nucl. Sci. 18(3) (1971) 1066-1067][a)].	F.J. Sacherer and T.R. Sherwood, IEEE Trans. Nucl. Sci. 18(3) (1971) 1066.
Reports	CERN Report	E. Hagebø and S. Sundell, The ISOLDE target system, *in:* The ISOLDE Isotope Separator On-Line Facility at CERN (eds. G. Rudstam and A. Kjelberg), CERN 70-3 (1970), pp. 63-80.	E. Hagebø and S. Sundell, CERN 70-3 (1970), p. 63.
	Internal Report	H.H. Umstätter, Calculations on septum magnets, internal report CERN MPS/Int. RF/63-3 (1963).	H.H. Umstätter, internal rep. CERN MPS/Int. RF/63-3 (1963).
Miscellaneous	Brochure	Electrical Research Association, Leatherhead, Surrey (1973), Thermal endurance.	Electrical Res. Ass., Leatherhead, Surrey (1973), Thermal endurance.
	Thesis	P. Müller, Mechanische Eigenschaften von Aluminiumoxid (Mechanical properties of alumina). Thesis (Braintown Univ., Braintown, NY, 1975).	P. Müller, Thesis (Braintown Univ., 1975).

a) The issue number may be included within ( ). For some periodicals which do not continue pagination from issue to issue, it is absolutely required [e.g. Sci. Amer. 233(4) (1975) 38-50].

b) *et al.* should not be used for less than three authors.

Fig. 4 Recommended style of references

3.7 Checklists

To help authors and typists when finalizing their work, for each of them a checklist has been devised. Fig. 5 shows the one for authors.

- ☐ *title* must be concise and informative
- ☐ *authorship:* author(s)' names, *affiliation* (standard addresses) and collaboration (if any)
- ☐ *abstract* briefly presenting the main ideas of the paper; covering purpose/method/results/conclusions/etc.; using specific terminology; detailed enough for the expert, but also intelligible to an outsider.
- ☐ *table of contents* (if any) clearly structured
- ☐ *introduction* mentioning background and purpose of the paper
- ☐ *numbering:* consistent, logical, clear numbering of chapters/sections, references, formulae, figures, tables, appendices (if any)
- ☐ *titles/captions/headings* must be concise and informative
- ☐ *figures and tables* must be clear and informative; good illustrative material
- ☐ *units/abbreviations/terminology* must be standard
- ☐ *conclusions* (forming a counterpart to the introduction: question-answer relation)
- ☐ *acknowledgement(s); copyright*
- ☐ *references* complete and standard; reference numbers in the text verified
- ☐ *approval* by Division Leader
- ☐ *model* of final document (if helpful); marginal notes for position of figures/tables (when attached separately)
- ☐ *last check;* approval for reproduction.

Fig. 5 Author's checklist

4. Outlook

The "Reference Manual for the Preparation of CERN Scientific Papers" and its accompagnning "Author's Short Guide" and "Typist's Short Guide" are presently being tested. This feed-back is essential for the preparation of an improved and up-dated version.

Whilst this "just-another-guide" has its justification for the European Organization for Nuclear Research, it is yet hoped that publishers, learned societies, research institutions, etc. could find a consensus to issue a single reference manual, which is valid and useful for all involved in publishing and processing of scientific literature; and last but not least, should be used to train future scientists and engineers at university level.

References

1) Europhysics Style Manual - a guide to authors of papers in Europhysics Journals, Europhysics News 3(4) (1972) 1-4

2) Guide for the Preparation of Scientific Papers for Publication (UNESCO, Paris, 1968; SC/MD/5)

3) Notes for Authors (The Institute of Physics, London, 1976)

4) Style Manual (American Institute of Physics, New York, 1973; rev. ed.)

5) A Manual of Style (Univ. of Chicago Press, Chicago, 1969; 12th ed.)

6) M. O'Connor and F.P. Woodford, Writing Scientific Papers in English (Elsevier, Amsterdam, 1975)

7) A. Jesse, Proposal for a Reference Manual for the Preparation of CERN Scientific Papers, internal report CERN DD/Int./76/5 (1976)

STYLE! WHY BOTHER?

Asa Kasher, *Editor, Philosophia (Philosophical Quarterly of Israel) Bar Ilan University, Ramat-Gan and Department of Philosophy, Tel Aviv University Ramat Aviv, Tel Aviv, Israel*

In the present paper I would like to point out what seems to be the betrayal of the editors. The latter takes the form of deliberate nurture and spread of an intellectual contiguous disease. A presentation of some of its symptoms will be followed by some suggestions concerning possible cures.

The common concept of style involves the author's conscious choice among alternatives offered by a language for the expression of what one knows, thinks or feels. However, what springs to a person's mind when he encounters a Style manual, a style sheet, or any garden variety of instructions for authors, is not which choices are offered to him but rather which choices are taken away from him. "In citing dates" - reads one style manual - "spell out the name of the month, precede it with the day and follow it with the year, as follows: 2 September 1675". Do not use the form "1st" or "4th" or the like." Another instruction, all too common, is to capitalize "the first word and each important word of the title" of an English book you mention. A journal of a certain royal institute (capital R, capital I) informs prospective contributors that the first line of each paragraph must be indented, except of the first line of the first paragraph of a chapter. The style sheet of a local quarterly devotes 750 words to instructions concerning the form of references, including directives such as the following: Titles of books should be in single quotation marks, in case they are in Hebrew; should be underlines, in case they are in Latin or Cyrillic; should be neither in quotation marks nor underlined, in case they are in Arabic.

"It is by style we are saved", said Henry James; alas, glancing through such lists of stylistic restrictions, it is all too natural for a prospective author to conclude that it is by style we are lost.

The most common argument in favour of stylistic uniformity is that effectiveness is gained by standardization. My counterargument rests on the observation that effectiveness is here crucially ambiguous. Devices and activities are not simply divided into effective and non-effective. Effectiveness judgments depend completely on the goals and means at hand. Now, when a paper in a journal is under consideration four different parties are involved, each entertaining its own goals and means. These are the publisher, the editor, the reader and the author. A publisher tries to maximize profits or in any case, to minimize costs. A possible consequence, e.g.: footnotes are printed at the end of the paper, causing every reader some inconvenience. An editor tries to maximize, say, reputation. Possible means for attaining such a purpose are, for exampl , extended solicitation of papers and omission of sections of discussion notes or book reviews. Natural victims: younger scholars whose papers are rejected or not published promptly, and the readers who are deprived of some professional services, such as book reviews and fresh contributions. A reader is usually interested in maximizing under-

Miriam Balaban (ed.), Scientific Information Transfer: The Editor's Role, 299-301.

standing of works which are relevant to his own work or interest. For that sake, auxiliary items, such as abstracts, key-words, all kinds of indexes, are very helpful, but someone should pay the price. Should it be a prospective author whose paper has been rejected because the space needed for it was spent on twenty abstracts? An author is always interested in maximizing the impact of his message and minimizing his efforts to produce the paper. His work is often too long, obscure, disorganized, even stupid. If it is too long, other prospective authors might suffer; if it is obscure, the readers will; if it is disorganized the publisher might be required to pay the extra expenses; and if it is stupid, the editor too is going to lose what he might have earned.

Since different goals and means are deeply involved, and consequently also different, or even incompatible criteria of effectiveness, the ruling parties cannot help it - they have to set aside some considerations, following a prevailing criterion of effectiveness. Assuming that in most cases both the publisher and the editor are interested in maximizing circulation within a certain audience, the power of attraction entertained by the journal is of central importance. But, as is well known, the appeal is in the eye of the beholder, and thus the author's convenience is overrode by the reader's desire to be free of any discomfort.

Since I accept the premisses of this argument, namely - that it is impossible to fulfil the incompatible requirements of all parties concerned, but I do not accept the conclusion, that is - I do not think that the reader's interests should prevail over the author's ones, let me try pointing out what went wrong in that train of thought, which led from a true premiss to a false conclusion.

Intuitively speaking, the job of the editor of a journal is to render a professional service neither to a community of writers who would like to publish, nor to a community of readers who would like not to perish, but to a certain professional sector or a learned community *at large*. The editor's job is, given the restrictions imposed on his activity by the publisher, to minimize the efforts needed, on the part of all concerned members of the community, for transmitting a mathematical result, a philosophical analysis, scientific findings, or what have you. Now, efforts are required at both sides of the channel and the distribution should be reasonable and just. It is obviously reasonable to ask the author tu supply bibliographical details, such as titles of books, names of their authors, but is it really necessary to mention the name of the publisher of each book and all the cities in which he manages to maintain major branches? Is it reasonable to assume that a reader will be able to follow a paper about Kant's philosophy but would not be able to figure out whether a title is that of a book and not of a paper, unless some quotation marks and underlinings are supplied?

"Style" - said the American philosopher Blanshard - is not the feather in the hat, it is the feather in the arrow." But even a sniper will often miss his target if he pays too much attention to the eyelets on the feather of his arrow.

The average professional, be it in philosophy or in physics, masters such proficiency in his field that he hardly needs most of the details appearing in the lists of references of papers he reads. The beginners need almost all the information supplied, but then their time is much less important so to speak, than that of the writer, who is - at least in philosophy and similar areas - the author himself and not one of his assistants or colleagues.

The gist of these remarks should be obvious by now. Stylistic instructions of two kinds: those that should be followed in order to enable a professional reader to put under scrutiny every detail of the content of the paper, and all the rest. A certain extent of formal shaping is very helpful - the title should always precede the paper itself, to mention an extremely obvious point - but beyond a small cluster of necessities authors should be exempted from any requirement which is not directly related to the processes of getting down on paper their thoughts, and furbishing it for the sake of clarifying the content.

There is one exception to this liberal editorial policy in matters of style. Authors must be consistent in their own practice, throughout one and the same work. Philosophers have developed a variety of arguments and devices concerning the rational reconstruction of human, intentional activity. Since the style of the present paper is bound to be telegraphic let me mention just one point out of the whole philosophical field. Our first attempt at understanding any deliberate activity on the part of any person, aims at explaining his acts by some rules of rationality, including maxims governing the amount and form of information given by a speaker to his audience. At this stage of groping for understanding, the reader has the right to assume that the author has a reason for saying what he says rather than anything else, in the form which he uses for expressing it, rather than any other form. It is, then, natural to assume, for example, that different occurrences of the same proper name denote the same person, unless there is sufficient evidence to the contrary. Similarly, when different marks are used for designating anything, the author should have had some reason for not using the same mark twice, and which would be the best reason if not that the different marks stand for different items? Thus, authors are naturally assumed to be consistent in their own practice, and for their own benefit - they better be.

What went wrong in the former argument in favour of the reader's convenience is, then, that the alternatives are not just two: either the author's interest prevails or that of the reader does. The third alternative is to consider the global professional effort, required by both authors and readers. The practical consequence of this approach are easy to draw.

When I have presented by argument to some colleagues, more than once raised eye-brows could have been detected. Doesn't style - I have been asked - has to do with the aesthetic qualities of your journal? Effectiveness aside, what about harmony, serenity, solidity, all meant to be produced by strict stylistic guidance? Well, beauty too is in the eye of the beholder. I do not see any serious reason to believe that aesthetic quality is gained by uniformity of trivialities such as indentations of first lines of first paragraphs of first chapters. Diversity is beautiful, isn't it? (Ask your wife about her dresses, if you fail to grasp the point easily).

"Subject without style" - said a famous philosopher, Collingwood - is barbarism; style without subject is dillettantism. Art is the two together." Too many authors provide content without real style; too many editors impose style without real content. These combinations of barbarism and dillettantism should, indeed, be desolated, paving the way for a fair and reasonable combination of effectiveness and creativity.

DISCUSSION on "The Fourth Edition of the CBE Style Manual"
by Philip L. Altman

"Style - Why Bother?"
by Asa Kasher

ANDERS MARTINSSON - to Prof. Kasher. My general remarks I can embody in very few words, namely: I strongly disagree. Many of those things that you specified, including those points where you caught the laughter of the audience, are well thought-out details which serve their purpose - perhaps not the author's purpose but the reader's purpose. Many of these stylistic details, the layout of the text in the publication, the placing and arrangement of figures, etc. really serve a very particualr purpose, namely saving the time and labour of the selective reader.

BASIL WALBY - I think that I should be delighted not to have to employ so many staff on the journals if we could adopt the laissez-faire position you suggest. Unfortunately the extra costs incurred in my whole production operation would completely counteract any editorial savings. This is why we request some standard form of presentation by the authors. It will be very interesting to see what happens if journals relinquish their responsibility for quality control at any or all stages of production.

RUTH WEISS - This time I speak as a somewhat unwilling editor. It is a job that keeps being thrust my way. I try to distinguish between editing of the sort which definitely contributes toward clarity and copyediting. If by editing the information that the scientist is putting out is going to get over to the reader more easily and more clearly, then I feel I am performing a function that has a very definite use. If I can improve language, reduce redundancy, increase clarity, then maybe at least I have done some good. On the other hand, I agree with you entirely. The fiddling little things that, as an editor, I am unwillingly led to do, (e.g. standardizing references) is, to say the least, unproductive work. We must distinguish between the main task of an editor (which has a lot of sense) and the subsidiary tasks which would link up with the plea of the authors for simpler instructions. I think we would all benefit if we would view these 'subsidiaries' in scale.

ASA KASHER - I am not an adherent of something that would be expressed by the slogan "no style at all", but I do believe in only minimal style requirements. By minimal I mean just real substantial requirements like the name of the author, coherent title and an address, and, of course, editing the wording and things like this to maintain clarity of expression, is of utmost importance in philosophy, as I guess anywhere else, so I have nothing against this inclination.

Now concerning computers, I have some reservation. I was recently instructed by my university that while grading student's work I should use only one figure grades, because in the computer just one character has been reserved for this purpose. Now I feel I should react very furiously to such things. If your computer instructs me to use certain characters, so to hell with your computer, not to hell with my work. Now if there comes the day in which editing will be done partly by computers I would suggest something like this. Not to put the whole burden on the authors who will be slapped back in their faces by our computer for not abiding regulations, but let us do it this way. Every author will be supplied forms where he would fill in his title, address, position, references then let your computer process the forms. Why should the author do it?

KNUT FAEGRI - I think the provocative ideas presented by Professor Kasher, and also by others regarding the necessity for style manuals and their value were of the greatest interest. I think we may suppose that guides for authors, style manuals, or whatever they are, are not produced by sadists who want to hurt authors. This may be one of the premises even if sometimes producers of guidelines say this hurts me more than it hurts you (this is always a very empty excuse). I think we may also agree that in some cases guidelines are too detailed; they contain ideas which many authors feel are completely irrelevant and only serve the self-gratification of editors or publishers in many cases, perhaps this has not been thought about but has been taken over simply because it has always been that way. So the question is, where is the balance? the factor which has to be taken into account is firstly economy. You can't produce a journal uneconomically. Secondly, it is legibility for the understanding. I think it was Dr. Mizoguchi who put forward a very interesting provocative question here, namely what does the word correct actually mean, what does better mean? It is the better thing we are interested in, not the correct thing, correctness in itself is really nothing. It is the thing we are aiming at, and if we could get some ideas on this I think it would be great. Dr. Helbig.

KLAUS HELBIG - Prof. Kasher was certainly provocative, but I think he was far off the mark. Many of his ideas, of course are correct, but it seemed to me, for instance, that he would prefer to do without the abstract he says that there should be a title, an author, and then there should be the text which would place the abstract apart from the layout. There is no question that the general reader reads the titles, the author's name, and the abstract. He rarely reads an article unless the abstract catches his interest, because if he did not work this way he would not be able to keep track of the many journals. I believe that many of the things Prof. Kasher does not like are done really to get the author's ideas across, to get the paper across. Only if he sells himself correctly will his paper be read. One of the intents of the style guides is to make the correct sales pitch from the author's standpoint. The editor himself would like his journal to be read, but if he wants to achieve that, he has to have a certain packaging, and this is one of our problems.

PHILIP ALTMAN - In response to the scientist who asked for a single style guide to replace all the guides and manuals available, I'd like to say that an international organization might be the first step in the production of a single style manual. I think you need an international organization where people can get together, discuss editorial problems and produce a single manual that will be acceptable to all. If nothing comes out of this conference other than the establishment of a permanent international organization, then one major barrier to cooperation will have been eliminated.

RUTH WEISS - At the risk of repetition may I perhaps strengthen these two pleas for uniformization. I am concerned, as I said, with editing which is thrust upon me. This has been not an established journal, but a new unit. I have looked with horror at all the conflicting instructions and I have come to the conclusion that I have to make my own decisions. I think it is a terrible thing that this should be so, and I am also grateful to this forum because there is an impressive need for standardization.

DAVID SHEPHARD - I'd like to come back to the author here. We did have a conference 3 or 4 years ago in which we had a session on the author's problems, and I think that many editors do recognize the problems of authors. One thing we haven't considered is cooperation between authors and editors. I would like to see some kind of cooperation between authors and editors, and I certainly would like to see authors making more use of editorial services. We have to have some control, we have to produce a package, but I think there is common ground somewhere in the middle between editors and authors. Editors are not quite as critical and as purist as some people think, and authors have much more to say than many editors think. I'd like to confirm what Dr. Altman said: maybe we can reach a consensus, and perhaps if we do have another meeting, we could have a session with authors taking part.

KNUT FAEGRI - I think that if the organization which Phil Altman here envisages can give a description of the best method, not the correct method, but the best method, I think we may perhaps hope for it being gradually adopted. Changing a little during adoption of course, but I think this is perhaps a better way of going than trying to lay down hard and fast rules which will meet with all kinds of resentments and with people saying this is not the way we do it. If we just describe the thing in nice terms like Mel Day did and say this is the most economical way of doing it and it is still quite legible I think we could perhaps achieve something and not in my lifetime, perhaps in yours, but I agree with you. That would be really a very fine thing we can do. Are there any other comments to this point.

ASA KASHER - I like the adversity and would want just to express another point. The situation might be different in philosophy than in the natural sciences. A philosophical article is never too structured - you cannot first present the experimental data and then the consequences and so on. Style is an important part of the content.

KNUT FAEGRI - Are there any other questions or problems emanating from this meeting? Well, if there are no more, I thank you all and the contributors for being brief so that we have managed to keep within our timetable. We all thank Phil Altman for having organized this session.

STANDARDIZATION OF WRITTEN INFORMATION IN SCIENCE AND TECHNOLOGY

Dorothy U. Mizoguchi
Technical Editor, Japanese Cancer Association, Kami-Ikebukuro 1-37-1 Toshima-ku, Tokyo 170, Japan

ABSTRACT

The Science and Technology Agency of the Japanese Government set up a committee in 1973 to draft standards for written communication of information in science and technology. This was part of a plan to develop a National Information System for Science and Technology (NIST), the purpose being to draft gúidelines for standardization of primary and secondary publications for a better information flow in science and technology. Guidelines have been drafted for 'Writing of Abstracts','Bibliographic Information' and 'Abbreviation of Journal Titles', work is in progress on 'Names of Institutions and their English Equivalent'.

In addition, the Standardization Committee of the Japan Documentation Society has drafted a standard for 'Layout of Periodicals' with the hope of making it a national standard.

These standards are discussed in relation to standards in other countries and international standards (ISO).

Although there are many international and national standards pertaining to written communication of scientific research results, they do not seem to be much heeded by the editors and publishers of scientific journals, or by scientists who write papers. These standards are classified as "Documentation" in the International Organization for Standardization (ISO), and many of these are adopted as national standards by ISO member countries. These "documentation" standards include "Presentation of Contributions to Periodicals," "Layout of Periodicals," "Bibliographic References," "Abbreviation of Titles of Periodicals," etc., which can be and should be used by academic societies and commercial publishers for scientific journals. Some journal editors are not even aware that there are such standards, even in their own country. This is also true in Japan, and the situation is worse because there are no Japanese standards on documentation. In Japan, there are only Japanese Industrial Standards (JIS) and Japanese Agricultural Standards (JAS), and a few standards on documentation included in JIS are international standards covering terminology and symbols that have to be used in the industry.

In order to correct this situation, the Standardization Committee of the Japan Documentation Society has been working hard to have documentation standards included in JIS, and it seems possible that this will be effected within a few years. One difficulty in having documentation standards adopted for JIS is the bureaucratic rivalry among the Government offices. Anything that has to do with academic societies must be in the hands of the Ministry of Education (recently name changed to Ministry of Education, Science and Culture). Everything to do with science and technology must be in the hands of the Science and Technology Agency. Japan Industrial Standards are drafted and published by the Agency of Industrial Science and Technology, although actual publication and sales are in the hands of Japan Standards Association.

Miriam Balaban (ed.), Scientific Information Transfer: The Editor's Role, 307-309.

Therefore, something to do with scientific journals, published by learned societies, must be approved by the Ministry of Education, but this Ministry has no interest in standardization.

With all these difficulties present, we have been aware that standardization in documentation is necessary for scientific journals, especially in the present age of international cooperation in scientific research and publication of its results. We can of course adopt many of the international standards but they are for English and other European languages, and there are many points that must be different when expressed in the Japanese language.

The Science and Technology Agency set up the National Information System for Science and Technology (NIST) to develop a system for better flow of informations in science and technology. As a part of this plan, the Agency set up a committee in 1973 to draft standards or guidelines for written information in science and technology. In other words, these would be a draft for future standards in documentation for scientific journals (primary publications) and secondary publications. The term "standards" cannot be used at present and the term "guidelines" is at present used. The Committee selected about a dozen subjects and started work with the aim of completing one subject per year. With the work starting in 1973, the Committee has drafted four guidelines, and work is on way on another at present. In drafting these guidelines, the Committee referred to international standards (ISO) and national standards of other countries, notably those in Germany, United Kingdom, and the United States of America, so as not to deviate from international practice.

The first to be taken up was the "Writing of Abstracts" because this was one of the most urgently needed. Scientific journals written and printed in Japanese cannot be read by European and American scientists, and they must at least be provided with English abstracts. Therefore, this guideline had to give abstract writing in both English and Japanese. The English part is much the same as the international standard (ISO/R214).

The next to be taken up was the "Bibliographic References". This is the most controversial subject and there was much debate on which was the best way to give references, in the text and in the reference list. Finally, the draft gave all different systems of giving references in the text and for listing the references. This guideline also gives the standard way for giving references in English and Japanese, and the English one is on the line of international standard (ISO 832).

The third guideline is on "Abbreviation of Periodical Titles." Since there is an international standard (ISO 833), there was no need to draft a new one for English and other European language abbreviations, and the guideline was drafted entirely for Japanese titles. These abbreviations do not apply to Japanese titles written in Romanized letters, but only to those written in Japanese (or Chinese) letters.

The fourth and the latest guideline is on "Names of Institutions and Their English Equivalent." This is also mostly on the names of Japanese institutions; what can or cannot be omitted, which and how to abbreviate, what to add, etc. The foreign institutions are to be given *per se*, as written in the original document, nothing more, nothing less, even if they are too abbreviated for easy understanding.

In addition, there is a guideline on "Layout of Periodicals" drafted by the Committee on Standardization of the Japan Documentation Society. It is on the line of the international standard (ISO/R8), but with modifications in the case of Japanese. This is now being offered as the first draft for the Japanese Industrial Standard.

Standards on documentation are not binding. You are not required by law to use these standards, and there are no fines or punishment for not using them. In contrast, industrial standards must be followed exactly, if industrial products are to be sold and used by the customer. This is the most difficult part in the dissemination of standards. Many people who must and should use these standards do not know their existence. Even if they did know of their presence, they will not use them if they do not like the standards. This is also true of the guidelines prepared by the Science and Technology Agency. The printed guidelines have been distributed to academic societies in the field of natural sciences, but are being hardly noticed. The Science and Technology Agency has held several meetings in various cities to explain these guidelines and the meetings have been well attended but not by the editors or editorial staff of society journals and mostly by librarians. The librarians are the most interested in journals because they have to process them and use them as reference tools. The fact that the utilization of scientific journals by librarians would become much easier if scientific journals are standardized is not understood by editors and editorial staff of scientific journals.

The only exception to the non-use of international standards by scientific journals is the standard on "Symbols for Quantities and Units" (ISO 1000), adopting the SI units. Majority of physics and chemistry journals use this standard, and some of biology and medical journals, but not all. There is even a Japanese standard (JIS Z8202), and standards on conversion tables for SI units (JIS Z8438 to Z8445). These units should be used by all scientific journals but, actually, the old system is still being used by many journals.

Another international standard that should be adopted by scientific journals is the "International List of Periodical Title Word Abbreviations" (ISO 833). This has been published since 1974 but many European and American scientific journals are not using this system yet.

Japanese scientists use European and American journals in their field as a model for Japanese journals. Therefore, even if they are told that they should follow international standards, like SI units, they will not use them because their counterpart journals in Europe and America are not using them. This is a very difficult business for those pushing standardization of format and style in scientific journals.

AN EDITOR'S VIEW OF STANDARDS AND STANDARDIZATION

J. Christopher Rigg
Centre for Agricultural Publishing and Documentation,
P.O.B. 4, Wageningen, The Netherlands

ABSTRACT

Some standards are good; others are mediocre. As means of communication, they are slow, expensive, poorly covered by secondary services and often poorly edited. Perhaps through lack of staff and finance, standardization bodies sometimes fail to encourage standards prepared by others. The process of standardization is sometimes based on unsatisfactory compromises rather than on scientific or technical evidence. Perhaps such deficiencies contribute to the poor implementation of standards on presentation of publications.

Future standards need to allow for new modes of preparation and production of publications. They need to be based on research among producers (authors, editors) as well as users (readers, secondary services, libraries). Authors and editors cannot be expected to have a complete set of standards to hand, so rules need to be consolidated and interpreted for particular groups of users, for instance through editors' association.

There are satisfactory international standards on the choice and placing of bibliographic elements, for instance in books (ISO-1086), journals (ISO-8), articles (ISO-R-215) and translations (ISO-2384). For abstracts (ISO-214), good rules exist; there are ISO standards for romanization (transliteration) of the main scripts of the world apart from Chinese and various other Asiatic scripts. On bibliographic references, ISO-690 is unsatisfactory. On indexing, good national standards exist. ICSU-Codata (1974) and the international scientific unions are working on guides for the presentation of numeric data; for statistical aspects the best source is ISO. Sources of recommended technical terms are indicated.

Free descriptors: presentation; publications; standardization; bibliographic elements; terminology; editors; authors; quantitative data.
bibliographic elements; terminology; editors; authors; quantitative data.

UDC 001.818+002:389.6(100)

Also to be published as a section of the Unisist Manual of Information Handling.

Miriam Balaban (ed.), *Scientific Information Transfer: The Editor's Role,* 311-319.

At the Unisist Working Group on Systems Interconnection 4 years ago, I reviewed the need for international standards in the preparation and presentation of scientific publications. The paper I then wrote gave a check on what progress has been made since and what still remains to be done. My present paper is in two parts: first general shortcomings of standardization; second a selection of aspects where standards may help those who make publications. Even for international standards, I can mention only a few, some of excellent quality, others mediocre.

1 Standards in general: some criticisms and remedies

To encourage the flow of scientific and technical information, we need standards of two types: minimum specifications, for instance of items of information required on the title page, and unification, for instance in range of paper size or image arrangements in microfiche. Perhaps also a third type, standards of perfection that cannot be attained but that can be aimed towards. Standards are difficult to make. They result from compromises and may leave much to be desired in quality. Yet in criticizing the product and the system, I do not criticize the hard-working servants of that system! Standardizers are often idealists. In practice, standards themselves are a poor means of communication and do not meet some specifications laid down for other types of publication.

1.1 Cost and inavailability

Standards are dear, often more than $1 per page, because the institutions tha produce them try to supplement their meagre government support by passing on costs of standardization. The result is low sales. Few abstracting services cover standards on information processing, exceptions being Bulletin Signalétique and РЕФЕРАТИВНЫЙ ЖУРНАЛ. ИНФОРМАТИКА.

Occasionally draft standards have been published cheaply and distributed widely in journals, for instance various United States standards including that on abstracts (ANSI-Z39-14) and the international standard on presentation of translations (ISO-2384, 1977).

Recently there has been unnecessary expense for users of ISO standards through transformation of 'recommendations' into 'standards'. Whether the tex differs is difficult to check without buying the new impression: for instance ISO-690 (1975) on bibliographic references contains the same 'printer's' [?] errors as ISO-R-690 (1968).

1.2 Introspectiveness

With Unisist aid, an information centre on documentary standards (ISODOC) has been established to monitor guidelines as well as official standards. It is to be hoped that it will help to overcome a tendency towards introspectivenes in standardization. It might also help in obtaining a better balance of interests in international working groups.

Surprising was a comment, for instance, in the foreword of a recent ISO draft abridged International Standard Bibliographic Description for monograph (ISO-TC46-1168), which claimed that two main systems were in use for bibliographic references: ISO-690 and ISBD (prepared by the International Federatio of Library Associations). Partly wishful thinking and partly confusion of library cataloguing with citation.

1.3 Obsolescence

International standards commonly take over 4 years from the drafting of first proposals to publication. There is therefore always a lag between what is technically feasible and what is recommended. There is also a danger that standards will represent conservative ideas rather than innovation.

ISO Technical Committee 46 'Documentation' deserves praise for the large program of revision in the past 4 years. Quality still leaves room for improvement: inept rules; unhappy compromises; perhaps some bias.

1.4 Competing standards

A recent example is the preparation by the British Standards Institution of a Glossary of Documentation. Perhaps through lack of funds, the British group has played little part in international work on ISO's Vocabulary of Documentation and Information(DIS 5127). It could ultimately have been cheaper for the British group to concentrate on the international project and to introduce some British pragmatics alongside the German thoroughness and French poeticism. However such duplication is not entirely harmful, since evolution is only possible if the species (here the standard) varies.

A welcome trend is to promulgate international standards as national standards, sometimes even untranslated as in the Netherlands.

1.5 Testing of standards

In a recent Unisist paper, Dr French (1976) of the ISO Secretariat in Geneva took up this problem, which was raised in the Unisist Working Group on Systems Interconnection in 1972.

Standards should contain a note inviting comment and criticism. Unfortunately little comment comes from outside the group involved in preparing and drafting standards. The best indication one can obtain of how much a standard is implemented is by research on usage, for instance checking how many journals mention their International Standard Serial Number (ISSN) and how many place it at the top right of the cover. In most countries, a standard does not have the support of law indicated in the preface of Soviet standards that "Anyone failing to implement this standard will be liable to prosecution".

Survey of implementation of standards requires research, for instance by university departments of library science, information and communication.

As an example if such a study, Kövendi (1975) examined the implementation of 9 ISO standards on presentation of periodicals. In his sample of 150 periodicals on library science and archives, some rules were implemented by almost all and others by hardly any. Of his sample, 87 provided indexes but only 26 met, even to some degree, the requirement of ISO-R-8 (1955) that issues containing indexes state so on the cover. Reasons for such omission must be examined.

One reason for failure to implement is that journals try to publish what their readers want and attach second priority to the needs of libraries and secondary services. In popular or practical journals, for instance, users may attach little importance to abstracts (Westwood et al. 1976).

ISO-R-690 comes off badly in Kövendi's survey. For reference to a whole book, 127 out of 137 journals included all the elements required but for reference to an article in a journal only 3 out of 134 included all the elements recommended: place and date of publication were usually omitted!

Surveys of order and punctuation of bibliographic elements in lists of references in journal articles (Olivecrona 1972; Rigg unpublished) show that style manuals and guides like US-CBE (1972), GB-RS (1965, 1974) and GB-RSM (1972) are more implemented (for US-CBE about 6%) than ISO-R-690 (about 2%). One reason is certainly the need for a format that allows reference in the text either by name-date or by reference number.

2 Standardization in publishing of scientific research

2.1 Conservatism in publishing

I must eat my words of 4 year ago about the influence of the computer on publishing in contrast to secondary services. The computer has now established its value in searching and compiling bibliographic data. At our office in the

Netherlands, we search data stored in California as a matter of routine.
But now we also have typewriters with magnetic storage and we are examining how to eliminate repetitive keyboarding of typescript and printed matter. The systems and equipment available for our work are flexible but will undoubtedl encourage restriction of typographic design to a few modules.
In automation, mathematical copy and languages with unusual characters will raise problems in choice of character sets.
Microform editions of journals have appeared. Most scientists are not enthu siastic about them. Such journals are not the revolution some expected. Microform is becoming a supplement rather than a substitute for printed matte
Publishers are showing their flexibility and standards for publications must also be flexible to exploit the new options.

2.2 Consolidation of standards

Standards have been dispersed but ISO and Unesco recently published a handbook of over 50 standards on information transfer (ISO-SH-1 1977). Some defin material requirements; some define requirements for particular types of work or particular types of publication; others define a particular feature like International Standard Book Numbers (ISO-2108). For Unisist, I prepared an annotated bibliography of standards (Rigg 1974, being reprinted with updating
The number of standards I traced on certain matters, for instance microform (about 500 standards) and drawings (about 200), exceeded all expectations. Surprisingly there is little guidance to the author or editor on illustration despite all the standards on technical drawing.
Probably about a tenth of the titles listed in 1974 have now been supersede and much new material has been published. The ultimate user cannot keep pace with such change and the rules can only reach him if they are consolidated into manuals and textbooks. Editors associations are an important channel bot for the implementation and the criticism of standards.
In a Dutch group, we attempted such consolidation for authors (and editors) as one of the proposals for the revision of ISO-R-215 (Rigg et al. 1976). We gathered from about 42 standards of ISO, hoping that the bare bones were essential but referring the author to sources if he cared to check.
The editor would need that manual and much more, dealing with many hundreds of standards and glossaries. Unisist is attempting to deal with that problem in a proposed manual of information handling, in which several of us are working on a chapter entitled Preparation and Presentation of Documents containing Scientific or Technical Information. Among the most comprehensive existing guides of this type is Butcher (1975). Besides books and journals, it deals with proceedings and other publications with many contributors, with multivolume works and with occasional series.

2.3 Bibliographic identification

For many of the workers in ISO, 'presentation of publications' is merely the choice and placing of bibliographic elements in a publication, often called 'title information'. This reflects the dominance of librarians and secondary services in bibliographic standards published in the 1960s. Professor Anders Martinsson has coined the term 'biblid' for a compact set of identification elements and this concept has reached the Draft International Standard 30 (DIS-30) approved in February in Stockholm. Unfortunately his concept has been somewhat emaciated in committee.
Bibliographic identification, library cataloguing and bibliography should be a continuum but there are many discrepancies in names, definitions and specifications of bibliographic elements between standards on presentation and those on bibliographic reference.
With these reservations, there are adequate international rules for presentation of books (ISO-1086), journals (ISO-8, in press), articles (ISO-R-215; proposals for revision to be presented in November at Paris), research reports (in press as DIS) and translations (ISO-2384).

Unisist has prepared guidelines on occasional series (Giniaux 1976)and on dissertations (Unisist 1976). Further guidelines are being prepared on composition of titles and on proceedings (for which the best existing manuals are Manten, 1976 and BS-4446).

Adequate rules seem to be lacking for annual reports, sheet maps and drawings, the latter two either as separate publications or as accompanying materials.

2.4 Guidelines on composition

Guiding the author on how he should write is dangerous ground for standardizers. The standard on abstracts (ISO-214, 1976) has successfully ventured into composition. But there is a widespread view that rules on how to write and what to include are matters for style manuals, of which plenty exist, and textbooks of technical writing (such as Woodford 1968). The gap is perhaps mainly in advice to the scientist trying to write in an international language: ELSE has made start in this matter. Many courses are being set up and there has been some exchange of experience.

2.5 Terminology and nomenclature

A separate technical committee of ISO deals with the principles of terminology and has done excellent work (for instance ISO-R-704 and ISO-R-860).

The scientist often thinks he knows the recommended terms of his speciality. But the editor ruefully discovers otherwise and must gather a large collection of glossaries of terms from a variety of sources to prove his point: for instance from scientific unions and associations, many of the 164 technical committees of ISO, United Nations agencies, and other multinational bodies.

Unesco has sponsored clearinghouses for thesauri and glossaries in Warsaw (PL-IINTE) and Vienna (INFOTERM). INFOTERM will shortly be publishing a massive bibliography (Verlag Dokumentation, Pullach bei Munchen, West Germany). Their output needs to reach the groups preparing international style manuals, as in ELSE and US-CBE.

2.6 Transliteration

Japan and ISO have recently switched allegience to the Kunrei system from the modified Hepburn system (ISO-3602). ISO-2805 (in press) provides rules for about 50 non-Slavic languages written in Cyrillic. Proposals to align ISO-9 (Slavic Cyrillic) with the unaccented system used in ISO-833 (periodical title-word abbreviation) and in the Unisist Reference Manual failed to gain a majority at Brussels in May 1976.

International rules are lacking for Caucasian, Ethiopian, Indian and southeast and central Asian alphabets and syllabaries. Supplementary Roman and modified Roman letters give some difficulty and are touched on in ISO-833.

It is a pity that standards do not stress the difference between character substitution and assimilation of word roots. For instance, ISO-R-843 transliterates the Greek word for house as 'oikos' but in scientific terms the Latinized form 'oeco-' or the simplified form 'eco-' should be used. In view of the range of systems in use, particularly for Cyrillic, an interchangeable typing head is better than character substitution (as in §1.1).

2.7 Numeric data

ISO-TC12 (ISO-31 Pts 0-13), ICSU-Codata (1974) and some of the scientific unions have completed much work since 1972. Not all have accepted the Système International in entirety and few of those who have done so realize all the implications. The switch is not difficult if reference data is available in the new mode. Unfortunately even some reference data published recently by international scientific unions and United Nations agencies is still in old modes.

We have seen a revolution in units, but we need a similar revolution in names of quantities. There are the rudiments of system in a few quantitive terms. The groundwork of a more systematic approach exists mainly in French and could be adopted in English. German and Russian raise more difficulties. This matter is under discussion in the International Union of Pure and Applied Chemistry.

Guides on tabulation and graphic representation of data are unsatisfactory, though some recent ones are better (GB-RS 1975; Maltha 1976).

Statistical aspects of data, in particular interpretation and indication of scatter, have recently received attention in ISO (2602, mean and confidence interval; 2854, mean and variance; 3301, comparison of means of paired observations; 3534, statistical terminology and symbols).

In February, Canada circulated a first proposal for the revision of ISO-690 (ISO-TC46-WG6-42) but for the moment editors can best stick to existing style manuals. There will be considerable difficulty in reconciling librarians (dealing with documents as a whole) with secondary services and authors (interested in the intellectual divisions of documents), and with the long-suffering but not always accurate typist. Editors need a uniform system and internationally IUB-CEBJ (1973) has agreed on one suitable for either the name-date or number system but their short form of bibliographic description is less informative than those of US-CBE (1972), GB-RSM (1972), GB-RS (1974) and O'Connor & Woodford (1975).

2.9 Indexes and filing order

IUB-CEBJ (1972) offers a pragmatic solution for order of references. Butcher (1975) and Anderson (1971) describe more of the intricacies of indexing, both in analysis of content and in form of entry. For more elaborate needs, BS-1749 (filing order), BS-3700 (indexes) and ANSI-Z39-4 (indexes) are good. For indexes of persons and corporate bodies, the International Federation of Library Associations has published several guides (e.g. IFLA 1975).

2.10 Proof correction

Perhaps proof correction is a dying art. The author must submit correct copy or have his reputation damaged as a reward for slovenly copy preparation. Nevertheless a proposal for unification of proof-correction signs is under study and with some allowance to local practice has been published as BS-5261.2 (1975).

3 Conclusion

Standard are the imperfect result of an imperfect system. Nevertheless the rules they contain merit careful consideration and need interpretation by groups within associations of editors. The best way of judging what weight to attach to particular rules is to help to draft them and criticize them. Don't be scared off by the long opaque document numbers!

References

Only the latest versions published are listed for standards. Published draft international standards (DIS) are listed unless they have been voted down. International organizations are alphabetized by acronym and national organizations first by two-letter country code (ISO-3166) and then by acronym; DIS is alphabetized with ISO.

The form of reference is a personal view of what ISO-690 should be. A space dash (n-rule) indicates a 'jump' of bibliographic level: contribution -- monograph (or conference) -- serial. Full stops separate the main bibliographic groups. The equals sign introduces a parallel title, either of a parallel text or a parallel edition. Brackets are used for data not stated in the publication, for instance ICSU-Codata's date of publication, for an issue number where it is not essential to identification, and for functions, for instance '(Ed.)'.

Anderson, M.C.; 1971. Book indexing. Cambridge University Press, London. iv+36 pp. -- Cambridge Authors' and Printers' Guides.
Butcher, Judith; 1975. Copy editing: the Cambridge handbook. Cambridge University Press, London. ISBN 0-521-20550-6; £6.50. xii+326 pp.: glossary; index.
French, E.J. (ISO, Geneva); 1976. A methodology for testing and evaluating standards with special reference to information, documentation and related fields. Unisist, Unesco, Paris. -- SC.75/WS/102. 11 pp.
GB(Great Britain)-BS(British Standard)-1749; 1969. Alphabetical arrangement and the filing order of numerals and symbols. 13 pp.
GB-BS-3700; 1976. Recommendations: the preparations of indexes to books, periodicals and other publications. 10 pp.
GB-BS-4446; 1969. Specification for the presentation of conference proceedings. 10 pp.
GB-BS--5261.1; 1975. Guide to copy preparation and proof correction. Pt 1. Recommendations for preparation of typescript copy for printing. 22 pp.
GB-BS--5261.2; 1976. Copy preparation and proof correction. Pt 2. Specification for typographic requirements, marks for copy preparation and proof correction, proofing procedure. 15 pp.: illus.
GB-RS (Royal Society); 1974. General notes on the preparation of scientific papers. 3rd edition. GB-RS, London, ISBN 0-85403-065-4. 31 pp.: figs. Supersedes 1965 edition.
GB-RS, Symbols Committee; 1975. Quantities, units and symbols. 2nd edn. 54 pp. Supersedes earlier texts of 1969 and 1951 and 1st edn (1971).
GB-RSM (Royal Society of Medicine); 1972. Units, symbols and abbreviations: a guide for biological and medical editors and authors. Edited by George Ellis. RSM, 1 Wimpole St., London W1M-8AE. ISBN 0-9501555-4-3. 36 pp.
Giniaux, G.; 1976-08. Guidelines for the presentation of publishers' series. Assisted by F. Cestac, H.E. Kennedy and C. Rigg. Unesco, Paris. 17 pp. Unisist-SC-76-WS--74.
ICSU(International Council of Scientific Unions)-Codata(Committee for Data in Science and Technology), Task Group on Publication in the Primary Literature; (1974). Guide for the presentation in the primary literature of numerical data derived from experiments. = Guide pour la présentation dans la littérature primaire des données numériques déduites des expériences. Unesco, Paris. 7 pp.: 35 refs.
IFLA(International Federation of Library Associations), Working Group on Organization of Headings for Voluminous Authors; 1975. The arrangement of entries for complex material under headings for personal authors. IFLA Committee on Cataloguing, c/o Reference Division, British Library, Great Russell St., London. 29.5 cm; viii+6 pp. ISBN 0-903043-8 £1.50; $4.
ISO(International Organization for Standardization)-DIS--8; 1975-12. Documentation. Presentation of periodicals. = Documentation. Présentation des périodiques. 5 pp.
ISO-R--9; 1968-09. International system for transliteration of Slavic Cyrillic characters. 8 pp.
ISO-R-30; Bibliographic strip. = Manchette bibliographique. 2 pp.
ISO-31.0; 1974. General principles concerning quantities, units and symbols. 14 pp.
ISO-DIS--31.1; 1975. Quantities and units of space and time. 14 pp. To supersede ISO-R--31.1 (1965).
ISO-DIS--31.2; 1975. Quantities and units of periodic and related phenomena. 7 pp. Supersedes 1958 edition.
ISO-DIS--31.3; 1975. Quantities and units of mechanics. 19 pp. Supersedes 1960 edition.
ISO-R-31.4; 1960. Quantities and units of heat. 9 pp.
ISO-DIS--31.5; 1975. Quantities and units of electricity and magnetism. 24 pp. Supersedes 1965 edition.

ISO-31.6; 1973. Quantities and units of light and related electromagnetic radiations. = Grandeurs et unités de lumières et de rayonnements. 10 pp.
ISO-R-31.7; 1965. Quantities and units of acoustics. 9 pp.
ISO-31.8; 1973-04. Quantities and units of physical chemistry and molecular physics. 13 pp.
ISO-31.9; 1973-04. Quantities and units of atomic and nuclear physics. 12 pp.
ISO-31.10; 1973-07. Quantities and units of nuclear reactions and ionizing radiations. 15 pp.
ISO-R-31.11; 1961. Mathematical signs and symbols for use in physical sciences and technology. 6 pp.
ISO--31.12; 1975-04. Dimensionless parameters = Paramètres sans dimensions. 6 pp.
ISO--31.13; 1975-12. Quantities and units of solid state physics. = Grandeurs et unités de la physique de l'état solide. 18 pp. *f* 34.65.
ISO--214; 1976. Documentation. Abstracts for publications and documentation. = Documentation. Analyse pour les publications et la documentation. 11 pp.
ISO-R-215; 1961. Presentation of articles in periodicals. = Présentation des articles de périodiques. 4 pp.
ISO--690; 1975-09-01. Documentation. Bibliographical references. Essential and supplementary elements. = Documentation. Références bibliographiques. Eléments essentiels et complementaires. 8 pp.
ISO-R--704; 1968-04. Naming principles. = Principes de dénomination. 15 pp.
ISO-833; 1974. Documentation. International list of periodical title word abbreviations. = Documentation. Liste internationale d'abréviations de mots dans les titres de périodiques. 38 pp. Supersedes ISO-R-833 (1968). Based on Unisist & ICSU-AB (1970).
ISO-R--843; 1968-10. International system for the transliteration of Greek characters into Latin characters. = Système internationale pour la translittération des caractères grecs en caractères latins. 7 pp. Being superseded by ISO-843 (in preparation, DIS approved Brussels 1976-05).
ISO-R--860; 1968-10. International unification of concepts and terms. = Unification internationale des notions et des termes. 15 pp.
ISO--1086; 1975-08. Documentation. Title leaves of a book. = Documentation. Feuillets de titre d'un livre. 2 pp. Technically almost identical to ISO-R-1086 (1969).
ISO-2108; 1972. Documentation. International Standard Book Number (ISBN). = Documentation. Système international pour la numérotation des livres (ISBN). 4 pp.
ISO-2384; 1977. Documentation. Presentation of translations. = Documentation. Présentation des traductions. 7 pp. Draft published in Translation News 1974-03 4(1)6-14.
ISO--2602; 1973. Statistical interpretation of test results. Estimation of the mean; confidence interval. = Interprétation statistique de résultats d'essais; estimation de la moyen; intervalle de confiance. 5 pp.
DIS-2805; 1974. Documentation. Transliteration of non-Slavic languages of the Soviet Union using Cyrillic characters. = Translittération des alphabets des langues non slaves de l'Union Soviétique utilisant des caractères cyrilliques. 17 pp.
ISO-2854; 1976. Statistical interpretation of data. Techniques of estimation and tests relating to means and variance. = Interprétation statistique des données. Technique d'estimation et tests portant sur des moyennes et des variances. 46 pp. *f* 75.60.
ISO-3166; 1974. Codes for the representation of names of countries. = Codes pour la répresentation des noms de pays. 22 pp.
ISO-3301; 1976. Statistical interpretation of data. Comparison of two means in the case of paired observations. = Interprétation statistique des données Comparaison de deux moyennes dans le cas d'observations appariées. 6 pp. *f* 16.05.
ISO--3534; 1977-07. Statistics. Vocabulary and symbols. = Statistique. Vocabulaire et symboles. 44 pp. Supersedes ISO-R--645 (1967) and ISO-R--1786 (1970) "Statistical vocabulary and symbols. First and second series" and

DIS-3725 (symbols).
ISO--3602; 19.. . Documentation. Romanization of Japanese.
ISO-DIS-3725; 1975. Statistics. Symbols. = Statistique. Symboles. 4 pp.
ISO-DIS-5127.1; 1977-03. Information and documentation. Vocabulary. Section 1. Basic concepts. = Information et documentation. Vocabulaire. Chapitre 1. Notions fondamentales. v+42 pp.: index.
ISO-DOC (International Information Centre on Standards in Information & Documentation); 1975-04. Standards in the field of documentary reproduction: international, American, British, French and German standards. c/o DIN, Berlin. 26 pp.
ISO-SH-1 (Standards Handbook 1); 1977. Information transfer. ISO, Geneva. ix+516 pp.: 53 ISO standards. ISBN 92-67-10017-3.
IUB-CEBJ (International Union of Biochemistry, Commission of Editors of Biochemical Journals); 1973. The citation of bibliographic references in biochemical journals). Recommendations (1971). -- Biochemical Journal 135:1-3. Also: Biochimica et Biophysica Acta 1973 329:1-3.
Kövendi, Dénes; 1975. The presentation of documentation, library and archives journals and the ISO standards. -- Unesco Bulletin for Libraries 29(9)206-219.
Maltha, D.J.; 1976. Technical literature search and the written report. Pitman, London. ISBN 0-273-00399-2. viii+175 pp.
Manten, A.A.; 1976. Symposia and symposium publications: a guide for organisers, lecturers and editors of scientific meetings. Elsevier, Amsterdam. ISBN 0-444-41419-3. x+155 pp.: refs; index.
O'Connor, Maeve (Ciba Foundation, London); Woodford, F. Peter (Institute for Research into Mental & Multiple Handicap, London); 1975. Writing scientific papers in English: an ELSE--Ciba Foundation guide for authors. Elsevier, Amsterdam. ISBN 90-219-4035-3. vii+108 pp.
Olivecrona, C.; 1972. Enhetliga manuskript: referensmaterial och kommentarer. Nordisk Publiceringsnämnd för Medicin, Biomedical Documentation Center, Karolinska Institutet, Stockholm. Loose-leaf.
PL-IINTE (Instytut Informacji Naukowej, Technicznej i Ekonomicznej, Al. Niepodległośći 188, 00-950 Warsaw); 1976. Bibliographic Bulletin of the Clearing-house at IINTE: supplement 1975 ... . Warsaw, Poland. 73 pp.
Rigg, J. Christopher; 1974. Unisist bibliography on the preparation and presentation of documents containing scientific or technical information. Unesco, Paris. SC.74-WS-21. 183 pp.
Rigg, J.C.; Maltha, D.J.; Manten, A.A.; 1977. Rules for authors preparing papers for learned journals or similar publications. iv+50 pp. ISO-TC46-WG7-88 rev.
US(United States)-ANSI(American National Standards Institute)-Z39-4; 1968. Basic criteria for indexes. 12 pp. $3.25.
US-ANSI-Z39-14; 1971. American national standard for writing abstracts. $3.50. Also: Journal of the American Society of Information Science 1970 21(5)351-357.
US-CBE(Council of Biology Editors); 1972. Style manual. 3rd edition. American Institute of Biological Sciences, Washington, D.C. 20016. xii+297 pp.
Unisist, Working Group on Primary Sources of Information; 1976-08. Guidelines for preparation of texts of dissertations for their presentation and publication. Paris. SC-76-WS--79. 18 pp. Prepared by G. Vanderborght with assistance from A.J.C. Wilson, G. Lorphèvre and J.C. Rigg.
Unisist-SC74-WS-20; 1974. Reference manual for machine-readable bibliographic descriptions. Compiled by M.D. Martin. Unesco, Paris. 71 pp.
Westwood, M.N.; de Hertogh, A.A.; Kretchman, D. (ASHS Publications Committee); 1976-04. Membership publishing behaviour and attitude towards ASHS (American Society of Horticultural Science) publications. HortScience 11(2)88-91. 1 fig.; 5 tables.
Woodford, F. Peter (Editor); 1968. Scientific writing for graduate students: a manual on the teaching of scientific writing. Rockefeller University Press, New York. x+190 pp.

PRESENTATION OF NUMERICAL DATA IN THE PRIMARY LITERATURE

EDGAR F. WESTRUM, JR., SECRETARY GENERAL CODATA

DEPARTMENT OF CHEMISTRY, UNIVERSITY OF MICHIGAN, ANN ARBOR, MICHIGAN 48109 USA

Abstract

CODATA (Committee on Data for Science and Technology) has devised a "Guide for the Presentation in the Primary Literature of Numerical Data Derived from Experiments". This report is concerned with the presentation of numerical data capable of verification by repetition such as those on the physical and chemical properties and behavior of material systems. General recommendations that apply to all branches of the physical sciences are made concerning the information that should be reported so that the data may be compared and correlated with those obtained from other studies. Some existing specialized guides that expand upon these recommendations are listed in the bibliography; others are under preparation.

It is not a style manual for writing scientific papers; rather it is a statement of the minimum information that is needed to ensure that the reader can understand the quantitative data, can assess their precision and accuracy, and can recalculate the results when values for auxiliary data change. Although the author of a paper has the primary responsibility for providing the reader with the type of information outlined in this Guide, the document also provides journal editors and referees with a set of consistent, considered criteria for judging the completeness and acceptability of papers insofar as the reporting of numerical

Miriam Balaban (ed.), Scientific Information Transfer: The Editor's Role, 321-328, *All Rights Reserved.*

quantities are concerned. Adherence to these principles will permit evaluators to consolidate the author's results with existing data and facilitate their incorporation in critical compilations. Appropriate guides for the geo- and bio-sciences are under production within CODATA.

INTRODUCTION

This historical role of the primary literature as the major channel for the communication of experimental results to interested readers is well established and has been clearly stated in a report by a CODATA Task Group:[1] "The remarkable progress of natural science has been possible by virtue of the systematic structure of scientific knowledge; namely, new research results are built upon the achievements of predecessors and this excessive accumulation of knowledge has led to the construction of the huge edifice of modern science. The essential factor in this construction process of science is a smooth communication of research results, as scientific information, among research workers, so that scientists are well informed of the achievements of their predecessors and can be rid of research duplications. Thus, scientific information serves as the key link connecting research workers. The applicability of the achievements of science and technology also depend upon this feature of scientific knowledge." There are, however, some serious difficulties with the primary literature as a source of information, and more specifically as a source of data. The first of these is the volume of literature published. There is ample statistical evidence of its continuing growth. This difficulty gives rise to certain specific problems. First of all, the volume of material presented for publication strains the capacity of many journals and leads editors to impose restrictions on the number of pages which the authors can devote to comprehensive presentation of their data results. Frequently, only summary tables, exemplary graphs may be included. The full exposition of measurement conditions, calibrations, etc. must be sacrificed. As a result, only a fraction of data are ever available to the user and those data which are presented cannot in many cases be adequately interpreted, analyzed, or compared to other results. It is not an overstatement to say that the primary literature is hostile to adequate data publication! However, this means that those concerned with the compression of the literature in the form of compilation and critical evaluation of data (sometimes called the "tertiary literature") often do not have the necessary raw material to do an adequate job of compilation, evaluation, and fitting the new data in to the already extant--sometimes conflicting--sources of data. As a matter of fact, Dr. David Lide, of the National Standard Reference Data System in the United States has estimated that between 50 and 90% of the data in the literature cannot be critically evaluated because the authors had failed to give sufficient information to allow its analysis and interpretation and comparison with other results. This is a serious indictment since the cost of the production of scientific literature is very high and to have much of it unsalvageable is inexcusable.

For these reasons, the CODATA (Committee on Data for Science and Technology) of ICSU (International Council of Scientific Unions) has addressed itself to the problem of "making the literature safe for data" if one may be permitted to perturb a phrase harking back to the days of World War I. But even the secondary literature is also hostile to data. By and large, the abstracts to be found in the secondary literature are concept abstracts rather than data abstracts.

THE GUIDE FOR THE PRESENTATION OF DATA....

To meet the needs in the core sciences of chemistry and physics and other similar laboratory sciences, a CODATA Task Group on the Presentation of Data in the Primary Literature in 1973 published a "Guide for the Presentation in the Primary Literature of Numerical Data Derived from Experiments". This Bulletin[2] in English has been made available to all the participants on the literature tables so that I do not need to recite to you its very useful contents. The definitive version is in English but it has been translated in French, Spanish, Japanese, German, Swedish, and Russian and has been already widely disseminated.

Let me tell you just a little about it. First of all, the Guide has only to do with the presentation of data. It does not tell the author how to write a paper. It does tell him the minimum information needed to ensure that the reader can understand the data presented. Remember that the primary literature is to a large extent a peer group communication device and a worker in a frontier field does not always feel obliged to explain to the non-expert readers the details of his endeavors. It also ensures that the author provides the reader with enough information so that he can assess the accuracy and precision of the data presented and that he can recalculate these data values when auxiliary values change.

Some of the fringe benefits of the Guide are that it does provide considered criteria for journal editors and referees as to what constitutes adequate presentation. We will be happy to make available copies to editors who wish to use them in sending them to authors who do not meet the minimum standards for the presentation of data. Finally, it should be noted that the Guide reflects the needs and the experience of data evaluators. Adherence to the suggestions in it will help to ensure the incorporation of the data presented in the compressed scientific literature.

THE CONTENTS OF THE GUIDE

First of all, the Guide is concerned with the description of the experimental procedures involved. This requires that the author give a suitable definition of the systems studied, that he explain the method used for sample characterization, that he enumerate the measurements made, that he tell something about the apparatus, about the experimental procedure and indeed about the performance of the measurement systems, i.e., the calibration, etc. The author is also obliged to tell the reader something about the reduction of data. This would include assumptions made about the nature of the experiments and the complete description of the physical models that the author is using for interrelating his data, the identification of adjuvant data used, as well as the identification of standards, fundamental constants, key values, and other items utilized.

Finally, the author is given guidance in the assignment of imprécision and inaccuracy to his data so that he can convey on the basis of his broader experience some notion of the ultimate reliability of the data.

All of these things are discussed in much greater detail in the Guide itself. But doesn't any scientific author instinctively know the answers to these questions? The answer appears to be "No!", judging from the statement made by Dr. David Lide earlier in this presentation. As a matter of fact, the Guide to which I have already made reference is not the total solution to the problem of data presentation. Other adjuvant and parallel endeavors are needed.

THE TOP OF THE PYRAMID

I like to regard the needs for the edification of authors in terms of a pyramidal stack of instructions. The Guide that I have just discussed is, in fact, conveniently referred to as the top of the pyramid for the core sciences--chemistry and physics--and related disciplines broadly utilized by scientists in many other areas. The Guide that we have been discussing is designed to deal in generalities with the needs of data presentation so that the data themselves may be salvaged and incorporated into the tertiary literature. But different disciplines will need to have their own "disciplinary guides" pertinent to the particular data problems in their field. Thus, chemical kinetics, thermodynamics, crystallography, spectroscopy, etc. will each need to have a disciplinary guide for the guidance of their authors. Fortunately, guides do exist for the disciplines that I have enumerated but there are many other disciplines for which we do not yet have such disciplinary guides. It is a continuing function of the Task Group to encourage international unions and other such bodies to create disciplinary guides as supplement to that of the Guide.

But there are many sub-disciplinary areas which are very important in the sense that a lot of data is produced in them and it is used very widely by other scientists. Not long ago, thermodynamic measurements were made largely by people who call themselves thermodynamicists. At present most thermodynamic measurements are made by people who call themselves, for example, a biochemist, a physicist, etc., etc. This makes it even more necessary to inform these authors adequately how the data that they take needs to be described in order to be meaningful to scientists of other disciplines who will utilize or compile the information. Thus we find a need for Sub-Disciplinary Guides for such fields as combustion calorimetry, differential thermal analysis, gas chromatography, etc. And so the data guidance pyramid broadens out at the base from the Guide at the top, into the disciplinary area, and then into the sub-disciplinary area. We would estimate that the availability of approximately 20 to 30 disciplinary guides and perhaps an equal number of sub-disciplinary guides would reasonably well serve in the general area of the core sciences today. You can judge from the references given at the back of the Guide(2) how far along we are toward the achievement of this goal.

THE THREE PYRAMIDS

CODATA has broadened its scope beyond the core science --not because it was running out of problems in the central area--but because it was felt that its expertise would be helpful also in other areas; in particular, in the geo- and astro-sciences, and in the bio-sciences. In a very real way these two groups of sciences taken together with chemistry and physics constitute rather different data problems. For example, in both the geo- and bio-sciences, one is confronted with non-numeric data as well as with space- and time-dependent data. These give rise to special problems. Whereas laboratory work in the geo- and bio-sciences may well be covered by much of what is to be found in the Guide(2) but in the field and observational aspects separate and distinct guidance may be necessary.

For this reason, CODATA has more recently set up task groups to prepare Guides in the other two areas. One such task group is that on the Presentation of Biological Data in the Primary Literature. This Task Group headed by Dr. P. L. Altman is charged with the creation of a Guide useful for the presentation of biological data. This Guide is now in the penultimate version and hopefully will be published soon as a CODATA Bulletin.[3]

I will not say more about it inasmuch as Dr. Altman is here and will in a few moments relate to you the progress which has been made in the production of such a Guide.

Still another Task Group on the Presentation of Geoscience Data in the Primary Literature headed by Dr. G. A. Wilkins has also produced a penultimate version of a "Guide for the Presentation of Geoscience Data". This Guide is broad enough to include not only the numerous aspects of geo-science but as well that involved in the astronomical areas of science. Hopefully, it too will soon be available in the form of a CODATA Bulletin.

No doubt other pyramids exist and in due time will be recognized and Guides devised to occupy their apex. In addition, disciplinary guides will need to be created for the diverse disciplines involved under each of them. Perhaps in a few years, it will be possible to collect together all of these Guides and make them available in a single volume for the convenience of authors everywhere. This is especially important as the inter-disciplinary nature of science increases in complexity.

OTHER MATTERS

Indeed there are many other concerns discussed in the Guide having to do with details on the importance of presenting original data as free from interpretation as possible in a form which it can be retrieved and used to test other theories and hypotheses than that which the authors had in mind.

Another important problem has to do with the question of the incompleteness of data as presented in the scientific literature. Here no single solution is available but some sort of a data depository--a supplementary access point which the author of an article and the journal may use--needs to be utilized if some of his valuable data and the conditions under which they were determined cannot be included in the primary publication. He can deposit a more complete file in such a depository and make reference to its availability in the journal article. Various other alternatives such as microform publication--either available on demand or actually inserted in an envelope in the cover of the journal itself have been provided by some journals and are available to others in the form of the NAPS (National Auxiliary Publication Service) in the United States and of VINITI in the U.S.S.R.

Data important to meet the needs of the evaluator and the specialist--as distinguished from the casual reader--may often be made available indefinitely through one of these mechanisms.

REFERENCES

1. "Study on the Problems of Accessibility and Dissemination of Data for Science and Technology", CODATA Bulletin No. 16 (1975), CODATA Paris, France, (Report of the CODATA Task Group on Accessibility and Dissemination of Data).

2. "Guide for the Presentation in the Primary Literature of Numerical Data Derived from Experiments", CODATA Bulletin No. 9 (1973), CODATA Paris, France, (Report of the CODATA Task Group on Publication of Data in the Primary Literature).

3. "Guide on the Presentation of Biological Data in the Primary Literature", Report of the Task Group on the Presentation of Biological Data in the Primary Literature to appear in late 1977 as a CODATA Bulletin.

BARRIERS IN THE FLOW OF NUMERICAL DATA FROM PRIMARY TO SECONDARY TO TERTIARY LITERATURE. A CASE HISTORY

A.S. Kertes
Institute of Chemistry
The Hebrew University, Jerusalem, Israel

The Solubility Data Project, an IUPAC venture, is concerned with preparing a comprehensive critical compilation of numerical solubility data in all physical systems. The work involves reprocessing and consolidation services, comprising compact compilation, rationalization and simplification, critical evaluation and selection of recommended experimental numerical data. The mechanism of the scientific work and the final output of the project both involve a number of innovations, demonstrated on a few typical cases, of which the most important is that career scientists, actively engaged in generating new experimental data, are the backbone of the project. The intellectual judgement involved, however, is only as good as the primary information available. This is exemplified by analyzing in detail a case-system of compilation, interpretation, synthesis and critical evaluation, carried out by subject specialists. If consolidating primary literature data are to be demands considered worthy of the effort of top scientists, the referee-editor institution that exercizes stewardship and control over primary scientific and technical information has to meet the challenge of present-day rigorous scientific requirements in order to improve the transfer of reliable numerical information from primary, to secondary and tertiary sources.

In the last ten years or so, international scientific unions affiliated with the International Council of Scientific Unions, and international engineering organizations affiliated with WFEO, the World Federation of Engineering Organizations, independently or in cooperation with agencies such as CODATA, the Committee for Data in Science and Technology, have initiated compilation and critical evaluation of numerical data on an international basis.

How to communicate and disseminate numerical data effectively in the chemical science and technology has been a problem of serious and growing concern to IUPAC, the International Union of Pure and Applied Chemistry for the last two decades. The steadily expanding volume of numerical scientific and technological information, the formulation of new interdisciplinary areas in which chemistry is a partner, and the links between these and existing traditional subdisciplines in chemistry, along with an increasing number of users, have been considered as urgent aspects of the information problem in general, and of the numerical data problem in particular.

Among the various numerical data projects initiated and operated by various IUPAC commissions, the Solubility Data Project - on which we now report - is probably the most ambitious one. It is concerned with preparing a comprehensive critical compilation of data on solubilities in all physical systems, of gases, liquids and solids. Both the basic and applied branches of almost all scientific disciplines require a knowledge of solubilities as a function of solvent, temperature and pressure. Solubility data are basic to the fundamental understanding of processes relevant to agronomy, biology, chemistry, geology and oceanography, medicine and pharmacology, and metallurgy and materials science. Knowledge of solubility is very frequently of great importance to such diverse practical applications as drug dosage and drug

Miriam Balaban (ed.), Scientific Information Transfer: The Editor's Role, 329-336.

solubility in biological fluids, anesthesiology, corrosion by dissolution of metals, properties of glasses, ceramics, concretes and coatings, phase relations in the formation of minerals and alloys, the deposits of minerals and radioactive fission products from ocean waters, the composition of ground waters, and the requirements of oxygen and other gases in life support systems.

The widespread relevance of solubility data to many branches and disciplines of science, medicine, technology and engineering, and the difficulty to recover solubility data from literature, lead to proliferation of published data in an ever increasing number of scientific and technical primary sources. The sheer volume of data has overcome the capacity of the classical secondary and tertiary services to respond effectively.

The Solubility Data Project developed a mechanism which involves a number of innovations in exploiting the literature fully, and which contains new elements of a more imaginative approach of transfer of reliable information from primary to secondary/tertiary sources. *The fundamental trend of the Solubility Data Project is toward integration of secondary and tertiary services with the objective of producing in-depth critical analysis and evaluation which are characteristic to secondary services, in a scope as broad as conventional tertiary services.*

While the proportion of secondary services - of the review article type - is generally increasing due to the rapid growth of all forms of primary literature, the review articles become more limited in scope, more specialized. The disturbing phenomenon is that in some disciplines, certainly in chemistry, authors are reluctant to treat even those limited in scope reviews exhaustively. There is a trend to preselect the literature, sometimes under the pretext of reducing it to a manageable size. The crucial problem with such preselection - as far as numerical data are concerned - is that there is no indication as to whether the material excluded was done by design or by less than thorough literature search. We are equally concerned that most current secondary sources, critical in character as they may be, give scant attention to numerical data.

On the other hand, tertiary sources - handbooks, reference books and other tabulated and graphical compilations - as they exist today, are comprehensive but as a rule, uncritical. They usually attempt to cover whole disciplines, thus obviously are superficial in treatment. Since they command a wide market, we believe that their service to advancement of science is at least questionable. Additionally, the change which is taking place in the generation of new and diversified numerical data, and the rate by which this is done, is not reflected in an increased third-level service. The emergence of new tertiary literature sources does not parallel the shift that has occurred in the primary literature.

Being the status of current secondary and tertiary services as they are briefly stated above, the innovative approach of the Solubility Data Project is that its compilation and critical evaluation work involve consolidation and reprocessing services when both activities are based on intellectual and scholarly reworking of information from primary sources. It comprises compact compilation, rationalization and simplification, and the fitting of isolated numerical data into a critically evaluated general framework.

The type of output generated by the Solubility Data Project can be demonstrated on the Water—2-Butanone system which belongs to the subdivision of liquid-liquid solubility systems. It has been prepared by Prof. S.S. Davis of the Pharmacy Department, University of Nottingham, U.K. in June 1975.

It is the standard practice that any given solute-solvent system consists of two essential parts: I. Critical Evaluation and Recommended Values, and II. Compiled Data Sheets. For the water-butanone system under consideration, there are a total of eleven pages: three pages of verbal text, tabulation and graphical presentation of critical evaluation, and eight data sheets.

Generally speaking, independent data sheets are given only to the best and endorsed data covering the known range of experimental parameters. Data sheets based on primary sources where the data are of a lower precision are given only when no better data are available. Experimental data with a precision poorer than considered acceptable are reproduced in the form of data sheets when they are the only known data for a particular system. Such data are considered to be still suitable for some applications, and their presence in the compilation should alert researchers to areas that need more work.

The typical data sheet, as shown in Fig. 4, carries the following information:

(i) components - definition of the system - , their names, formula and Chemical Abstracts registry number;

(ii) reference to the primary source where the numerical information is reported. In cases when the primary source is a less common periodical or a report-document, published though of limited availability, abstract references are also given;

(iii) experimental variables;

(iv) identification of the compiler, his affiliation and the date of compilation;

(v) experimental values as they appear in the primary source. Whenever available, the data are given both in tabular and graphical form. If auxiliary information is available, the experimental data are converted also to SI units by the compiler.

Under the general heading of Auxiliary Information, the essential experimental details are summarized:

(vi) experimental method used for the generation of data;

(vii) type of apparatus and procedure employed;

(viii) source and purity of materials;

(ix) estimated error and classification of the quality of experimental data;

(x) references relevant to the generation of experimental data as cited in the primary source.

The Critical Evaluation part gives the following information:

(i) a verbal text of evaluation which discusses the numerical solubility information appearing in the thirteen primary sources located in the literature, when the first publication appeared in 1898. The first page is shown in Fig. 1. The evaluation text concerns primarily the quality of data after consideration of the purity of the materials and their characterization, the experimental method employed and the uncertainties in control of physical parameters, the reproducibility of the data, the agreement of the worker's results on accepted test systems with standard values, and finally, the data fit to generally accepted graphical tests, as shown in Fig. 2;

(ii) a set of recommended numerical data, including values for the upper and lower critical solution temperatures. Whenever possible, though not in the case under consideration, the recommended set of data includes weighted average and standard maximum deviations, and a set of smoothing equations derived from the test experimental data endorsed by the evaluator;

(iii) a graphical plot of recommended data, as shown in Fig. 3.

The Compilation part consists of data sheets of the best experimental data in the primary literature. A typical sample of the data sheet is shown in Fig. 4. For the system under consideration only eight out of thirteen primary source data qualify for individual data sheets. The quality of the data in the remaining five primary sources were considered poor or unreliable, thus not reproduced in the form of independent data sheets. However, as emphasized before, they are fully discussed in the Critical Evaluation text.

Fundamental to the philosophy of the project is the recognition that the basic element of strength is the active participation of career scientists in it. Consolidating primary literature data and producing a truly critically-evaluated set of numerical data, and synthesizing data in a meaningful relationship, are demands considered worthy of the efforts of top scientists. Career scientists, who themselves contribute to science by their involvement in active scientific research, are the backbone of the project. The scholarly work is commissioned to recognized authorities, involving a process of careful selection in the best tradition of IUPAC. This selection in turn is the key to the quality of the output. These top experts are expected to view their specific topics dispassionately, paying equal attention to their own contributions and to those of their peers. They digest literature data into a coherent story by weeding out what is wrong from what is believed to be right. To fulfill this task, the evaluator must cover all relevant open literature. No reference is excluded by design and every effort is made to detect every bit of relevant primary source. Poor quality or wrong data are mentioned and explicitly disqualified as such. In fact, it is only when the reliable data are presented alongside the unreliable data that proper justice is done. The user is bound to have incomparably more confidence in a succinct evaluative commentary and a comprehensive review with a complete bibliography to both good and poor data.

Much care is taken not to produce the output of the project hastily, when the most obvious problem is the sheer size of the project. There are bound to be some inconsistencies between the practices of the individual scientists. We believe that some of the inconsistencies can be reduced to a minimum by enforcing strict and detailed guidelines, recommendations and sets of criteria for decisions on difficult and controversial issues. However, the most reliable way to reduce, or even eliminate, inconsistencies in the delicate process of critical evaluation and in the responsible decision-making process of recommending numerical data, is to provide the expert evaluator with every bit of information in the primary source he is dealing with.

And this sets the stage for our earnest appeal to the referee-editor team.

We, as individual and institutional users of primary literature, need the wholehearted support of the referee-editor team in solving our scholarly problems. We believe that a change in the users needs should influence editorial responsiveness, and that scientific editors are uniquely able to control and assure the quality of numerical data that they distribute through their basic primary literature. In calling particular attention to the ever-growing need for quality consolidation of the rapidly expanding store of numerical data vital to all branches of science and technology, we wish to make a few fundamental recommendations:

(i) to recognize the archival status of numerical data;

(ii) to assume responsibility for a more effective transfer of numerical data information;

(iii) to accept increased responsiveness to users' requirements by recognizing the fact that a detailed description of experimental procedure and technique per se is an essential part of the scientific communication in which intrinsic numerical data are reported.

COMPONENTS:	EVALUATOR:
(1) 2-Butanone; C_4H_8O; 78-93-3 (2) Water; H_2O; 7732-18-15	S.S. Davis, Pharmacy Department, University of Nottingham, Nottingham, U.K.

CRITICAL EVALUATION:

The system water-2-butanone has been examined by many authors - we discuss here the thirteen papers reporting measurements. The first study by Rothmund (1) reported an upper critical solution temperature. He also observed a tendency to a lower critical solution temperature but he could not realize it because of the formation of ice. The mutual solubility data of Rothmund have been reported widely in the literature; however, it should be noted that the quoted values are interpolated from experimental observations. The figure below shows that Rothmund's data do not agree well with subsequent measurements. Marshall considered that the discrepancy between his values and those of Rothmund's was due to the presence of an unknown impurity. He noted that Rothmund did not state that he submitted the 2-butanone to any purification.

The data of Marshall (5), Jones (6), Park and Graham (7), are rejected since it is difficult to estimate the error or source and purity of materials in these cases. The data of Evans (8) are more acceptable but refer to one temperature only.

The data of Ginnings et al. (9) are in good agreement with other more recent studies (10,11) at low butanone content; however, the agreement at high butanone content is less satisfactory. The study by Berkengeim (12) is rejected since the experimental values are at variance with all other studies.

The data of Shell Chemical Corporation (10) are classified as tentative since the experimental procedure was not described. The studies by Randall and McKenna (13), Campbell et al. (14) and Siegelman and Sorum (11) are accepted as being reliable.

Siegelman and Sorum (11) noted that their data agreed with those of Marshall (5) from 293 K to approximately 373 K; however, the enhanced solubility of 2-butanone in water and vice versa at temperatures below 293 K found in their work is at variance with Marshall's data for the same temperature region, but is in close agreement with the solubilities found by Ginnings et al. (9). Siegelman and Sorum (11) considered it probable that the results reported in other previous investigations (1,5,7,12,13) were in error. They noted that the work of Campbell et al. (14) agreed with the combined data of Marshall (5) and the Shell Chemical Corporation investigators (10) whereas their results were in disagreement at temperatures near to the upper consolute point. In addition the mutual solubilities reported by Rothmund (1) were shown to be incorrect by consideration of the UCST. The data of Siegelman and Sorum (11) indicate that the UCST is 412±0.5 K and 44.9±0.2% by weight 2-butanone and those of Campbell et al. (14) and the Shell Chemical Corporation (10), 416 K and 45% by weight 2-butanone, would indicate that Rothmund's data of 424.8 and 44.2% by weight 2-butanone are probably in error.

Campbell et al. (14) estimated that the lower critical solution temperature, if it were realizable, would lie at about 238 K and the critical composition would be 75% by weight 2-butanone.

Few workers have studied the effect of pressure (15,16) and it is difficult to estimate the accuracy of these data because of insufficient information in the published papers. The results of Steiner and Schadow (16) are useful information, however, the accuracy of this work is questionable as judged from the system 3-methyl-2-butanone.

A table for the mutual solubilities of water and 2-butanone at different temperatures based on the combined accepted data of Randall and McKenna (13), Campbell et al. (14) and Siegelman and Sorum (11) is given below, together with the appropriate figures.

References

1. Rothmund, V. *Z. Phys. Chem.* 1898, *26*, 433.
2. *International Critical Tables*. McGraw-Hill: New York. 1933, Vol. 3, p. 386.
3. Landolt-Bornstein. *Zahlenwerte und Funktionen II*. 1962, *2B 3-408*, 22263.
4. Stephen, H.; Stephen, T. *Solubilities of Inorganic and Organic Compounds*. Pergamon: New York. 1963, Vol. 1, Part 1, p. 398.
5. Marshall, A. *J. Chem. Soc.* 1906, *89*, 1381.

Figure 1

COMPONENTS:
(1) 2-Butanone; C_4H_8O; 78-93-3
(2) Water; H_2O; 7732-18-5

EVALUATOR:
S. S. Davis, Pharmacy Department, University of Nottingham, Nottingham, U.K.

CRITICAL EVALUATION:

MUTUAL SOLUBILITY
2-BUTANONE - WATER

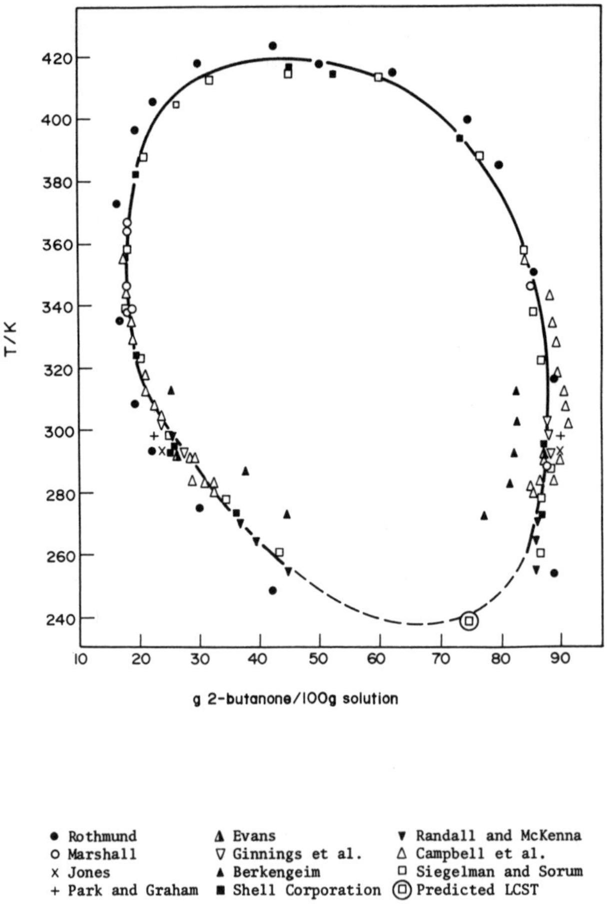

Figure 2

COMPONENTS:	EVALUATOR:
(1) 2-Butanone; C_4H_8O; 78-93-3 (2) Water; H_2O; 7732-18-5	S. S. Davis, Pharmacy Department, University of Nottingham, Nottingham, U.K.

CRITICAL EVALUATION:

References (cont.)

6. Jones, D.C. *J. Chem. Soc.* 1929, 799.
7. Park, J.G.; Graham, H.E. *Ind. Eng. Chem.* 1933, *24*, 134.
8. Evans, T.W. *Ind. Eng. Chem. Anal. Ed.* 1936, *8*, 206.
9. Ginnings, P.M.; Plonk, K.; Carter, E. *J. Amer. Chem. Soc.* 1940, *62*, 1923.
10. Shell Chemical Corporation. *Methyl Ethyl Ketone.* 1950, p. 67.
11. Siegelman, I.; Sorum, C.H. *Can. J. Chem.* 1960, *38*, 2015.
12. Berkengeim, T.I. *Zavodskaya Lab.* 1941, *10*, 592. *CA* 1946, *40*, 6961.
13. Randall, M.; McKenna, F.E. *J. Amer. Chem. Soc.* 1951, *13*, 4859.
14. Campbell, A.N.; Kartzmark, E.M.; Falconer, W.E. *Can. J. Chem.* 1958, *36*, 1475.
15. Timmermans, J. *J. Chem. Phys.* 1923, *20*, 491.
16. Steiner, R.; Schadow, E. *Z. Phys. Chem.* 1969, *63*, 297.

MUTUAL SOLUBILITY OF 2-BUTANONE (1) AND WATER (2)

RECOMMENDED VALUES

T/K	g(1)/100g solution		$x_{(1)}$	
260	42.5	86.0	0.156	0.606
270	37.0	87.0	0.128	0.626
280	32.5	88.0	0.117	0.647
290	28.0	89.0	0.088	0.669
300	24.5	90.0	0.075	0.692
310	22.0	90.5	0.066	0.704
320	20.5	90.0	0.060	0.692
330	19.0	89.0	0.055	0.669
340	18.0	87.5	0.052	0.636
360	17.5	83.5	0.050	0.558
390	21.5	76.5	0.064	0.448
410	30.0	63.5	0.097	0.303

LCST (predicted) 238K, 75% (2), $x_{(1)} = 0.428$

UCST 416K, 45% (2), $x_{(1)} = 0.148$

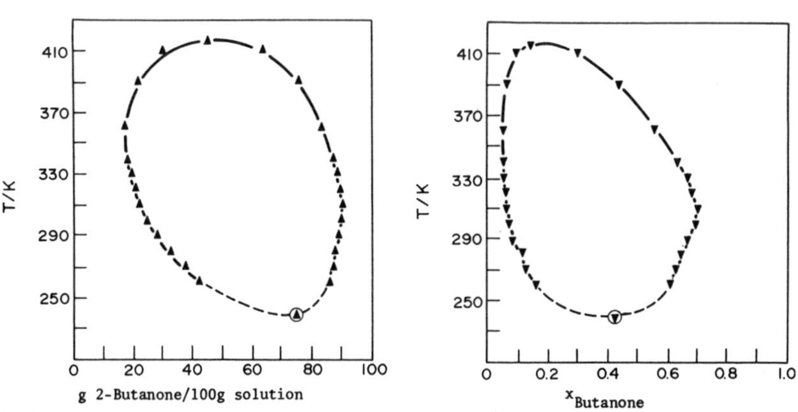

Figure 3

COMPONENTS:	ORIGINAL MEASUREMENTS:
(1) 2-Butanone; C_4H_8O; 78-93-3 (2) Water; H_2O; 7732-18-5	Campbell, A.N.; Kartzmark, E.M.; Falconer, W.E. *Can. J. Chem.* 1958, *36*, 1475.
VARIABLES: Temperature	PREPARED BY: S.S. Davis, June 1975

EXPERIMENTAL VALUES:

T/K	g(1)/100g solution	
261.6	43.8	86.8
278	34.3	87.0
298	25.5	88.4
323	20.6	87.1
338	18.3	85.8
358	18.6	83.9
388	21.1	76.9
413	31.7	59.8
415.6	45.0	45.0

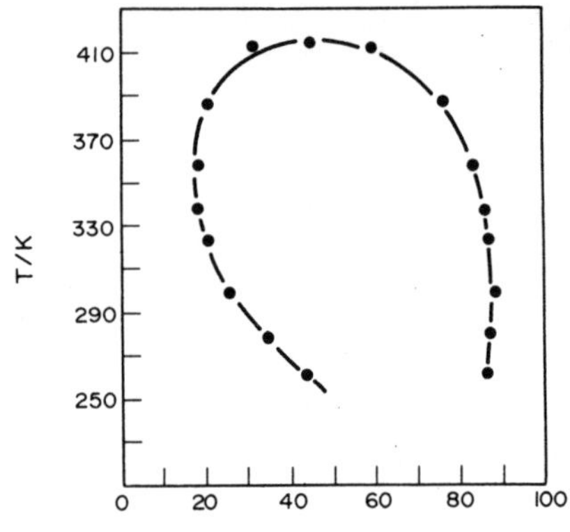

AUXILIARY INFORMATION

METHOD:
Refractive index and density

APPARATUS/PROCEDURE:
Isothermal solubility determinations conducted in three different thermostats. Mixtures were made up to approximately known over-all compositions such that the two equilibrium layers of more or less equal volumes would form at the temperature in question. Each layer was sampled after 24 hr stirring and assayed by refractive index and density.

SOURCE AND PURITY OF MATERIALS:
All chemicals were of highest purity.
2-butanone d^{25} = 0.79979

DATA CLASS: B

ESTIMATED ERROR:
Solubility ±0.2%
Temperature ±0.05K (author)

REFERENCES:

Figure 4

THE CODATA BIOLOGISTS' GUIDE FOR PRESENTATION OF NUMERICAL DATA

Philip L. Altman, *Chairman*
CODATA Task Group on Presentation of Biological Data in the Primary Literature
Federation of American Societies for Experimental Biology
9650 Rockville Pike, Bethesda, Maryland 20014 U.S.A.

A CODATA (Committee on Data for Science and Technology) Task Group representing 10 different fields of biology has just prepared the final draft of a "Biologists' Guide for the Presentation of Numerical Data in the Primary Literature." It borrows much from the "Guide for the Presentation in the Primary Literature of Numerical Data Derived from Experiments," but its recommendations are adapted for application to most of the biosciences. The general recommendations of the "Biologists' Guide" are intended to facilitate the use, evaluation, and comparison of data. They cover the description of experiments and observational procedures, the treatment of data derived from them, and presentation of the final numerical results. In terms appropriate to the biosciences, the important aspects of numerical data presentation related to preserving the data and promoting the usefulness of the quantitative results of scientific research are set forth.

CODATA, the Committee for Data on Science and Technology, was established by the International Council of Scientific Unions in 1966 to promote and encourage on a worldwide basis the production and distribution of collections of critically selected numerical values of properties of substances of importance to science and technology. As it was originally conceived, the organization concentrated primarily on physics and chemistry, but at the recommendation of ICSU it broadened its scope several years ago to include non-numeric data and time-and-place-dependent data, and so it now embraces the bio- and geo-sciences as well and is especially concerned with data of interdisciplinary significance. The scientific work of CODATA is largely done by task groups, ad hoc working groups, and advisory panels. The task groups composed of 4 to 10 specialists, work through meetings and correspondence. They perform definite tasks, the results of which are published in the *CODATA Bulletin*. One of the task groups is that on Presentation of Numerical Data in the Primary Literature. In 1973, it prepared a guide on the subject, emphasizing data that may be verified by repetition, such as physical and chemical properties and processes. The Guide has had wide dissemination internationally through Unesco and as *CODATA Bulletin* No. 9. As a result of the success of this particular group, a new CODATA task group on the Presentation of Biological Data in the Primary Literature was formed in 1975. This task group consists of 10 different individuals representing the various fields of biology, including zoology, botany, microbiology, functional and analytical biology, environmental biology, animal physiology, plant physiology, nutrition, biochemistry and biophysics, and pharmacology.

Miriam Balaban (ed.), Scientific Information Transfer: The Editor's Role, 337-339.

The first draft of the "Biologists' Guide" was prepared in March 1976 and distributed for critical comment to the 10 members of the Task Group. On the basis of changes recommended by these scientists, a second draft was prepared early in June 1976 and sent to the Task Group members for additional comment. This second draft was also distributed to the 30 biologists who attended an open meeting on Presentation of Biological Data in the Primary Literature at the Fifth International CODATA Conference in Boulder, Colorado, late in June of 1976. The suggestions for modification from those responding were incorporated into a third draft completed in December 1976. This third draft was widely distributed to 240 scientists representing the national and scientific-union delegates to CODATA; to a select group of editors of biological journals; and to the members of sections, committees, and commissions of the International Unions of Biochemistry, Pure and Applied Biophysics, Pharmacology, Nutritional Sciences, Physiological Sciences, and Biological Sciences. The suggestions received from these scientists have been considered by the Task Group which met on 18 April at the Royal Society of London to prepare the final version of the guide. The published document, representing the best consensus of valid recommendations, should be available for distribution late in 1977 or early in 1978.

Although the "Biologists' Guide" borrows much from the Guide for the Presentation in the Primary Literature of Numerical Data Derived from Experiments, its recommendations are adapted for application to most of the biosciences. The general recommendations of the "Biologists' Guide" are intended to facilitate the use, evaluation, and comparison of data.

The "Biologists' Guide" covers the description of experiments and observational procedures, the treatment of data derived from them, and the presentation of the final numerical results. The recommendations are encompassed in 3 rules as follows:

Rule 1 prescribes that an adequate description of the procedures used to obtain the numerical data be provided. The major points to be considered in establishing the quality of the information about how the observations or measurements were made often determines the acceptability of the results. These points constitute 1) identification of the organisms used or observed--whether animal, plant, bacterium, fungus, or virus--in accordance with the rules of the appropriate international code of nomenclature; 2) a definition of the system studied, sufficiently precise to ensure reproducibility of the investigated properties within the limits of experimental error, and its unambiguous identification for the purpose of relating the results to those for the same or similar systems; 3) a description of the methods and procedures used in obtaining the results; 4) a description of the apparatus and/or the chemicals used; and 5) the performance of the measurement system, that is the sensitivity and resolution achieved in the measurements.

Rule 2 applies to the reduction of observational and experimental data. This is achieved by describing the derivation of the reported results from the actual measurements. The important components of this process are assumptions made about the experiments or observations, a complete description of the physical models and the mathematical models derived from them, and auxiliary data (experimental results and physical constants taken from other sources).

Rule 3, the presentation of numerical results, states that the results are to be reported separate from interpretation, and in such a manner that the uncertainties of the data can be independently assessed so the reader can reanalyze the data in terms of a hypothesis different from that considered by the author. In order to fulfill this rule, citable results must be made available in explicitly titled tables that are separated from the discussion of the work. An acceptable alternative to a complete table is the compressed presentation of unsmoothed data. In addition to the primary results, tables of smoothed data which permit interpolation may be included. A statement of reliability noting the various sources of error should be described and evaluated in clearly defined terms. There should be a clear separation between the random error (precision) and the systematic error (accuracy) of the results.

It is strongly recommended that, unless overriding reasons indicate otherwise, the symbols and units recommended by the International Standards Organization or by other international organizations should be used and identified. Of course, SI units and their associated symbols and conventions should be used, if at all possible. Tables of data and figures should be logically complete units, independent of the main text, and with explanatory notes where necessary.

In terms appropriate to the biosciences, the guide attempts to specify the important aspects of numerical data presentation related to preserving the data and promoting the usefulness of the quantitative results of scientific research. The recommendations in the "Biologists' Guide" are consistent with the usual exhortations to authors for brevity and clarity in their papers. It makes it possible for journal editors and referees to judge the completeness and acceptability of papers insofar as reporting of numerical quantities is concerned by using a set of consistent criteria.

These general recommendations are supplemented by a bibliography that provides further guidance in the reporting of numerical data.

FOUR KINDS OF STYLE

Wendell Cochran, *Geotimes, Falls Church, Virginia U.S.A.*

and Mary Hill, *U.S. Geological Survey, Menlo Park, California U.S.A.*

ABSTRACT

The English word 'style' has four distinct meanings as used by authors and editors:

1. Editorial (or stylebook) style concerns primarily matters of consistency in such things as capitalization, abbreviation, punctuation, use of numerals, and the order or components within bibliographic citations.

2. Typographic style involves the typefaces, sizes, measure, and other design features in a particular publication.

3. Literary style (or, as E.B. White put it in The elements of style, 'the sound that words make on paper') relates to euphony, rhythm, avoidance of obtrusive alliteration and accidental rhymes, and the like. It also concerns personal preferences and habits of expression, and so is sometimes called 'personal style'.

4. Usage style deals with definitions, connotations, idiom, and such distinctions as those between 'eager' and 'anxious'.

'Style' may be the most common piece of jargon used by writers and editors; it may also be the most confusing. Many people talk about style, but they are not always talking about the same thing. As one result, communication between authors and editors is often inefficient. For another, many stylebooks are badly organized, hard to use, and difficult to compare.

In order to examine this problem (and, we hope, reduce the confusion) we have examined about two dozen 'stylebooks' and divided 'style' into four kinds:

1. Editorial style (also called 'house style').
2. Typographic style (or 'design style').
3. Literary style (or 'writing style' or 'personal style').
4. Usage style.

Editorial style primarily concerns consistency, on the theory that readers are disturbed at finding 'grey' in one sentence and 'gray' in the next. Among the most common problems of editorial style concern spelling ('spelled' or 'spelt'; 'gage' or 'gauge'); abbreviation ('November 1' or 'Nov. 1'; 'Brown & Company' or Brown & Co.'); capitalization ('the Chairman' or 'the chairman'; 'Federal Government' or 'federal government'); punctuation ('8, 9, 10...' or '8, 9, ten...'); the syntax of citations ('Gowers, Sir Earnest [1962]. The Complete Plain Words' or 'Gowers, Sir Earnest. Pelican Books, Baltimore [1962]'). And so on.

Clearly, editorial style is the province of the editor although it may be strongly influenced by the dialect of the specialty concerned; an editor of a journal concerning the history of past civilizations may prefer the spelling 'archeology' but will be reluctant to force that form on readers accustomed throughout their careers to seeing it spelled 'archaeology'.

Miriam Balaban (ed.), Scientific Information Transfer: The Editor's Role, 341-343. *All Rights Reserved.*

Typographic style refers to the design of the publication; a given journal almost always displays its name on the cover in the same size of the same typeface; uses only one typeface (with attendant consistency in size, leading, and measure) for the body type; uses the same headings for the same departments...

Like editorial style, typographic style rarely concerns the reader, although an author who uses angstrom units may find that the journal's body type does not include the character Å and so the editor must at each occurrence change it to the word 'angstroms'. Of course the editor -- better yet, the designer -- should when choosing the body type take into account any special characters or diacritical marks that occur often in the journal's subject matter.

Literary style can be an intensely personal thing. Its extreme form is poetry, in which an editor dares interfere with the author's copy only to the extent of pointing out to the poet something that might be an error. Prose may be another matter. Some authors have efficient, distinctive styles, and in general editors should recognize the fact and not tamper with their work. However, at least one American journal tries very hard to made consistent the literary style of all its articles, and so its editors re-write every article contributed to it (the authors are allowed to correct errors of fact but only rarely may change anything else).

Many authors -- probably the great majority -- have no distinctive style, or at least not the kind that should be encouraged. Some of those simply haven't written enough to develop a personal style (and if they have, it may be unacceptably verbose), and some unconsciously mimic the passive, jargon-ridden prose found so often in English-language journals since the first World War.

If the author does have a distinctive and efficient style, the editor should not change it. If changes are necessary the editor should be an expert parodist or -- preferably -- should point out the specific problems to the author. ('Not clear' or 'poor style' may approach the ultimate in insults.) At any rate someone should (as Fowler said of the split infinitive) remove all signs of the struggle.

If the author produces only dull, turgid prose, the editor should be firm enough to reject it and also competent enough to suggest tactfully how to make the necessary repairs. (A grey area: 'writing style' sometimes occurs in stylebooks not as literary style but as a part of either house style or the grammer part of usage style. At any rate, few compilers of stylebooks can resist adding it in one form or another.)

Usage style deals with grammar, syntax, idiom, and shades of meaning. As Mark Twain once said, the difference between the right word and almost the right word is the difference between lightning and the lightning bug. Many books are devoted exclusively to usage, notably (in English) Modern English usage (second edition, revised by Sir Earnest Gowers, 1965; Oxford University Press) and The careful writer, a modern guide to English usage (by Theodore M. Bernstein, 1956; Atheneum, New York). An ordinary language dictionary may be a great help, especially if it distinguishes between synonyms, but lack of context poses danger. However, some dictionaries include sections -- usually as appendices -- on style of one kind or another.

Now that we've distinguished four kinds of style, it is easy to see why the word 'style' alone can be confusing, and why so many stylebooks are so hard to use. An editor or referee criticizing a manuscript should do more than complain about style in general, but should in each instance specify the kind at fault.

Most of the stylebooks we examined included rules on editorial style. Very few in wide use specified typographic style (the most notable example being the University of Chicago's Manual of style for authors, editors, and copywriters, 12th edition, 1969; University of Chicago Press); most others including typographic style are intended for one journal, or for a very few published by the same organization. Most stylebooks included some rules on 'writing style' and also at least some pointers on usage. Books on literary style alone are few; books on usage alone abound.

Bearing in mind those different kinds of style, the editor can also judge more precisely the many stylebooks on the market, and more easily choose the one to meet his (or her) specific needs. For example, a stylebook written for another journal may be perfectly acceptable, provided that the editor can accept or choose to ignore any rules on typographic style. One stylebook may be strong on hints for literary style; another may be good only for usage. The editor can choose the one, or the combination, that best meets his needs.

Finally, in communicating with authors and fellow editors, he can be more precise in the primary business of editors -- communication.

30

ECONOMICS AND ORGANIZATION OF SCIENTIFIC PUBLISHING

Paul Nijhoff Asser, Chairman

stm

INTERNATIONAL GROUP OF SCIENTIFIC, TECHNICAL & MEDICAL PUBLISHERS

'AN AUTONOMOUS GROUP AFFILIATED TO THE INTERNATIONAL PUBLISHERS ASSOCIATION'

INTRODUCTION TO MORNING SESSION

Paul Nijhoff Asser
STM Secretary, Keizersgracht 462, Amsterdam 1002, The Netherlands

Introduction to Session and to STM

Ladies and Gentlemen,

I would like to introduce, and draw attention to, the following elements of this morning session on the economics and organization of scientific publishing - which subject will be dealt with in more detail by nine other speakers.

May I first introduce myself; then draw attention to you, the audience; after that, summarize what I hope and trust will become clear to you in the course of this morning: that is, that it is good to be a publisher - even if youreconomics and your organization are not perfect, to say the least.

Myself, first, then, not because of my inborn immodesty but because I want you to know what the mysterious letters STM mean in the programme of this session. STM stands for International Group of Scientific, Technical & Medical Publishers, an association of at present nearly 120 publishing houses (commercial and institutional) in twenty countries. I have been that association's full-time secretary since 1970, after having spent the previous 23 years in academic bookselling and publishing.

The STM Group is fortunate in having a great number of the high-quality international publishing houses of the world among its members; looking over the list of 300 participants in this conference I counted 23 or nearly 8% of the total connected with houses such as Masson of Paris, Universitetsforlaget of Oslo, John Wiley & Sons and McGraw-Hill of New York, Addison-Wesley of Boston, Cambridge University Press of London, Elsevier of Amsterdam, and smaller but equally distinguished houses such as Dr. W. Junk Publishers of the Hague, The Weizmann Science Press of Israel, the British Medical Journal, London, the Institute of Physics of Bristol, and so on, all of whom are members of the STM Group.

More information about this organization is available from me and I have even put some of our recent publications on this table which the most eager among you may take home: our members list, a recent Newsletter, a recent report of a seminar of which we organize two per year, and a

Miriam Balaban (ed.), Scientific Information Transfer: The Editor's Role, 347-349.

short history of our Group.

By now you may be curious about what, apart from organizing these two unspecified seminars and publishing a newsletter, the STM Group is actually doing. Let me just tell you what our purpose is, and the rest will follow: our aim is "to assist publishers and their authors in their activities in the dissemination of the results of research in the fields of science, technology and medicine" which implies that publishers in order to become members of STM have to publish primary stm literature - and that includes social and behavioural sciences.

To pursue this aim, the STM Group, or rather its secretary, has a number of duties one of which is the collection and dissemination of information on key problems concerning st and m publishing, including the processing, storing and retrieval of information by electronic means - and now I will stop and draw your attention to the second element I mentioned, and that is you, the audience.

You describe yourself, collectively, as scientific editors. Not every single one of you is, or wishes to be described as, a scientific editor, but since your interest in that curious profession has brought you all together here, I believe that it is fair to assume that you at least have an affinity to, or like, the editorial function.

And what we are here for, I like to believe, is to define that function. Now Miriam Balaban has been very effective in that she has succeeded in highlighting just about every aspect of this function which even if she had not done so would be a many-splendoured thing. But she has apparently been looking for (because she has *found*) persons who could present any single one of the myriad wonderful, brilliant and ever-active and evermoving, lovely aspects of the author-publisher-reader relationship.

I, therefore, suggest that we, editors and publishers alike, replace Miriam's name by Myriad since she is the symbol of ourselves: she is the combination of all these wonderful, brilliant, ever active and evermoving, and, above all, lovely aspects of what we, in our amateurish way, mistakenly prefer to call our profession.

In any publishing house worth its salt the editorial department is despised by all: management, production, marketing, accounting, design just to mention the other main departments. The more the editor is despised, the better relations he will have with the best authors, and the better the publishing house will flourish. The best publisher, therefore, will try to find the most despicable editorial staff. Yesterday at John Ziman's session I tried to picture the publisher's role in his nice scheme where between authors and readers he painted a picture of the editorial function which really moved me. You know why? Because it proved that the editorial function - apart from just printing - is, in my opinion, the essential one in publishing since publishing started.

Now enough about you, before I start getting emotional.

The third element is the subjects to be dealt with this morning. My friend Peter Bakker of Junk Publishers in the Hague, a small but prestigious house in the field of biology, entomology and ophtholmology will give you a broad survey of all the grandeurs et misčres of a publishing house and of the editor's place in it; then Michael Harris of John Wiley in New York - who by the way is a member of the STM Group Executive and the Chairman of the STM Copyright Committee - will say something about why a good editor could be (but is not always) a good publisher, and after those two commercial publishers, a not-for-profit one, Michael Bowen of the American Chemical Society, will tell us why the strong need for cost effectiveness, business management skills and other needs encountered in commercial houses will be, and are, welcomed in not-for-profit houses as well - to which I might add that the need for copyright protection, without which no publisher could exist for long in the western culture; is equally present in for-profit and not-for-profit publishing organizations alike.

The other speakers will be introduced after the coffee break. I intend to allow no nonsense. The time schedule is there to be adhered to, not to be delayed. Those who came in late have missed at least the beginnings of an inspired introduction. I have not been made Chairman for nothing, and I will not spare the rod if necessary. So, keep your ears peeled and check your watches. We are now exactly 35 minutes behind schedule. Let's make a concerted effort to have that book be published on the scheduled date - I mean, to have our lunch at 13.00 and not at 13.01. If I do not succeed, please do not forget that that may be one of the reasons why publishing managers despise editors, and why publishing, and editing, remain such unruly and wholly wonderful professions.

THE ORGANIZATION AND ECONOMICS OF PUBLISHING AND THE FUNCTION OF THE EDITOR

Simon Peter Bakker
Junk Publishers, P.O.B. 3713, The Hague, The Netherlands

There is a Dutch saying 'so many heads, so many minds' and this certainly applies to the subject I have to talk to you about under the title *The organization and economics of publishing and the function of the editor*. To start with, there are many different kinds of publishing companies, such as commercial and university presses, publishing houses of learned societies, and state publishing houses. Depending on their size, they have different types of organization. Also, there are several kinds of editors, for instance the journal editor, the advisory editor, the acquisition editor, the desk editor, and so on.

Because of this great variety, I have decided not to treat the organization and economics of the several types of publishing houses, mentioned or forgotten, but I have chosen one of those as an example. It is a company that has just been founded called 'STM Inc.'. As most of you belong to the tribe of editors, I will put special emphasis on the various editorial duties, while talking about the organization of this company.

Almost all existing publishing houses were one-man-enterprises in the very beginning, and thus at this moment STM Inc. is a very small company — a one man show — and the founder unites in his person all functions: he is the publisher, the acquisition editor, the desk editor, the marketing manager, the salesman, yes, he even is the one who invoices and sends out to the customers the items ordered.

Before he is able to start any publishing activity, he must have made up his mind about the fields to be covered and the type of publications to be issued. And by 'type of publications' I do not only mean journals or books, but also the various levels — such as post-graduate, student textbook, primary research, university research and so on.

As the seat of the firm, one should, in principle, choose a place that presents the best economic possibilities such as availability of qualified personnel, the lowest salaries, fiscal climate, and so on. In Western Europe, for instance, the disadvantages of the high salaries, the high costs of housing, mailing, and the fiscal climate are such, that the competitive position of the publishers in that region is very much affected. It was not without reason that one of the biggest STM-members has recently founded a publishing company in Ireland!

Miriam Balaban (ed.), Scientific Information Transfer: The Editor's Role, 351-357.

To establish a good basis for the continuation of the company, this publisher has decided to start with the publication of a scientific journal. Here we meet the first specimen of an editor, the editor-in-chief of a scientific journal.

I will not deal with the financial management of journals in detail but I just want to indicate why our publisher has chosen to start with a journal rather than to publish books. Although the investment needed for a journal is much larger than the amount of money needed to produce a single book, it is more attractive to invest in journals. If a journal sells well, it solves many financial problems, as the pre-payment procedure enables publishers to work with other people's money.

With the editor-in-chief, our publisher comes to an agreement on, to mention the most important items, the scope and level of the journal, the number of pages to be published, the speed of publication of the articles submitted, and on the composition of the editorial board. One important detail of such an agreement should always be a clause for a rotation scheme for the members of the editorial board, in order to prevent a growing rigidity in the years to come. (A much used rotation scheme is that one third of the initial members is replaced after three years, one third after four years, and one third after five years; each member will serve on the board for three years.)

As a good publisher, our man will make sure that he meets the editor-in-chief, and possibly the members of the board, once a year and he will provide them with information on the number of subscriptions per region and the statistics on the origin of the papers. The editor-in-chief will inform him of the supply of manuscripts, the number and percentage of manuscripts accepted, referee procedures, and so on. With the help of this exchange of information they will discuss the steps to be taken in the near future, which could be an increase in the number of pages to be published, additional promotion activities during congresses, etc.

After having set up his journal, our publisher will realize that this will be too small a basis for building up a publishing house. Therefore, he will try to set up other journals and to supplement the journals with the publication of books.
To this end he will need the help of one or more advisory editors. You can imagine that the editor-in-chief of his journal could be such an advisory editor.
I don't think it will be necessary to explain the task of an advisory editor in detail: the term is self-explanatory. The advisory editor not only reviews the quality of the manuscripts received, but he will also advise the publisher on potential authors, subjects of interest and so on. The relationship between such editors and the publisher is a remarkable one most of the time: it is not based on a business relationship, but has its firm roots in a more personal relation.
I can't illustrate this better than by the fact that most advisory editors are, in the first place, not aware of their 'official' function and, secondly, that they receive hardly any allowance for their counsel.
The moment the publisher contacts an author and invites him to write a manuscript, he functions as an acquisition editor. As long as the company will

remain small, this will be a normal thing, but when STM Inc. starts growing, our managing director will decide to hire acquisition editors, who will each have responsibilities for a certain part of the fields he has chosen to cover. The task of an acquisition editor is to set up a publishing programme (which needs a lot of research), to acquire authors and editors and to guide them until the manuscripts are submitted. The position of an acquisition editor has always been a difficult one. In Europe, acquisition editors are mostly scientists with a degree — ironically spoken 'fallen scientists', in the USA they usually have a trade background, that is, they are 'risen college travellers'. I use these — not very kind — expressions because they clearly illustrate what we expect of an acquisition editor: he should have a good scientific background to ensure that on the one hand, he will be able to discuss the various projects very thoroughly with the authors, while on the other hand he should realize that publications will have to be sold profitably, which makes his work seem to be a 'contradictio in terminis'. In my opinion we should never underestimate the function of an acquisition editor: Isn't he the one who is responsible for the image or part of the image of a publishing house?

As soon as the manuscript is ready for publication, we meet the copy or desk editor. He checks the manuscripts on completeness (text, tables, figures) and will — if necessary — revise the manuscripts from a linguistic point of view with responsible casualness. Due to his job, the copy editor has in many cases a special relationship with the authors or editors-in-chief. Sometimes it appears that they have a feeling for publishing and in that case they might become acquisition editors. In most cases, however, they identify themselves too much with the authors and pay too little attention to the financial aspects of publishing and, therefore, it will always be an exception if a copy editor becomes an acquisition editor.

Up to now I have only treated the organizational ties between the publisher and the various editors, and I somewhat neglected the organization itself and its economic aspects. I shall now turn to these points.

Let us go back to the point where STM Inc. was founded. At that time I said that in his person the founder unites all functions, and I mentioned some of them. It is quite easy to draw an organization scheme for this firm:

STM inc

FOUNDER

+

secretarial help

In the beginning, it will be easy for this organization to do all the work, but after a while our man will need help for the daily routine duties. If he is wise, he will stop looking after the distribution of his books himself, and will entrust the job

to one of the firms which have the necessary facilities for it, like storage, invoicing, bookkeeping, shipping, etc. To this end, he could also co-operate with other publishers. In this connection it would also be suitable to sell his publications abroad by means of co-editions with foreign publishers. An advantage would be that a share of the financial risk will be borne by his partner or partners.
One great disadvantage is that the title will be issued under someone else's imprint, and this may have a negative effect when the publisher tries to build up contacts with authors in that region. The selling publisher might try to have a co-edition published under a common imprint, but, for obvious reasons, the buying publisher usually does not accept this.

If our publisher is successful in having the 'physical handling' done somewhere else, very little will change in the simple and also cheap form of the organization sketched above. If he is not succesful in this respect, the distribution of his titles by his own company has far-reaching consequences for the organization. In this case it could look as follows:

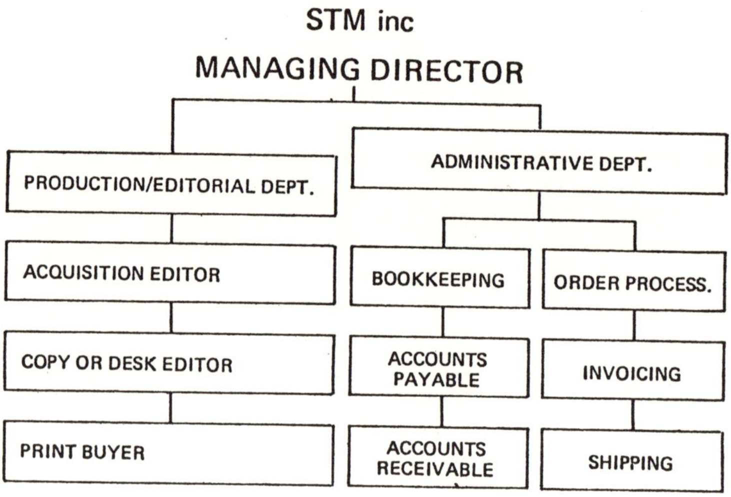

By this time, our company will be facing a big economic problem, namely that of the overhead costs. The words 'overhead costs' are a collective term for all costs that are necessary to run a firm, but cannot be charged to a specific product. Such costs are, for instance, salaries, housing, and administration.
For example, in Western Europe salaries to be paid by the employer are really a heavy burden. If we fix the gross salaries (which include amongst others social security and pension costs) in the Netherlands on 100%, comparatively those in the USA are 75% and in the United Kingdom 50%. In my country, the costs of executing an order, range per invoice Dutch Guilders 15.00 (US $ 6.00) for a rather small company to Dutch Guilders 25.00 (US $ 10.00) for a large company. This and the ever increasing costs of postage have forced publishers a couple of years ago to start with the introduction of a 'handling charge'. The above should illustrate that it is wicked to think that publishers make a profit on their charges for handling!

Now that I have brought up the 'touchy' subject of costs, I would like to treat the aspects of the production costs by means of a simplified calculation scheme. In order to show you the reality, I have taken as an example a title which is now in production at my company.

CALCULATION SCHEME

Title: NEKTON
Author(s): Aleyev
Editor(s):
ISBN: 90 6193 560 1 Format: 16 x 24 cm

Text: 27¾ sheets of 16 pages. Total 444 pages.
Number of figures: 247; 21 tables Number of coloured figures:
Print run: 2000

Text:	Coloured figures:	Cover:
No. of copies:	No. of copies:	No. of copies:

		Calculation	Invoice
A	**setting**		
	Typing at f per page		
	IBM composer at f per page		
	Monotype at f 50,— per page	22.200,—	
		600,—	extra for tables
B	**plate-making**		
	Four-colour lithos		
	Black/white screen blocks 55 figs	2000,—	
	Line drawings 192 figs	3840,—	
C	**printing**		
	Text at f 180,— per sheet	5000,—	
	Photo pages		
	Four-colour lithos		
	Dust jacket	1300,—	incl. design
	Paper used for text:		
	wood-free m.c. 90 g/m² at f p.s.	5000,—	
D	**binding**		
	Cloth/dust cover at f5.80 per copy	11.600,—	
	Stitched/sewn at f per copy		
	Insertion(s)		
E	**various**		
	Correction charges extra corr.	3000,—	
	Editor's fee	3000,—	
	Translation charges		
	Drawing charges	600,—	
	Author's royalty		
	production costs	58.140,—	

The 2000 copies of this title cost Dutch Guilders 58,140.— but as we need a 100 copies for review purposes, author's copies, editor's copies, etc., we have to base ourselves on a print run of 1900 copies to find the price per copy, which is Dutch Guilders 30.60. In order to get the sales price we have to multiply this amount by four and this gives us as a result Dutch Guilders 120.— (DG 122.40). I will now show you why we need a sales price like this:

Sales price		DG 120.—
Discount for booksellers 33 1/3%	40.—	
Royalty 10% of sales price	12.—	
	52.—	52.—
		68.—
Costs of sales 20%		13.60
		54.40
Cost price		30.60
		23.80
Depreciation 20% of cost price		6.12
		17.68
Overhead on production 15%		5.40
		12.28
Interest 15% of cost price + overhead		6.10
Profit		6.18

Although this investment is considerable, I already told you that the investment in a new scientific journal will be many times more than this. The return on investment of a book is faster than that of journals, but the sales possibilities are exhausted earlier. If a journal sells well — and with a good quality periodical the 'break-even-point' could be reached within three to five years — it will be one of the pillars of a publishing house.

Until now, I have dealt with the main aspects of what I would like to call the 'input', that is, from the acquisition of a manuscript up to the final product. Now we will have to try to sell it. It will be clear to you that we have already taken into consideration the sales possibilities when we took the decision whether to publish or not, but I want to give you a sketch of the physical means we have at our disposition while trying to sell the STM publications. Before I do so, however, I would like to present a scheme in which the organization of the sales department has been added.

'Direct mail' is the most important means the STM publisher has to his disposition to bring his titles to the attention of prospective customers. Investigation has shown that more than 70% of the sales of books is initiated by the leaflet on the particular title. In order to make the best use of this means, one needs extensive address-material. In practice, it is necessary to have mailing lists that are very specialized, as the high mailing costs make it impossible to scatter advertising material.

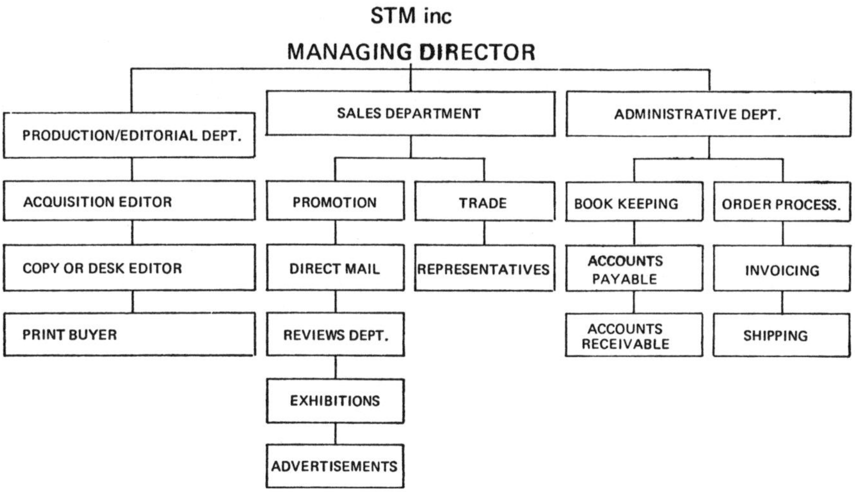

Depending on the groups to be reached — such as libraries, booksellers, private customers —, split up according to specialism, the publisher will send out leaflets, special lists or general catalogues.

The same investigation has shown that reviews of titles in scientific journals are also of some importance. Sending out reviews copies is always a delicate subject, because the publisher would like to have some guarantee on the publication and especially on the time of publication of a review of his books, but such guarantee will hardly ever be given. 'Science' or 'Nature' for instance, publish reviews of only approximately 15 - 20% of the number of titles received. If a title is reviewed positively in such a journal, this will certainly have a stimulating effect on the sales, provided that the review is published not too long after the actual title has been issued.

Exhibiting titles during congresses or symposia is a third way to reach the interested public. The influence on sales of this means should not be overestimated, but the public relations effect for both the publisher and his authors should not be neglected either.

As a fourth way I would like to mention advertising. There is no use in going too deeply into this, but it will be clear to you that if the review is preceded by an advertisement in the same journal, this could have some good results.

Although, as a rule, many authors and editors have the feeling that the promotion of their titles is insufficient or even bad, I feel right in stating that this is certainly not the case in general. It has been my experience that the promotion departments of publishing houses are usually the least neglected departments. Should a book sell badly, this is, in my opinion, not due to a lack of promotion but to the quality of the book itself, and it is the publisher who will have to take full responsibility instead of the author or editor.

IS A GOOD EDITOR A GOOD PUBLISHER?

Michael Harris, *Wiley-Interscience Publishers, Vice President*
605 Third Avenue, New York, N.Y. 10016

There is a relatively simple answer to the question of, "can a good editor be a good publisher?" The answer is yes, it is no, it is maybe. It all depends on the editor, on the function of the editor, and that of the publisher. At first glance I must say I don't understand why a good editor would want to be a good publisher, or to be a publisher at all, why he would want to join that relatively distinctive group of people who have been characterized in so many fashions.

I recently read, and I commend this to you, a paper by Richard de Gennaro the librarian of the University of Pennsylvania, about publishers in the February 1977 issue of the American Library Journal. He describes publishers as vultures who live off the flesh of librarians who themselves constitute an innocent and helpless breed. They are victims of publishers. They cannot resist the wiles of publishers, particularly the habit of publishers of placing a number on a book. If the book is part of a serial or series, or is in any way consecutively numbered, that book is guaranteed according to Mr. De Gennaro, to be purchased by every single librarian because the librarian cannot leave himself or herself in an exposed position.

The article is profusely illustrated, and depicts in cartoons the publisher as a kind of ogre with an enormous mouth voraciously devouring the innocent breed of librarians. The article goes on to say that the publishers devour the librarians, their treasuries and simply trade on their helplessness by publishing journal after journal for no good reason at all and creating new journals which leave librarians in the position of having to obtain more.

Now why an editor would want to join that kind of life,a publisher's life, is beyond me. I would like to parenthetically remark, although I'm straying a bit from the subject on the discussion which took place earlier in this session. These were the complaints which have been levied against publishers. De Gannero made the complaint that publishers have made a practice of publishing too much about too little, and therefore the big problem today in publishing is to either reduce the number of journals or to find some means of conveying scientific information in a fashion which will eliminate many of the journals which are published today.

There is a mood among some scientists, and certainly among some of the organizations that deal with science, to criticize the publication of journals. Without going into any detail I must say that it is strange to me that organizations which support scientific research - such as the National Science Foundation is the United States - maintain that no research is worth undertaking unless it is published, and even goes so far as to offer page charges to authors in order to encourage them to get their materials published. How can an organization of that kind at the same time complain about the extent of publication? It seems to me that there is some inconsistency in the point of view or that other institutions that are responsible for the publish-or-perish syndrome - and this includes educators and scientists and scientific organizations who use publications as the measure of the quality of a scientist - should object to journals which have to carry an increasing traffic.

Miriam Balaban (ed.), Scientific Information Transfer: The Editor's Role, 359-361.

That is a side comment, and I must say that if such a problem does exist, perhaps it is no easier to cope with than the original problem that confronts science. That is - what research should be engaged in, and how to determine what research should be financed and supported. Science has had great diffifulty to find answers to that question and the problem of the growth of scientific journals and the growth of publications of papers, to me at least, is merely a reflection of the basic problem of science in trying to ascertain ways and means of distinguishing between research which should be supported and research which should not be supported.

Back to the subject of the editor and the publisher. Here I think it is necessary to go into a couple of definitions. In the first instance I am talking exclusively about scientific publications - scientific, technical and medical books, and not general trade publication which, although I've been in publishing for some years, remains to me a mystery. I am talking again largely about books and journals as the main form of publication of scientific and technical information.

Books and journals are two different aspects of scientific publication which require somewhat different talents for the editors engaged in them. The talents of an editor of a primary research journal and the editor of a book program are not mutually exclusive, but they are not identical. And although some talents are shared, there are distinctive differences between the responsibilities of the editor of a scientific journal and the editor of a book program. The book editor generally is the full-time employee of a publishing house, a member of the staff. In the United States and in some other parts of the world, the book editor, as I use the term, is essentially an acquisitions editor, and editor who runs a publishing program whose responsibilities are to know the field in which he publishes, to select the right authors, the right subject, at the right time. He negotiates although he does not necessarily sign the contract with the author, and his responsibility is to get the book from the author in satisfactory form and content and within a reasonable period of time. The book goes to the production department and there after the marketing, sales, distribution and promotion people are involved in the production and the ultimate sale and distribution of the book.

The book can be a text or a monograph or treatise, and the editor's skill determines to a large extent the relationship he has with the author and his advice to the author on the form in which the book should be printed. He works in one or more broad areas of scientific inquiry, and some editors cover several areas. He must know the authors and he must know the market, and he is charged with the responsibility of bringing in books which will produce a profit.

The journal editor is generally not a staff member or an employee of the publisher. He is a professional scientist working in his chosen field in which he is actively engaged. He is responsible for obtaining manuscripts or establishing editorial skill for obtaining the best possible contributions to his paper. He examines the contributions generally in order to ascertain that they are within the scope and meet the quality that he desires. He selects and sends manuscripts to referees who send their comments back with the manuscripts, and ultimately the editor makes the essential decision about the quality of the work and whether it deserves publication.

This position is somewhat different from that of the book editor. The publisher in both cases essentially has an executive or managerial responsibility. It is easy enough to become a publisher. I think the problem of a publisher is to remain a publisher, and this is a problem with which he is seized virtually every day at his operations. The publisher essentially has full responsibility for all of the operations of his publishing program and the publishing house, and at times he may seem inordinately concerned with his profit and loss.

Certainly economics and economic viability of his program are essential and perhaps the most significant part of his responsibility. He determines the fields in which he wants to work as a publisher and selects his staff to perform various functions. He is responsible for all the elements involved in the success of his program including the quality of the work to be published although much of that is delegated to the editor, the potential market, assessing the market for certain kinds of books in various fields. He is responsible for pulling together all the elements of a successful publishing program,production, marketing, distribution and so on. In short he has full executive responsibility for book and journal program. The publisher and the book editor share many common concerns. They are concerned about the market, the quality of the book, the profitability of the book.

Journal editors also are concerned with those but much less so with profitability. Being an outside employee, he is more concerned with circulation and the prestige of his journal. The publisher is distinct from the editor though. He spends much more time and is much more concerned with sales and marketing, distribution and promotion. Basically the difference between and editor and a publisher, a book editor and a publisher, are matters of degree. An editor is much closer to an author than a publisher is, but a publisher is also very much interested in maintaining good relations with authors although he certainly cannot substitute for an editor in relationships with authors.

I think all this leads us to saying that a good editor can be a good publisher. The editorial choice is the very basis of the success of a publishing company, and I myself believe, as a publisher, that the most important function in any publishing house is the function of an editor because it is his choice, his selection, that determines the quality, the reputation, and in the final analysis, the success of the publishing company.

The good editor can be a good publisher, but not by definition. He must have training and talent for some of the executive responsibilities which I have mentioned and particularly those which require managerial responsibility. The editor is a natural source for training to be a publisher, and at least in practice, from my own experience, it is basically editors who have become publishers. Now I admit that's quite different in other types of publishing but in our kind of publishing, it is the editor whom we look forward to ultimately becoming the publisher. I can see nothing incompatible between the two positions. I think that the question of whether the editor wants to become a publisher depends on what he desires because I believe that a career editor is a very honourable occupation that should be appropriately awarded, and there is no reason in the world why a person who wants to remain an editor which I think is an important single function in publishing should not have every encouragement, including prestige and monetary award in maintaining his position in that capacity.

FINANCING SCIENTIFIC SOCIETY JOURNALS IN THE USA

D.H. Michael Bowen
Director, Books and Journals Division, American Chemical Society,
1155 Sixteenth Street, Washington, D.C. 20036

ABSTRACT

Publishing scientific journals in not-for-profit organizations like the American Chemical Society is big business with a difference. Though the profit motive is absent, there are nevertheless many factors which contribute to a strong need for cost effectiveness and business management skills. US journals published by societies normally are largely dependent on subscription revenues, but physical science and earth science journals also receive substantial revenues from "page charges". Some ACS journals also receive large sums from advertising, a helpful source of income, but one that is difficult to manage.

In this presentation on the management of scientific society journals in the U.S.A., I will of necessity refer most often to the journals of the American Chemical Society. ACS journals are not necessarily typical of American scientific and scholarly journals in general (nor are they necessarily atypical!) and American journals, of course, have some characteristics that different from journals in other parts of the world. I will try to point out some of these differences as we go along.

My main plan will be to examine, in turn, the main components of revenue and expense that characterize the finances of U.S. scientific journals and to indicate their relative importance to the overall financial picture of a journal.

REVENUES

The main component of revenue for the great majority of scholarly journals everywhere is from subscribers. Revenue from members of a scientific society--who generally are offered preferential (i.e., low) subscription rates--is less significant these days than revenue from nonmembers, usually institutions. Within the ACS, a member usually pays one fourth of the price a nonmember has to pay. As a consequence, we still have around 10,000 member subscribers to some of our journals, but the number diminishes yearly, as the price has to rise. A major problem attendant on giving preferential rates to members is "cheating," by which an institution obtains its copies of a journal by subscribing in the name of or through a member. In the ACS we try to minimize cheating by (1) having all new members sign a pledge that they will restrict journals purchased at the member price to their personal use; (2) putting a special cover on copies mailed to members, on which the personal use restriction is restated in highly visible type.

Miriam Balaban (ed.), Scientific Information Transfer: The Editor's Role, 363-370.

The Internal Revenue Service (the U.S. government agency that is responsible for collecting income taxes) is looking askance at the granting of price privileges to society members. How the practice will evolve is, in the U.S., likely to be quite dependent on IRS actions with respect to scientific societies.

Some U.S. commercial publishers also offer special prices to individuals and/or society members, usually only if the individual's institution already has a subscription at the regular (higher) rate.

Nonmember subscription revenues form the backbone of revenue support for most scientific journals. Most publishers consider that the support from non-member subscribers is relatively price insensitive. As costs go up they simply raise the price to nonmembers. While this has proved more or less successful from the point of view of the "bottom line," it is a policy that eventually must be self-defeating, as declining circulations indicate.

Advertising revenues are rarely a factor in scientific journals, especially basic journals. Yet when the scientific discipline corresponds to a market for products or services, the potential for large revenues may be present. Even then, since advertising does not sell itself, a sales organization that is both knowledgeable and aggressive is needed for the potential to be realized. In the U.S., advertising revenues to scientific societies may well be taxable as "unrelated business income." The ACS basic journals receive advertising revenue in excess of $1 million a year. As I will show you in a few minutes, most of that comes from just a few journals.

Page charge revenue seems to be a U.S. specialty. Historically, page charges were started in physics journals and were philosophically dependent on the contention of the U.S. government (the largest single research sponsor in the U.S.) that research is not complete until it is published in a recognized scientific journal. The use of page charges is confined to non-profit journals, since U.S. policy prohibits U.S. government grantees from paying them to for-profit journals. In many cases the page charge--typically $30-80 a page--is "voluntary," and payment is "expected" rather than required. In the ACS just one half the pages in basic journals have the page charge paid voluntarily. The rest are waived upon request submitted by the author. Nevertheless, this brings in more than $1 million a year, about 25% of the revenues for a typical journal. Some societies assess a mandatory page charge (although U.S. government authors are not permitted to pay into mandatory page charge systems.) Yet other societies collect the majority of potential page charges by allocating the majority of available pages to articles on which the page charge is paid and subjecting unpaid articles to a fate of slower publication.

Page charges are likely to remain an important source of revenues to U.S. society publishers in the physical sciences for the near term, perhaps through 1985. The consequence of dropping them, even though they tend to be unpopular with scientific authors, is higher subscription prices.

Reprint revenues are a significant source of net revenue for ACS, though not necessarily for other scientific publishers. We do not give authors free reprints, even if they pay the page charge. We view reprint sales to authors at the time their papers are published as a vanity press. Indeed, the paper used for reprints may be better than that used in the journal issue itself. More than 90% of all authors purchase reprints, and the minimum number available is 100. As the price rises the number of reprints purchased has fallen somewhat, but it is still a $750,000 a year business for us. Sales of single reprints, by contrast, is a mere $4000 a year.

Back issue revenue seems to be a declining business, but can still bring in significant revenue, especially for well-established journals. To a growing extent sales of hard copy back issues is being replaced by microfilm and microfiche sales. Our present policy is to reprint sufficient hard copies for five years of normal sales, and to discard older hard copies from expensive storage. We also will not reprint any copies more than one year old.

Permissions and royalty fees are at present miniscule. Copyright developments over the next few years will dictate whether this sort of revenue can grow to respectable levels. Most of our present revenue in this category come from royalties from book publishers who republish copyrighted journal articles or from commercial suppliers of reprint/photocopies, such as the Institute for Scientific Information in Philadelphia, Pa.

There are some other potential or real sources of revenues:

- Sales of audiotapes of scientific meetings.
- Book sales--in ACS a fast growing business now worth $1 million a year.

Both of these areas require special expertise but should be within the capability of a journal publisher.

- Investment revenues from funds derived from subscription sales made in advance of production.

The mix of these revenue components can be very different, depending on the type of journal:

A. The typical basic journal (Figure 1) is highly dependent on subscription revenues and to a lesser extent on page charges.

B. The atypical "advertising" journal (Figure 2) gets a lot of its revenue from advertising and is much less dependent on subscription revenue. But it poses problems of a different kind--not the least of which is the danger of being dependent on the continuation of advertising revenue, a volatile form of financial support.

EXPENSES

On the expense side, most journals have considerable costs tied up with people:

1. Scientific editors who are paid modest stipends.
2. Authors of certain types of material (e.g., reviews) who may be paid honoraria.
3. Manuscript review editors who act fulltime to operate the peer review and manuscript logistics operation.
4. Technical editors, who copy edit and help produce the journals.
5. Administrative and support staff, who rejoice in the ability to add "overhead" to the entire system.

Printing and distribution costs represent the bulk of the expense on the average scientific journal, whether published commercially or by a non-profit society like ACS. Major components of P&D costs are:

1. Composition--i.e., setting into type.

2. Illustrations, paste-up, camera work, plate-making etc.

3. Press make-ready and press running, binding.

4. Affixing mailing labels, mailing (and wrapping).

5. Paper

6. Postage

We will analyze later the relative importance of these components and how that varies with the type of journal. Suffice it to say for the moment that all these costs have been rising, and costs for paper and postage have been skyrocketing.

Subscription fulfillment is a necessary cost of business--to maintain the necessary information on subscribers and to produce mailing labels. This function can be, and often is, farmed out to a specialized firm. In the ACS it is done internally. Costs can be substantial (in the ACS more than $2 per subscriber per year) if you keep an extensive amount of information about members and subscribers on file, as we do. Then there are the costs connected with:

1. Producing reprints.

2. Storing and mailing back issues.

3. Manufacturing and selling microforms.

4. Circulation promotion. (The bulk of these costs are incurred on "advertising" journals where maintaining a large circulation is important for advertising purposes.)

5. Miscellaneous other expenses, such as travel, office supplies, telephones, etc.

6. General overheads for accounting and other administrative expenses.

Now we shall analyze how the various components of expense play a large or small part in determining the overall expenses of various types of journal.

Figure 3 shows a basic journal's dependence on production and distribution costs.

Figures 4 and 5 compare the breakdowns between a typical basic journal (Figure 5) and an atypical "advertising" journal (Figure 4). The basic journal shows the heavy influence of composition costs. Such a journal has few subscribers and a short press run, thus paper and postage are relatively unimportant. The "advertising" journal in figure 4 has over 30,000 paid subscribers and a great deal of advertising. It is not surprising, therefore, that paper looms large among its expenses, and on such a journal savings in paper are eagerly sought.

Over all scientific journals, however, it is composition coststhat catch the eye most. It is for this reason that we in the ACS have spent much time and effort in trying to reduce composition cost by changing from hot metal composition to photocomposition ("cold" type).

FINANCIAL STRATEGIES FOR THE FUTURE

We feel we are still too dependent on subscription revenue from hard copy journals, and we hope in the future to have microforms and royalty payments shoulder more of the load. It is most unlikely that revenue from either page charges or advertising can be increased significantly.

On the expense side we intend to continue to attack compostion costs through the adoption of new photocomposition technology. We also will purchase our own paper, rather than rely on the printer to do that for us. We will also look for "secondary" products and services which could be produced from the "primary journal data base" that is a byproduct of the method by which our journals are produced. We also intend to continue our experimentation with new formats for journals--the synopsis or summary journal is one such example--in the hope that we can once again produce a useful and inherently less expensive information source that may reattract the individual subscribers we have lost over the years.

Figure 1
REVENUE BREAKDOWN
"Page-charge" Journal (*Jl. of Org. Chem.*)
1976 Figures
Total $: $840,000

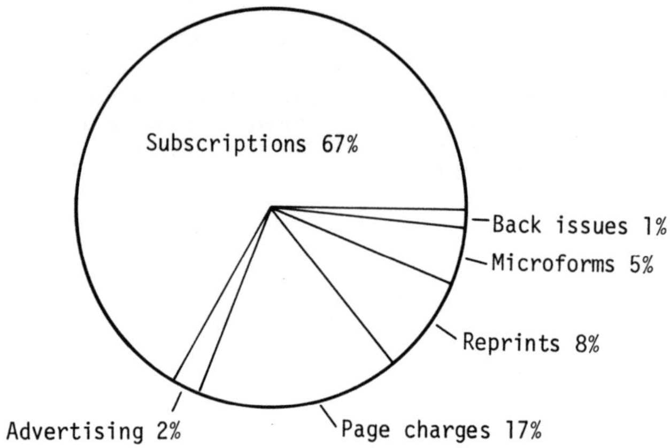

Figure 2
REVENUE BREAKDOWN
"Advertising" Journal (*Analytical Chemistry*)
1976 Figures
Total $: $1,150,000

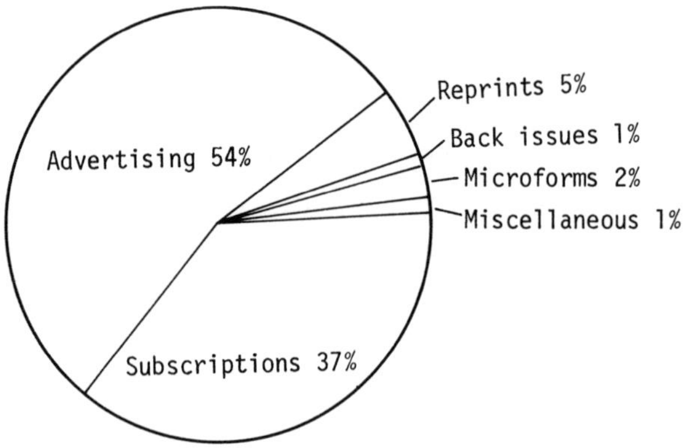

Figure 3
EXPENSE BREAKDOWN
Typical basic journal (*Jl. Agric. & Food Chemistry*)
1976 Figures
Total $: $180,000

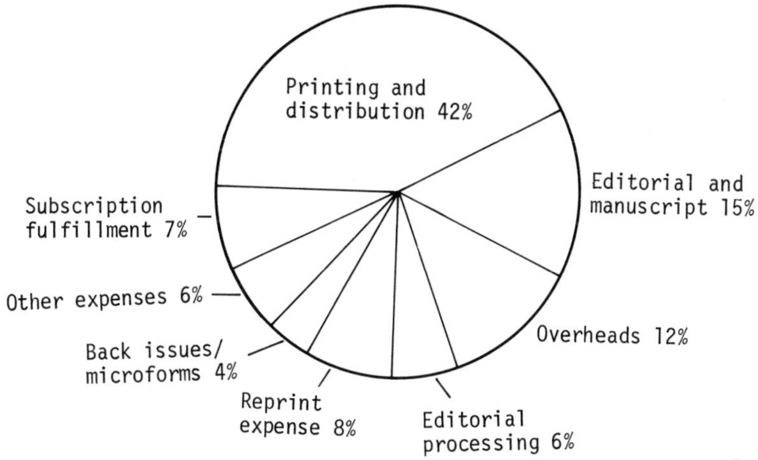

Figure 4
PRINTING AND DISTRIBUTION EXPENSE
Large-circulation journal (*Analytical Chemistry*)
1976 Figures
Total $: $512,000
Editorial page expenses only
Circ.: 30,967

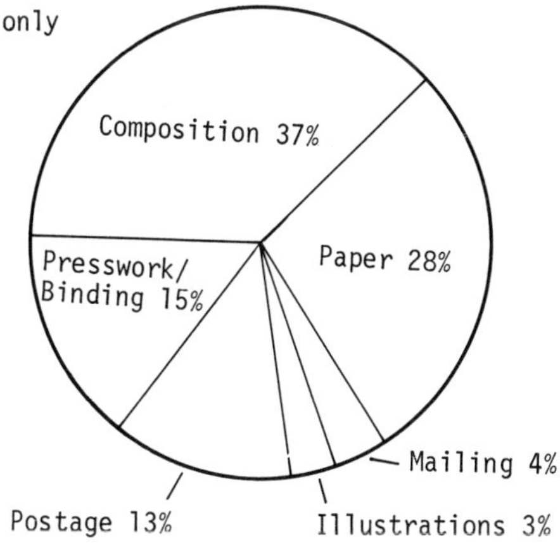

Figure 5

PRINTING AND DISTRIBUTION EXPENSE

Small-circulation journal (*Jl. of Chem. & Engng. Data*)

1976 Figures

Total $: $43,000

Circ.: 2031

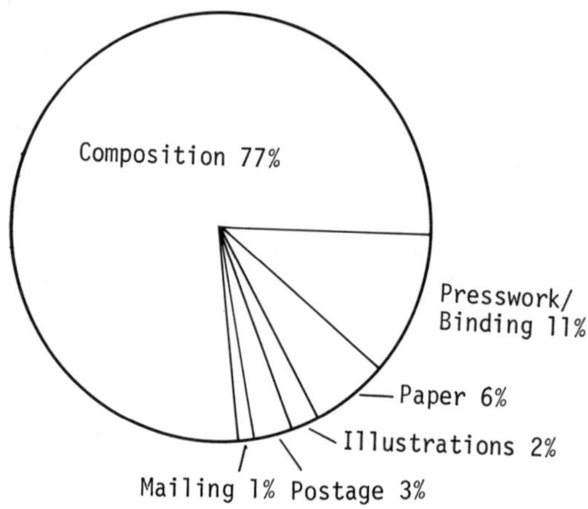

DISCUSSION on "The organization and Economics of publishing and the function of the editor"
by Peter Bakker

"Is a good editor a good publisher?"
by Michael Harris

"Financing scientific society journals in the USA"
by D.H. Michael Bowen

ALFRED FISHMAN - I wonder if I might address the three comments that Mr. Harris made parenthetically because I think they permeated the entire discussion. The first had to do with the question of support of research and the maxim "if it's not worth publishing, it's not worth doing". I agree with this dictum; research has little value if the results are not reviewed critically, dispassionately, and made available for examination by others who are expert in the field. The second is the concept of publish or perish. I think that this concept is a bit out of date, at least in the bio-medical sciences. No longer is the height of a stack of publications a measure of the level of academic performance. A faculty member is not promoted simply because he has written a great deal. A review or an abstract or a best seller of any kind has little value with respect to academic advancement. Moreover, publication in certain journals - especially those which practice rigorous criticism - has much more meaning than in others. Finally, I would like to address the question concerning the use of the support of research as a mechanism for preventing the flood to the publishers and editors. Thus, "flood" stems in large measure, from the ease of getting things published. So much that is valueless is published. The controlling mechanism is not the funding of research - which has its own goals - but a poor review mechanism. What we are seeing in the biomedical sciences is a flood of uncritical publications simply because there is a journal or book for every manuscript.

MICHAEL HARRIS - First of all, I doubt if there is any genuine distinction between commercial publishers and society publishers and the selection and determination of what should be published. I'm a commercial publisher, publish twenty journals. We have a peer review system, which is identical to that used by the professional societies and use in many cases some of the same people for review as are used by the professional societies. I doubt that there is a great distinction between the two. I think there may be a distinction between the willingness to take a new journal because a society generally looks from the point of view of its entire membership and the problems of the society, whereas a commercial publisher maybe much freer to decide to create a new journal in a newly developing field which has not yet established such a large body as to develop its own society or even a section within an existing society. So I doubt very much that the distinction is a very great distinction between the practices of commercial publishers, not in all cases to be sure, but in general I don't think there is such a vast difference.

The point that I really wanted to make with regard to the question of research is this. Some people maintain that publishers should, and this goes for commercial societies, for professional societies, not-for-profit societies, as well as commercial publishers. Quite a number of people and institutions have maintained that publishers are not really fulfilling their duty determining what should be published. They are publishing too much, they are publishing material which is irrelevant, that in any given journal there are only one or more articles or certainly a small number of articles that are interesting to the journal reader. That may be so. My point was basically that I don't think that publishers once having established a peer review system once having involved the scientific community and having people in the profession say yes this is what we are publishing, I don't see how a publisher can then exercise any further judgement about what should be published. And I make a point that for me at least, that is the same problem as is involved when a number of scientists go to an institution for a grant. What research do you support? How so you support that research? It is a very difficult problem and one for which so far as I'm aware no suitable formula that is satisfactory to most people has ever been found. So I think it is with publishing.

My only point about the publish or perish syndrome is that while it may be dying in some sectors, it is still very true in others and that there is in some institutions a lot of pressure on people to publish in order to gain tenure. In some cases in order to achieve prestige or for other purposes of that kind. And I'm not at this moment arguing for or against it. All I'm saying is that these are pressures which tend to build up the number of papers that are submitted, the number of things that are written and contribute to this vast ourpouring of information which ends up being published. And I must say that I don't feel very distressed by that. I think that in the final analysis librarians and individuals have to determine for themselves and no one else can determine for them, what is important to them, what they should read and what they should spend their time on.

STEPHEN LOCK - I would like to ask the panel to what extent strong journals should subsidize the weak in an entire publishing house. We started off at the BMJ with twenty specialist journals and with the philosophy that we would break even and make a little over - the little over being used to start new journals and to subsidize the weak. What happened in fact was that some strong journals emerged almost immediately, such as cardiology, thoracic medicine and things like this, and research journals in the field of education and venereal diseases always remained weak. Eventually the strong journals got fed up with subsidizing the weak and they insisted that they all stand on their own legs. I just wonder what the panel's philosophy is about this.

PAUL NIJHOFF ASSER - You must have missed the STM seminar on journal publishing in Denmark where it was said that an ailing journal which an example was given should not be allowed to survive artificially. It was at least the opinion of the majority of the audience at that time.

MICHAEL BOWEN - Yes, we have looked at this quite a bit and actually have had a situation where far from the strong subsidizing the weak, we have some of the weak subsidizing the strong. Which needs some explanation. Of course you mean weak and strong financially. We have a policy now that every single journal has the same financial goal and that is, as I said before - to break even plus a bit extra. The bit extra is something up to five percent. It hasn't been pinned down to an exact number. In reality of course this doesn't really happen. Now there are no glaring changes from this policy right at the moment in our stable of publications. What has been done essentially here is if a journal cannot attract a sufficient number of subscriptions at a given price to break even we generally raise the price so a number of our smaller, weaker journals are in fact priced higher per word or per page than the big journals which in our particular stable are the stronger and the best journals intellectually. We have some discrepancies yet because some of the very big journals are underpriced, but because they are very big journals, their absolute price appears in the eyes of some people to be too high.

BASIL WALBY - Michael Bowen said that the levying of page charges was some sort of honor system.

MICHAEL BOWEN - Our journal editors do not know who pays the page charge and who does not, because the decision to accept a manuscript is made before any mention of page charges is directly made to authors. Our editors are kept away from that sort of information altogether. They suspect but they really don't know who pays and who does not.

CYRIL BELSHAW - I am interested in this question of page charges because in addition to page charges in some of the social science journals, there are now submission fees. That is, before an article is accepted to go to referees the author of the article has to pay a fee which is intended to discourage certain kinds of articles. Now in my kind of journal I'm opposed to this for certain reasons, but the reasons involve international communication. Essentially that is, it appears to us in our kind of field, that it is almost impossible to to get people to pay these kinds of fees on an equitable basis where the funds from research and the personal income are very weak or almost non-exist nt or where there are financial restrictions to the payment of such things across international boundaries. Now, in the natural and medical sciences it appears that this is going ahead so I would like to ask the people who are concerned with this what is the effect on the international flow of manuscripts?

PAUL NIJHOFF ASSER - I think that I have only heard once that Michael Bowen has said that the page charge system exists mostly in the United States of America and hardly exists in Europe. I heard once that because of this, manuscripts were submitted to European publishers but whether this is true I have no way of ascertaining. Perhaps you know something about this.

MICHAEL BOWEN - Well, you'd have to look at the proportions. I don't even know if you could do an experiment like this and find out significantly. You have to look and find how many papers came from outside the U.S. before page charges became common, which was in the sixties some time, and how many are coming in now. I think we consistently had about ten percent of all the papers in our journals from outside the U.S. and we do know the propensity to pay off the foreign authors who now publish in our journals. That is no worse than the propensity to pay by American University professors without federal funding.

YEHUDAH WERNER - I do agree with Mr. Harris that there is no difference in principle between journals published by learned societies and journals published by commercial firms as to their selectivity. It is not the flag that counts. But much depends on technical details: there is a natural difference in selectivity between a journal which has a fixed number of pages and selects papers to fill these from among a large number of manuscripts and a journal which reviews and approves manuscripts and then publishes as many pages as is necessary in order to get these manuscripts printed. The selectivity will be lower. Secondly there is a natural difference in selectivity between a journal where the author is bound to submit manuscripts to one editor who then has his anonymous referees and a journal where the author may submit his manuscript to any one of twenty or fifty persons listed on the cover, and that one person will then be his selected and perhaps well known editor and reviewer. So there you may have differences.

MICHAEL HARRIS - Very briefly, there may or there may not be. I think it depends on the journal and the publisher and the practices of the publisher. I can tell you that in my own case for obvious reasons we have had to simply say that we are not going to permit journals to grow any more-with greater or lesser success-because we've had to say at times that there is a fixed number of pages, and somehow you must find a means of coping with it. Sometimes we found that to work, and sometimes we found that unreasonable. I find it hard to argue about the matter because I think it really depends on the individual cases.

KLAUS HELBIG - I would like to comment on Dr. Werner's last remark. I don't believe that a learned society can print papers "because they are there". I edit a journal which is owned and published by a learned society and I'm strictly told "the budget does not allow to publish more than eight hundred pages a year". Whether you are a commercial publisher or a learned society somewhere you have to pay the bill. Being a learned society does not mean you can publish simply every paper that comes in: one might accept it, just put it on the shelf, and publish it when its turn comes. But since we cannot publish more than eight hundred pages a year, we would simply limit the number of papers by the delay building up.

PETER BAKKER - I think this might be true for non-commercial publishers, but with a commercial publishing house, if there is a big offer of manuscripts th commercial publisher could decide to increase the number of pages. He doesn' like to do this in the middle of the year because then he has a problem with his subscription rates and things like that, but he could decide to publish twenty percent more pages in the next year, depending on the marketing resear But that's an easy decision and many publishers have taken such decisions during the last couple of years. Now there is a tendency to restrict the numbe of pages.

MICHAEL BOWEN - What we try to do in the American Chemical Society is to try to look ahead and see what the manuscripts are likely to be, and of course we depend very heavily on the editors and the advisory boards for a knowledge on how that particular field is going. If there was likely to be a pressure of manuscripts in the field of biochemistry, for example, it wasn't very difficult to work out that there really was an increase so we allotted more pages. So even though we are not-for-profit, we set budgets ahead of time. If we have a budget for the year, we simply do not like to exceed that budget. We have on occasion gone to our board of directors and said we need so many more pages for such and such a journal. It's going to cost so many thousands of dollars. Because we know that we are going to have to publish these papers eventually anyway, we want to publish them as quickly as possible, but by and large we have to do it through thinking ahead and budgeting and not off the seat of our pants so to speak.

PAUL NIJHOFF ASSER - I think in this respect there is again not much difference between most of the good quality commercial houses and most of the good quality society publishers, and again I think this proves that they are more or less similar. There is in this respect the matter of adaptability, of fle ibility and of course also on both sides what the market will bear.

MARK KELLER - There was only one remark about reprints and no one has commented on it. The remark was that reprints are a form of vanity publishing and therefore the author can be charged enough for them so that the journal makes a profit on them. To me it seems that when an author has gone to all the trouble of writing his paper and submitting it and going through the agony perhaps of revising it twice to please the editor and the referees, and at la gets his paper published and needs to let a wider world know that his paper has been published than the readers of the journal - he is being exploited if the journal then also tries to make a profit on him by charging him more f the reprints than their production costs. And I think it would be much faire simply to charge for reprints whatever the cost of production is. That's the policy that we have followed. Maybe we've been unwise.

MICHAEL BOWEN - I think what I said was that we view reprints as a vanity pre I think the experience bears out that we are correct. I mean it's a rough ol world, but one way or the other we've got to get the money in. The fact is i we didn't get it from reprints we'd get it from subscribers. Authors themselves very rarely pay for reprints. It's their institution or their source of support by and large that pays for reprints. There are exceptions to this I'm sure. And the reprints are of course sent all over the world to advertis the expertise of the author and of his institutional sponsor. So it's public relations for them. We don't feel too badly about it overall. (Laughter).

MICHAEL HARRIS - I thought I'd just add to that to show how difficult life can be in trying to make generalities. I am a commercial publisher and I'm out to make money. We do give free reprints to our authors, and if they want a certain number more we do charge them. But because we don't want to be crassly commercial we take as an example the price schedule of the not-for-profit American Chemical Society. (Laughter).

IRVING AMRON - The American Society of Civil Engineers is a not-for-profit organization. During the last 15 years, our costs of publication have gone up about two and a half times, in the same way as our dues income went up, while the publication support dollars remained constant through these years. The number of copies of each of our 15 or 17 journals during the years increased, independent of the increase in the subscription rates. And if we take the support that the members of the society are giving our publications, they dropped from 11 to 3½% so that this in no way influenced the number of subscribers to our journals. If the journals are of high quality, the subscribers will continue to pay for them. We are publishing a translated journal from the Russian and we've increased the subscription price approximately five times during the past ten years, and the subscribers have remained constant. If they want to buy it, they will pay for it, and they understand the need for paying more.

Concerning page charges. The American Society of Civil Engineers in principle has been against the use of page charges even though we are in the United States. It was mentioned here that the number of foreign authors is about 10%. We have the same percentage independent of page charges or anything else. There are some constants which remain the same in any type of publishing.

MICHAEL BOWEN - I would like to ask Dr. Amron a question. There is an allocation of dues in the American Society for Civil Engineers that goes to the publication program. Is that correct?

IRVING AMRON - Yes. Our publications program includes advertising which is a large percentage of our income. This is a separate branch of our publishing of Civil Engineering. All other expenditures are paid for by the subscribers themselves. As I mentioned, the dues increased two and a half times, the subscription rates approximately the same, as well as our expenditures. Therefore the dollars that were given to us for covering our losses were covered constantly over the years because we watched the way our costs were rising. Our subscription rates were rising. Relative costs remained the same over the years.

THE PUBLICATION OF SCIENTIFIC JOURNALS IN FRANCE

Jean-Paul Abadie
Director of Scientific Publications, Masson Publishers,
120 Boulevard Saint-Germain, 75280 Paris, France

ABSTRACT

The international reputation of certain French journals seems to have lost some of its prestige in recent years.

Essentially, the principal reasons for this decline are the following :

1 - Bad organization, and insufficiently-developed editorial criteria on the part of these journals, which interferes with the rigorous selection of articles.

2 - Inadequate knowledge of the principles of medico-scientific writing, which prevents foreign readers from rapidly grasping the significance of research papers.

3 - The decline of the French language as a medium of scientific communication.

4 - The attitude of French scientists themselves, who have accelerated this decline by reserving their best work for foreign reviews.

Nevertheless, this description should be attenuated, taking into account the following factors :

- The decrease in the number of subscribers is still insignificant.

- Great improvements are being made in the editorial organization of journals and in the presentation of articles.

- Above all, the French scientific community is giving evidence of a renewed desire to support French publications.

Miriam Balaban (ed.), Scientific Information Transfer: The Editor's Role, 377-383.

A general feeling has gained ground in the last 20 years that the advanced French scientific journals have lost some of their prestige. This impression has some basis if one judges by the number of French articles cited in other journals.

Whether this feeling is justified or not - and we will come back to this point later - it seems now to be well-established and I thought it would be interesting to analyze the factors leading to this state of affairs, and based on this analysis, discuss the policies that might be instituted so that French journals can regain their place on the international market of scientific information.

1 - The principal reproach made to French journals is their lack of selectivity : they sometimes publish articles of doubtful originality and disputable scientific value. In effect, there is no doubt that many French journals have been slow in developing the editorial structures necessary for rigorous selection - and some of them have even not yet solved the problem.

There are several explanations for this phenomenon :
- Since France is a small country, with a limited number of scientists - as compared to the United States, for example - it is safe to say that almost all scientists working in the same field know each other. And the concentration of laboratories in Paris only accentuates this phenomenon. An Editorial Board member therefore, finds himself in a delicate position in refusing or criticizing an article submitted by one of his peers.

- In certain fields, the number of scientists - and especially the outstanding ones - is so limited, that an adequate editorial structure can only be established if the French Editor-in-Chief enlists the participation of foreign colleagues. But this solution is possible only if French research in the particular field is especially competent, and if the Editor himself has the necessary international stature to make such contacts.

- Finally, for certain journals, the number of articles submitted and their quality are inadequate for real selection : rigorous selection would threaten the very existence of the review, for it would then no longer have enough papers to publish.

2 - A second criticism frequently levelled at French journals concerns even the orgazination and form of their articles. While I am less well-informed of the problem in physics and chemistry journals, I know that the criticism is well-founded for medical and biological reviews.

It is universally accepted that the organization of a research paper should follow the now-classic outline of distinct, prescribed sections, and that special rules apply in writing this type of article. Unfortunately French authors are still guilty of confusing a research paper with a literary essay.

The foreign scientist therefore is disoriented when trying to read certain French articles whose subtilities of grammar and organization only aggravate his poor knowledge of our language.

A British physician, Dr. Farfor, - who has written a great deal on the subject - quoted in one of his articles the following comment of an American friend : "Finding the message in a French research article is like looking for train information in a timetable written by Proust " !

In fact, no serious effort is made to permit the non-French-speaking reader to rapidly grasp the essential findings and their significance. In addition, the English summaries, where they exist, are too often so sketchy as to be perfectly useless.

3 - I come now to the third factor in the declining influence of French reviews : the language in which they are published. And in approaching this subject - I can't help but think of that famous French cartoon which appeared at the height of the Dreyfus Affair.
The first scene shows a French family happily united around the dinner table. In the second scene, the dining room has become a battlefield - with the members of the same family engaged in a furious fight. The caption reads : SOMEONE DARED TO RAISE THAT QUESTION !

My question - should we publish in French or in English - is likely to rouse the same passion. I shall therefore discuss this question very cautiously - later on - and will limit myself presently to the following observation : If all French journals had achieved international standing, with articles written according to the norms of scientific communication, and especially, if they had been given priority in publishing the best work of our best scientists - then the language of publication would not have the same importance.

In this regard, we have some significant statistics available : While less than 25 % of the articles published in all scientific journals, world-wide, are cited more than 5 times in the 5 years following publication, this rate rises to 50 % for articles published in French in multi-lingual reviews (principally, anglo-Saxon). In addition, while the percentage of articles in French in French journals which are never cited may exceed 50 %, this rate is only 20 % at the maximum for articles published in French in multi-lingual reviews.

4 - A final factor in the difficulties of French journals is the attitude taken by French scientists when faced with the initial problems.

Instead of making an objective analysis of these problems (problems of quality, of form, and possibly language) - and adapting French journals so that they could meet international competition - French scientists, including the best of them, especially those who had studied or worked in the United States - turned rather to foreign periodicals to publish their work.

It must be said in their defense, that 10 or 20 years ago, the necessity for these modifications - which seems self-evident now- appeared perfectly absurd to both editors an publishers.

The example set by these precursors became rapidly widespread Paradoxically, it was not unusual to see the most fervent defenders of the status-quo - the very authors who would not tolerate from a French journal the slighest doubt about the scientific and editorial quality of their articles - meekly submit to the editorial verdict of foreign reviews and humbly agree to make all the required modifications. And you can imagine their amazement at the number of reprint requests they received once the article was published !

This trend has been intensified by some of the official organizations responsable for French scientific policy : articles published in very selective reviews, particularly Anglo-Saxon journals,earn their staff members more points than articles published in less selective reviews, that is, practically all the French journals.

Thus, the ever-increasing flight of articles abroad reinforced the decline of French journal, and in consequence, the desire of French authors to publish in foreign journals. In the face of this viscious circle, it is surprising that French journals have been able to survive at all.

During the same period, similar publications - German, Scandinavian and Dutch - gave proof of their extraordinary capacity for adaptation, often with less important human, scientific and financial potential.

French publishers are often accused of not displaying the same marketing flair as their European colleagues. But what do you do when an important foreign library, subscriber to a journal for 40 or 50 years, suddenly decides not to renew its subscription ?

Nevertheless, and fortunately, French journals still have great possibilities, and having made this severe analysis, we shall go on to discuss the available solutions.

As a preliminary though, - since my harshness may have shocked some of my scientific compatriates - and understandably so - I must emphasize that the picture is not all black :
First of all, many French journals have been able to make the necessary adaptations and transformations to hold their place on the international market : among the Masson reviews we can cite the Journal de Physiologie, the Nouvelle Revue d'Optique which has just become the Journal of Optics, the Annales de l'Institut Pasteur, Biomedicine, and Biochimie which has replaced the Bulletin de la Société de Biochimie.

In addition, our subscription analysis shows that :

1 - the loss of foreign subscribers is for the moment still slight and certainly insignificant, given the variations observed over the last 10 or 20 years.

2 - North America is by far our most important market, placing before the French-speaking countries - mediterranean or Est-European - which constitute traditionally an easier market for French journals.

3 - Our progression on new, non-French-speaking markets, like Japan, is constant, and Japan is the second foreign market for the Bordas/Dunod/Gauthier-Villars Group, and the 6th for Masson.

These elements should enable us to see the "decline" of French journals in a more-balanced perspective.

Returning now to the available solutions :

1 - The most urgent task to attack is the form of French scientific articles. Manuscripts can no longer go to the printer in the form in which authors submit them, and it is indispensable that everyone : authors, editors, Editorial Board members and publishers decide to respect the rules of scientific communication, that is :
- meaningful titles using functional words
- informative summaries, clearly presenting the research study; why and how it was conducted, the results and their significance
- structured organization
- rigorous demonstration
- presentation of data
- concise style, as simple as possible, in basic French, understandable even to foreigners.
- detailed legends for figures and tables so that they can be understood without referring to the text.

These rules are borrowed from Dr. Farfor, and like him, we are convinced that a research paper presented according to these principles will have a greater success, even if written in French.

2 - The next step is obviously publication in English. But, I must insist on the fact that this solution can be considered only if the organization of articles is beyond reproach : a confusing, though interesting article will be as difficult to understand in English as in French. Otherwise, I fear that the Anglo-Saxons will realize even more clearly that their difficulty in reading articles in French is due not to their poor knowledge of the language, but rather to the poor organization of the articles.

Assuming then that articles are perfectly written, the first stage in the "anglicization" of our journals is the translation of the title and summary. The English summary, moreover, should be more detailed, with possible references to essential tables or figures. In the same spirit, legends of tables and figures should be bilingual.

I am willing to bet that even at this stage, most readers will manage to sift out from a research report all the aspects which interest them, as easily as if it were an English text.

And I would advise a French author - with insufficient mastery of English - to limit himself to this stage, rather than write his article in a ridiculous English which can only diminish the scientific aspect of his work.

In addition, English translation is impractical, first because of the difficulty in finding qualified, specialized translators, and secondly, because of the excessive costs for the budget of our reviews.

Writing in English should be undertaken only by those whose knowledge of scientific or medical English is good enough, or by those who have an Anglo-Saxon colleague available to do the translation.

In any case, I feel it desirable to add an Anglo-Saxon representative to the Editorial Board, who would ensure the linguistic correctness of manuscripts. We recommend too that he be a resident of England or the United States, for in our experience the English of an Anglo-Saxon living in France is contaminated in less than 5 years.

Just one word on this linguistic problem : I would like to emphasize again that even if English has become the imperative vehicule of scientific thought, it should only be used under appropriate circumstances : writing no matter what, no matter how, in English, can only have an even more disastrous effect on the image of our publications and of French research in general.

Some French scientific personalities feel bitter about the substitution of French by English; they underline the cultural and even political aspects of the language, and are already predicting as a consequence the decline of French thought. I believe, on the contrary, that if French is to remain the language of communication for 230 million people, it is indispensable that the French intellectual community be able to continue to make itself felt, in the necessary form, and possibly in the required language, but above all, wherever possible, through the medium of a French journal of international stature.

This last point has a fundamental, strategic importance in a world where scientific quality is not always the sole criterion for accepting an article. And this leads me to the last point of my expose.

3 - For finally, if we are to re-establish the position of French journals, there must be concertation on all levels of the French scientific community, to develop common guidelines for official policy-makers, for scientists, and for publishers alike. This is of prime importance, the mainspring without which all the rest is merely wasted motion.

Some recent developments in this direction lead me to believe that there is reason for optimism. For example, in meetings held a few months ago, bringing together official research organizations like the I.N.S.E.R.M. (1), the C.N.R.S. (2), the B.N.I.S.T. (3), with scientists an publishers, it was possible to identify the problems and define a coherent publication policy as follows: It was decided to promote the creation of what we call "lighthouse" reviews - that is, journals covering a rather wide scientific sector and capable of reaching an international audience.The sectors covered might be : physics - chemistry - cells biology - biochemistry - plant physiology - animal physiology - ecology - oceanology - geology - economy - linguistics - computer science, etc. These reviews could result from the merger of already-existing journals or be newly-created by special grants.

In the framework of co-production contracts, responsability for scientific content would be assumed by an official research group, an association, or a French or foreign professional society, while a private publisher would ensure production and

(1) National Institute fo Health and Medical Research
(2) National Center for Scientific Research
(3) National Bureau for Scientific an Technical Information

distribution.
French scientific authorities, especially the B.N.I.S.T., would contribute financial support, signing long-term development contracts with the publisher.

But if these projects are to succeed, French scientists must again agree to offer their best work to these new French journals, even if in the beginning, their circulation is not as great as the corresponding foreign reviews. In effect, we are asking them to bet on the future.

French publishers, who also share responsability for the decline of French journals, have finally been given the means to ensure a better penetration of our reviews : the new structure and reorganization of Bordas/Dunod/Gauthier-Villars - and the creation by Masson in less than 2 or 3 years of 5 or 6 publishing subsidiaries throughout the world - the last in New York - are but two examples.

The recent launching of the "Nouveau Journal de Chimie", published jointly by the C.N.R.S. and Gauthier-Villars, with the participation of the best French and foreign scientific talent, is the first result of this new policy. I hope that others will follow soon.

INTEGRATED SCIENCE PUBLISHING: AN EXAMPLE FROM AUSTRALIA

Basil J. Walby, *CSIRO Editorial and Publications Service, 372 Albert Street, East Melbourne, Victoria 3002, Australia*

If we consider journal publishing in science from the point of view of authors and readers we find that their demands are not necessarily the same as those of publishers or systems analysts. We find that the latter are generally seeking ways to reduce the cost of publishing by preparing truncated journals, greater author involvement in preparation of material, use of microfiche or of untried methods of disseminating information tailored to meet the presumed needs of the reader. But what do scientists tell me they require? Much the same as before. They want journals they can handle and browse through - they do not like squinting through microfilm readers - they like reprints which are more convenient to handle than whole copies of journals, they do not want a mass of irrelevant references spewed forth by a computer in response to their imagined requirements. Publishers have found and are continuing to find reaction to their attempts to force new types of publication on their subscribers. Even librarians who aver to a preference for microforms are spending very little of their budgets on non-book material. To take an extreme example, CSIRO librarians insist on seeing a printed version of their "Annual List of Subscriptions" in spite of having the material on microfiche produced directly from the Organization's computer. If the proponents of microforms won't use them, what hope is there that simple scientists will turn avidly towards them?

Costs, however, continue to rise and will go on rising. So what steps can we take to ensure that our publishing budgets are used to best effect?

First, let us take a close look at our published material over the past year. Are we satisfied that all of it justified the space we gave to it or indeed whether it should have been given any space at all? Our referees must be more watchful than ever that non-original or repetitive material does not get published in the primary literature. Let us stop the Ph.D. thesis being used as a source of several indifferent articles. We must ensure that undigested data be relegated to accessory publication for lodgment either in data centres or with the editor for supply to those people who really need it for further work. I would dare to suggest that it is time for the taxonomists of the world to rethink their rules of publication. Is it very sensible to insist that material should be reproduced typographically in full when the number of people who may wish to study a particular species may be no more than half a dozen spread over several continents and probably over just as many decades?

Rationalization of publishing is certainly necessary. How else will we make the best use of editorial, production and distribution resources? Some rationalization may also stop the unhappy business of seeing last year's rejected manuscript appear in this month's edition of a less-than-prestigious journal floated in the hey-day of speculative publishing. Australia with its small population has felt the effect of overpublishing to a considerable

Miriam Balaban (ed.), Scientific Information Transfer: The Editor's Role, 385-390.

extent. The position has been aggravated by high wages and high costs of living. Many Institutions and Societies have been forced to reduce their publishing activity because there are just too few local consumers. Instead of working together to set up publishing cooperatives with a specialist staff of editors, designers and salesmen they have either sold out their interests to foreign publishers or have given up the unequal struggle altogether. One might have thought that they would have looked at the example offered them by CSIRO where just such a cooperative exists.

It says much for the enlightened leaders of CSIRO in the late forties who saw that strength lay with the big battalions and created at that time a publishing organization that was as self-sufficient as possible without being too bureaucratic. Each year CSIRO publishes about 15 regular journals, rather more than that number of irregular serials, a dozen or so new books and sundry newsletters, brochures and Divisional annual reports. This makes it the largest indigenous scientific publisher in Australia. The budget for 1975-76 was $A1.36 million. The main categories of expenditure are shows in Figure 1. This output is achieved by a tightly knit publishing group which is illustrated schematically in Figure 2. The number of personnel engaged in the activities shown here is indicated at the bottom of each of the hexagons. There are in addition to this the Publisher, the Printing Manager and the Administrative staff which altogether make up a total of some 78 people. The editorial and manuscript control activities are housed about a mile away from the printing establishment. Although a courier from the printery visits the editors three times each working day, the telephone lines still run pretty hot. The close links with the printery enable us to adhere to strict production schedules which on average require the publication of one issue of a major journal each week of the year. Table 1 shows the output for a typical recent year.

TABLE 1: Australian Journals of Scientific Research 1975/76

Journal	No. Issues	No. Pages
Aust. J. Agric. Res.	6	954
Aust. J. Plant Physiol.	6	820
Aust. J. Biol. Sci.	5	682
Aust. J. Bot.	6	922
Aust. J. Chem.	12	2864
Aust. J. Mar. Freshwater Res.	3	506
Aust. J. Phys.	5	668
Astrophysical Supplement	(4)	306
Aust. J. Soil Res.	3	386
Aust. J. Zool.	4	632
Zoology Supplement	(8)	784
Total	50 (12 supplements)	9524

Manuscripts are controlled centrally by the front office. Over 1250 submissions for the national journals arrive each year, each having to be checked for completeness, correct number of copies and readability (particularly the existence of duplicates of illustrations) and then acknowledged.

Figure 1

Breakdown of expenditure

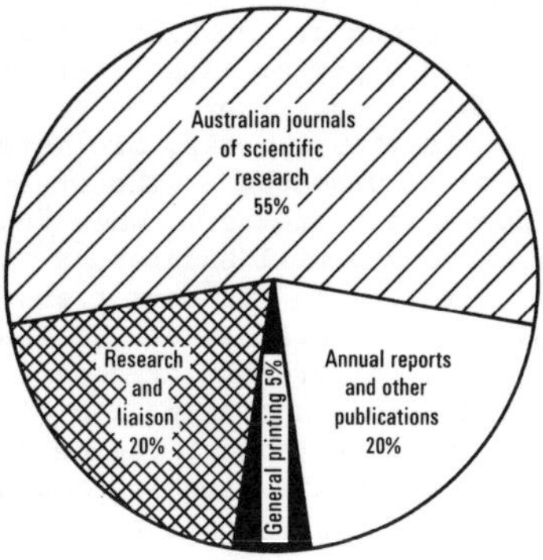

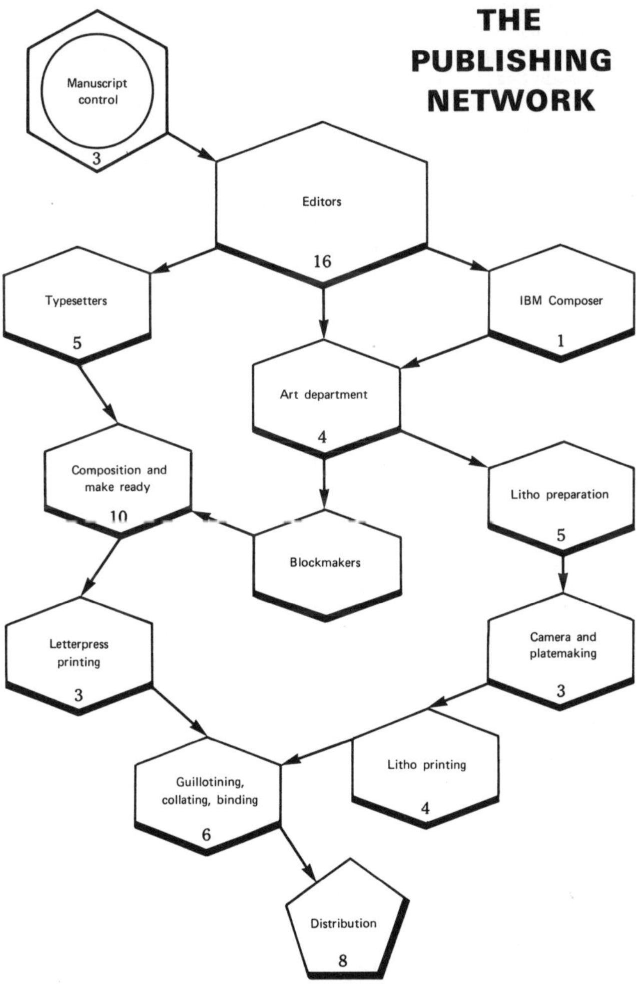

Figure 2. The publishing network

They are then passed to one of the senior editors responsible for either one journal or a group of journals. The editorial service, like all Gaul is divided into three parts. The first two of these handle only the primary journals, the third has perhaps the most difficult task, that of dealing with the liaison publications, books, brochures and the Organization's Annual Report which has to be produced at the greatest speed in time for the budget session of Parliament in September. When you realize that our financial year ends on the 30 June you will understand that problem.

But to return to our senior subject editors. They act as secretaries to the Advisory Committees of the journals. The Committees help these managing editors in the process of choosing referees for the submitted papers. Referees are chosen not only from within Australia and New Zealand I might add. It is vital to enlist the help of scientists from the whole world in a country like Australia with its small research population many of whom may be working in the only local group dealing with a particular topic. The size of Australia itself and its distance from even its nearest neighbour makes us very dependent upon a quick turn round of manuscripts by referees and within our own office. The success of this may be judged by the time to publication from receipt of manuscript to appearance of the article in a journal. The median time is 6 months - half of which represents the time to acceptance and includes any reqriting that is necessary as a result of advice from the referees or the Committee. The close proximity of specialist editors makes for an efficient and much prized copy editing facility which cannot be attained by those who publish journals in splendid isolation. This convinces me of the need for more group or cooperative publishing (to which I have referred above) particularly by small societies that cannot afford to have more than one copy editor (often underpaid or not paid at all). Our interdisciplinary editorial team contains physicists, chemists, zoologists, a botanist, and a mathematician, all within shouting distance of each other. This pays considerable dividends and authors appreciate the careful attention their manuscripts get within the publishing office.

I should like now to turn to production matters. The need to conserve funds has always been important, but this started to become much more acute as long ago as 1966 when we were faced with publishing a considerable amount of radiophysical data resulting from work done in the University of Sydney and the CSIRO Division of Radiophysics. It was at that time that the Astrophysical Supplements were born. It was clearly pointless to bother to reset data coming from a line printer. In addition there were also many maps and photographs which represented a considerable investment in blockmaking. Photolithography was clearly the answer. And this is how we started to move away from our traditional letterpress operations. We invested in our first IBM selectric composer at the end of the 1960's. It was used initially for various series of CSIRO Technical Papers. By 1971 a new problem had arrived in the form of an increasing number of long papers in insect taxonomy and systematic botany. IBM-composed photolithographic supplements were introduced to contain the costs of composition of these publications. By 1974-5 only 29% of all pages printed by our Organization were set in hot metal. The Australian Journals of Scientific Research (the national journals which we publish in collaboration with the Australian Academy of Science) were responsible for most of that 29%. The cost of external hot metal setting (and we rely on outside typesetters for 50% of this) had risen by 40% in two years between 1973 and 1975. A 48 x 30 pica page of complicated physics set by commercial typesetters in 10-point monotype now costs $45 - an amount we certainly cannot afford. Hence all our most complex mathematical setting is done in-house by our own experienced and dedicated typesetters and compositors. We have recently driven our composer operator to the limit by giving her a new mathematical journal to set. No commercial strike-on composition company was willing to quote for this type of work but we have been led to believe that the cost would be little different from that for hot metal if anyone could be persuaded to take it on. But dependence on one very experienced composer operator is clearly unsatisfactory. Hence we have been

327

PHYSICAL SCIENCES, ENGINEERING AND TECHNOLOGY (cont.)

Engineering and Technology (cont.)

Civil engineering (cont.)

Continuous dewatering of Victorian brown coal Stanmore, B R and Boyd, A R Natl.Conf.Publ.Inst.Eng.Aust. 76/6:76-9 11 refs (1976)

Demineralized brown coal as a alternative to current hydrocarbon resources Bowling, K McG and Rottendorf, H Natl.Conf.Publ.Inst.Eng.Aust. 76/6:86-91 18 refs (1976)

Enamelled exhaust silencers Bork, H Aust.Corros.Eng. 20(6/ 7):19,21 (1976)

Oil shale as a source of hydrocarbons in Australia Samson, D A Natl.Conf.Publ.Inst.Eng.Aust. 76/6:11-16 18 refs (1976)

Operation of standpipes in hydrocarbon catalytic cracking plants Leung, L S and White, E T Natl.Conf.Publ.Inst.Eng.Aust. 76/6:42-5 11 refs (1976)

Resources, environment and soil materials Ingles, O G Lect.Ser.Div.Appl.Geomech.CSIRO 31:20pp. (1976)

Sugarloaf creek project - flood studies Water supply division. Investigation, planning and research department Melb.Metro.Board Works: Monographs in series W-0056:30pp. 3 refs (1975)

Building

4207 "Designing to a cost" and its effect on cost planning Douthwaite, A D Build.Econ. 15(2):94-5 (1976)

4208 Forest product research Muncey, R W R Aust.For.Ind.J. 42(7):39,41,43,45-6 (1976)

4209 In search of the "Ideal" construction contract Dessewffy, A G J Build.Econ. 15(2):69-78 18 refs (1976)

4210 Methods of preparing value/time graphs for building contracts Cole, L J R Build.Econ. 15(2):82-4 4 refs (1976)

4211 New shed ideas catching on Hicks, P Dairyfmg Dig. 23(3):16- 17 (1976)

4212 Optimisation of plywood sheeting under buckling loads Hanley, D P Aust.For.Ind.J. 42(7):53-4,57 5 refs (1976)

4213 Professional ethics and the quantity surveyor Hodgetts, M F Build.Econ. 15(2):85-6 (1976)

4214 Quantity surveyor in the building industry Kinlay, G Build.Econ. 15(2):79-81 (1976)

4215 Shrink packaging develops Aust.Nat.Clay 7(3):2 (1976)

Figure 3. A page from *Australian Science Index*.

experimenting with photocomposition both within CSIRO and outside. In our case photosetting has to compete on cost grounds alone with traditional composition techniques as we are not in general producing databases for scanning or searching purposes. There are exceptions of course like *Australian Science Index* which is a record of all work published in Australian scientific serials. This particular monthly is generated entirely up to the camera-ready stage by computer. It is keyboarded by staff of the CSIRO Information Service directly into CSIRONET the Organization's computer network, edited in Melbourne on Datamedia Elite 1520 terminals and then output in Canberra as microfiche and as bromides for subsequent photolithography. Figure 3 shows a typical page from *Australian Science Index*. Until the middle of 1976, this journal was produced in a typewritten format. Indexes took up to six months to produce. The index for 1977 will be generated from the computer record in a matter of two weeks.

Most local phototypesetting is limited to fairly simple text which can be made available in a multiplicity of type faces and type sizes. The eyes of most printers' representatives glaze when we mention our need for something equivalent to four-line mathematical setting. Our own Cyber 76 computer with its third-generation photo-typesetter (a COMp 80) can produce very adequate tabular work and simple listings. I have some examples here for you to see. And recently a new editorial package has been written which enables editors to generate simple text through the system. This is not entirely satisfactory from our point of view as the costs of holding text in store pending author's corrections is prohibitively expensive. We need something other than an on-line system to give us the flexibility we need in journal or monograph publishing. To prove the point that cheap phototypesetting of complicated material is possible, we have been sending some material to that well-known low cost centre New York City. I assume that the keyboarding is done by typists working in sweat-shop conditions and that the text itself is set between the hours of 3 and 5 a.m. by an otherwise idle computer. There appears to be no other way of obtaining a product of such high quality at such an incredibly low cost. It equates to about $A12.50 per page - less than half of the notional cost in Australia if equivalent facilities were available. This then represents the challenge I have offered to our computer programmers. Any wholesale changeover to computer typesetting on our part will require a massive retraining program amongst our editors and typesetters not to mention much stricter control over our authors. The latter will have fewer opportunities to change their minds when checking their proofs. Just as the earlier changeover to direct paging tightened up editorial control so will the newer processes encourage greater preparation and disciplined thinking.

Finally I must return to contemplating the future. We have always attempted to produce what the customer has asked for by harnessing every relevant piece of technology as it becomes available. Nevertheless, the day may come when we cannot afford to give the sort of service we have been renowned for. Many publishers seem to be content to pass the cost of preparing material for press back to the customer. The insistence on sending in camera-ready copy which is now being extended by the arrival of synopsis journals available on microfiche or in miniprint form is just an extension of this. The journals which have levied page charges have been more honest in making scientists include the cost of publication in their research budgets. Government subsidies will not continue indefinitely even in countries like Australia where science has been supported for years from public funds. The day of the "free lunch" has ended. The consumer will either pay for what he wants or he will get something less than perfect. Some of us are fighting strong rearguard actions to preserve standards but we will probably not win in the end.

ECONOMICS AND ORGANIZATION OF SCIENTIFIC PUBLISHING: THE ESRC MODEL STUDY

Anders Kylin
Dept. of Plant Physiology, University of Lund,
Fack, S-220 07 Lund 7, Sweden

ABSTRACT

Already from its start, the ESRC (Committee of the West European Science Research Councils), now a standing committee within the ESF (European Science Foundation) paid attention to the chaotic situation within science publishing - more than 4600 journals in the area were found in Western Europe alone. A model study on possible rationalization gains was initiated in the field of plant physiology, where three major journals kindly made their lists of subscribers available for analysis. An amalgamation of the three journals would increase both the scientific and economic potentials of the publication system. Scientifically, each published contribution would immediately become available for practically all consumers of information in the field, whereas each of the present journals cover only 35-50% of the market. Economically, the longer printing series would mean savings in the order of 2 million Danish kroner a year (about £ 190 000 with the present exchange rate); which is almost 50% of the present annual turnover for the three journals at the time of the investigation. At the same time several problems are involved in an amalgamation. One must find bodies willing and capable of taking care of the scientific and economical responsibilities. The "former ownerships" may be complicated and have to be resolved; between one and four owners were noted for the three different journals investigated. Also the starting costs and the building up of reserve funds must be taken into consideration. All in all, it is the hope that the model study can provide a key, whereby science research councils with publication grants on their program can give rational advice to journals in their field of interest - to subsidy-taking as well as to independent ones.

INTRODUCTION

The ESRC (Committe of the West European Science Research Councils) was founded in 1972 as a body for cooperation between national research organizations within the natural sciences. In 1975 it became a Standing Committee within the wider organization ESF (European Science Foundation). Already from its start, the ESRC took an interest in the rather chaotic situation for publication of scientific results. As a first measure, an inventory was made of the number of journals in the natural sciences in Western Europe, and one arrived at the figure of at least 4600 (ESRC Review n:o 8, 1974). This is a clear indication that the publication structure may be inefficient, so that the situation should be analyzed further, from both scientific and economic points of view. Scientifically, a responsibility for the ESRC in this respect follows since all of the organizations within the ESRC sponsor

Miriam Balaban (ed.), Scientific Information Transfer: The Editor's Role, 391-398. *All Rights Reserved.*

research activities; economically, it follows since many of the organizations pay direct subsidies to maintain scientific journals.

To pursue the analyzes, the ESRC in the autumn of 1974 formed an ad hoc Group for the Rationalization of Scientific Journals. During its first sessions, the ad hoc group found that it had immediate access to practical data from journals within the field of plant physiology. It was decided to take up a Model Study within this area in order to get solid knowledge about the advantages and difficulties involved in structural changes of the pattern of scientific publication.

THE COMMUNICATION POINT OF VIEW. LEVELS OF SCIENTIFIC COMMUNICATION

Scientific research is pointless unless the results are communicated, so that they can be utilized by others. Seen like this, publication is the end of the research process, and it becomes important that scientific results are disseminated as widely as possible and with regard to the interests of the receivers.

Within the responsibility of a national research organization, there are several levels of information about scientific results to consider. Data from applied sciences or investigations primarily concerned with local problems (for instance inventories of natural resources and the like) have their immediate "consumers" inside the nation. Further, there is a need to keep in touch with "interested amateurs", who form an important link in the popular feed-back, which is necessary if science is to be accepted as a useful part of a modern democratic society. But there is also the level of international communication between scientists - the level which allows results from abroad to be utilized in one's own country in exchange for data from one's own laboratories.

It is the latter level, which falls within the responsibility of an international organization like the ESRC. From this point of view the field chosen for the Model Study is appropriate, since the primary problems of plant physiology - the life processes of plants - are general in character, so that good results and data obtained in one nation should be of interest also in other countries.

STRUCTURAL CONSIDERATIONS. THE SCIENTIFIC ASPECT

The great number of scientific journals in Western Europe - 4600 according to the inventory of the ESRC (ESRC Review n:o 8, 1974) - indicates too many and too small units from an economic point of view. In the field of plant physiology there are at least six:

1) Journals with a main influece by learned societies:
1.1) Journal of Experimental Botany (U.K.)
1.2) Physiologia Plantarum (Scandinavia)
1.3) Physiologie végétale (France)
2) Company owned journals
2.1) Planta (DBR)
2.2) Plant Science Letters (The Netherlands)
2.3) Zeitschrift für Pflanzenphysiologie (DBR)

Furthermore, there are a couple of journals - New Phytologist and Annals of Botany, both U.K. - which in the main contain material from plant physiology. However, these two journals prefer to regard themselves as journals of general botany. There are also several journals covering only parts of plant physiology like photosynthesis and phytochemistry; and journals in related subjects like soil science that overlap into the area of plant physiology.

For practical studies on the scientific and economic consequences of split-up in small units, the ad hoc group had access to the three journals in category (1). The data have been published in connection with the Oxford Conference of the ESCR in 1975 (Science Journals in Europe. An analysis by the ESRC ad hoc group on the rationalization of science journals. Appendix III, enclosure 3. - In Friborg, G. (ed.) Science Research Councils in Europe II, pp. 143-178. ISBN 91-546-0192-4. ESRC/Editorial Service of

NFR, Stockholm, Sweden 1975.). Some of the main features will be shown and discussed here.

Based on the address lists over subscribing libraries at the shift of year 1974/75 (l.c., Table I), the following features were found:

Journal	Number of subscriptions at full price
J. Exp. Botany	1532
Physiol. Plantarum	1477
Physiologie vég.	838
Total subscriptions	3847

Please note: Membership subscriptions are not included in the figures since they do not carry the full price

Total library subscribers = (Total subscriptions) - (Libraries that buy more than one of the journals) = 2901.

Probably, the number of total subscribers can be regarded as fairly close to the potential library market for plant physiology. This assumption is corroborated by the fact that the main journal in the field, the American "Plant Physiology", had at the same time a little more than 2700 library subscribers.

Looking upon the journals as media for scientific communication, a few conclusions can immediately be drawn:

I) Research data published in one of the investigated journals reach only 30-50% of the potential consumers directly.

II) By creating a journal on European basis rather than on national basis, it should be possible to make the scientific results available to all potential users almost directly after publication.

Some additional remarks should be made here. The method to investigate the situation - an analysis of subscription lists - should be of general applicability. The result will have to be discussed from case to case. In plant physiology, the data reveal surprisingly little overlap between the journals - only about 1/3 of the subscribing libraries keep more than one of the journals investigated. A comparison with the American "Plant Physiology" gives at hand that this is regarded as the main journal - the "must" for every organization in the field. In addition, libraries may buy one or more "secondary" journals, and national aspects may then often enter the field. The French journal is greatly preferred by the French-speaking countries, the U.K. journal by the Commonwealth, whereas "the rest of the world" is mainly split up between J. Exp. Bot. and Physiologia Plantarum.

STRUCTURAL CONSIDERATIONS. THE ECONOMIC ASPECT

The two conclusions of the preceding section imply (A) that the split-up of the publishing structure in Europe leads to unsatisfactory dissemination of scientific results and (B) that there are cases where publication on European basis would greatly enhance the efficiency of scientific communication.

Clearly, the creation of a completely new journal on "European" basis, in addition to the existing ones, would only increase the split-up and the negative effects. Such a new journal would not change the existing structure of publication, it would only act as a competitor to the existing journals. Furthermore, the undertaking of a new journal would be economically unsafe, since the stagnation of research money will mean shortage of both manuscripts (raw material) and library money (buyers) over the period to come.

For these reasons, the ad hoc group decided to investigate the reverse strategy - what will it mean in economic terms if the existing journals merge to form a European unit with world-wide dissemination?

One condition appeared immediately necessary for such a strategy. The total amount of publication space must be kept unchanged, otherwise the members of learned societies behind the journals would not wish to cooperate. Furthermore, from the point of view of the ESRC, it was natural to build up the comparisons

A) Technical production cost for Physiologia Plantarum alone

B) D:o for (Physiologia Plantarum + Physiologie végétale)

C) D:o for all three journals together.

This was so, since the French journal is subsidized by the CNRS, the French natural science research organization; and since Physiologia Plantarum, although independent of subsidies, is working in close contact with the Nordic Publication Board for Natural Sciences, a cooperating body for the research councils of the Scandinavian Countries. Contrariwise, the research organizations of the United Kingdom do not include scientific publication in their direct sphere of interest, so that Journal of Experimental Botany is only indirectly accessible for considerations from the ESRC.

In order to make the data comparable, the potential publication volume of each journal was recalculated to "standard pages". Furthermore, technical production includes not only printing but also technical editing (or copy editing), distribution and marketing, which are normally undertaken by a publishing company. Quotations from a printing company and a publisher were solicited for these items. The complete data are again published in "Science Research Councils in Europe II".

The following data are derived from Tables III and V of Appendix III, enclosure 3 to the report on "Science Journals in Europe" in the publication cited.

Structural level[1]	Number of Copies[2]	Number of Pages[3]	Printing Cost (Dkr[4]) Total	per Copy	per Page
A	1400	1400	385 000	275	0.196
B	2000	1900	557 000	278	0.146
C	2900	2700	875 000	302	0.104

Structural level[1]	Technical editing-distribution-marketing (Dkr[4]) Total	per Copy	per Page	Total technical production (Dkr[4]) per Copy	per Page
A	240 000	172	0.122	447	0.318
B	265 000	132	0.070	410	0.216
C	310 000	107	0.040	409	0.144

1) A = Physiologia Plantarum
B = Physiologia Plantarum + Physiologie végétale
C = Physiologia Plantarum + Physiologie végétale + Journal of Exp. Botany

2) Number of copies calculated as (combined subscriptions minus subscribers to more than one of the journals).

3) Pages given in "standard size" with 6000 signs per page.

4) At the time of writing a British pound is a little more than Danish kroner 10:-.

The conclusions are again obvious, although it was a surprise that so much is to be gained economically by cooperation.

I) Short printing series (a low number of copies printed) make scientific information expensive (printing cost per page is high).

II) By amalgamating two of the journals under study, the cost for the scientific information would decrease by more than 30% per page. If amalgamation of all three journals were possible, roughly 55% of the technical production cost per page can be saved. The production cost per copy does not change much.

Again some additional remarks are appropriate. Qualitatively, it is a general conclusion that economic savings can be made by longer printing series, and the method used for the study is of general applicability. Quantitatively the results must be studied and discussed from case to case. Also, it is a matter of course that the actual data will vary with time, due to inflation and similar phenomena.

STRUCTURAL CONSIDERATIONS. THE EDITORIAL ASPECT

At this point, the ad hoc group realized that it had also to take the editorial aspect into consideration. The split-up structure and the small units characteristic for most of the scientific publication in Europe has made it traditional and possible to have editors that work "idealistically". Small journals can handle only a small number of incoming manuscripts, and an institute or a single scientist can feel enough rewarded by the contacts created by the editorship.

However, the "idealistic" solution means in practice that single scientists give a hidden subsidy to the journal. The subsidy is paid by loss of research time, and when a journal grows this burden becomes too heavy to be born by single institutes or single individuals.

The necessity to switch from "idealistic" to "professional" handling of the editorship will as a first approximation depend on the number of incoming manuscripts. For the three journals under study this number was approximately:

Physiologia Plantarum	350 manuscripts a year
Physiologie végétale	70 manuscripts a year
Journal Exp. Botany	150 manuscripts a year

Percentage refused manuscripts varies between 20 and 40.

In the main, J. Exp. Botany and Physiologie végétale work with "hidden" editorial costs. Physiologia Plantarum has changed to a system with open costs for editorship, where scientific editing is regarded as a form of advanced scientific teaching. To get reasonable equivalence with the normal (teaching + research) career, the journal pays for 0.75 man-year editing duty and 0.50 man-year research duty. The editing time allows a mean of four hours of work with each manuscript. The research time compensates the institute that houses the journal at the same time as it gives the individual reasonable chances to qualify for promotion in the scientific career. The system presents also the best guarantee that the editorial staff will maintain the scientific qualifications that ensure a competent handling of the manuscripts and, thereby, a good reputation for the journal.

The growth of a journal raises also the question of subspecialities in the specific field. The primary expertise in a subspeciality lies with the referees, but it is also desirable that the scientific editors do not work too far from their own speciality. It then becomes desirable to collect manuscripts from different subfields of a science in the hands of editors with their main scientific competence in different areas.

During the discussions of the ad hoc group it was, finally, pointed out that the scientific editors together with the referees have a key role of the input to the system for scientific information. It is well known that the number of scientific papers printed grows much more rapidly - by a factor of about ten - than their utilization. The only remedy for this is increasing awareness at the input level - during the continuous international prepublication seminar that in the ideal case is constituted by the cooperation between authors, referees and editor.

A couple of conclusions can be formulated from the debate in the group:

I) The traditional pattern of scientific editors that work on "idealistic" basis necessarily keeps the size of the journals too small for efficient scientific communication and for efficient use of printing economics.

II) The interests of the journal, of the institute that houses the journal, and of the individual doing the job, all speak in favour of professional scientific editors with a combined duty of editing and active research.

Both conclusions are general in character. For the specific case the cost for several alternative editorial organizations were computed. The variations included.

a) the different alternatives (A, B, C) for journal amalgamation as given above;

b) different assumptions about the possibilities to obtain research money from outside the budget of the journal;
c) different alternatives to organize the editorial service according to scientific subspecialities.

The different alternatives and the costs implied at the shift of 1974/75 are published as Table II of Appendix III, Enclosure 3 to the report on "Science Journals in Europe" in "Science Research Councils in Europe II". For the present purposes it is sufficient to state that the scientifically most ambitious - and thus also the most expensive - alternatives will be used for the computations of the total balance sheets below.

THE BALANCE SHEET

From the three preceding sections it is evident that the scientific efficiency would increase and the cost for the technical production of scientific information would decrease by a change in the publication structure, so that small units could amalgamate into reasonably big ones. The primary bottle-neck for such a procedure is the arising need for professional scientific editors, who are at the same time active research workers. With small units it has been possible to keep the editorial cost hidden, with increasing size and demand of time it must become open. The question then becomes to what extent the savings on the side of technical production can cover an increased open cost for scientific editing.

In the Model Study, balance sheets were worked out for several different alternatives in editorial organization. For the present purposes it should be sufficient to show the outcome for the most expensive - that is, the scientifically most satisfactory - editorial alternative for each of the three structural levels under consideration.

Structural level[1]	Number of copies	Library cost (Dkr) for present units[2]	Total revenue at unchanged price (Dkr)
A	1400	690	965 000
B	2000	900	1 800 000
C	2900	1280	3 710 000

Alternative	Total cost (Dkr) (Technical production+scientific editing)	Calculated "profit" (Dkr)
A	795 000	170 000
B	1 112 000	688 000
C	1 615 000	2 095 000

[1] As for the meaning of the structural levels and the currency used, see notes to the table given under "Structural considerations. The economic Aspect".

[2] Prices for 1975 were used for the Model Study.

Evidently, enormous theoretical "profits" - or from another point of view "savings" - would arise even at the scientifically most complete and satisfactory organization of a combined journal. As compared with the old structure, the new type of organization would (a) increase the efficiency of scientific information, (b) decrease the costs for technical production, and (c) allow a better editorial function. Which would be the practical problems involved if one wishes to favour such a development?

SCIENTIFIC RESPONSIBILITY

All the three journals investigated are sponsored by learned societies. This means that the society takes the responsibility for the scientific standard of the journal. The society appoints editors and members of the editorial board, and members of the society are normally willing to act as referees. One cannot directly presume that a specific society is willing to sponsor a compound journal. Neither can one presume that a specific society is willing to give up its own responsibility and accept another society on the same level as the sponsoring organization. It became a natural task for the ad hoc group to find possible solutions for the case of the Model Study.

Interviews with different national organizations related to plant physiology made it clear that they were interested to form a Federation of European Societies for Plant Physiology, modelled after the example of the biochemical organization FEBS. One might hope that such a European organization would be willing to take over the scientific responsibility for an amalgamated journal and that this could be accepted by the mother societies of the ingoing journals. After negotiations at Ronneby, Sweden, and in Paris, France, an interimistic FESPP was formed in Copenhagen, Denmark, in November 1975; basically according to the material given as Appendix II and Appendix III, Enclosure I, of "Science Journals in Europe" in "Science Research Councils in Europe II". The interimistic organization is now acting as an independent body. It expects to have its first general meeting in July 1978 and by then transform into a permanent body. It will then be up to the parties concerned to decide to what extent the case used for the Model Study suits their particular needs.

THE PRICE FOR INFORMATION - THE LIBRARY PROBLEM

The figures given in the section on "The Balance Sheet" indicate a great theoretical "profit" if the price for total amount of scientific information would be kept unchanged after an amalgamation and build-up of the editorial resources. This is unreasonable from the point of view of the libraries, and one of the desirable effects of structural rationalization is to make it possible to disseminate scientific information more effectively than now. From general market considerations it appears that at least half the theoretical "profit" should be used to lower the price for the information. The practical decision will depend also upon the type of considerations that are dicusssed in the section below.

OWNERSHIP RESPONSIBILITY

The ownership responsibility presents problems, which are related to but not identical with the problems involved in the scientific responsibility. As for the present ownerships, the three journals of the Model Study proved quite different. In the case of Physiologia Plantarum the situation is simple - the Scandinavian Society for Plant Physiology owns the journal. In the case of the British journal both the Society for Experimental Biology and the Oxford University Press are involved. As for the French journal, it is in itself the result of an amalgamation between three different earlier scientific journals, and a publishing company is also involved. The Scandinavian and the British journals are self-supporting and give a surplus, whereas the French journal carries a subsidy from the CNRS. How are these interests and responsibilities to be combined at the same time as the interests of a new scientific sponsor on the European level, the FESPP are considered?

The solution suggested by the ad hoc group is to form a Foundation, with representatives for the ingoing journals as well as for the organization that takes over the scientific responsibility - in the case given the FESPP.

As an example and as a starting-point for the discussions, draft statutes were formulated and presented as Appendix III, Enclosure 2, of "Science Journals in Europe" in "Science Research Councils in Europe II".

The British journal did not want to pursue this part of the investigation. Furthermore, the possible future scientific sponsor on a European level, the interimistic FESPP, must have time to turn into a permanent organization and to formulate its own policy (cf. the section of "Scientific Responsibility" above). However, for the purposes of the Model Study, the ad hoc group found it of interest to gain some insight in the practical problems that may arise. For this purpose a series of negotiations between the French and the Scandinavian interests were sponsored in Copenhagen,

Denmark, during the spring 1976.

The negotiations led to a draft agreement. A journal where the authors decide whether they want to publish in English, French or German was envisaged, as well as a referee procedure with one referee from outside the nation of the author(s) and one from inside the language used in the paper. The price must recognize the need for the libraries to obtain the scientific information at a lower price per page than now, so that the real "profit" was envisaged as much lower than the theoretical one of the "Balance Sheet". The real "profit" should first of all be used to build up a reserve fund, and a possible dividend be shared in proportion to the "ingoing" pages from the parent journals. In the budget one must also take measures against possible losses of subscribers during the first year.

Practical solutions could thus be found for all these points. The main problems that were indicated concern complex and unregulated ownerships behind the ingoing journals; and how to finance the period of transition between the old system and the new, until funds have been built up.

CONCLUDING REMARKS. SUGGESTIONS HOW TO PROCEED

The Model Study has provided the ad hoc group of the ESRC with valuable experience on the problems and difficulties connected with the split-up structure of scientific publication in Europe to-day. The mandates as given at the start of the group and as complemented at the Oxford Conference of the ESRC cover the situation. The first problem involved in future practical work appears as one of communication: how can the ESRC and its ad hoc group make known that their services are available, and what further aspects would the presumptive users of the services want to have elucidated? There is also the problem that the specific field chosen for the Model Study may have obscured the general applicability of the results - even though several groups have made contacts, others may wrongly have felt that plant physiology was the only concern of the investigators.

To attack these practical problems of communication and to get a response to its activities, the ad hoc group would like to send out its report to be commented by different editorial and scientific bodies on the European level. The group would appreciate the views of organizations like ELSE (the editorial organization for the life sciences), EDITERRA (for the earth sciences) and STM (for the commercial journals - at least small publishers should in all likelihood have felt the structural problems in the same way as the members of the ESRC). It would also be of interest to listen to the opinions of scientific organizations like FEBS and FESPP, and there are also the Taxonomy Groups within the ESRC itself. Finally, it might be of interest to compare the policy behind the publication grants of different member organizations within the ESRC itself. Several of the organizations have discovered the problems involved in coordination of the funding for multinational journals, at the same time as the Model Study points to the need of multinational cooperation. Could the policies for publication grants be parallelized between some of the member organizations of the ESRC?; and would it be useful to compare, for selected journals, what cooperative measures that may be taken after a quantitative analysis in the terms of the Model Study?

DISCUSSION on "The Publication of Scientific Journals in France," by Jean-Paul Abadie

"Integrated Science Publishing. An Example from Australia," by Basil J. Walby

"Economics and Organization of Scientific Publishing: The ESRC Model Study," by Anders Kylin

MICHEL GIBB - I would like to put a short question to Prof. Kylin. What about the language problem. In what languages can manuscripts be provided.

ANDERS KYLIN - We discussed that question rather closely and of course the journal like the present journals will be open for publication in English, German and French. It will be up to the authors to decide what language they want to use, but there will be abstracts in English in all non-English papers. And we'll also use their cooperation to help in refereeing. Our general idea about refereeing in a journal like this is that each paper should be reviewed at least by two scientific workers of high standard, but we would like to choose them so that one referee speaks the language in which the paper is written and one referee from outside the nation from which that paper has been produced. We think that we create reasonable assurance about linguistic correctness and readability at the same time as we can be reasonably assured that national aspects will not play any dominant role in the judgement of the paper.

JEFFREY WATSON - I'm very sympathetic to Mr. Abadie's statement about the serious cultural concerns that arise from the loss of French as a preferred language of science, since we have the very same concerns in our own country. I'm sure that the experiment to encourage French authors to publish their works in French journals will be followed very closely in Quebec where unilingual French journals really have similar very serious problems. The policy followed by the Canadian Federal Government journals is to publish original research articles in the language of choice of the authors and to have to abstract in both French and English. However interestingly enough, very few articles are received in French and this is unfortunate I think. Most French-Canadian authors feel they need to publish in English to be read and this is little problem to them since often they're fluently bilingual. However, it does raise a cultural problem which has not been resolved. And all articles incidentally received from French authors are all in English and we don't find we have any problems with atrocious English. In fact they're usually received in impeccable English. Most of the articles that we receive in atrocious English are from people who are English speaking. (Laughter). Where scientific communication is not totally scientist to scientist as it is in the primary literature, then we run into another problem and that is one of providing the scientific article in both languages. And I must say here that translation is almost impossible because of the difficulty of finding competent translators. We find this is a really serious problem. And we find that for most translated works scientists go to the original English version rather than read what the translator thinks the author was saying.

However, we have another problem and that is one in areas of broad interest such as the one I'm involved in which is aquatic science. And there, there is really a need to provide to an interested lay audience - basically a repackaging of scientific information that has been published in the primary literature. And this has to be suitably repackaged and suitably written in understandable fashion. And available to the lay audiences in the two mother tongues that we have. Although as I said before this translation problem is difficult, I don't think it is insurmountable. And we have in the past few years made great strides in providing written versions in both languages and it has been acceptable by these audiences.

DIMITRI VIZA - I was very impressed with Mr. Abadie's thorough, and I would say unprejudiced, analysis of French politics pertaining to French publications. Living in France, I have some experience with these policies and I wonder how French reports are affected by the policy of the French government and agencies of giving large grants to French-language publications. If I'm correct, they will give, through their agencies, large grants to any publication provided that it is published in French, regardless of what rubbish is published there. As a corollary, English publications in France do not get any governmental support. Now does this not encourage mediocrity in French publications?

JEAN-PAUL ABADIE - It is very easy to answer to that because I would say that up to now the policy of the French agencies like CNRS for example, was to give grants to journals, to non-French journals, to French journals where it was not written in French. For example Nouveau de Journal de Chimie which was just launched some months ago, the only words in French is the title and that's all and you won't find one article in French and so I think that the most important factor for an agency to give out subsidies is not a lingual problem. It is first a quality problem. If the journal has a good editing organization and if the quality of the papers will be good enough then it will get a grant, otherwise it won't and then the lingual problem is a secondary one.

DIMITRI VIZA - I am sorry to contradict you, but I have personal experience with governmental agencies denying or suppressing funding for my own journal for the sole reason that it does not contain articles in French. (Note added in proof: This policy is still in force as of February 1978).

JEAN-PAUL ABADIE - Well then it was the last of the old policy. (Laughter) I totally agree with you, but now things have changed and now I can tell you because now I was with one official of the BNIST*, which is a very important agency in France for editing, and the real policy now is the one I gave you. First quality and then language.

ANDERS KYLIN - I look on the situation from a point of view of a man who is Swedish speaking. I mean a lot of people in the world don't have English and don't have French or don't have German and we are still scientists. In my generation we've got French, English, German, but this is not so in the new generation. Due to different social pressures, schools have to teach other things than language and I'd say that more young Scandinavians know English and have just sniffed at a second language, but no more than that. I had a very practical example a few weeks ago. A paper written in French came back from the referee with a comment that it was an excellent paper. But unfortunately in my Institute only I and one more could read it. And the situation is just as bad with German. I sent that away to the author of course, and the paper came back translated back into English. This is life, if you want to reach people that really cannot read French, how do you do it? You have to write in English because for the non-English, non-French, non-German speaking nations which are quite a few, linguistic study is just dwindling due to other social pressures.

ARIE MANTEN - To Mr. Abadie. Elsevier publishes quite a number of journals. The more recent ones have only English as their official language. A statistical survey of those journals points out that only two to five percent of the articles published in those journals comes from the French language area. The more established journals, established ten years or more ago, normally use English, French and German as official languages. Peculiarly enough those journals do not have a higher percentage of articles from the French language area, which indicates also that perhaps primarily other criteria determine

*Bureau National d'Information Scientifique et Technique

where the Frenchman is submitting his paper. It may be that he is direct himself most of all to those people who have the greatest influence in his career, and that is the people in France. By extending this argument, this implies that if those people having a significant influence on a scientist's career adopt all the criteria for selection, for instance a better style of publication, this is the way to improve your situation and then publication in English may be of less importance. To Mr. Kylin, national journals tend to have a certain percentage of contributions which are only of national interest. If you are merging three journals what is going to happen to the articles that do not really deserve European, not to speak of world-wide, distribution?

ANDERS KYLIN - In this specific case I don't think this is a problem. Plant physiology in itself is international. I would say that you could read more about that problem if you write to the Swedish Natural Sciences Research Council and ask for this which is called Botanical Publication in Sweden, because when you come to general botany it is quite another thing. There are a lot of publications that have only national interest and this means sorting out the journals according to other principles, but you can take it from there.

ILANA FERBER - I'd just like to remind us all of the famous proverb and the famous equation in it. One picture is worth a thousand words and it solves a lot of economic problems. From several impressions I've had here during the few days I've been listening very intently to what was being said, is that not enough attention is being paid to good graphic presentation and illustrations in various journals. And I'm sorry to say that many of the people presenting their papers here are not giving graphic presentation to the audience.

LARRY LESTER - To Basil Walby. You stopped with the promise to provide us with some information about the financial structure of the organization. I assume that most of your journals do not even break even. I'm basing myself on that assumption. Perhaps you could comment on that and have a few cracks at our chairman about financially non-viable journals and the value of publishing them.

BASIL WALBY - Thank you for asking that question. We certainly do not break even on our publications. The budget I told you represents exactly 1.16% of CSIRO's budget. This is not a high proportion of a research budget. With that we are publishing at the moment nine national journals of science apart from all the material emanating from our own research departments. So when we look at our financial structure over all I don't expect to make a profit. It would almost be an indecent thing to do I think to make a profit within our organization. When we add our actual income to our national income, and our national income is very high, CSIRO until quite recently has been the Australian National Science Library. Consequently we have been charged with selecting material from all over the world. We do this by a very large exchange program. We reckon that we actually save something like one million dollars in subscriptions by offering our journals to eminent institutions overseas. We don't get many commercial journals that way obviously. But it means in effect that we are running this whole publishing operation on a subsidy of about four hundred thousand dollars. Which I personally think is pretty good. When I put prices up next year, I hope I can reduce that deficit even more.

LOUIS COHEN - The first point on the combination of journals - the attempts to combine journals. When the European Physical Society was founded some eight years ago, one of the reasons for having such a society was to try and combine the European journals in some way which could counterbalance the American journals in physics which of course play a dominant role in that they publish well over a third of the world's physics literature. This perhaps is one of the reasons why English is the main language in physics, perhaps more so than any other of the sciences and why indeed English was the working language of the European Physical Society from the very beginning. The degree of success achieved so far in eight years of the European Physical Society has been minimal. There have been rules for recognition of Europhysics journals which give basic requirements for membership of editorial boards, referee procedures and so on. But now after eight years, there is beginning to be an attempt to get some sort of combined structure. There is a special problem, however, with the journals for physics teachers. We came to the conclusion that it was impossible to give recognition for such journals in the EPS framework for the reason that you couldn't expect the school teachers to be sufficiently fluent in English to be able to accept English as the working language at that level.

One final point if I may make a comment on a different topic. This is a very large subject of the government subsidy of publications. Now this takes a variety of forms and it is not always obvious that government is subsidizing a journal. This I assume is largely talking about the institutional publications rather than the commercial publications. But the fiscal system that operates in a country can give a subsidy in a way that is not immediately apparent. In the United States where the not-for-profit system operates it looks as if there isn't a subsidy. But in fact the acceptance of page charges by the NSF constitutes a government subsidy. In the United Kingdom where we have a rather different system of non-profit making organizations, indicated by so-called charitable status, we can make profits as long as the money is used for approved purposes. In that respect the government is subsidizing us because the government is not taking the taxes. In France and I know in many other countries, particularly the smaller countries in Europe, there is a direct subvention and we've heard that in Australia there is a direct subvention. So this is an extremely interesting topic - to what extent do the national governments subsidize publications and indeed to what extent ought they to subsidize publications.

JOSEPH PALTI - The fact that English is the dominant language in scientific publication is just a fact of life. We all have to live with it. Another fact is that many European authors feel grieved that they are not cited properly in North American journals even if their papers have been abstracted. And now I think there should be more of a dialogue on this between those publishing in English and those finding it difficult to publish in English and also our English speaking colleagues should show a little more interest in getting more papers published in English, the international language.

Now I have two suggestions for this. One is for a paper not written in English to be translated. You know the author has to make an effort to get it translated. The journals that have page charges should consider dropping their charges for this sort of paper, because the author has to go to considerable expense to get it translated. And the other suggestion is the other end of this. Many more European institutions, not English speaking institutions should have a resident English speaking editor and I know this is the case in some Dutch institutions. It is the case here in some institutions. I think this should be adopted in more scientific institutions in Europe.

R.E.M. VAN DEN BRINK - I should like to make a suggestion for the Second International Conference of Scientific Editors because we have dealt with the logistics of being publishers or being editors. We have seen all the cost structures, the rationalization possibilities, the mergers, etc., but the most important thing is the change in the needs and demands for scientific information, and therefore I have only one question. Speaking not about logistics but about strategy, it is necessary to have sufficient data, sufficient figures. I think when we take science not too broadly, three billion dollars are involved in scientific information transfer. But we don't know - I don't know all the things behind that. Speaking about strategy, changement in scientific information must be based on hard figures. And my question to the secretary of the STM is are they now preparing with their very important background of 150 members (and I think that with three hundred publishers you will have nearly eighty percent of the total of information transfer). Are you preparing the figures on which basis perhaps the next time we can speak about strategy. Without having a strategy or demand the logistics I'm afraid are not so important.

PAUL NIJHOFF ASSER - I must apologize, we have not been preparing any statistics yet apart from separate subject fields, for instance in journals, but we are aware that our coverage of the total system of scientific information transfer is far from complete. Even if you should only look at the society and the commercial publishers then you would also need to have information from government publications such as our Australian friend and I think this is very important if you think for instance that on the subject of scientific journals there is absolutely no statistic available on the total impact. Nobody knows how many scientific journals there are. We just heard that there are 4,600 in Europe alone. But world-wide the estimates run from 30,000 to 100,000 and there are no definitions. There is absolutely nothing to start on. The statistics given to us by, let's say, central, national, statistical offices are quite deficient because they do not distinguish between newspapers and scientific periodicals. And this is all a ridiculous state of affairs. We are far more advanced in book publishing as you well know, Mr. van den Brink, but even there it is very difficult and in science I think it is nearly impossible. But we will start and we have started especially in the journals field by investigating various problems such as border crossings issues, etc., and we have found quite a few I think representative figures. But it doesn't mean they can be proved to be true for the total picture. I think that's enough about it now, but we will certainly go on as we grow and have a better survey of the total field.

SURVIVING INFLATION?
SOME RECENT FINANCIAL PROBLEMS OF A SPECIALIST MEDICAL JOURNAL

E.H. Hare, M.D., *Editor, British Journal of Psychiatry,*
17 Belgrave Square, London, SW1 8PG, Great Britain

Many learned journals are run by amateur ('honorary') editors who, being unfamiliar with business affairs, may have problems in dealing with the effects of rapidly increasing inflation on their journal accounts. This is particularly so where the journal is owned by a learned society who, nurtured in a tradition of service and perhaps of Empire, have come to accept as a matter of course, if not of policy, that their journal does not aim to make a profit and that its yearly accounts will and indeed should show a moderate overall loss.

Figure 1 shows the inflation in Britain over the past century and Figure 2 shows the galloping inflation which began in 1966. From Table 1, one can infer that the British Journal of Psychiatry (formerly the Journal of Mental Science) has in general been a financial liability to the learned society which owns it (The Royal College of Psychiatrists, formerly the Royal Medico-Psychological Association); and the fact that these losses have been recorded in the annual reports of the Society under the heading "The cost of the Journal" seems to indicate that the losses were accepted as the Society's tribute to the dissemination of learning. But Table 1 also shows that, when the value of the pound is held constant, the "cost of the Journal" has been gradually increasing; and Table 2 shows that the average deficit during the decade 1964-73 was, at the 1975 value of the pound, more than £ 10,000 a year (say $20,000). In 1973 the Council of the newly created Royal College of Psychiatrists - at that time rather impecunious and about to be burdened with the mortgage interest of its new residence in Belgrave Square - requested the editor to look into the finances of their Journal to see how far the rate of loss might be reduced.

It has been the practice of the College of Psychiatrists to choose its editors for their professional standing and their interest in literary and learned matters. When I was appointed editor in 1973, I had had no experience at all either of editorial procedures or of journal management. I should like to say at once that, in as far as I was able to improve the financial position of the British Journal of Psychiatry, I have been much indebted to the valuable information circulated to its members by ELSE (the Society of European Life Science Editors) and, through them, by the CBE (Council of Biological Editors).

An immediate way of saving costs was to stop the issue of free off-prints to authors. The cost of 50 free off-prints to authors had amounted (in terms of the 1975 value of the pound) to about £ 2,000 a year. Stopping free off-prints seemed a reasonable step because most authors have access to research funds and these usually include a sum to cover publication; and the development of photocopying has often made it cheaper for an interested reader to take a copy than to send a postcard for a reprint. We received no significant complaints. Next we abandoned our practice of beginning each

Miriam Balaban (ed.), Scientific Information Transfer: The Editor's Role, 405-413.

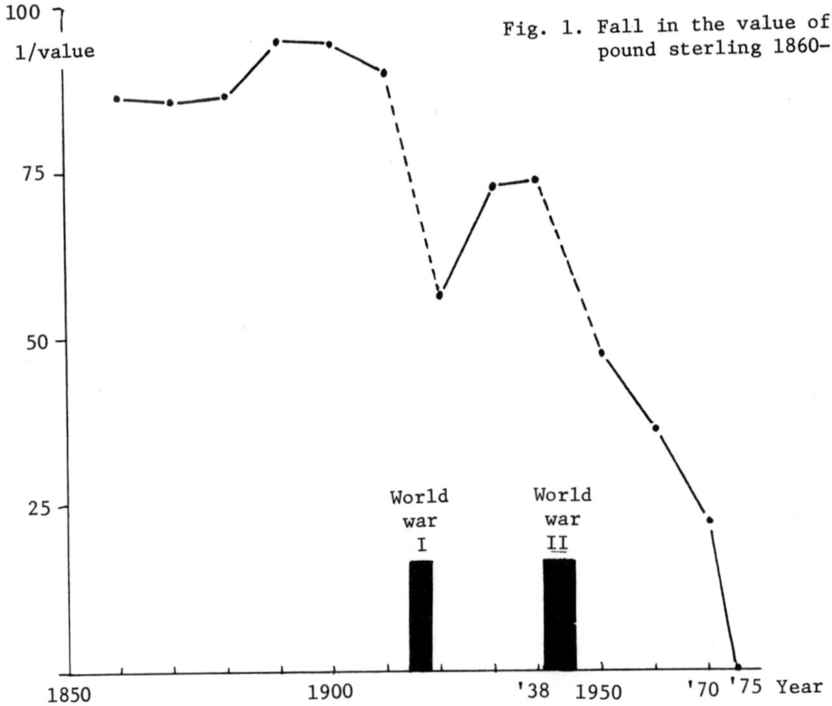

Fig. 1. Fall in the value of the pound sterling 1860–1975.

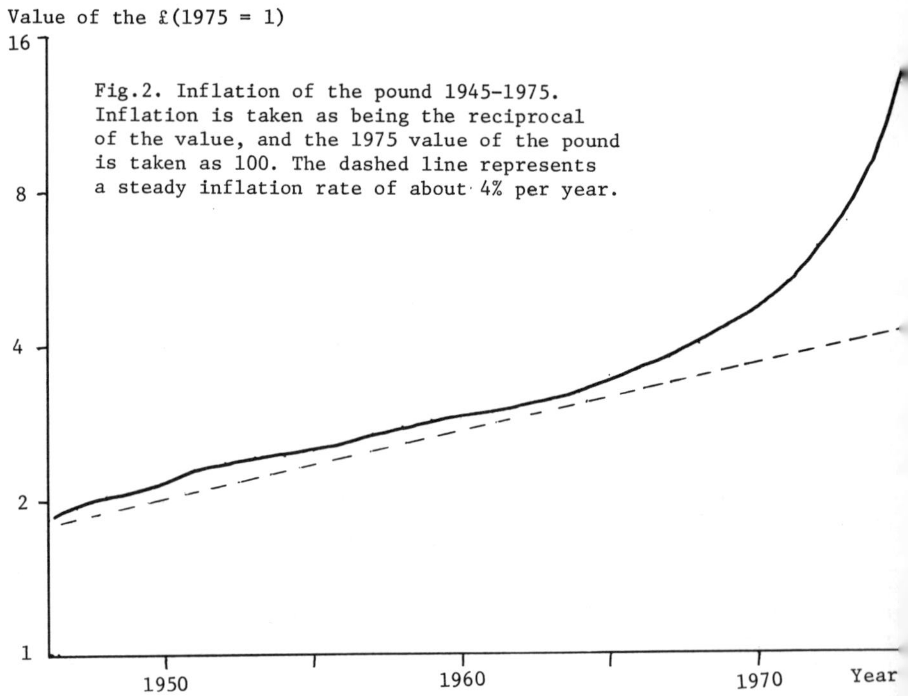

Fig.2. Inflation of the pound 1945–1975. Inflation is taken as being the reciprocal of the value, and the 1975 value of the pound is taken as 100. The dashed line represents a steady inflation rate of about 4% per year.

TABLE 1: "The cost of the Journal" 1855-1965 (British J. of Psychiatry)

Year	Deficit £	Deficit (at 1975 value of the £)
1855	68	800
65	113	1,200
75	158	1,700
85	268	3,300
95	295	3,900
1905	323	4,200
25	928	6,200
34	750	6,300
45	946	4,300
55	1,162	3,500
1965	6,375	14,300

TABLE 2: "The cost of the Journal" 1964-73 (British J. of Psychiatry)

Year	Balance £ (deficit)	Balance (at 1975 value of the £)
1964	1,200	2,800
65	6,400	14,300
66	5,600	12,300
67	6,100	12,800
68	8,900	17,700
69	4,300	8,100
70	6,900	12,500
71	7,000	11,600
72	800	1,300
73	9,700	13,800

journal article on a right hand page - a practice which had been adopted largely because it facilitated the production of offprints - and this at once saved some 7 or 8 pages of paper per issue and so several hundred pounds per year.

The most obivous way of increasing a journal's income is to raise the subscription rate. Raising the subscription rate had, very properly, always been a matter of concern to the Journal Committee for one could never know how far its hoped-for effect might be more than offset by a fall in the number of subscribers. But the subscription rate had to be raised, cautiously at first and then more boldly - or so it seemed. Between 1966 and 1976 the subscription increased from £ 6 to £ 35, that is, by a factor of nearly 6. Yet when the effects of inflation are taken into account, the sbuscription rate had not even been doubled (Table 3). Table 3 also shows that an apparently reasonable increase in subscription rate (e.g. between 1971 and 1974) might not in fact compensate for the effect of inflation. A doubling (in real terms) of the subscription rate seems not unreasonable, in view of the continued heavy losses of the Journal over many years, and it has not in fact been associated with any fall in the number of subscribers.

TABLE 3

Subscription rate
(British J. of Psychiatry)

Year	Rate, £	Rate (1975 value of £)
1961	6.00	15.40
1968	7.50	15.00
1970	10.00	18.10
1971	15.00	25.00
1974	20.00	24.75
1975	25.00	25.00
1976	35.00	29.00

Perhaps the next most obvious source of income for a medical journal nowadays is the sales of advertisements. Income from advertisements has been published in the annual accounts of our Journal since 1890, but our editors have rarely concerned themselvers with the cost incurred in printing these advertisements. They did do so, however, during a period between the wars. Although the sums are trivial by modern standards, Table 4 shows that, during this period, the cost of printing advertisements overtook the receipts from sales. This curious circumstance - that the Journal was in effect paying for the privilege of carrying advertisements - was never mentioned in the annual reports of the aditor, and we do not know how long it persisted because no figures for the cost of printing advertisements are given in the reports for subsequent years. Although the present printers of the British Journal of Psychiatry have in their monthly invoices provided precise details of the cost of printing advertisements, these had not been carefully studied until, in 1973, I happened to notice that the monthly cost of printing advertisements seemed to exceed the monthly receipts from their sale. This led me to look more closely into the matter and to go back as far as I could in the files. Table 5 shows that during the 6 years 1968-73, the cost of printing advertisements exceeded the receipts from their sales by an average (at 1975 value of the pound) of more than £ 3,500 a year. I do not know if the British

TABLE 4

Advertisement account 1922-32 (British J. of Psychiatry)

Year	Balance, £ (surplus or deficit)
1922	+ 6
24	+ 8
25	+ 4
26	+ 7
27	- 15
28	- 18
29	- 19
30	- 19
31	- 26
32	- 14

TABLE 5

Advertisement account 1968-76

Year	Balance surplus/deficit	Balance at 1975 value of the £
1968	600	1,200
69	1,700	3,200
70	(1,100)	1,800
71	3,500	5,900
72	3,200	4,900
73	2,800	4,000
74	0	0
75	6,400	6,400
76	16,700	13,500

Journal of Psychiatry has been unique in this respect, but the reason for the loss was of course that our charges for advertising had not kept pace with inflation and therefore with the increasing costs of printing. Although I have no doubt that our Journal Committee had no one to blame but themselves, yet it seemed to us that an alert advertising manager, if only to maximize his own commission on sales, would have considered more carefully whether the page charges he recommended were at a proper level. Our losses on advertising, when taken with the moderate surplus that might reasonably have been expected, probably account for about two thirds of the total deficit experienced by the Journal during the decade 1964-73, and these losses were therefore the single biggest factor in "the cost of the Journal". Closer consultations with a new advertising manager have now led to a more satisfactory situation (Table 5).

There are certain other aspects of the finances associated with advertising which may be noted. (a) Advertising managers deal with advertising agents who generally charge a commission of ten per cent on gross receipts. For reasons which are now lost in the past, the British Journal of Psychiatry had for many years been paying its advertising agents a commission of fifteen per cent. I do not think any of our staff were clearly aware of this until it was pointed out to us by our new advertising manager who, under some pressure from me, has since persuaded the agents that it would be fair to reduce their rate of commission to the normal ten per cent. (b) The weight of advertising matter in a journal may have a significant effect on postal costs. Thus in 1975, the postal costs of distributing the Journal were £ 17,000; and if the Journal had carried no advertisements at all, its lesser weight would have incurred only £ 12,000 postal costs. In other words, our net receipts from the sale of advertisements in 1975 (that is, after allowing for the cost of printing) were hardly more than the savings there would have been on postage if the Journal had carried no advertisements at all. It seems to me that the extra cost of postage due to advertisements should be set against the advertising revenue and therefore should be taken into account when considering the appropriate page charge for advertisements. (c) Exchange advertisements with other journals. In view of the evidence[1] that promotion campaigns are unlikely to achieve much for an established journal, we felt that the purely financial value of such exchanges (of which we had a number) must be doubtful. But we were also disappointed to find, in a survey of our exchanges, that in at least one case where we had been publishing advertisements for another journal, that journal did not appear to have been publishing ours; and when we sent a letter to this journal requesting copies of the advertisements which they had undertaken to publish for us, the letter was unanswered. We have therefore cut down on exchange advertisements.

An increasing item of journal expenditure in Britain is postal costs, which in recent years have risen considerably faster than the general rate of inflation. The really important factor is the weight of the journal package, which should be considered in respect of the various levels which determine the postal rate. For example, the present rate for a package of 255g is 20p inland and 22p overseas; and the corresponding rates for a package of 245g are 17p and 12p. For the British Journal of Psychiatry, with a home circulation of 5,500 and an overseas one of about 3,000, a journal package of just over 250g will cost over £ 5,000 a year more in postal rates than would a package of just under 250g. The simplest way to reduce the weight of a journal package is to use lighter paper. In mid-1976, we changed from $70g/m^2$ to $50g/m^2$ text paper. Quite apart from the saving in postage, lighter paper costs less. Table 6 shows such savings in cost during 1976, in spite of the fact that both postal and printing rates increased during that year. When our Journal Committee decided on the use of $50g/m^2$ paper, I was apprehensive that the resulting journal, which would be lighter, slimmer, and flimsier, might be viewed askance by readers as a falling off in standards. But in the event it passed largely unnoticed, the only comments being of approval that the Journal was now less heavy to carry and more supple in handling.

The effect of all these changes on the balance of the Journal accounts since 1973 are shown in Figure 3.

There must be many other ways of saving on journal costs, and new possibilities occur all the time. We have recently been studying the economy of webb offset printing on paper of $40g/m^2$ or lower weight. The fashionable format of quarterly or monthly learned journals - and one which probably reflects the available techniques hitherto associated with the print run of a relatively small circulation - has for long been that of quarto size with

TABLE 6: Effect of Paper Weight on Costs, 1976

Item	Jan.–June	July–Dec.
Paper weight (g/m^2)	75	50
Weight of package (g)	303	224
Postal costs	£10,200	£8,100
Printing costs	£29,300	£27,100

Figure 3. The annual balance of account of the British Journal of Psychiatry 1947–1976, in terms of the 1975 value of the pound. The journal was issued quarterly until 1961, bimonthly 1961–64, and monthly since 1965.

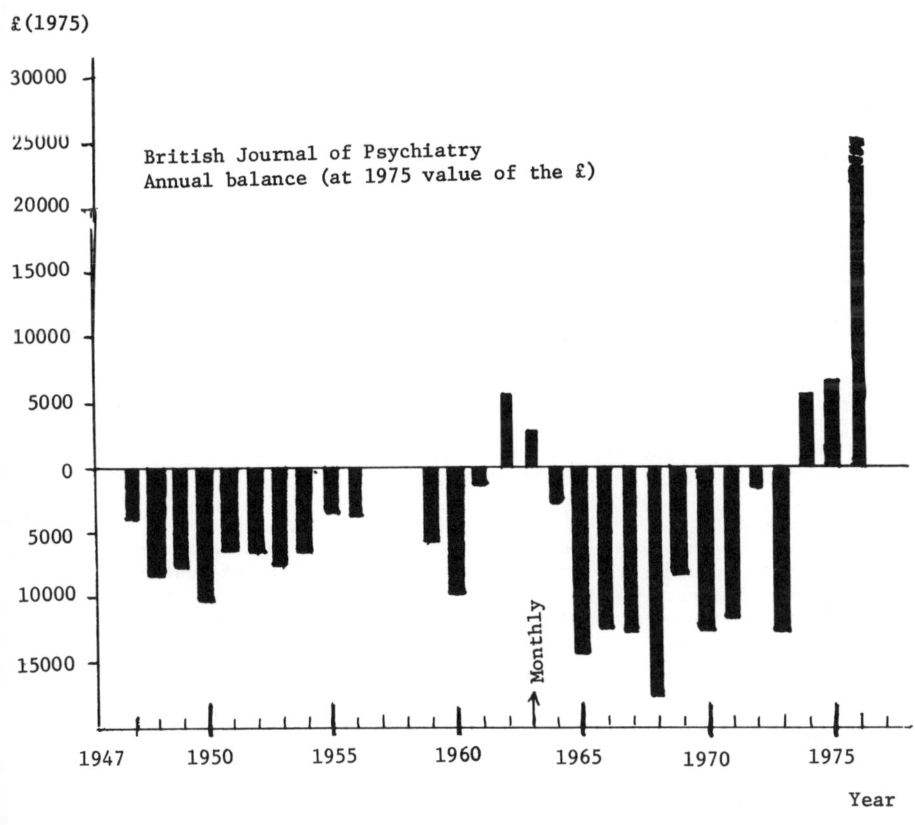

stitched binding and a spine (on which, incidentally, much useful information can be put). The webb offset technique is at present (I understand) most efficient on A4 size of paper and, for the average journal issue, this implies metal binding and no spine. For our journal, it seems that the cost of printing by webb offset would be lower than our present method of printing on single sheets by something between £ 10,000 and £ 20,000 a year. Would readers take kindly to such a change in format? Or would they be ready to accept that a monthly has nothing to be ashamed of if it appears in the same format as Nature or the British Medical Journal?

Apart from these considerations, the costs of a small learned journal will depend to a large extent on the way in which its running is organized. If much of the work of sub-editing and general management is done by honorary officers in their spare time, the costs to the learned society which owns it will be less than if the journal is run and managed by a publishing firm. But there is probably a limit to the extent to which a journal can be run by honorary officers and non-professional clerical staff. To date, the British Journal of Psychiatry has always been run in this way; but in 1974 a large publishing firm offered to take over its management. Careful study of this firm's proposals left us in no doubt that our College would be some tens of thousands of pounds worse off annually than under the present system, and we therefore decided to go on with our cottage industry. But the increasing complexity of the operations, and the increasing difficulty of honorary officers finding sufficient time under the conditions of the National Health Service, may soon make it hard for us to continue in this way.

As a final point I should explain how the financial accounts of our Journal have been presented by its owners over the past 120 years. The costs of supplying the Journal to each of the members (now 5,500) are included under the heading of Expenditure, but the Income never includes a contribution from the membership fees towards these costs. It is not difficult to see how such a state of affairs arose. When a learned society is formed, its journal is designed primarily as a means of communication between its members, who receive the journal as part of the privilege of membership. What is then published in the accounts is simply a statement of the cost of producing and distributing the journal. If the journal proves a success, it will be bought by non-members. The society will charge non-members an appropriate subscription fee and these sbuscription fees will then appear in the accounts as part of the journal income. For a long time however the costs of production are likely to exceed the income, and the resulting deficit will continue to be labelled as "The cost of the Journal" to the society. If the circulation of the journal continues to grow, the time will come when the income from subscriptions and advertisements will, if economically managed, balance or even exceed the total cost of journal production. But long before that happens, the learned society is likely to have forgotten, or to have decided to overlook, the circumstance that if an appropriate part of the membership fees were allocated to the journal income to cover the cost of supplying members with the journal, then on that basis there would probably have been for some time a substantial surplus in the journal account.

The present scrutiny of the finances of a small specialist medical journal was begun under the impression that the journal was bearing an unacceptably high loss. The losses have been made good - temporarily at any rate - but I have been left with the rather uncomfortable realization that, if an appropriate part of the membership fees were credited to the journal income, then even in its recent poor years it was in fact making quite a sbustantial surplus. This might seem to raise some ethical problems. How far is it proper for a society to use its journal surplus for non-journal purposes? How far should a journal surplus be used for paying its honorary editors an honorarium, or for keeping the subscription rate low, or for expanding its promotion of knowledge and learning? The editor is probably not the right person to answer such questions. But if he is asked, his first impulse is likely to be one of caution. More than anyone else, he learns to appreciate what a complex ecological system a journal is and how any change in the system, whether by design or accident, may have long term effects which cannot be foreseen. In other words, he is likely to have in mind the fat kine and the lean kine of Pharaoh's dream and so to believe that the first call on any years of plenty is to provide against possible years of famine ahead.

Reference

Economics of Scientific Publications. Council of Biology Editors. Washington, D.C. 1973. Pages 3,4.

EDITING A PHOTOGRAPHICALLY REPRODUCED MATHEMATICS JOURNAL

Earl J. Taft
Editor, Communications in Algebra
Department of Mathematics,
Rutgers University, New Brunswick, N.J., U.S.A.

ABSTRACT

This paper describes four years of experience of editing Communications in Algebra, a mathematics journal photographically reproduced directly from the author's manuscript, published by Marcel Dekker, Incorporated of New York starting January 1974. Some of the good experiences have been comparatively rapid publication after acceptance of paper by the referee, cheapter cost of production, no mistakes due to typesetter, no corrections of galley proofs and no difficulties with papers in foreign languages and even foreign alphabets (e.g. Cyrillic). The bad experiences include the difficulties encountered by some authors in preparation of manuscripts for direct reproduction and the related poor physical appearance of the published article. For the most part, there has been a steady flow of good quality papers with an international representation. It is felt that Communications in Algebra is able to effectively compete with other algebra and mathematics journals of a comparable quality which are published by traditional typesetting methods.

1. Introduction

This paper will be a description of the origins, short history and experiences of Communications in Algebra, a rapid-publication mathematics journal published using photographic reproduction methods by Marcel Dekker, Inc. of New York. My point of view is that of the Executive Editor of Communications in Algebra.

2. Origins

Marcel Dekker, Inc. is a scientific publishing house whose background was primarily in biological and chemical sciences. Later they had engineering publications and had started a physics program during the 1960's. When I first became connected with them about 1970, they were just beginning a series of monographs and lecture notes in pure and applied mathematics. At that time, they also had recently started a small program in statistics, with which I have not been involved. At that time, they were publishing various communications journals in non-mathematical areas. In 1972, when I was acting as advisor for their mathematics series, they suggested a communications journal in mathematics. Since I was to be the editor, it was decided to do it in algebra, in which my special research interest lies. A communications

Miriam Balaban (ed.), Scientific Information Transfer: The Editor's Role, 415-418.

journal in statistics had recently been started, and seemed to be doing well. Communications in Algebra is a journal for the publication of original research articles in all branches of algebra. However, classical number theory was excluded, since there are ample outlets for this elsewhere.

It should be noted that journals in mathematics published by direct reproduction are fairly rare. It is more common for certain Lecture Note series in mathematics to be directly reproduced, such as the series of Springer-Verlag, Inc. and of Marcel Dekker, Inc.. However a quick glance through a complete collection of current journals would reveal less than a handful which are directly reproduced.

3. Beginnings

The organization of the Editorial Board occurred during parts of 1972 and 1973. Of about 60 invitations issued, 41 accepted. This is a high rate and, in general, the response was enthusiastic. Concerning the reasons of those declining to serve on the Editorial Board, besides the usual reasons involving lack of time, the most prevalent argument concerned the proliferation of mathematics journals, both specialized and general. At that time, there were several algebra journals, including the established Journal of Algebra, the Journal of Pure and Applied Algebra, and several journals centering around topics in linear algebra. These latter linear algebra journals are not comparable since they branch into analysis and applied mathematics areas. All of these other journals are published by traditional type-set, so that the main justification for a new journal was the photographic process leading to more rapid publication than is possible with the other algebra journals. Of the 41 acceptances, one resigned before the first issue appeared because he disliked the photographic process and subsequent physical appearance of the articles. This point will be discussed in greater detail later. One other resigned for health reasons, and the number of Editorial Board members stabilized at 39, where it remains today.

In 1973, papers were solicited. The response, both from the general public and from the Editorial Board members themselves, was very high. The first issue appeared in January of 1974. It had been decided in 1973 that we would start with 6 issues per year, but before the first issue even appeared, it became apparent that 12 issues per year would be appropriate. Hence, volume 1 appeared during the first six months of 1974, and volume 2 during the last six months of 1974. Since then, one volume per year has appeared.

4. Manuscripts

The author is required to have his paper typed in camera-ready form. A rather detailed list of instructions for preparation of manuscripts for direct reproduction appears on the inside rear cover of the journal. Although this list appears formidable at first glance, the only instruction that really matters is that there should be clean, clear even impressions on good quality paper. In other words, the paper should reproduce well when photographed. Most of the other instructions concern style, e.g. references, footnotes, title, etc. and are there only to encourage a uniform appearance. If such details are not precisely followed, the article will still be photographed if it reproduces satisfactorily.

5. Editorial Procedure

The manuscript is submitted to a member of the Editorial Board, who is ultimately to decide on its fate. The member may referee it himself or use the services of an outside referee. If the member accepts the paper, he forwards it to me for publication. Papers are not supposed to be sent directly to me, but sometimes they are. In such cases, which have become rare as the system is more widely understood, I will either return it to the author suggesting that he send it to an Editorial Board member, or send it to a member or to a referee myself, depending on the circumstances of the paper. The published article does not contain the name of the Editorial Board member who transmitted it, an idea proposed by various members of the Board itself.

6. Publication

I submit the papers for the various issues to the publisher in New York where it is arranged for direct reproduction. This involves some cutting and pasting, and some minor adjustments of position of addresses, footnotes, completion of references, etc.. It is then sent to the printer for reproduction, assembly and distribution. The time from the receipt of camera-ready copy by the publisher to the distribution date is approximately three months.

7. Backlog

The backlog was fairly steady for several years at around six months. This refers to the time from date of submission of the paper until date of publication. In 1976, this backlog increased to about one year. Although this is still quite favorable in comparison with other algebra journals, most of whose backlogs are in the vicinity of two years, it nevertheless starts to go against the idea of rapid publication. Hence, in 1977, 14 issues will appear. Also, each issue which had contained about 100 pages will now contain about 120 pages, which is near the limit permitted by binding considerations. This will cut the backlog to around 9 months. In the future, it will be possible to further increase the number of issues per year in order to decrease any backlog which is considered too high.

8. Mathematical Quality

In general, the mathematical quality of the papers appearing in Communications in Algebra has been high. It is certainly comparable to the best of the existing algebra journals. The Editorial Board, which is international in scope, is very distinguished. Of course, with 39 members, some unevenness is inevitable, but things have generally proceeded smoothly and there have been no major hassles.

9. Physical Quality

The physical quality of the published articles is generally very good for photographed articles. There have been no complaints from authors or readers. Although the process has become quite sophisticated, it naturally will not appear as attractive as articles published by standard hot-type methods. The final appearance is directly proportional to the physical appearance of the original manuscript. I occasionally receive accepted papers which have to be returned to the author to be retyped into camera-ready form. This happened more frequently during the early stages of the journal, and has become very infrequent more recently. It is most prevalent in countries where access to the instructions is not available or where the nature of the typewriters and paper is poor.

10. Reactions

Author reaction to direct reproduction cannot be fully judged since no accurate assessment is possible of how many potential authors decide not to submit papers because of the care needed in the preparation of the manuscript. I have had no direct comments of this nature although it is surely a factor in some authors' thinking. Probably we do not lose many papers because of it. A good sign of acceptance is the large number of good papers coming from authors who had previously published in Communications in Algebra and are not deterred from doing so again. With one exception, no author has refused to retype a manuscript which was not in acceptable form. The exception was a paper accepted before the first issue actually appeared. It was a very long paper which would have had to be retyped in Germany, and it was inconvenient for the authors' institution to do so. It seems that whatever negative feelings are created by the necessary process for preparing manuscripts are minor in comparison with the desire for rapid publication. A small number of authors submit a roughly typed manuscript, and then have it retyped if the paper is accepted. Another positive factor in some authors' thinking is the lack of galley proofs to correct. The author is assured that what he submits is what will appear. Of course, he is also then responsible for any errors - typographical as well as mathematical. Another minor advantage of cold-type, but one which demonstrates its versatility is the absence of problems concerning language. All languages and alphabets are equal to the camera. We have had a paper appear in Russian in Cyrillic letters treated as the other papers.

11. Conclusion

In conclusion, most authors do not seem to be bothered by the care needed in preparation of manuscripts. The guidelines seem widely understood, mainly that the manuscript should reproduce well when photographed. Communications in Algebra is doing well mathematically and also seems to be accepted by the mathematical public with respect to the final appearance of the published articles. The publisher has subsequently started a similar Communications journal in Partial Differential Equations, and one is now being planned in Numerical Analysis for 1978. One might predict that as the technical reproduction processes get better and public acceptance grows, publication of journals by direct reproduction will become a widespread phenomenon.

PAGE COMPOSITION COSTS OF PHILOSOPHY JOURNALS: A REPORT ON A STUDY

Robert N. Beck, *President*
The Association of Philosophy Journal Editors
Editor, Idealistic Studies
Clark University, Worcester, Massachusetts, U.S.A.

and Richard H. Lineback, *Secretary/Treasurer*
The Association of Philosophy Journal Editors
Bowling Green State University, Bowling Green, Ohio, U.S.A.

ABSTRACT

This paper reports the main conclusions reached in a study of cost-cutting alternatives in page composition for philosophy journals in North America. Composition costs are singled out for study because they are a major cost factor for most journals. Sponsors of the study were the National Endowment for the Humanities, the American Philosophical Association, the Canadian Philosophical Association, and the Association of Philosophy Journal Editors.

The major conclusion of the study is that cost-saving composition alternatives are available to editors if they choose to adopt them. To show this, the study surveys major factors in composition costs and indicates the expense to journals of each of them. Also surveyed are various kinds of composition equipment available to editors. Suggestions are made to show how, if the options are adopted, editors can make savings approaching 50% of their current average composition costs.

The paper concludes with some observations on the financial practices of philosophy journal editors.

The purpose of this paper is to report on a study of journal page composition costs which we recently conducted[1] on philosophy journals in Canada and the United States. Underwritten by the National Endowment for the Humanities,[2] the American Philosophical Association, the Canadian Philosophical Association, and the Association of Journal Editors, our project arose out of felt needs of American journal editors faced with increasing financial pressures. In fact, our society took official action requesting its officers to explore alternative solutions to cost problems. One approach would have been to try to secure outright grants to supplement journal budgets. But this turned out not to be a viable option for us in the United States, partly because of the scarcity of funds for such purposes and partly because of the magnitude of the problem. The alternative we adopted was to study possible means by which journals might cut their costs and thus in effect help themselves. Since one of the major expenses for most philosophy journals is page composition, we decided upon a study to determine with some precision the availability of less expensive methods of composition and printing. Included in our purview were the capabilities of alternative methods, for example,

Miriam Balaban (ed.), Scientific Information Transfer: The Editor's Role, 419-426.

their ability to handle logic symbols, languages other than the native tongue of the keyboarder, and so on; the costs, both initial and recurring, which they involve; and something of the nature of the services -- time spans, scheduling, etc. -- which alternative systems could provide. Also of critical importance was the question of how much individuality in printing styles editors would sacrifice to meet financial pressures.

The major aspects of our work were three: (1) to use as many relevant inputs as possible, including questionnaires to members of our Association and contacts with other individuals and groups with similar concerns; (2) to collect as much information as possible on recent technological innovations and composition machinery from the printing industry; and (3) to try to organize this information in such a way that journal editors could learn and take advantage of it.

Since our study was primarily directed toward the problems faced by North American editors, the precise financial details discussed in our work may not be relevant to the problems of editors in other countries around the world, or to their solutions. We might say, however, that for editors in our country who feel inclined to adopt our recommendations, savings of nearly 60% (roughly in dollars, from $12 to $4.80 per page, based on 1976 cost estimates) could be effected. On the other hand, some report on our work may be suggestive, if not an actual model, for editors in other countries. Accordingly, we shall organize the remainder of our brief report to you under three headings: cost factors in page composition, alternative technologies, and general observations.

I

Although many, if not most, of the following page-cost considerations may be known to editors, it may be nevertheless useful to list and comment on them systematically. Once again, we should remind you that our estimates are determined relative to the American printing industry. At the same time, however, we want very much to stress the fact that these cost considerations are to a large extent within the control of editors and are influenced by editorial decisions. What we as editors do and decide has a direct bearing on the page composition costs we have.

Providing Clean Copy.

Perhaps the single most important element which will allow rapid keyboarding of manuscripts is clean copy for the person doing the keyboarding. Copy that is of poor quality to begin with or copy that has been badly marked and edited reduces input speeds. If an editor is willing regularly to provide unusually clean copy, he can negotiate a reduced rate for page composition. Good clean copy can easily save 10% of composition costs.

Proofing.

Many composition firms include in their basic per page composition charge the cost of a professional proofreader proofing the galleys before an editor sees them. If an editor who carefully reads the galleys anyway is willing to assume responsibility for catching and marking all typographical errors, page composition costs can be reduced by approximately 20%.

Reducing Foreign Language Content.

Keyboarding information in languages other than the native tongue of the keyboarder slows down the person doing the keyboarding. This is the case for a variety of reasons, including the person's lack of familiarity with foreign spellings, the necessity to look up hyphenation points in a foreign language dictionary, and the need to insert diacritical marks.

Using A More Efficient System.

Although most composition systems are similar, a more efficient system may do the same job at a significantly lower price. Since composition houses must distribute the costs of their equipment over the pages set and since the life of equipment is relatively short, a significant amount of the per page cost is for equipment. By choosing a composition firm with an efficient system that is heavily scheduled as opposed to a system that is less efficient or less used, an editor can save up to 15%. And if one chooses a vendor with a comparatively inexpensive strike-on machine or a direct input phototypesetter, one saves up to 30%.

Footnotes.

Footnotes generally increase the cost of page composition. In order to minimize page composition costs due to footnotes an editor may reduce the number of footnotes used, place them at the end of the article, or place them in the running text. Footnotes placed at the bottom of each page are the most expensive. Depending on the process used, one footnote at the bottom of a page usually costs between 10¢ and 25¢ more than it would cost to put the footnote at the end of the article or in the text.

Changing Type Size.

Minimizing the number of times that changes are required in the size of type reduces composition costs. Frequently the same aesthetic effect can be obtained by putting the text in bold type rather than in a larger size. Because of the way most systems are set up, it is cheaper to bold face than it is to change the size being utilized.

Alterations.

Author and/or editor alterations should be avoided as much as possible. Such alterations increase the cost of page composition regardless of the system used. If an editor is willing to hold such alterations to an absolute minimum, page composition costs may be reduced by signing a contract that provides for special charges to cover any such alterations that need to be made. If the alterations have been suggested by the author the practice now followed by some members of the Association of Philosophy Journal Editors may be used, which is to bill authors for any charges due to their alterations.

Symbols.

Composition costs may also be lowered by reducing the use of special symbols, such as $\forall$, $\exists$, and $\setminus_c$, as much as possible. In order to have a special symbol on the page, it is usually necessary to keyboard two or three characters for each symbol. Some special symbols also require that a new type font be inserted into the phototypesetter. Thus, special symbols reduce input speed and frequently reduce the rate at which the phototypesetter operates.

Fonts.

The use of commonly available fonts reduces page costs. Although fonts for phototypesetters are much cheaper than for Linotype or Monotype machines, the acquisition, storage, and utilization of a special type font for a given journal will increase per page costs. For example, a $200 type font distributed over 400 journal pages per year and over a two year period would add approximately 25¢ per page to the composition charge. If an editor needs and insists upon custom fonts, the additional charge per page would be several times this much.

Common Font.

If it is possible, the same page format as other journals being produced by the same composition firm should be used. Such a practice reduces the machine set up time and allows the operator to proceed at a faster pace.

Type Size.

Reducing the size of type from 10 point type to 9 point type will increase the amount of information on each page by approximately 10%. Assuming a given amount of copy, this change in type size will reduce the number of pages to be printed by the same percentage. Although the cost of printing is not directly a function of the number of pages, there is a close correlation. Hence, reducing type size will in a vast majority of cases reduce printing costs.

Margins.

Another way of putting more information on a page is simply to use more of the page for text and less for margins. Consider for example the following extremes. If one used a column of text 3 1/2 inches wide and 6 inches long on a 6 x 9 inch page, approximately 21 square inches of the 54 square inches available would be used. In contrast, if one used a column 4 1/2 inches wide and 7 1/4 inches long on the same 6 x 9 inch page, 32.6 square inches of the available 54 square inches would be used. This is a 55% increase in the amount of information on each page. In other words, for each quarter inch added to line length, one adds 6% to the information on the page. For each quarter inch that is added to the length of the page, one adds an additional 4% to the information presented. Thus, even though an editor may not want to make a major change in the dimensions of the text, relatively small changes can effect significant savings.

Leading.

Reducing slightly the amount of white space between the lines of a paragraph and between paragraphs can mean a significant savings. For example, if one is currently using 10 point type, reducing the leading between lines and paragraphs by one-half point would increase the information on the page by approximately 4%. Although 4% may not sound like much, when it is combined with slightly smaller type and slightly narrower margins, it is easy to increase page content by 25% or more. The number of pages required for a given issue will be reduced proportionately and the amount saved on the printing will approach this percentage.

Running Heads And Page Numbers.

For journals currently using both a running head at the top of the page and a page number at the bottom of the page, the content of the page can be increased by putting the page number on the same line with the running head. This change reduces paste-up time and allows space for an additional two or three lines of text. An additional three lines of text means 8% more information on the page.

Page Size.

Most printers have presses which can produce a given page size more efficiently than any other. If one uses the size for which the print shop is set up, one can get the lowest possible price. Otherwise, some paper is wasted and the assembling and binding process must be adjusted to your particular size. On the basis of the contacts made in this study, a 6 x 9 inch page seems to be the page size preferred by most printers in the United States.

Cooperation.

Virtually all composition and printing firms contacted agreed that there would be significant savings if a group of philosophy journals would use one firm. Estimates covered a rather wide range, but it seems safe to conclude that five or six journals could save approximately 15% each by using the same composition house and printer. If this number could be increased to some twenty-five journals, it is probable that an additional 10% could be saved. The reasons for these savings are various. For example, if all of the journals were printed on the same stock, stock could be purchased in greater quantities and at a lower price. Covers could be printed for eight journals at a time. Cooperation also allows the composing and printing facilities to be scheduled more efficiently. In addition to saving money, philosophy journals by cooperating together could demand better service than each can demand by itself.

OCR (Optical Character Recognition).

If journal editors are willing to assume the responsibility for typing their manuscripts on an IBM Selectric typewriter using a special typing element and if they are willing carefully to proof this copy before sending it to a processing center, composition costs can be significantly reduced. The basic reason for this is that the composition firm does not have to keyboard the information or proof it. This step by itself can reduce composition costs for an average journal by approximately 40%.

II

A second area of study for us -- and one which we believe is important for all editors to know -- is the changing technology of the printing industry. That this technology is changing, and rapidly, is a fact about the United States and, we are inclined to believe, a fact about most industrialized nations. Editors who keep abreast of these changes can turn their knowledge into cost savings for their journals.

Very roughly put, printing technology has moved from the older hot metal letterpress through photocomposition to current computer-driven photocomposition and offset printing. Actual printing use seems to have followed these developments. In the United States, the use of hot lead has declined significantly, largely because of the higher costs involved in this process. A second reason for the decline is the increasingly high quality of materials being produced by photocomposition and offset presses. To be sure, the question of comparative quality between materials printed using letterpress and hot type as contrasted with materials produced by offset printing and photocomposer has been much debated. Letterpress is one of the oldest, most widely used, and highest quality printing processes. The combination of photocomposition and offset printing has existed only for some twenty years and has come into general use only in more recent years. In the beginning -- and with some justification -- the combination of photocomposition and offset printing was characterized as "quick and dirty." With recent improvements in both photocomposition and offset printing, however, there now appears to be little difference in quality. We should note parenthetically that quality differs not only between printing processes, but between machines utilizing the same process. Thus there is low quality as well as high quality letterpress work. The same holds true for offset printing. Rather than making general comparisons between the two processes one might better be advised to make specific comparisons between a given letterpress and a given Linotype machine, on the one hand, and a given photocomposition machine and a given offset press on the other.

The question of the quality of letterpress versus offset printing was raised by the members of the Association of Philosophy Journal Editors at their meeting in December, 1975. As a result of that discussion a brief survey was conducted in the following manner. A paragraph from a journal produced by letterpress and a paragraph from a philosophy journal produced by photocomposition were cut out and pasted on the same poster board. The next twenty-five philosophers to visit the display table of the Philosophy Documentation Center at the American Philosophical Association convention were then asked to look very carefully at the two samples and to indicate which of them was the higher quality printing job. Six philosophers said that the letterpress paragraph was of higher quality, five said the two paragraphs were of equal quality, and nine said that the paragraph set by photocomposition was of higher quality. Although this survey did not control a number of related factors, it suggests that the average reader sees no significant difference in the quality of print produced by the two systems.

Another indication of the acceptable quality produced by the combination of photocomposition and offset printing is the number of commercial publishers and university presses that are using this newer printing process. Again at the American Philosophical Association convention, twelve exhibitors were asked what typesetting and printing processes were used to produce their books. Six of the twelve confessed that they were merely salespersons and had no knowledge of the process used to produce their books. The other six were more helpful. St. Martin's said that they use both methods. Harper & Row indicated that they were currently using both methods, but that they were moving towards photocomposition. Macmillan indicated an exclusive use of letterpress and McGraw-Hill claimed to use mostly letterpress. In contrast, Cornell University Press indicated that it was using some photocomposition, while Harvard University Press indicated exclusive use of photocomposition. The fact that photocomposition and offset printing are being used by commercial publishers and university presses to produce quality publications (as contrasted for example with newspapers) suggests that this newer printing process produces satisfactory quality at the least.

A special, recently developed feature of photocomposition is that it can be used in conjunction with an optical scanner. Although there are many optical scanners on the market, one of the more common models can read pages produced by an IBM Selectric typewriter which utilizes a special typing element. The typing element has the standard alpha numeric characters on the top so that the text can be read in the normal way. However, in addition there is a unique bar code under each character which the optical scanner can read.

This system allows an editor to have a secretary type the manuscript according to his instructions. The editor could then proof and correct the copy before forwarding it to be read by the optical scanner. The sheets are then fed into the optical scanner, which in turn reads the text and stores it in an appropriate mode. Print commands which cause headers to be printed in larger and bold face type and which cause book titles to be printed in italics are usually added by an experienced operator utilizing a terminal.

The conclusions we reached for North American editors were direct, perhaps even obvious. Photocomposition and offset printing offer the better cost-effective alternative; and when or if an editor can utilize optical scanning devices and engage an offset printing firm under good contractual conditions, cost savings can be remarkable.

III

Financial studies, including our own, are seldom only that: they usually address, or at least reflect, other types of questions as well. In this third and concluding part of our report, we shall share with you a few of our general observations and impressions. For the sake of brevity and directness, we shall adopt a question and answer format.

(1) To what extent are less costly modes of publishing journals available? To what extent are they being used?

Less costly modes of publishing journals, we can immediately answer, are sufficiently available to meet journal needs. But we must qualify this statement in a very important way, for a good part of the meaning of "less costly" depends on editors themselves and on the decisions they make about what they feel they must have. Currently some 40% of American philosophy journals are using photocomposition, which is a less costly mode of composition. However, for an editor who feels that letterpress is required, there are certainly fewer less costly alternatives.

(2) Are there behavioral or institutional factors inhibiting the move to less costly alternatives?

Although study of the "behavior" of editors was not directly part of the project, it may be of interest to pull together the data we have on the question. Queries concerning possible cost saving methods in our questionnaires were frequently answered in the negative. And it was not only journals that are "wealthy" who did not want to avail themselves of savings. In contrast, we have learned that one financially sound journal has just changed to photocomposition, partly as a result of the Association's discussions.

It has been suggested that an editor's position and the philosophical qualifications he brings to it do not lend themselves easily to certain business considerations. Editorial work is scholarly and professional; financial considerations, while necessary to survival, are secondary to professional goals. This hypothesis certainly requires further testing. However, we have some information of which any hypothesis should take account. The number of negative responses to the cost saving possibilities suggested in our questionnaire is significant. Other answers, such as that a change would be considered only if a savings were substantial as against, say, a number of small savings, may be significant. We have also learned that changing printers for economic advantage and asking for annual bids are not common practices.

(3) Finally, what percentage of the philosophical journals are facing the same problems? If there are some who are less affected, what are the major reasons?

With but one exception in our survey, all journals seem to be facing very similar problems: costs rising faster than income, major dependence on institutional subsidies, and subscription lists primarily made up of libraries. Since there are general pressures on both institutions and libraries to cut costs, journals are being affected. To be sure, journals seem to be affected differently by these pressures. The major variables involved in the differences appear to be the following: (1) the quality of management practices, (2) the amount of salary costs borne by the journal budget, and (3) the size of the subscription list. In general, journals which are high on (1) and (3) and low on (2) are less affected by cost pressures; those which are the opposite tend to be highly affected.

NOTES

1. Our study is dated February 15, 1976, and was submitted to our sponsoring organizations in the spring of that year.

2. Under NEH grant AS-23525-75-503. The conclusions of our work represent, of course, our own judgments and not necessarily those of our sponsors.

DISCUSSION on "Surviving Inflation? Some Recent Financial Problems of a Specialist Medical Journal," by E. H. Hare

"Editing a Photographically Reproduced Mathematics Journal" by Earl J. Taft

"Page Composition Costs of Philosophy Journals: A Report on a Study," by Robert N. Beck and Richard H. Lineback

KLAUS HELBIG - I have a question to Prof. Taft. How do you print your equations? Are they typed or are they handwritten?

EARL TAFT - Well the equations are prepared by each individual author. I'm not sure I completely understand your question. If I don't answer it, please ask again. In general, the symbols that are available on the typewriter are typed in and those that are not, are usually entered in ink or something that will reproduce satisfactorily.

KLAUS HELBIG - I ask because one normally runs into quite some trouble between a small e used for distance and the symbol of the base of the natural logarithms, the capital E for expectation and the inclined E which is an ordinary symbol; all these kinds of things, which I have always hesitated to leave to authors in order to avoid mix-up. Also some use the 1 for the one, and others have a special one.If you do not have italization, how do you distinguish the 1 which is used for one from the symbol l, and so one. Of course, if you instruct the authors very closely you can rely that after all this is his responsibility in your journal. I have always found in our journal that authors have to be very carefully watched in this field.

EARL TAFT - Yes well, let me say to you that in mathematics I think usage is fairly standard so that I have not found this a major problem. The one difficult point that I have found is the tendency of some authors to use too many symbols and numbers in titles of articles. This is prevalent in certain branches of mathematics. I have sometimes made suggestions to authors so that they try to avoid the use of symbols and numbers in the titles of articles.

JANINA BARD - I have a very simple question for Dr. Hare. Is his fifty gram paper opaque?

EDWARD HARE - I have a simple answer, yes.

JANINA BARD - I was told that below about seventy it becomes very difficult for the reader to reprint an abstract journal and we were advised not to go below seventy.

PAUL NIJHOFF ASSER - It depends of course what you print on it. But generally it depends on the chemicals used in the production of the paper. If there are no more questions I can summarize perhaps very shortly.

This morning we have been moving away from publishers towards editors again as we should because this is an editors' conference, but this session was devoted to the publishers problems and I think the editors we had have coped quite successfully with publisher's problems. We heard four publishers as well and as I said they have analyzed their problems and I think the editors have looked a bit further and we have tried to solve a lot of publishers' problems for them. At the end we have heard three people and these perhaps are very appropriate because they symbolize something. The mathematician perhaps is the symbol of the correct organization of our thoughts and the psychiatrist perhaps made us think of the pleasure principle in what we are doing, and finally we have had the philosopher, without which I think science would not exist. After all science has its roots in philosophy and has been repeated many times. This morning we cannot have science without publishing and we cannot have publishing without editors.

INTERNATIONAL, NATIONAL AND SPECIAL PROBLEMS

Wolfgang Löhner, Chairman

INTRODUCTION TO SESSION: INTERNATIONAL, NATIONAL AND REGIONAL PROBLEMS

W. Löhner
Division of the General Information Programme
UNESCO, 7, Place de Fontenoy, 75700 Paris, France

When I compare Recommendation 11 of the UNISIST Feasibility Study concerning the responsibilities of scientific editors and read there that "it has often been the practice in the past for editors to work in isolation, responsible only to the scientific society or institution which employs them", I wonder whether it would be fair to say that this past is over.

It is my impression that, despite all the efforts, within Unesco and outside, scientific editors still work under conditions which are not adequate anymore to ensure an efficient and rapid transfer of information.

Unesco tries to deal with the complete information transfer chain starting with the author, going on to the editor and then further to publishers, libraries, documentation centres, information analysis centres until the user. Looking at the scientific editor from this point of view, one could say that only the author might operate in a more isolated way.

Publishers, librarians, archivists, secondary services have - at least in the so-called developed countries - more or less powerful organizational structures which enable them to influence quite efficiently governments and intergovernmental organizations with the result that their interests are - to a large extent - being taken care of.

Editors, and in particular scientific editors, seem to have difficulties, for various reasons, in building up a mechanism or mechanisms which would enable them to play a more active and decisive role in information handling.

Unesco, under its UNISIST programme, has in the past, and will continue to do so in the future, initiated and lent its support to several activities intended to strengthen the role of scientific editors:

1. At the national level, Unesco has organized courses in scientific editing in Indonesia (1973), the Philippines (1974), Nigeria (1976) and India (1976). The basic objective of these courses was to familiarize the editors in the countries concerned with the latest international practices pertaining to various aspects of editorial work. Some of these courses could be considered also as a regional activity since participants from neighbouring countries participated as well.

In evaluating these courses, one always finds that it is of great value to the individual editor to be able to exchange ideas and experiences and also the proof that the problems of scientific editing are not unique to each journal. The problems presented in all these courses were the same:

- scientific journals are insufficiently funded
- <u>or</u>, have much too small a circulation
- <u>or</u>, receive no manuscripts
- <u>or</u>, take months to be typeset
- <u>or</u>, suffer from enormous postal delays and are consequently of virtually no interest to the scientists of the region as vehicles for the transfer of scientific information.

Miriam Balaban (ed.), Scientific Information Transfer: The Editor's Role, 431-432.

The trouble is that the problems mentioned are inter-dependent and, taken together, form a vicious circle.

A course for scientific editors can only serve to make those who are concerned with scientific information aware of the problems and, in particular, of their inter-dependence. It thus forms the first step towards a solution, but other activities will have to follow. They include:

- a suitable journal policy
- the regular provision of funds
- the development of adequate typesetting and printing capacities
- the development of publishing house-type support for editors
- permanent training courses on specific problems.

For obvious reasons, it seems desirable to "organize" editors at the national level first and courses on scientific editing have been appreciated very much as a first step. Unfortunately, only a small number of editors can participate in these lectures whereas the "hunger" is tremendous. Therefore, Unesco is now planning the publication of "Guidelines for Editors of Scientific and Technical Journals" of which I received the manuscript only a few days ago.

With the courses already organized and the Guidelines, we hope to achieve a multiplier effect which would enable Unesco to hand over the training of scientific editors to national authorities without too heavy a financial burden.

2. At the regional level, Unesco's efforts have to be critically evaluated. The original idea was to form regional or sectorial professional groups which would provide useful mechanisms for the discussion of common technical and economic problems and ensure a qualitative improvement in primary publications. Furthermore, Unesco promotes a closer liaison between scientific editors and the abstracting and indexing services, the information analysis centres and the libraries in order to improve the overall effectiveness of the information transfer chain.

To this end, Unesco has initiated the creation of editors' associations in various disciplines in Europe: Life Sciences, Earth Sciences, Chemistry, Physics, Engineering, Mathematics, and in a combination of disciplines and other regions: Earth Sciences in Latin America, Earth Science and other natural sciences in Southeast Asia.

A proposal was formulated at the Training Course in Nigeria to create an Association of Scientific Editors in Africa.

Looking at these attempts, opinions on the successfulness of those projects are very much divided. In the life sicences and the earth sciences, ELSE and Editerra respectively have - it seems- firmly established their associations and long-term plans of action are being implemented. Also in the physics field, the European Physical Society has a well-founded scheme of standards for the Europhysics Journals. Chemistry and Mathematics journals have not, or did not want to, succeed in creating an organizational regional structure. This is by no means a criticism, just a statement and I would welcome any reasons which could be given for this failure. Or is this not a failure at all? Are there good reasons why they have not been organized? Not being an editor myself, I am open to any constructive proposals which will be evaluated sine ira et studio.

3. At the international level, I should like to refer to the abovementioned statements. It has always been the ultimate target of Unesco - with respect to editors' activities to support an International Federation of Scientific Editors' Associations. However, Unesco would be acting against its mandate and the interests of scientific editors were it to try to impose any international structure on them. To avoid any misunderstandings, I must point out that we will encourage any proposals in this direction provided that there is ample proof of the firm will and active participation of editors which in turn are the only guarantee for the survival of such a mechanism.

EXPERIENCES OF THE NORDIC PUBLISHING BOARD IN SCIENCE (NOP)

Synnøve Irgens-Jensen
Nordic Publishing Board in Science, NAVG, Munthes gt. 29, Oslo 2, Norway

ABSTRACT

The four largest Nordic countries (Denmark, Finland, Norway and Sweden) have a population encompassing a little over 22 million people. Although each country has its own distinctive language, a large part of the citizens of each country has a fairly good understanding of each other's languages.

Previously, each country had completely separate publishing policies, and in total a great number of scientific journals. Within zoology, for instance, there were not long ago as many as 50 Nordic journals competing for manuscripts, subscribers and government aid.

NOP, or the Nordic Publishing Board in Science, was formed in 1968. Its main goal was to give economic support to Nordic scientific journals serving the international market, and publishing mainly in English. Since the formation of the society, several fusions of journals have taken place, and work is steadily going on to streamline the whole publication structure of the area to obtain bigger, viable units. NOP is not opposed to supporting popularized science, and at the moment gives financial aid to two such journals where articles appear in one of the Scandinavian languages.
To work for continual improvement of the NOP journals, seminars are regularly scheduled for editorial workers from journals of alternating scientific disciplines.
The governmental research councils for natural sciences in the four countries are the sponsors of NOP, and its activities are controlled by a Joint Committee on Nordic Cooperation in Science (NOS). Two members are appointed from each country and decisions on policy are decided at biannual meetings of the appointees.

Miriam Balaban (ed.), Scientific Information Transfer: The Editor's Role, 433-439.

The Nordic Publishing Board in Science, or NOP for short, is a joint body established since 1968, by the governmental research councils for natural sciences in Denmark, Finland, Norway and Sweden.

As a map of the northernmost part of Europe demonstrates, the Nordic countries are small, and the total population is slightly in excess of 22 million people Nordic cooperation is by tradition strong. The political systems are very similar, and language barriers hardly exist. True enough, Danish, Norwegian and Swedish are different languages, but the similarities between them are so great that there are no difficulties for the peoples in these countries communicating, each using his own tongue. With Finland the situation is particular. The Finnish language is totally different from the others. However, 5-10% of the Finnish population have Swedish as their mother tongue, and a large part of the rest of the people have become familiar with Swedish as part of their education.

Although Nordic cooperation has been significant in many fields, it took a remarkably long time to realize advantages to be gained by cooperating in the publication of scientific journals. Until the establishment of NOP each country had its own journals in every conceivable scientific discipline. The result was small editions and small, mainly local readership. Although some journals were in English, one may say that the majority had little impact on the outer world.

Remnants from this nationally-operated publication system still show their resistance to change in the four Nordic countries taken together.

NOP then has as its goal to promote common Nordic scientific journals through fusion of already existing national journals, and through the creation of new ones in fields where there is a need. The end result should be that each country will then be able to abandon many of its smaller journals which are increasingly more costly to operate; and one substitutes a Nordic forum for a national one.

So far, this work has resulted in 19 common Nordic journals which are tabled below.

NOP-supported journals in 1977

International, devoted to primary research:	Printed edition	No. of pages	% Nordic authors
Acta Chemica Scandinavica A + B	2400	2160	100
Astronomy & Astrophysics	1500	3880	4
BIT	1400	500	40
Boreas	800	240	72
Entomologica Scandinavica	800	320	80
Grana	700	224	33
Hereditas	1400	880	100
Lethaia	1800	320	18
Lindbergia	1000	185	51
Lithos	1100	320	35
Mathematica Scandinavica	1900	750	87
Nordic Hydrology	800	320	61
Oikos	1400	550	55
Ornis Scandinavica	1000	224	78
Physica Scripta	1400	800	60
Scand. Journal of Statistics	1000	192	90
Zoologica Scripta	1000	320	59
Popularized Science:			
Astronomisk Tidsskrift	3800	200	100
Nord. matemat. tidsskrift	2100	148	92

Editors are spread all over the area, and the office of editor moves from one Nordic institution to another. A condition for successful fusion of journals is to clearly define the scientific level of the resulting journal, be it devoted to original research results, review articles or popularized science. The intended group of readers must be defined. So far, of the 19 NOP-journals, 17 are written in English with an eye to international readership. Only 2 are in popularized science, mathematics and astronomy. These publish articles in the Scandinavian languages Danish, Norwegian and Swedish.

Within NOP we try to be as flexible as possible. In one field, astronomy, we support a European journal "Astronomy and Astrophysics". Fusions on a European scale are also encouraged for other fields where the Nordic countries seem too small a basis. Our bryological journal, "Lindbergia", is thus a Dutch-Nordic venture, and further European cooperation in this field is expected in the future.

In two fields where we may say we have been most successful in fusing journals mathematics and astronomy, there now exist only two journals for each discipline - one devoted to original research, the other a popularized version.

In other disciplines, especially the biological ones, reductions on such a scale are impossible. Biological species are very different in Southern Denmark and Northern Norway, whereas the stars and the mathematical formulae are common. However, NOP sees it as part of its task to work for the rationalization of the whole publishing structure in the Nordic countries. A working group has been started trying to reduce the number of zoological journals. The most eager proponents of reduction envisage that out of today's 48 journals we can create 6-8 new common Nordic journals, and still cover all fields of zoology. A similar rationalization-group for botany will also be created in the future.

When I mentioned the flexibility of NOP, I also want to add that any country is free to withhold its support from one or more journals it considers irrelevant to itself. We have thus inside NOP both tri- and bilateral arrangements.

Authors may be from any country, not only Nordic. For some journals, actually in excess of 50% of the articles are produced by scientists of other countries. For bio- and geo-science journals, however, the interests that are emphasized are naturally on Northern conditions and species. The NOP-journals systematically use international referees or Nordic referees with an international reputation, thus guaranteeing the scientific standard of the journals.

NOP is always ready to aid any of the journals it supports by hiring outside technical assistance to help with printing-methods, distribution problems and so on.

Arrangement

NOP is governed by a board consisting of 2 members from each of the four parent research councils. Biannual meetings are held in alternating countries. A secretariat is presently situated in Oslo.

Although economic support of scientific journals and the rationalization of the whole Nordic journal structure is NOP's main goal, NOP also takes on other tasks. One such is meetings of editors and editorial staff of journals of separate disciplines. In 1976, a conference was held for zoological editors, in March of this year another conference for botanists, and in May a similar conference is planned for geologists. At these meetings, which so far have been termed very useful by the people involved, there is a chance to discuss common problems, to learn new ways to tackle them and foundations for cooperation and fusions can be laid. At these meetings lectures intended for education and information are held, and workshops where groups get together and work on and discuss actual manuscripts. In 1978, NOP will consider still another venture, - a course for university instructors in writing and publishing techniques. This is a subject which has been sorely neglected at many Nordic universities,with the consequence that the editorial work on papers written by Nordic scholars has often been needlessly heavy. It is hoped that with the education of instructors at the Nordic universities, courses on these topics can be instituted there and further create scholars more aware of the problems regarding style and public presentation in what they write.
Fig. 1 shows the setup and the tasks of NOP, how money flows from the parent research councils through NOP, there to be distributed to the different undertakings of the organization.

Finally, a few words about economics. Of course NOP's activities are rather costly. Yet the individual countries can begin to see that as NOP grows, their economic burdens and responsibilities on the national level diminish.

Since the countries are of different sizes, it is found fair to divide the economic burdens every year proportionally to each country's gross national product for the preceding year. This means that in 1977, for example, we have the following division of contributions:

Denmark	22%
Finland	17%
Norway	18%
Sweden	43%

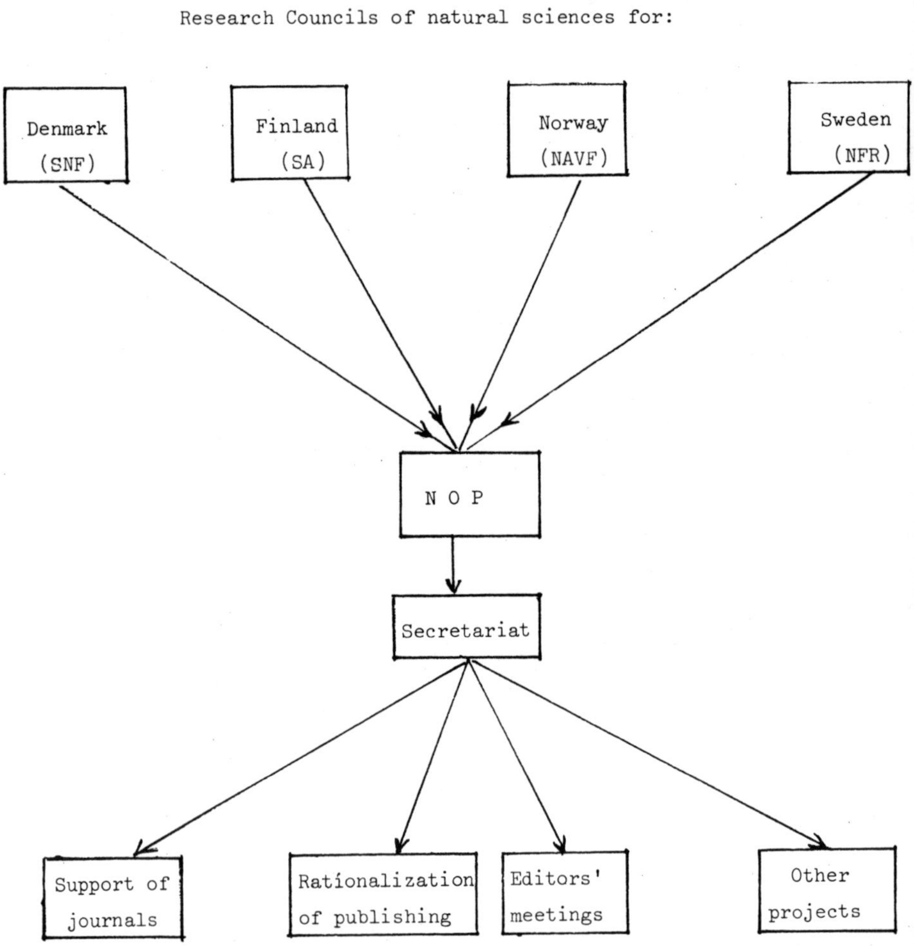

Fig. 1. The makeup and function of NOP.

An advantage here is that if bad times should befall one country, it will not see its scientific output curtailed, as the other countries will then take up the burden.

Our journals are encouraged to aim at being self-supporting. The subscription price should be set with this goal in mind. However, some disciplines are small, and NOP realizes that the goal of self-support for them will be hard to achieve. With the exception of newly-started journals, NOP is reluctant to give support in excess of 25% of total costs. A journal must prove its viability, something which may often be hard in times of growing competition from other journals, happening at times of diminishing funds for libraries and research institutions.

To summarize our ventures and our economy, we can look at 1977's proposed budget given in Norwegian crowns:

Support to journals:	nkr.	1.250.000,-
Rationalization of publishing structure:	"	40.000,-
Editors' conferences:	"	50.000,-
Other projects:	"	30.000,-
Secretariat:	"	110.000,-
	nkr.	1.480.000,-
	= US$	280.000 and
	DM	670.000

Approximately 85% then goes directly to journal support.

To summarize our success, we can mention that of the 2443 journals mentioned in the Science Citation Index for 1974, 3% were Nordic. The average number of references to an international journal article was 13, to a Nordic journal article, however, 16.

CENTERS FOR SCIENTIFIC PUBLICATIONS

Dorothy U. Mizoguchi
Technical Editor, Japanese Cancer Association, Kami-Ikebukuro 1-37-1 Toshima-ku, Tokyo 170, Japan

ABSTRACT

The Center for Academic Publications, Japan, was founded in 1971 to undertake the mechanical part of journal publications for scientific societies. As of October, 1976, the Center undertakes publication of 45 journals, some of them in Japanese, some in English. Eight of these journals are monthlies, and the remainder are bimonthlies, quarterlies, semi-annuals, and annuals. Acceptance of papers and their refereeing are in the hands of academic societies but all editorial matter, including copy reading, proof reading, and business dealings with the printer, is done by the Center staff. The printed journal is mailed by the Center. This system minimizes the number of staff necessary for academic societies, and also helps in the standardization of journal format. There is also a Business Center for Academic Societies which undertakes the business of collecting membership fees, mailing of notices and journals, preparing a list of members, etc. The Center undertakes such business of about 80 academic societies, as of October 1976.

Administration of a learned society has many problems. Large societies with a membership of over 10,000 can afford to employ sufficient number of staff for various business of the society and a technical staff for editorial work. In small societies, with a membership of a few hundred or less, employment of a full-time staff is difficult, and editorial and publication work are usually in the hands of scientists themselves. Scientists are usually not an expert of editing and publishing, and this work detracts them from their research and educational activities. Usually, scientists undertaking such editorial and publishing activities serve for one or two years, so that they have no time to become acquainted with the editorial work, and they are not interested. Of course, there are scientists who are interested in such a work and try to learn the works but such scientists are very rare - at least in Japan.

The University of Tokyo Press had been undertaking publication of several journals from such small societies and, after learning the difficulties of editor scientists, decided to set up a center for scientific publications and a business center for academic societies to take the work load off the shoulders of scientists, and also to improve the quality of scientific journals, especially when published in English. This plan was carried out, and the two centers were established in 1971.

As of October 1976, the Center for Academic Publication, Japan, has a staff of 11 and undertakes publication of 280 journal titles from 50 scientific societies. The Business Center for Academic Societies has a staff of 22, and a large computer, and undertakes the business of 65 societies, with a

Miriam Balaban (ed.), Scientific Information Transfer: The Editor's Role, 441-443.

total membership of around 65,000.

Center for Academic Publications, Japan

The author sends his manuscript to the Society addressed to the Publication Center, the Center confers with one of the members of the Editorial Board of that Society, and sends this manuscript (and a copy) to two Referees designated by the Editorial Board. The Referees send back the manuscript to the Editorial Board, and the Board decides on its acceptance or rejection. The Editorial Board holds a meeting at appropriate intervals and decides on the papers to be accepted and published in the next issue. The accepted manuscripts are placed in the hands of the Center and its staff does all the work necessary for publication; copyreading and checking for style, sending out graphs and illustrations for tracings and block making, and finally marking instructions to the printer. In the case of English language journals, some manuscripts are sent out for checking or correcting English. The first proof is sent to the author but also proof-read by the Center staff. After 2 or 3 proofs, the journal is printed and distributed. Mailing of journals is undertaken by the Business Center. Orders for reprints of each paper are filled by the Center and reprints are sent out with a bill.

When a new journal is to be published, or when a new society is established and wishes to publish a new journal, the Center suggests the format and style, cover design, etc., and editorial board of the society decides on these points. In deciding the format and style, the Japanese scientists generally follow the style of journals published by the societies of the same scientific field in Europe and U.S.A. Therefore, if such European and American journals, and also international journals, do not use the standard units and symbols, the Japanese scientists refuse to use them, even if the Center indicates that standard units and symbols should be used.

As in Europe, the scientific publications in Japan are complicated by the publication in two languages; one in Japanese and one in English. The difference between Europe and Japan is that the two languages, the Japanese and English, are so vastly different. In Europe, at least both the native and English languages use the Roman letters. In addition, scientific writing in Japanese invariably contains Roman letters; for example, atomic symbols, names of chemical compounds, symbols for quantities and units, numerical formulae, abbreviations in Roman letters, etc. Therefore, the editorial staff of the Center must be familiar with the editing in both languages, must be able to proof-read in both languages, even when scientific journals are published in Japanese.

The Center therefore holds seminars and lecture meetings for their staff, as well as for editorial staff from other scientific societies, in addition to the usual in-house training.

Several good results have emerged from the publication of scientific journals from the Center. It has made it possible to carry out cooperative editing - a concerted editing of several journals in different fields. To do this, however, format and style, symbols for quantities and units, abbreviations, etc., must be unified. At the time of the start of the Center, there were some journals which used periods after each unit symbol, and some that did not. There were some journals which used plurals, and some that did not. Symbols of physical and chemical quantities varied from one journal to another The use of italics and capital letters and punctuation marks also differed among several journals. This meant that the editorial staff must have a complete list of these unit abbreviations, symbols, etc., for each journal and have that list to refer to in editing the manuscript and in proof reading. This proved too much for the staff, and the Center called a meeting of the editorial board of four large societies to discuss possible unification; the Pharmaceutical Society of Japan, Japanese Biochemical Society, Agricultural Chemical Society of Japan, and Institute of Applied Microbiology. After four meetings, some agreement on unification was made and each society incorporated these unified rules into their "Instructions to Authors." This unified rule has been and is being slowly adopted by the journals of smaller societies, the majority of which do not have "Instructions to Authors." Drawing up of such "Instructions to Authors" has been another task of the publication center for societies which never had one before. The one thing that could not be unified was the system of bibliographic references; each society, or each field of science, has its own system and none of the societies wanted to change. Therefore, this remains varied. This is true for all the journals in the world. Take, for example, the American Chemical Society, which publishes about 20 journals. There are several systems of giving references in the text and in writing the bibliographic references. If one were to translate Japanese into English or type a manuscript for several of the American Chemical Society journals, one will have to refer to each journal for the correct way of giving references in the text and in writing the bibliographic references.

Although the ISO standard on SI units and recommendations for the use of their multiples and of certain other units (ISO 1000) has been adopted by many countries of the world, including Japan (JIS Z8202-1974), there are still many scientific journals in Europe and U.S.A., as well as international journals, which do not use these SI units. As stated earlier, Japanese scientists follow the style and format of journals in Europeand U.S.A. in the same field, and will not use correct international standards even when recommended by the publication center if European, American, and international journals do not use them.

Changing the format and style of a scientific journal is very difficult, even in a good direction. It is hoped that such journals would adopt the correct usage of standard symbols of quantities and units, and, even if they cannot unify the presentation of bibliographic references, at least they can use the correct international abbreviation of periodical titles (ISO 833).

Business Center for Academic Societies

This is a Center that undertakes the business side of scientific societies. Membership cards, membership fees, and mailing cards are all processed by a computer. Based on the membership cards, list of members can be prepared and printed as needed. Bills for membership fee are sent out and the fees collected by the Center, which reports the state of income to the society. Sending out of society notices, annual meetings, and other document distribution are all undertaken by the Center. The Center charges the Society about $1.00 for each member for this work.

In addition to these usual business of scientific societies, the Business Center also undertakes the planning and business of annual meetings or conventions, international meetings, lecture meetings, etc., with attendant work of sending out notices, printing of programs and abstracts, distribution of materials, and arrangement of meeting halls.

QUANTITY AND QUALITY IN ISRAELI SCIENCE

Shaul Katz
Department of Sociology and Social Anthropology,
The Hebrew University of Jerusalem, Israel

ABSTRACT

Science in Israel has two inherent difficulties - distance from centers of science and a limited dimension. Although from its inception, science has been conceived as an important component of the national upbuilding, there was a clear policy of preference for publication in international forums rather than in local ones. Thus local scientists were channeled to compete with their research products in the international market. Publication abroad and constant entrance into new fields of research were among the most important means of cultivating science at appropriate standards.

Israel has a peculiarity in its volume of scientific activity. De Solla Price says that the scientific population of Israel as measured by its publications is "more than a factor of three higher than that which one would expect by the standards even of such highly developed scientific nations as Britain, Switzerland, Hungary and Czechoslovakia".[1] However, producing a quantity of about one percent of the world's total scientific papers in natural sciences, medicine, agriculture and technology[2], does not necessarily imply quality. Cultivating high standards in the conduct of scientific inquiry is problematic because of the distance and the peripherality of the country to the Western scientific centers[3]. In the following, the questions of quality and quality control mechanisms in Israeli science will be examined mainly with the aid of journals, which have served the scientists of this country for more than half a century.

The Holy Land has been for centuries, and especially from the 19th century onward, the most, or one of the most explored countries.[4] However, as Humboldt or Darwin were not in any sense part of the early history of science in Spanish America[5], so, famous and other lesser European and American naturalists and archaeologists had no social or cultural ties with the Land of Israel which they explored. The articles stemming from their explorations were not published in local journals, and when published in journals abroad there were no signs of relationship to local interests.

The founding of a Hebrew University in the old-new Homeland was motivated by the central place of science in the Zionist plan or utopia. It was both due to its moral and intellectual promise for a substitute, or as a complement to the old religious tradition, and due to its practical and technological promise as a central tool for making the desert bloom[6]. There were great hopes and expectations that in Jerusalem there would be established an outstanding center of excellence as a modern realization of Isaiah's words: "For out of Zion shall go forth the law and the word of the Lord from Jerusalem."[7] While university institutions were established as teaching institutions outside western and central Europe in such places as Norway[8], Ireland, Asia and Africa[9] in the 19th century and earlier in the United States[10], the Hebrew

Miriam Balaban (ed.), Scientific Information Transfer: The Editor's Role, 445-455.

University was first destined to be a research university. Prof. Chaim Weizmann, later the first president of the State of Israel, was a central figure in the Zionist movement and one of the founders of the Hebrew University. Weizmann, a well-known research chemist, presented the aims of the Hebrew University in these words:

> "We have no attempt to extemporize a pretentious replica of one of the great teaching universities of the West. We have made up our minds that it is for our University to win its spurs and build up its reputation by the distinctive value of its contributions to the common stock of knowledge"[11].

The founders of the Hebrew University saw an advantage in delaying teaching until the staff could reach high academic and scientific standards[12].

Despite the great national hopes and the apparently vast resources of first-rate Jewish scientific manpower, the Hebrew University opened in 1925 with 3 very modest research institutes - Jewish Studies, Oriental Studies and Chemistry - and other few small research units, all with limited possibilities. Only very few of the eminent Jewish scientists came to reside in the Holy City and to participate actively in the Zionist revival[13]. The Hebrew language was given a special position in the new institute. As Weizmann put it: "We feel in our innermost being that if this University is to express and, by expressing us, look at the world from our (national) angle and make a distinctive contribution to civilisation, it must be through the medium of Hebrew."[14] Indeed the use of Hebrew as the "medium" for teaching which began some years later was not without difficulty for the teachers. There were more than a few anecdotes about university professors reading in Hebrew their lectures which were written in Latin characters.

As far as research in the natural and allied sciences was concerned, the reality with the Hebrew "medium" was different. The research results were not published in Hebrew nor in any local journal written in an international language of science. It was in obvious contrast to the practice which had been adopted in most of the European as well as non-European countries, where publishing in the vernacular or establishing a local *Proceedings of the National Academy of Science* in French or English, had been conceived as an important component of the national upbuilding process.[15]

The question to be asked here is how did it occur that despite the nationalistically-loaded circumstances within which the establishment of the Hebrew University had been embedded it did not imply the demand for home-based scientific literature. It seems that two elements were involved here. One was the nature of the scientific work as had been conceived by the scientists themselves; the other was the test for the "usefulness" of such a work as conceived by the scientists' immediate administrative environment and the lay public.

Those scientists who came here, came already as established professionals experienced in working in established scientific centers mainly in Central and Western Europe.[16] Since their arrival was, first of all, for the purpose of doing basic research it is reasonable to assume that they naturally did not tend to give up their being read by colleagues. Writing in Hebrew about research in natural sciences, in the 1920's, apart from heaping difficulties on the potential writers themselves,[17] meant giving up any chance for receiving competent response for one's professional work; and receiving competent response for one's professional work is a central component of the scientific enterprise.[18] The same applies, although to a lesser degree, to publishing in a local journal (see below).

The justification for conducting pure research in the natural sciences, resided with the nature of their expected "products". Chemistry and micrbiology were the first subjects in the natural sciences which had been developed in the Hebrew University. It seems that the reported results with potential application in local medicine, agriculture and industry[19] could have been considered as satisfying national expectations.

The counterexample of Jewish studies in Jerusalem may support this explanation. The Hebrew University scientists did write their scientific articles in Hebrew, and they did establish journals for Judaica and allied fields in Jerusalem.[20] But by writing Hebrew they did not lose professional competent readers. On the contrary, Jewish and non-Jewish scholars naturally could read (and write) Hebrew as Hebrew had been the substance matter of their studies. More than that, by writing in Hebrew and in local journals, which in the course of time became international by nature[21] they gained the wider Jewish audience; teachers, writers and lay scholars who certainly considered even the mere Hebrew articles, written in Hebrew and published in Jerusalem, as a "proof" for the "usefulness" of their endeavours.

Thus the handful of scientists who continued here their previous lines on research or who opened new lines, and whose main task has been the doing of research, continue here to compete with their research products in the world market of science. Although being few, isolated and remote from scientific centers, by submitting their papers to the Mathematische Zeitschrift, Kolloid Zeitschrift, Nature, or The Journal of Bacteriology, as they did before, they put themselves under the universal criteria employed by the editors and referees of the central scientific journals who had to evaluate Israeli research works.

Publishing in European or American journals - which for some measure, could be something of routine or convention to those scientists who came here "fully baked" - became the norm of those not few, new Israeli scientists, who were recruited for academic research from local autodidacts or semi-professionals, at the beginning;[22] and from local graduate students, later,[23] during the last five decades of scientific activity in this land.

Before statehood, and especially after it, there had been some attempts to found Israeli scientific journals, mainly based on contributions from Israeli scientists in agriculture and natural sciences, medicine, mathematics and other fields. Generally, these attempts to publish bilingual, - Hebrew and English, - or English journals, did not gain much success. Publishing in English rather than in Hebrew required a public apology[24].

The most conspicuous long-lived scientific journals which have been published in this country before statehood were the medical journals like Folia Medica Orientalia which were associated with the endeavours of Prof. Arieh Feigenbaum, and the Palestine Journal of Botany (and Horticultural Sciences) associated mainly with the efforts of the late Prof. H.R. Oppenheimer. Both journals were first published in the 1930's when there was an increase in scientific output of regional nature, and when local and other Jewish scientists disconnected their ties with the German scientific centers and literature. This transition from German to English as the main publishing language is very conspicuous if the publication lists of Israeli scientists are examined. This transition coincided with the general process of the decline of German as a central language of science and scholarship in the years around World War II. These journals or their "derivatives" were incorporated after statehood in the publications of the Weizmann Science Press.

After founding of the State of Israel in 1948 the Research Council (now the National Council for Research and Development, NCRD) decided that "the time had come to issue a scientific publication in Israel, which would report on main scientific results or investigations of regional import and at the same time give scientists here the opportunity to announce briefly the results of fundamental research in all fields of science, in the form of letters to the editor. The Council's decision to publish such bulletins was also due to the difficulties encountered in connection with the publications of scientific papers and journals abroad, owing to increased scientific output in all countries during the last years, and to the priority given to scientists of the country in which the Journal appears."[25]

The Bulletin of the Research Council of Israel, first published in English, and the first volume also in Hebrew, was edited by Miriam Balaban. The journal was subsequently subdivided into the Israel J. of Botany, Chemistry, etc. and published by the Weizmann Science Press, a national scientific publishing house of the NCRD and Israel's universities. A third long-lived series was the journal Ktavim for agricultural research first published both in

Hebrew and English, and after some years turned completely to English as the Israel Journal of Agricultural Research. All these had local editorial boards.

An obvious exception in the arena of scientific publishing in Israel is the Journal d'Analyse Mathématique founded as a result of the private initiative of the late Prof. Binyamin Amira in 1951.[26] This journal is considered in the mathematician circles as having first rate good standing. From its very beginning the Journal d'Analyse Mathématique recruited contributions from foreign contributors and based its activity on its international editorial council composed of well-known mathematicians from all over the world. The connection between international standing of the journal and the international editorial council seems to be conspicuous. But this subject deserves more fundamental study taking also into account the regionality versus universal characteristics of the material to be published and the fact that periodicals were assisted by foreign referees.[27] As follows from the above-mentioned description local scientists usually preferred to send their magna opera to European - and after the 1950's - especially to American journals.[28] But it must be noticed that local publication in English at least ensured that the papers could be read and easily abstracted abroad.[29]

II

It was important for local scientists to publish abroad thus submitting themselves to the foreign quality control mechanism since Israeli institutions required international recognition and approval. Eminent Jewish scientists and scholars from various countries participated actively in the scientific administration and governing bodies of the Hebrew University, established 1925, the Daniel Sieff Research Institute, established 1934 which became the Weizmann Institute of Science in the middle 1940's. Prof. Albert Einstein for example presided for a period in the Academic Council of the Hebrew University[30]. These Jewish scientists and scholars insisted, explicitly, on excellence in research while considering new appointments. In the course of time, with the consolidation of these institutes, the direct vast academic administrative powers which had rested with foreign scientists diminished. But this mode of centrally organized international orientation was not affected. Some of the functions of the Senate's foreign members were substituted by the practice of consulting foreign referees for assessment of candidates for academic posts.[31] This does not mean that this policy has always succeeded. Indeed, from timt to time claims have arisen that the Hebrew University, for example, had not fulfilled expectations of being a center of excellence. Sometimes the claims were more bitter[32]. It is only to suggest that the pressure to publish internationally, at least prevented the local isolated scientists, from adopting local standards, that would probably have been mediocre or worse.

The pressure, or better the incentives, to publish abroad did not diminish with time. On the contrary: the trend of emphasizing scientific peer-oriented research, strengthened, especially after 1948 statehood.[33] It occurred not only in the old veteran Technion, in the new teaching universities[34] which had been established in the 1950's and 1960's, but also in non-university research organizations. A process of academization has passed upon the whole system, but its marks were apparent especially in the applied research sector. In medicine, agriculture and technology there were obvious signs of emphasis on publication abroad instead of publishing in local media.[35]

In the 1950's and especially in the 1960's some foreign research-financing agencies began to be active in Israel. In certain years these agencies, private and governmental, financed up to a quarter of Israeli scientific activity.[36] Both pure and applied local scientists benefited from the foreign control by referee panels who considered the requests for grants. Such panels, like editorial boards, were active in assessing the "end product" of the research projects observed. But their uniqueness rests with the introduction of foreign quality control by selection and guidance at the launching point of the research process, concentrating on its intellectual and material inputs, rather than the evaluation of the end products as post factum, namely the research paper or book.

There are good indications for the positive role of such fund evaluation procedures in raising the quality of local scientific work.[37] The introduction of the relatively massive foreign financing sources affected especially applied science in Israel where in many quarters there were no rigorous formal or non-formal institutional quality control mechanisms, up to that time, if at all.

III

Nevertheless, publishing abroad, and even in respectable journals, does not necessarily imply evidence for another trait, usually required from national systems of science: the permanent renovation, or, in other words: being at the frontiers of science.

Examination of the journals in which Israeli scientists publish their articles reveals that some 9 percent (in 1955) and 16 percent (in 1965 and in 1975), of the Israeli scientific output, had been published in new journals.[38] Apparently, sometimes new journals provide an outlet for the growing pressures on old journals and sometimes the counter-phenomenon takes place: veteran journals revolutionize their editorial boards and content, without changing name and cover. However, within this important reservation, at least part of the new journals do reflect new gatherings in new frontier fields and frontier specialties. Consequently, the systematic sharing in new journals indicates entrance into new areas of scientific investigation. Therefore, the participation in new journals is interpreted as a good indication that a reasonable part of the growth of science here comes from permanent renewal.[39]

There is a difference between a growth of scientific fields in a country due to scientists who are to a lesser or greater extent among pioneers and founders of new specialties or due to bandwagon phenomena. It is suggested that at least a fair portion of the growth of Israeli science is due to pioneering work.

A good support for this suggestion emerges from the examination of the share of Israeli scientists in the world "population" of publications of the sort of Annual Review of or Recent Advances in (Table I). A contribution to such collections is usually the result of an invitation from an international editorial board. Hence, such an invitation may be regarded as an indicator to international recognition in the standing of the would-be contributor in his specialty. The quota of Israeli authors among the population of the Review authors, is suggested here as an indicator for the excellence of some Israeli scientists in their corresponding new specialties. It is to be stressed that these authors are distributed more or less equally among the various scientific disciplines.

TABLE 1. Share of Various Countries in Authorship of about 4,000 Scientific Papers of the *Annual Review* Type (in percentage)

United States	50.0
Great Britain	13.5
France	2.5
Sweden	1.8
Switzerland	1.8
Holland	1.2
Japan	1.0
Belgium	0.9
Denmark	0.6
Italy	0.5
ISRAEL	0.5
All other countries	26.0

Source: S. Katz: Science in Israel - Figures and Evaluation (in Hebrew), National Defence College, (Jerusalem 1965).

To sum up: Israeli science - although placed in an area, peripheral to the principal scientific centers - was successful in establishing and perpetuating fair standards in the content, and in the conduct of scientific inquiry. This was accomplished, first due to competition for a place in international journals of science and for grants, and secondly, due to renovation as part of the permanent growth.

The irony is that, at least from the point of view of the original founders of Hebrew science, this was achieved, at least partly, at the price of cultural, societal and economic losses: cultural - because of giving up of Hebrew as the language of publication as a central component of the new or renewed national culture;[40] societal - because a central sector of the Israeli society has "built in" orientation to peers abroad, a phenomenon which stands in some contradiction with the perception of the nation-upbuilding interests; and economic - because of difficulties to recruit scientists to engage in research whose output is a product of process rather than a paper, or to turn to local practical research problems whose potential output is "only" papers in a local journal.[41]

The accentuated international orientation has also its consequence within the Israeli scientific system itself. This is the looseness or practically the absence of local scientific community, if the test for the existence of such a community is the existence of some arrangement for "inside" self governing and "outside" representing the overview upon all or most aspects of the local scientific enterprise and its practitioners.

There are some central bodies, like the Israel Academy of Sciences, The National Council for Research and Development, or the Committee of the Presidents of the Universities. All of them have some responsibilities on some sector of the national scientific system. They draw their authority principally due to their financial and administrative powers. But there is no organizational framework which draws its authority from the voluntary gathering of the practicing scientists and represent the overall scientific intellectual and social interests inside and outside the scientific system.[42] It is suggested that this situation is wholly or partially a direct result of the limited interest which Israeli scientists have in their immediate local colleagues and local scientific societies. With the emphasized orientation on the world, especially American, scientific centers of excellence, and in limited internal control roles, there is only very weak standing, rewards and prospects for investing time and effort in representing the scientific interest on the national level.

It seems that in days of zero growth of scientific activity in this country, and in changing outside and inside environments, the absence of such representation imposes difficulties on properly dealing with Israel's 1% portion in the world production of science. The main problem of Israeli science is still how not to become a head to foxes of the periphery, but remain a tail to the lions of the center.

NOTES AND REFERENCES

1. Derek J. de Solla Price, "Measuring the Size of Science", The Israel Academy of Sciences and Humanities Proceedings, Jerusalem, 4(6)(1970)98-111.

2. According to de Solla Price, ibid; Shaul Katz, The Growth of Science in Israel, National Council for Research and Development, Jerusalem, 1970 (translation from Mada, 13(1968)82-87; Science and Technology in Israel 1975/76 (in Hebrew), Jerusalem, NCRD 10-77 (1977) p.26.

3. About centrality and peripherality in science and about the special problem Israel faces in connection with this problem, see: J. Ben-David, "Scientific Endeavor in Israel and the United States", The Amer. Behav. Sci. 6 (1962) 92-96.

4. Shaul Katz, "Aharon Aharonsohn, Reishith Ha'Mada Ve'Reishith ha'Mehkar Ha' Haklai Be'Eretz Israel" (Aharon Aharonson, The Beginning of Science and of Agricultural Research in Land of Israel") Kathedra, 3(1977) 3-29.

5. Marcel Roche, "Early History of Science in Spanish America", Science, 194 (1976)806-810.

6. Official history of the Hebrew University and the development of the University idea in shorter or longer versions, may be found in the yearbooks of the Hebrew University. An early version of such presentation is The Hebrew University Jerusalem 1925-1926 (Jerusalem: Workmens Printing Press, April 1926), 48 pp; The Hebrew University of Jerusalem 1925-1950, semi-jubilee volume. (Jerusalem; Goldberg's Press, 1950), See also Norman Bentwich (ed.), Hebrew University Garland, A Silver Jubilee Symposium (London: Constellation Books, 1952); Norman Bentwich The Hebrew University of Jerusalem 1918-1960 (London: Weidenfeld and Nicholson, 1961).

7. See ibid no. 6, and especially; Martin Buber, Berthold Feiwel, Chaim Weizmann, Ein Judische Hochschule, (Berlin: Judischer Verlag, 1902) reprinted by the Hebrew University of Jerusalem, (Magnes Press, 1968); Yom Hagenu (Our Celebration Day). The details of the cornerstone ceremony of the Hebrew University in 1918 (Jerusalem: Azriel Press, the year of 5818 (1918)), 28 pp., Inauguration of the Hebrew University of Jerusalem (text in Hebrew and English, Jerusalem: 1925), 115 + 11 pp.

8. Hans Skoie, "The Problems of a Small Scientific Community: The Norwegian Case", Minerva 7(1900) 399-425.

9. Eric Ashby, Universities: British, Indian, African (Cambridge: Harvard University Press, 1966).

10. Lawrence R. Veysey, The Emergence of the American University, (Chicago: University of Chicago Press, 1965).

11. Chaim Weizmann, in Inauguration of the Hebrew University of Jerusalem, pp. 23-26.; For Weizmann's views about the aims of the Hebrew University see Chaim Weizmann, Trial and Error, (New York: Harper and Brothers, 1949), Chapter 28, pp. 215-329 and other places; see also Hugo Bergmann, "Dr. Weizmann's conception of the Hebrew University", Chaim Weizmann (ed: Paul Goldman) (London: Victor Gollancz, 1945) pp. 93-99. Reuven Cahana in his paper "Emdoth Ha'Idealogia Ha'Dominantith Klapei Mada, Madaanim Ve'Professionalim Bi'Thkufath Ha'Yishuv" (Attitudes of the Dominant Ideology to Science, Scientists and Professionals in the Yishuv Period") evaluates the various zionist ideas about the Hebrew University and other

research and institutes of higher education; Prof. J. Ben-David claims that science in Western Europe was considered largely as "cultural consumption". J. Ben-David, Fundamental Research and the Universities: Some Comments on International Differences, (Paris: OECD, 1967) p. 46.

12. The Hebrew University of Jerusalem 1925-1926,(April 1926) p. 7.

13. ibid., p.12.

14. Chaim Weizmann, ibid., p. 11.

15. In his review of the rise of the scientific journal, Meadows says that "...right from the beginning journals tended to publish in the vernacular of the country of origin...because of the increasing emphasis on the vernacular scientists became afflicted by the Tower of Babel problem that still exists today. The difficulties were felt particularly by scientists in countries that were minor producers of science, for, if they wrote in little-known languages, their contributions were almost certain to be overlooked...One way round this, was to learn the language of a major-producing country and to write all articles in that language. Thus Russian scientists most frequently wrote in French and the practice was continued for some time, even after the October Revolution",(A.J. Meadows, Communication in Science, (London: Butterworths, 1974), pp. 69-70). Basalla, while discussing the transition from what he calls "Colonial Science" to "independent scientific tradition" says that: "Should national pride dictate that the contributions to the new journal be printed in the mother tongue when that language is not familiar to Western Europeans, or should some concession be made in order to gain European readers? This was the question faced by the founders of scientific periodicals in Japan, in China and (in the case of Rumania) in Central Europe." George Basalla, "The spread of Western Science", Science, 156 (5 May 1967) 611-622.

16. The first six professors were S. Adler formerly of the University of Leeds and Liverpool and later awarded F.R.S.; M. Fekete formerly private-dozent in the University of Budapest; A. Fodor formerly Ausserordentlicher Professor in the University of Halle; A. Fraenkel formerly Ordentlicher Professor in the University of Kiel; I.J. Kligler formerly with the Rockefeller Institute of New York; O. Warburg Professor at the University of Berlin. There were in 1930, 35 senior and junior scientists in the natural sciences and mathematics. On the whole there were at the Hebrew University 10 professors and associate professors and 50 other junior scientists. According to information in the Hebrew University Jerusalem 1929/30 (Jerusalem: September 1930).

17. See for example: F.S. Bodenheimer, A Biologist in Israel (Jerusalem: Biological Studies Publishers, 1959) pp. 42-43.

18. M.J. Mulkay, "Sociology of the Scientific Research Community" in Ina Spiegel-Rösing and Derek de Solla Price (eds.), Science, Technology and Society (London and Beverly Hills: Sage Publications, 1977) pp. 93-148.

19. The Hebrew University of Jerusalem 1926-27, ibid., pp. 7-9, and in subsequent reports.

20. In the 1920's the periodicals that have been published by the Hebrew University, were Madaei Ha'Yahaduth, Kiryat Sepher and Tarbiz Zion, and Leshonenu which began to be published some years later. All of them are still being published and are central organs in Jewish and allied studi

21. "One of the great merits of the Institute of Jewish Studies in the Hebrew

University is that it has transferred the centre of research in Jewish studies to Jerusalem and that it has made Hebrew the medium for it. Works in this feild had for the main part been written in foreign languages, principally German...periodicals, dedicated to Jewish Studies, were found in German, French, etc. and many of them enjoyed a lengthy existence... Those few which were written in Hebrew, mainly by scholars from Eastern Europe, did not receive that measure of appreciation which they deserved. Since the founding of our institute a great improvement has taken place in this respect. It is impossible today to ignore what is being produced in this language, and the scholars who write in it can now feel assured that the number of his readers will not be smaller than if he had written in a foreign language." S. Assaf, "Jewish Studies", *The Hebrew University of Jerusalem, 1925-50,* ibid.

22. See publication lists of for example, A Eig, M. Ashner, Z. Avidor (Klein) and O. Theodor. They all were technical assistants at the Hebrew University and later became professors of this institute and others (Technion).

23. The most known (famous) among the first graduates of the Hebrew University are the late Prof. Aharon Katchalsky-Katzir and Prof. Ephraim Katchalski-Katzir.

24. See for example: Aryeh Feigenbaum, "Ventures into Medical Journalism in Israel", *Isr. J. Med. Sci.*, 1(5)(1965) 1056-1064.

25. S. Sambursky, "Science in Israel and the National Research Council", in *Hebrew University Garland*, ibid., pp. 111-117.

26. S. Agmon et al., In Memoriam of Benyamin Amira, *Journal d'Analyse Math.*, 22(1970) xii-xvi.

27. Personal knowledge about the practice accepted in the various editorial secretariates of the Weizmann Science Press.

28. Compare: A.A. Manten "political Events Influenced Volume and Languages of Scholarly Publications", *J. Technical Writing and Communication*, 5(2)(1975) 143-153.

29. Three out of four scientific journals published in Iran in the late 1940's were written in the vernacular. The same applies to 5 out of 8 in Turkey and 8 out of 22 in Egypt, in the same period. All 7 scientific and professional periodicals of Israel were written in English. See: UNESCO: *List of Scientific Papers Published in the Middle East*, no. 2, (Cairo: Science Cooperation office in the Middle East, 1948). As a result of statistically analysing the reference lists of scientific publications in some research field and of direct inquiring among the authors involved, Manten comes to the conclusion "that authors who want their message to reach the world-wide professional community can only hope to be successful if they publish in English" A.A. Manten, "Language Barriers in Livestock-Husbandry Research", *Livestock Production Science*, 1 (1974) 421-430).

30. Among the members of the Academic Council of the Hebrew University in 1927 there were Dr. C. Adler, Philadelphia; Prof. S. Alexander, Manchester; Prof. Besredka, Paris; Prof. S. Brodetsky, Leeds; Dr. M. Buber, Heppenheim; Prof. R. Ehrmann, Berlin; Prof. A. Einstein, Berlin; Prof. Eisenmann, Paris; Prof. S. Freud, Vienna; Prof. J. Hadamard, Paris; Prof. E. Landau, Goetttingen, Prof. L.S. Ornstein, Utrecht; Prof. S. Weil, Berlin. "The Professors of the University [in Jerusalem] formed a section in the Academic Council to be consulted before a decision is taken" (*The Hebrew University Jerusalem 1927*, ibid., p. 18).

31. Selig Brodetsky, "Development of the Hebrew University", Hebrew University Garland, ibid., 118-125.

32. Brodetsky, ibid. and Leon Roth both for a period headed the University (Leon Roth, "Philosophy of the University and the Jewish Mind", ibid., pp. 65-72) criticized the situation in the Hebrew University in the early 1950's Prof. Chaim Weizmann him self had criticized the situation in this university in the late 1920's and early 1930's. See: various places in Sylvia Flowers, The History and Development of the Weizmann Institute of Science Thesis submitted to the University of London, January 1975.

33. The Hebrew Technion, (now Technion - Israel Institute of Technology) was established in 1925 as a technical college. See: Arthur Blok, "Early Days at the Technion", Technion Yearbook, 1952-53, pp. 110-113, 198-199, and also, see: Tholdoth HaTechnion. Be'Reishitho 1908-1925 (The History of the Technion in its Beginning 1908-1925) (Haifa: The Technion, 1953). The research in this institute has developed gradually mainly according to the personal inclinations of some of the teachers. In the 1940's there was created in the United States the American Technion Society with some of the most prominent American scientists and technologists. With the aid of the latter the Technion began to emphasize research. See: Rosa and Sydney Goldstein, "Technion Today and Tomorrow", Technion Yearbook, 1951, pp. 121-125, 230-232.

34. For the development of the various scientific institutes in Israel see: Shaul Katz, (ref. 2); Bar Ilan University was established in the early 1950's as a teaching (religious) university. See: Arieh Ben-Joseph, "Al Reshithah shell Universita Bar-Ilan" ("About the Beginning of Bar-Ilan University") in Bar-Ilan Universitah Be-Hibanuthah (Bar-Ilan University Upbuilding 1953-1967) (Ramat-Gan: University Bar-Ilan, 1969) and research activity developed there gradually. The same applies to the University of Tel Aviv. See for example the emphasis on teaching in the address of the rector of this university (Yedioth Universitath Tel-Aviv, July 1963, pp. 4-5,)or the address of the newly appointed president of the university (ibid., December 1963, pp. 2-3, 8-11).

35. For the change in medicine see: Lydia Aran and Joseph Ben-David, "Socialization and Career Patterns as Determinants of Productivity of Medical Researchers", Jour. of Health and Soc. Behavior, 10, 1, (March 1968), pp. 1-15; For agriculture see: Shaul Katz and Joseph Ben-David, "Scientific Research and Agricultural Innovations in Israel", Minerva, 13(2) (summer 1975) 152-182. As pertaining to technology at least half of the Technion publications in the period of 1948-1954 were published in local periodicals (see Catalogue of Scientific Papers by Israeli Scientists 1948-1954 (London:Embassy of Israel, 1954) as against less than 25 percent in 1968/1970. (See lists of publications in the Technion's Research Report 1973 (Haifa: Technion, 1970) .

36. Moshe Eshel, Mimun Mehkar Be'Israel MiMkoroth Memsheleth Arzoth-Ha-Brith (The Financing of Research in Israel by US Sources) (Jerusalem: National Council for Research and Development, December 1968).

37. While only 25% of the submitted research projects awarded by local and foreign judges the mark "80 and more", in 1973, some 65% of them gained this mark in 1975 assessment. United States-Israel Binational Science Foundation Synopsis 1974-76 (Jerusalem: 1977) Table 2 and pp. 28-29.

38. The Israeli scientists' rate of participation in new journals was calculated by counting the number of papers which appeared in volumes 1 to 10 proportionate to other volumes in a certain period. The publications of the

biological and chemical departments of the Weizmann Institute of Science were counted in this case. They produce some quarter of the total production of Israeli scientific publications in the natural and allied sciences. The publications were drawn from the lists of publication in WIS Annual of Scientific Activities in the respective years.

39. The new periodicals in which Israeli scientists from Weizmann Institute of Science published their articles were as follows:
in 1955: *Acta Cryst., Biochemistry, Biopolymers, Immunochemistry, Inorg. Chem., J. Rep. Fertil., Lab. Comp., Radiol. Med. Nucl., Transfusion*
in 1975: *Acc. Chem. Res., Biochem Behav., Biomed Mass Spec., Bioelectrochem. Bioeng., Biophys. Chem., Brit. Polym. Chem. Struct. Comm., Cell, Chem. Inst., Differentiation, Europ. J. Immun., Exp. Hematol., Instr. Radiat. Immunogenetics, J. Phys. Chem., J. Neurosci. Res., J. Phys. Oceanogr., J. Suprammol. Struct., J. Ster. of Mech. Ageing Develop., Microb. Ecol., J. Mol. Evol., Mol. Cell Endoc., Mol. Photochem., Nucl. Acid, Origin Life, Org. Mass. Spectrum., Phycology Plant Sci. Lett., Prostaglandins, Spectros. Lett., Throm. Res., Transplant Proc.*

40. See for example claims about damage which the making of English as the local language of science brings about on the cultivation of national culture. (N. Ben-Moshe, "Hithcahasuth la'Ivrith Bi'Mdinath Israel", (The Denial of Hebrew in the State of Israel"), Moznaiam, 38, 3 (1974), pp. 147-152.

41. This phenomenon is being discussed for a long period in local newspapers and public forums. It has been examined in the field of agricultural research. See: Shaul Katz and Joseph Ben-David, ibid. ref. 35.

42. The Israeli "Association for the Advancement of Science" may be taken as an example for the claims here presented. This association was established in 1954 by Prof. E.D. Bergmann, with the explicit aims of representing "the voice of science" in political decision-making "outside", as well as facilitating scientific cooperation and advancing scientific excellence "inside". (See: E.D. Bergmann "Ha'Igud Le'Kidum Ha'Mada Be'Israel" (The Association for the Advancement of Science in Israel", Technica Ve Mada (Technics and Science), 17, 5-6 (153-154) (February 1954), 1-2.) Practically this is a non-prestigious association without any influence "outside" or "inside", serving mainly in its Biannual meeting in disseminating popular science. The case of the magazine Mada (Science) is similar. It was established by Miriam Balaban and Prof. Ephraim Katzir and others in 1956. This Israeli version of Scientific American intended among other things to crystallize the local scientific community (see: the Introduction to the first volume, Mada, vol. 1, no. 1, 1956). However after success during the first years to attract articles from the country's foremost scientists, in later years, the journal enjoyed the participation only to a limited degree.

EDITEAST (Association of Editors in Science in Southeast Asia, Australia and Oceania) AND ITS ROLE IN SCIENTIFIC PUBLISHING IN THE REGION

Amando M. Dalisay, *Treasurer, EDITEAST*
National Research Council, PTRI Building, Bicutan, Tagig, Rizal, Philippines

I wish to present briefly an overview of Editeast and its potential role within the region. Potential because Editeast has been organized only recently and its activities have barely begun.

Editeast is the result of a pressing and continuing interest among science communicators in the region to develop a machinery for communication among themselves and as much as possible to link up with science communication systems in other regions. The proliferation of primary scientific journals in several countries of the region, notably Malaysia, Australia and the Philippines, has provided strong motivation for organization in order to coordinate their scientific publishing activities and contribute in some way to the flow of essential information among scientists and researchers in the various regions.

Unlike the earlier existing associations of this kind, which deal primarily with particular fields in pure science, such as Editerra and AESE for the earth sciences, ELSE and CBE for the life sciences, and the strong abstracting services in North America, the new regional organization is aimed at covering all the branches of science. This is a deliberate attempt to conform to the structure of scientific activities within the region and to provide a broad basis for membership in the organization.

Founded at a meeting of scientific editors in Bandung, Indonesia in November 1975, it has three main purposes: (1) to promote improved communication in science by cooperation among editors of serial publications, including primary journals, abstracting and indexing services, and similar media, (2) to help improve the education and training of scientific editors in the English language within the region, and (3) to encourage cooperation in improving the standard of scientific publication among the countries of the region.

The principal sponsors of our regional organization are UNESCO, operating through the regional office in Jakarta, and the IUGS in cooperation with IGU and IUGG. The godparents at its inauguration are Prof. Martinsson of the ICSU Abstracting Board and Mrs. Jager of the International Union of Geological Sciences who are both present at this conference. The absent guests in the initial ceremony were the different country national research councils and scientific communities, particularly the various national geological survey agencies who made possible the presence of their representatives at the Bandung meeting.

From the beginning, the regional organization faced several difficulties. Among these is the bilingual nature of scientific communication in the different countries of the region, although English is the official language.

Miriam Balaban (ed.), Scientific Information Transfer: The Editor's Role, 457-458. *All Rights Reserved.*

But each scientific editor must face this problem squarely in his country's publications and overcome this in meeting high-quality standards for scientific publishing all over the world. Secondly, the matter of distance between the countries of the region and the inadequate communication systems among them have served to compound the problem of establishing a satisfactory communication system primarily in the interest of science. Thirdly, the part-time or secondary nature of scientific editing as practiced in the various countries of the region poses hazards (of the scientific profession) as well as danger of lack of continuity and proper maintenance of editing of existing scientific publications in the region. This is a real challenge to the improvement of existing standards of publishing as well as a pressing need for the appointment of full-time, fully qualified editors for the primary scientific journals, which are published largely by main scientific bodies or private scientific organizations of a national character.

In spite of the above - and of the urgency to overcome these difficulties, Editeast must develop its short as well as long-run program of activities, the main outlines of which were laid down at the Bandung meeting. Unfortunately, our first president, Mr. Townley of Canberra, Australia passed away early last year and the governance of Editeast affairs was left to three other officers who are representatives of Malaysia, Indonesia and the Philippines in the Council of seven members.

The following is a glimpse of our future activities in the region:

1) The organization of national scientific associations of scientific editors or similar organizations in the different countries of the region or a primary activity, not only because of the need for the construction and support of improved scientific publications, but also because of the importance of a national communication and retrieval system. The librarians and archivists, particularly the special libraries catering to the needs of the various scientific disciplines, are already organized, but the scientific editors and their librarian associates will have to form a more close-knit association among themselves.

2) The essential coordination and collaboration work of national scientific bodies, including research councils and/or academies of science with the various scientific communication systems in other regions will help bolster the development of a regional communication system which will in turn result in more effective and economical means of publishing and exchanging scientific information among themselves. Editeastwill help promote and encourage this approach.

3) The training of young scientific editors or promising candidates for full-time editors of existing primary journals is an essential activity. To bring this about, Editeast needs the cooperation and support of ICSU and UNESCO as well as the existing regional organizations.

SCIENCE PUBLICATIONS IN HONG KONG

S.T. Chang
The Faculty of Science, The Chinese University of Hong Kong, Shatin, N.T., Hong Kong

PRINTING INDUSTRY

According to the special supplement published by one of Hong Kong's leading English newspaper "South China Morning Post" in April 1, 1977, Hong Kong has quietly become an important printing centre for the world in recent years. She now exports the products of her printing presses as far afield as Australia, Britain and the United States. Export earnings from the printing industry topped $200 million in the year 1976, it represents a healthy growth of approximately $100 million at 1971 rates.

Hong Kong has some 1,800 printing firms, employing more than 22,000 workers.

PUBLISHERS AND PUBLISHING ASSOCIATIONS

Hong Kong has approximately registered over one hundred commercial publishers engaging in publishing work. Less than half are reckoned as bona-fide publishers as many of them publish only one or two books or magazines then vanish totally out of the scene. New names appear now and then, but they are difficult to keep track of.

Two educational publishing associations are being officially registered in Hong Kong. Both the Anglo-Chinese Textbook Publishers' Organization and the Hong Kong Educational Publishers' Association vest their interests primarily in textbook publishing. The former specialized in English textbook publications and their adaptations for local use. Its membership comprises some major publishers from the U.K. and Commonwealth countries such as Oxford University Press, Longmans Green, Heinemann, Books for Asia Limited and others. On the other hand, the latter is organised by local Chinese publishers with Ling Kee as a leading member who advocates Chinese textbook publications. Both associations maintain a close relationship with Education Department of Government. There is a Tebtbooks Committee in Education Department. Its main function is to give schools more positive guidance in selection and use of textbooks which are defined as those books that are to be considered for inclusion on textbook lists of Kindergarten, Primary and Secondary Schools (excluding From 6). Curriculum packages and other instructional materials for use by student are also defined as textbooks.

In addition to the above mentioned bona-fide commercial, other English-language publishing in Hong Kong is carried on in the following ways:

(1) By newspaper publishers. In Hong Kong, there are three independent English-language newspapers. Each newspaper publisher often issues local guide books, almanacs or studies on local history.
(2) By Government Printers and by literature bureaux.
(3) By Universities and University College, Institutes and Centres of Studies.

Miriam Balaban (ed.), Scientific Information Transfer: The Editor's Role, 459-462. *All Rights Reserved.*

(4) By local history societies and research institutions.

SCIENCE PUBLISHING IN GENERAL

It is interesting to know that while printers in Hong Kong offer top quality work at competitive prices and enjoy prosperity in commercial and textbook printing, the output of Hong Kong's science publications in general shares only a meager fraction of its total production.

The nature of this phenomenon may be attributed by:

(1) that Hong Kong has been developed in its main capacity as a trading port. Trade is the primary interest and concern of its people and Government alike. At the governmental level, over 3,000 titles have been published by Government Printers throughout the years. Only .02% of the items were either directly or indirectly related to applied science fields under Meteorology, Agriculture and Fishery. Other publications were mostly concentrated in legal matters and business.

(2) that science publications, like other importations, are freely accessible in Hong Kong. These include materials published in English from all over the world and books in Chinese from the Chinese mainland and Taiwan alike.

Therefore the need for such a demand as to publish Hong Kong's own English version is small indeed. For publication of science materials in Chinese language in Hong Kong it faces practically the same dilemma.

SCIENCE PUBLISHING AND RESEARCH BY INSTITUTIONS OF HIGHER LEARNING

There are two Universities in Hong Kong. The Hong Kong University and The Chinese University of Hong Kong. The Hong Kong University was founded in 1911. A list of about 120 titles which the Hong Kong University Press has in print represents a major part of its scholarly publishing over the last two decades. Science, Engineering and Mathematics occupy approximately 10% of its total production.

The Chinese University of Hong Kong was established in 1963. A University Publications Office with a small staff was set up in 1968 in centralizing publishing activities of the University with sufficient flexibility to respond to growth and change. Now after nine years of operation, the office seemed to have attained a certain degree of maturity with scholarly publishing and the necessary experience for further implementation and growth. The University is therefore considering the praticability of establishing a university press on the foundation of the Publications Office.

Since 1968 the annual number of titles published by the Office has grown from 20 to 88, including research papers, journals and texts in the field of Biology, Chinese Studies, Computer Science, Economics, Education, English Literature, Geography, History, Law, Linguistics, Mass Communications and Physics. Due to its diversity in publication, science shares only about 10% of its total production.

General to the philosophy of both Universities which are committed to the integration of Western and Chinese Cultures, naturally they make the promotion of Chinese studies, Hong Kong studies, Social Sciences and Humanities their primary and continuing concern. But both Universities in Hong Kong recently seem to have kept in sight of their greater effort to advanced research in science. A good example of this is that, at the Chinese University of Hong Kong which I am associated with, the science research has been greatly developed and vigorously pursued since the Faculty of Science was housed under one roof, Science Centre, in 1972. Cooperative research and concentration on a few selected areas of each discipline have been emphasized and encouraged. However, each member of the staff has the freedom to continue his research according to his interest and background. Accordingly, numerous projects have been undertaken and a great number of them have been completed. Results are reflected not only in the large number of papers appearing in international journals (ca 405), but also in the ten books published during 1972-77 by local and overseas publishers (5 by Chinese University of Hong Kong Press; 1 by Sino-American Publishing Co. H.K.; 1 by World Book Co. H.K.; 1 by Clarendon Press, Oxford; 1 by Springer-Verlag, Berlin; and 1 by Academic Press, New York/London).

Furthermore, some interdisciplinary research projects within the two research units, Chinese Medicinal Material and Food Protein Production from wastes, at the Chinese University of Hong Kong have also been launched. The Chinese Medicinal Material Research Unit incorporates staff members from Science and Sociology Department who have been actively engaged in research in Chinese medicine for some years already. The basic objective is to undertake laboratory investigation of certain rationally selected Chinese medicinal herbs in order to establish their true therapeutic value. The Unit on Food Protein Production from Wastes consists of the staff members from Science, Building Office and Business Management Department. There are two main streams of research in this Unit; (1) Production of food protein from sewage wastes, and (2) Production of mushroom from cotton waste and used tea leaves. Both streams of research are correlative to convert wastes through microorganismic activities into food protein.

Another example of science research and publications is demonstrated by a Machine Translation Research at the Chinese University of Hong Kong. The research work was initiated in late 1969 in studying the po-sibility of automatic translation from Chinese to English by computer techniques. The first machine translation system, CULT (Chinese University Language Translator) was designed and tested in October 1972. Since January 1975, the machine translation system has been used regularly in translating the Chinese mathematical journal Acta Mathematica Sinica published quarterly by the Academy of Sciences in Peking, China. Many wellknown international university libraties, mathematical institutes and research organizations have become primary subscribers. The extention of the translator to other scientific and technological subjects has also been studied. Two issues of Acta Physica Sinica from Peking have also been tested by machine translation and they reveal the distinct possibilities of applying the language translator to other scientific and technological fields.

FUTURE PROSPECTS

Because of Hong Kong's geographical location and its unique political and economical relations with China, Hong Kong has the provision and know-how in becoming a true bilingual publication centre for the Western World. With this hope and concept in mind, the compilation of Chinese-English scientific terms has been well underway.

As it is understood, a university teacher's role lies in both teaching and research. These two functions are hardly separable from each other because they are complementary, interacting and mutually reinforcing. Therefore, the faculty of Science will continue to expand and improve its research facilities and to stimulate and encourage individuals and groups of teachers to do research eagerly and willingly. Therefore, there is no doubt that scientific publications in Hong Kong will increase more and more.

Another point which may worthwhile be mentioned here is that local publication of science textbooks has shown an upward tendence in recent years. Because they are (1) lower in price, (2)fit well to local educational systems and (3) available for a great demand.

To say the least, the future of science publishing in Hong Kong will look even more promising as Hong Kong exchanges more and more with the international world in scientific ideas, programs and projects.

ACKNOWLEDGEMENTS

I deeply appreciate the assistance from the following people with whom I have consulted: Mr. John Taam, Principal Inspector (Science), Education Department of Hong Kong Government; Mr. Leon Comber, Chairman of Anglo-Chinese Textbook Publisher's Organization; Mr. Au Bak-ling, President of Hong Kong Educational Publishers Association; Mr. Benny Szeto, Manager of Sino-American Publishing Co.; Mr. Rupert Li, Director of Swindon Book Co.; Mr. G.W. Bonsall, Director of Hong Kong University Press and Mr. Adrian Teng, Assistant to Director of the Chinese University of Hong Kong Publications Office.

DISCUSSION on "Experiences of the Nordic Publishing Board in Science (NOP)" by Synnøve Irgens-Jensen

"Science Publications in Hong Kong" by S. T. Chang

EUGENE GARFIELD - Do you have any idea about the percentage of publications from the Nordic countries that are being published outside this group. What percentage of the total publication output of Norway and the other countries is published in the international journals?

SYNNØVE IRGENS-JENSEN - You mean how many Nordic researchers publish outside their countries? No I don't have any figures on those, but I suppose the majority publish inside.

EUGENE GARFIELD - What are the languages of these journals?

SYNNØVE IRGENS-JENSEN - Well the seventeen primary journals are in English, and only the two popular science ones are in either Danish, Norwegian or Swedish.

MICHEL GIBB - This is an excellent initiative and we are going to try and do the same in the European community.

I was quite fascinated by the remark about the translation from Chinese into English. I first have a purely technical question. Has translation been made in the other direction - from English to Chinese? I ask this in the general context of the search for a lingua-franca. Our organization is frantically trying to develop automatic translation, and I must say I was full of admiration to find that it had been possible already to translate full texts. I think it is a remarkable achievement.

SHU-TING CHANG - I understand that they translate the text only from Chinese to English. So far the center is not yet able to translate from English into Chinese. There is a quarterly mathematical journal which translates whole texts into English. And now they try to translate physics and biology too. There are about forty journals in the biological sciences in China.

SOCIAL SCIENCES

Joseph Ben-David, Chairman

FUNCTIONAL DIFFERENTIATION OF SOCIAL SCIENCE JOURNALS

Peter Lengyel, *Editor, International Social Science Journal* *
UNESCO, 7 Place de Fontenoy, 75700 Paris, France

ABSTRACT

The publishing crisis has put the future of many social science periodicals in jeopardy. Yet a large number of titles continue to appear, of which a sizable proportion adhere to a typical formula which is poorly adapted to real communication needs. Three directions for desirable innovation are suggested: authoritative, international "state-of-the-art" periodicals in major disciplines, systematic coverage of the processes (rather than display of the products) of social science and servicing with regard to sources of primary data. Through functional rationalization along such lines, a considerable potential clientèle, not well catered to under present circumstances, may be reached to support ventures which will need powerful backing and promotion.

Despite the so-called publishing crisis, a large number of social science serials continue to appear. Just how many is very hard to assess: estimates range from about 2,700 to 27,000 in the world at large, according to the criteria of inclusion adopted. The lower total represents very strict and narrow definitions of what constitutes a scientific or scholarly periodical and which disciplines or areas of interest are subsumed under the general heading of social science. As these criteria are relaxed, so one moves up the scale until a point is reached where practically all serials dealing in one way or another with socio-economic phenomena, at whatever level of seriousness, originality or reliability, are included.(1)

Given the nature of the reality with which the social sciences seek to deal, it is hardly possible to say whether any given number of journals is excessive or insufficient. The only realistic approach is a qualitative one which takes into account the shifting needs and techniques of scholarly communication, the output of primary data and of research, the languages in which serials are published, the division of functions as between different periodicals and similar matters. Systematic analysis along these lines on a world-wide scale, while probably worth undertaking in principle, would be a vast task which, to the best of my knowledge, has never been even partially attempted. All I can do here, consequently, is to offer certain observations of an impressionistic nature based on fairly wide, but obviously by no means exhaustive acquaintance with the world's scholarly social science press (that is to say, those at the lower end of the scale I referred to above) and upon some fourteen years of experience in editing one of the few systematically international and multilingual social science serials published today, Unesco's

---

* Opinions expressed here are those of the author and do not necessarily reflect those of Unesco or the editorial policy of the ISSJ.

Miriam Balaban (ed.), Scientific Information Transfer: The Editor's Role, 467-471.

quarterly International Social Science Journal, appearing in identical English, French and Spanish editions, with selections also in Arabic. Two issues of this periodical, incidentally - one in 1966 and one in 1974 - have dealt specifically with problems of communication in the social sciences in what I believe to be a pioneering fashion.(2)

What is very striking, even upon a quite superficial examination of scholarly social science periodicals brought out in the majority of countries, is that the range of different types of serials is much narrower than might be imagined. In view of the vast variety of what they seek to cover, one might expect corresponding variety of types of periodicals embracing, for example, weeklies, monthlies and quarterlies, journals of reference and those rendering more mundane professional services, different formats and presentations, including the use of illustrations and so on. But this is not the case. The profile of the typical social science periodical can be delineated somewhat along the following lines. It is a quarterly, expensively printed but conventionally laid out, identified with one of the classic disciplines in a given country (e.g., American Economic Review), devoted to a problem field (e.g., Town Planning Review), to area studies (e.g., Pacific Affairs) or to specific cross-disciplinary or interdisciplinary technical questions (e.g., Social Indicators Research), directly sponsored by an institution or body (such as a professional association) active in the line of endeavour concerned. Our typical journal contains essentially discursive articles on its subject, along with some book reviews and miscellaneous information items, only occasionally also offering other material, such as notices of employment vacancies, the regular coverage of certain meetings, lists of theses, etc. As a rule, the typical periodical appears to have a circulation of between 1,500 and 5,000 copies, of which about 70-80% within the country of origin. Its editorial arrangements include a board or other device for collective responsibility, in name at least, but the donkey work of editing often falls on one person or a small team for whom this activity may be a sideline and who frequently have but rather inadequate support facilities at their disposal. Despite the highly perishable nature of much social science material, risks of overlap and duplication and the eternal perplexities of border-lines, the typical social science periodical observes a leisurely pace and is not designed for the easy retrieval of information. It often reports findings which are three or even five years old, is not analytically indexed and keeps its distance from primary sources of data, specializing in secondary material representing a finished product, signed by one or more authors as their own research or methodological contribution. Clearly, such a typical journal best serves the ritualistic functions associated with advancement in academic careers, guaranteeing that those who publish shall not perish, and that what is published has the stamp of acceptable current quality, or else conforms to the traditional pattern of learned journals in the humanities, concerned primarily with interpretation and not so much with the cumulation of knowledge, policy or operational concerns.

One may harbour one's doubts about the value of a number of such typical journals as media of communication. Indeed, as professional norms and the external manifestations of institutional prestige change, their very raison d'être, already largely bypassed or marginalized as they are in the real process of scientific communication today, may cease altogether to exist. If careers are not, in part at least, made through publication and if institutions cease to regard periodicals as important components of their public image and relations, considerable shifts are likely to occur. I propose to devote the rest of my time to considering what seem to me to be desirable directions in the future functional differentiation of social science periodicals.

The first such direction is the emergence of authoritative, international "state-of-the-art" (i.e. review) periodicals in each of the major disciplines and other established lines of professional endeavour. In so far as the well-known disciplinary core journals do set world standards and attract contributors from many countries, these already exist to some degree, but they at present generally aspire to promote the progress of the discipline, in other words to present original contributions largely by and to an élite, rather than to cover the state of the art systematically wherever and however it may be practised. Since this élite has other ways of communicating, such as the widespread practice of circulating preprints, attendance at conferences or word of mouth, some of the functions of core journals have become largely redundant while others, which they might assume, are so far going begging. Regular reporting on what is being taught in different countries, what emphases are emerging, which research groups are most active, perhaps in several major languages, would perform an invaluable service to practitioners everywhere; their appeal would probably be rather wide inasmuch as they would become indispensable sources of general reference with a place in every self-respecting library, as well as on many a private bookshelf.

"State-of-the-art" periodicals would essentially mediate between national communities of social scientists, bridging geographical, linguistic and cultural barriers. As such they could suitably extend to, or be supplemented by, material dealing not so much with the products as with the processes of social science. The decentralization which has marked the past two decades, and the implantation of social science teaching, research and practice in settings where they did not previously flourish, has thrown a series of problems relating to performance, content, training, status and range of professional activity into sharp relief. New patterns are constantly emerging. The social sciences are, and will certainly continue to be, highly culture-bound. Their assimilation to, and the way in which they are shaped by those cultures which did not originally give rise to them but where they have now been received and will presumably follow uncharted paths needs regular monitoring. The whole complex of dynamic processes which touch upon the dialectics of science in society, of the scientist as a citizen, of the uses of knowledge, of institutional forms, political conditions, even legal arrangements, needs ample airing. Certain journals, like Minerva, have for years concentrated on such matters, but so far the bulk of the social science press, and especially periodicals appearing in the Third World, where the acculturative problems ar most acute, have tended to stick to the more traditional patterns, displaying products rather than discussing processes.

A third direction is towards servicing with regard to the sources of primary data. Certainly almost any kind of information, from archaeological evidence to the latest findings of biology and from the popular romances of the 18th century to personnel management in a fruit-canning plant, can be raw material for social scientists. It is therefore not possible to serve all interests equally well, or to serve many specializations at all. Nevertheless, every State, as well as a host of regional and international organizations and private institutions, such as banks and research bodies, today turns out a formidable array of primary data on a more or less regular basis. Studies have shown that social scientists, as a whole, are not very well informed about these sources of data and that, even for the senior discipline of economics, closely tied as it is to a constantly-evolving data base, "there are as yet no comprehensive secondary information services sufficiently frequent and up to date to satisfy economists' needs."(3) As a result, it appears that economists have recourse to printed media and informal channels (including mimeographed versions of articles prior to publication) in about equal proportions.(4) Reliable guides to primary data sources will be in increasing demand as more and more data are stored in computers and can be retrieved easily by those who know where to locate it. Such guides would be different from abstracting services, which mostly refer back to secondary sources and, in so far as they would not actually contain data but merely indicate where they are to be found, they would avoid problems of rapid obsolescence and could concentrate on reliable listings, regular updating and the reporting of new sources, technical innovations and similar matters while sticking to the traditional journal format.

The present situation of the social science press is nobody's particular doing. It has been compounded out of historical accretion, of convenient routine and patchwork novelty, of captive audiences and subsidized publishing, of vanity, adventure and parochialism. As a system of communication it is, by common consent, an exasperating nightmare, even if its very eccentricity and oldfashioned individualism occasionally prove captivating. One must hope that its fringes will continue to ferment, that experimental periodicals will spring up here and there as tokens of creativity and forums of dissent. Meanwhile, however, something should happen to the mainstream. All the three directions I have indicated as offering opportunities at this juncture - authoritative international state-of-the-art journals, systematic coverage of professional processes and the dialectics of science in society, and guides to primary socio-economic data sources - are precisely within that mainstream. They could even be considered as contributions to the very stability of the social sciences threatened, as these continually are, by centrifugal forces which are particularly strong in periods of low consensus, such as the one we are living through currently. But journals which are to take up the challenge of these opportunities cannot be launched and run in an amateur manner from shaky bases. They require solid sponsorship and elaborate contributor networks. Some, like the guides to primary socio-economic sources, should prove viable on a national scale, especially in larger countries, but others would have to seek coverage of several countries, of a region, or even of the world at large. Their optimal characteristics, including the technical and economic aspects of production, distribution and pricing, format, periodicity, multiple language editions, as well as editorial policy, the mix of content and quality control, would all need careful consideration and the examination of what alternatives may be feasible.

The question then arises: who are to be the powerful backers and promoters ? Whose interests will be sufficiently engaged to motivate such difficult departures ? Undoubtedly, the influence of scientific communities, especially the large and powerful ones, would need to be thrown decisively behind innovation. Judicious merging of existing journals and other communication devices, resulting in a better complementary fit between serials and their sponsors and altogether broader horizons could impel renewal. An even greater spur to enterprise and investment, however, might derive from a reassessment of the potential clientèle for well-conceived social science information organs. The current situation, in fact, amounts to rationing information at one level while creating massive redundancy at another. The proliferation of illogically differentiated products tends to discourage the consumer who then seeks to protect himself by skipping and sampling. Thus, many a working social scientist today is faced with needs which, he considers, only a major library can satisfy, and relies largely on library subscriptions for his journals. Others, like managers, civil servants, social workers or journalists, whose interest in the social sciences is diffuse and oblique, hardly know where to turn in order to obtain consolidated coverage which is neither too detailed nor too general, too technical or too popular, too abstract or too narrowly applied. In all probability, therefore, several hundred thousand persons in the world are not being properly catered to by the existing social science press, and would be ready either to switch their present habits or become regular readers of suitable new periodicals. Equally, there are many resources upon which a wide audience would gladly draw if only it were offered convenient access to them. All this represents a considerable pool of consumers to be tapped. Perhaps the current crisis in publishing and the rethinking which it is stimulating may prove a blessing in disguise in the medium run: necessity, once again the mother of invention, might rationalize in a domain which is far from rationally structured at present.

NOTES

(1) These estimates are based on Maurice Line and Stephen Roberts: "The size, growth and composition of social science literature". International Social Science Journal, Vol. XXVIII, No.1 (1976), p.125.

(2) Vol. XIX, No.2 (1967), "The social science press", and Vol. XXVI, No.3 (1974), "Communicating and diffusing social science".

(3) John Fletcher: "Secondary information services in economics". International Social Science Journal, Vol. XXVIII, No.3 (1976), p.569.

(4) ibid., p.364. It may equally be noted that the centralization of information on data sources within nations is very imperfect. Several surveys of primary national socio-economic data networks have been commissioned by Unesco in 1976 and 1977, and the first set will be published in the International Social Science Journal, Vol. XXIX, No.4 (December 1977).

THE NEED FOR A COUNCIL OF SOCIAL SCIENCE EDITORS

James L. McCartney, *Editor, Sociological Quarterly*
Department of Sociology, University of Missouri
Columbia, Missouri 65201 U.S.A.

ABSTRACT

Although social scientific journals have some problems in common with other journals--rising costs and the explosion of information--they have many unique problems. These include smallness, narrow bases of support, exceptionally high rejection rates, short tenure for editors, long publication lag times, and no common style manual. Substantively some of the social sciences are experiencing a fragmentation and proliferation of perspectives, both theoretical and methodological. At a time when many social scientists feel that publication outlets should be increased, most journals are barely able to maintain their present size.

The most critical problem of social sceintific journals is organizational: they are isolated publishing enterprises, and only a few are coordinatively managed. This disconnectedness is a serious handicap in that they cannot easily take advantage of economies of scale that might be offered by technical innovations in publishing.

Editors in the social sciences need a council that can facilitate information sharing, provide a context for limited cooperative ventures, and initiate a discussion among publishers about critical information needs in the social sciences.

> . . . the paper journal is becoming an evolutionary dead-end, partly because of their cost, but also because of the size of the literature. One could, perhaps, liken the present journal system to the dinosaur which also consumed vast resources to support a large, unwieldy mass (Senders et al., 1975: 52).

The editor of a journal might be likened to a conservationist who desperately is trying to preserve an endangered species. Periodicum scholasticum eruditum has shown itself highly adaptable for a couple of centuries, but now, through processes of natural selection, is being pushed toward extinction. Editors of social scientific journals feel particularly threatened, for our journals are small, widely dispersed, and especially vulnerable to the jackal-like attacks of economic inflation. I will argue in this paper that editors in the social sciences must band together and evolve a form of council, not simply to preserve the species, but to explore new strategies for vitalizing the dissemination of information in and about the social sciences.

It has been prophesied that the paper journal most likely will not survive until the end of this century, partly because of the exponential

Miriam Balaban (ed.), Scientific Information Transfer: The Editor's Role, 473-484.

increase in the volume of literature, partly because of the alarming increases in the costs of producing and distributing journals, and partly because of the diminishing support from libraries and individual subscribers. We are advised that the future holds radically different modes of communication for scientists, including the electronic journal (Senders, 1976). Despite the technical feasibility and the economic plausibility of such innovations, there are many reasons--intellectual, emotional, organizational, and demographic--to suspect that social scientists will expend much effort to preserve journals beyond their "natural life."

In fact many social scientists are trying to create new journals and expand the size of existing ones.[1] The reasons for this expansionist movement are several:[2] (1) The demand for more publication outlets is partly a consequence of the publish or perish syndrome and of the demography of occupations. Most social scientists are employed in academe. The social sciences have had the largest increase of new Ph.D.s of any science field--doubling their Ph.D. production from 1965-1970, and showing the highest projected increase (20 percent) for 1975-1980 (King, 1976:38). There is fierce competition for the few positions available, and publishing still is considered the *sine qua non* for entering and moving up in an academic career. (2) Intellectually we are witnessing a proliferation of perspectives on social life, both in terms of new approaches--theories and methods--and in terms of the formation of small groups of scholars interested in studying special facets of social life. Whether scholars are exploring a particular theory, a new method, or a special area of investigation, their impulse is the same: "We need an outlet for our ideas!"

Most likely these demographic and intellectual pressures for expansion interact, and some might argue that by cutting down on the number of Ph.D.'s in the social sciences, we can solve the problem of journals. But this is a short-sighted view. A more optimistic scenario could conceive that by preserving or extending the apparently chaotic system of journals, social scientists eventually may resolve some of their theoretical and methodological problems and differences, and perhaps provide a base of information upon which informed policy can be made. Implementing this more favorable scenario, however, will take much more planning and coordination than we now have among the hundreds of persons and associations that publish social scientific journals.

Unique Problems of Social Scientific Journals

Social scientific journals face many of the same problems as journals in other science fields, but the spiraling costs of production and distribution have exposed dramatically the unique vulnerabilities of social scientific journals. Journals in the natural sciences have a more diversified basis of funding, and have relied on page charges and advertising revenue to a much greater extent than social scientific journals (King, 1976:242. See Table 3, *infra*.). Social scientific journals rely almost exclusively on subscriptions and society dues, and must try to pass their increasing costs on to academic institutions and scholars, both of whom are ill-prepared to absorb these costs.

Since the U.S. government currently is bearing an increasing burden of the costs of publishing journals in the natural sciences (mostly through page charges), they actively have been seeking relief, primarily through technical innovation. The National Science Foundation, for example, has been studying the feasibility of Editorial Processing Centers as an intermediate solution (Rhodes and Bamford, 1976), and there is some enthusiasm for the electronic journal as the mode of the twenty-first century (Senders et al., 1975; Senders, 1976).

The search for technical solutions proceeds on the premise that all scientific journals are facing common problems. Many of us in the social sciences, however, feel that our problems are unique. Whether we are at a more primitive stage of development than journals of natural sciences, or simply different from them, our differences present serious obstacles to taking advantage of the proposed solutions. Two examples may clarify this:

a) The economy of scale afforded by Editorial Processing Centers presumes that a minimum number of small journals (more than 4) would need to jointly operate such an enterprise (Aspen Systems, n.d.:35). But social scientific journals are scattered, with only one or two coordinating structures that publish more than a few journals.[3] Working out a manageable organizational structure to coordinate several independent publishers would be no small task.

b) Although an electronic journal technically might be feasible within the decade, the state of the art of indexing in the social sciences is so primitive that much information might be irretrievable in a wholly electronic system. Concepts and terms in the social sciences are several orders of magnitude less precise than those in the natural sciences.[4] Often our terms are part of the ordinary language, with all of the attendant confusion. Considering this substantial variation in the use of concepts and terms, many social scientists feel that their best retrieval method is subscribing to, or tracking, pertinent specialized journals, which they can peruse and scan.

Although the contingencies I have identified limit our search for solutions, they may not be insuperable. Even so, we cannot expect technology to save us. After elaborating a profile of social scientific journals, in the next section, I follow with the argument that organizing a council of editors permits us to plan for, and perhaps beyond, the fate of our prophesied "dead-end."

A Profile of Social Scientific Journals

Compared with journals in the natural sciences, social scientific journals are large in number, but small operationally; they are scattered and apparently highly inefficient; in scope, they have been generalist, mostly, but now are becoming much more specialized; there is much confusion about terminology, and even on presumably simple matters, such as style, there is no common guide; financially they are precarious, with little flexibility in their sources of funding. I will consider each of these dimensions more fully.

Number of journals. There are more social scientific journals in the U.S. than in any other of the eight major fields of science. Although social scientists write only 11 percent of the scientific articles, they publish 35 percent of the scientific journals /approximately 679 journals as of 1974 (King, 1976:82-3)/. This large number of journals is mostly uncoordinated. The major professional societies in the social sciences publish and set policy for only a few journals (see fn. 3).

Size. Social scientific journals are small. Comparatively the "average" journal in the social sciences is much smaller than the typical "small" journal in the other sciences (see Table 1). The average journal in the social sciences publishes fewer issues per year, fewer articles per issue, fewer pages per year, and has fewer subscribers than journals in the other sciences.

Efficiency. Social scientific journals apparently are highly inefficient. Rejection rates are strikingly high /(more than 80 percent on the average) (Rhoades, 1975; Zuckerman and Merton, 1971)/, and appear to be the converse of many science fields. Turnaround time for reviewing manuscripts is relatively long (Beyer, forthcoming), and once a paper is accepted for publication, the author waits a long time, often a year or more, before it is published. Some comparative data (Garvey et al., 1970) show that the dissemination of scholarship in the social sciences takes an average of six months longer, from the initiation of the research to publication, than scholarship in the physical sciences.[5]

Scope. Most social scientific journals have a generalist rather than a specialist focus, and increasingly this is preceived as a problem.[6] Along with this generalist scope, social scientific journals are moderately interdisciplinary, and for many topics (e.g., political sociology), disciplinary distinctions are not finely drawn. Social scientists increasingly are dissatisfied with their generalist journals, and despite the formidable economic problems of sustaining a journal, there is demand for many more specialized publications.

Terminological and Stylistic Variation. One of the most perplexing problems for authors and editors in the social sciences is knowing which language conventions to adopt. Principally because of the theoretical and methodological heterodoxy, as I have noted above, there is little consensus about the use of terms and concepts (cf. Beyer, forthcoming). Social scientists are handicapped in developing a short-hand mode of communication. Our articles, therefore, are longer [two or three times so (King, 1976:212)], and necessarily more elaborate than the compact articles in the natural sciences. Even on presumably more consensual matters such as style of references and footnotes, there are many variations. Social scientists do not have a style manual,[7] and even within a single discipline, one may find several different patterns of referencing.[8]

Financing. Like most journals in the sciences, those in the social sciences face bleak financial prospects. The situation is more complicated for social scientific journals for several reasons: first, because of their small size and their lack of coordination, they cannot take advantage of the economies of scale that are available to larger, coordinatively managed journals. Second, their funding pattern is less flexible. They are almost totally dependent upon subscriptions and society dues for their revenues: they receive miniscule amounts from advertising and almost nothing from page charges. (Some of these patterns are shown in Tables 2 and 3.) As a result of this narrow base of support, social scientists have few options when considering how to absorb their higher costs of production.

In this brief profile of social scientific journals I have emphasized their special handicaps, and yet all scientific journals face serious challenges to their viability. Perhaps nothing we do in the way of planning can prolong the life of journals for more than a decade or two at most. But clearly the need for a system of information for the social sciences will not disappear. If we hope to develop a system of information that will provide sufficient outlets for significant scholarly work, so that it may be disseminated efficiently and effectively, then we seriously must consider how we may begin planning in the context of a highly fragmented system of decision making.

Reasons for a Council of Social Science Editors

Many editors and other concerned persons in the social sciences recognize the need for sharing ideas about their common problems, but they have lacked an organization that would serve as a forum for their discussion. In the past few years, several groups of editors have begun to meet informally, often at the annual meetings of the large professional societies. The societies appear not to have instigated or planned these events, and in that sense, the move toward seeking solutions has a grass roots character.

In their discussions editors envisage many areas of cooperation. A few of their ideas apparently could be implemented with little difficulty, such as exchanging journal subscription advertisements. Other projects, such as preparing a common style manual, are more ambitious, and would require some kind of coordinating structure. Table 4 shows the responses of some social science editors to my questionnaire about areas of cooperation that seem most feasible at this time.

To achieve our cooperative objectives, several of us have begun to organize a council of social science editors. A council of editors cannot circumvent economic inflation and the explosion of information simply by organizing and discussing matters, but it can serve as a catalyst for the planning that must inevitably be done to preserve a system of information. As a grass roots organization, without previous organizational obligations and ties, the council can proceed in a number of ways to mobilize editors in the social sciences.

In another paper (McCartney, 1976) I have identified three major purposes that a council of editors can serve: information sharing, cooperative ventures, and policy consideration.

Information sharing: An immediate need for editors in the social sciences is to find out what other editors are doing and have done in the past that ought to be adopted. Since they have relatively short tenure as editors,[9] and there are not many centers of coordination and management of journals to act as reservoirs for the accumulated wisdom about editing, editors in the social sciences must learn their trade quickly.

Probably the easiest way to accomplish the sharing of information is through a newsletter. Thus editors could share information about innovations for saving money (using fourth class postage for mss, not returning mss to authors), share ways of improving their operational efficiency (using the telephone to locate willing reviewers), and share ideas about sources of income (how to plan a subscription campaign, how to solicit advertising).

Another important venture would be to compile regularly some statistics about the state of journals: their circulation, the volume of manuscripts, their operational costs, etc. These data collected over a period of time, would permit us to make informed assessments about the state of health of our journals, and assist those who must make policy decisions about them.

Cooperative ventures: One cooperative venture that will benefit authors, editors, and those who must prepare manuscripts is the creation of a manual of style. No doubt such a product will have to be multi-parted to accomodate the differing traditions of usage in the several social sciences. But certainly there are areas of agreement that will make this venture worthwhile. Since there is much interdisciplinary work now in the social sciences, it would be wasteful to have the disciplines begin the laborious process of constructing independent style manuals.

Since many editors are willing to explore limited forms of cooperation (Table 4), a council might serve as a clearinghouse, bringing together those who wish to exchange advertising or subscription lists, or who wish to solicit advertising jointly. Some editors feel that it would be advantageous to join together and find a printer who would give them a better price to jointly produce their journals. Like-minded editors can find each other more easily through a council, although it would not manage or produce such arrangements.

Policy: Consistent with the scattering of journals in the social sciences, decisions about policy concerning publications are located in a multitude of bodies. Even the largest of the professional societies in each discipline do not speak for the entire discipline. Clearly this fragmentation poses problems, but it serves other purposes. Social scientists, for example, seem to be most fearful of centralized decision making in their professional societies, and they seem to prefer to keep these organizations relatively weak.

Their reasons for fearing centralization partially may reflect their political inclinations, but doubtless they also reflect genuine substantive concerns. In the absence of a theoretical and methodological paradigm, or consensus, at this time, social scientists are suspicious of any signs that one theory is being favored in any way by the large professional societies. The selection of an editor for the most prestigious general journal in a field is thus a matter of great importance, even though the tenure of the editor is relatively short. The editor may feel strong pressures to avoid any appearance of bias in favoring a point of view with which he or she has been identified (see Zelditch, 1976:924, 1094).

A council of social science editors will have to be careful to avoid raising fears that their activities will shape, constrain or limit substantive directions in other than "natural" ways. Some social scientists already fear that the gate keeping functions of editors give them too much control over the direction of ideas,[10] and the existence of a council of editors may give the appearance that we are trying to create a super body to control social scientific journals. A council will have to walk the treacherous divide between appearing to do too much and not doing enough to address the publication crisis in the social sciences.

If a council of editors must be cautious about appearing to set policy, it nevertheless must be vigorous in raising questions about policy that other groups should address. Some of the following questions appear to be important, but they are by no means an exhaustive list: Do we have a sufficient number of outlets for publishing scholarship in the social sciences? Should there be more specialist or more general focus journals? Where do journals fit in the system of information of the future? Can we find ways of reducing the time between the initiation of research and its dissemination to users? Which groups should take the initiative in experimenting with other forms of disseminating scholarship in the social sciences? How can journals help bridge the gap between scholars and users of social scientific knowledge?

By raising these and other questions and providing a forum within which concerned parties can explore alternatives, a council of editors can give its greatest service.

Organizing the Council

Already we have a steering committee for a council of social science editors. It has communicated mostly by mail and has not yet resolved the many problems that attend the creation of a new body. The preceding outline of the purposes of this group, therefore, must be regarded as tentative, and reflect my personal interpretation of our discussions so far.

The organization of the council is still very much an open matter and we seek the widest possible participation in this effort. At the moment, the following organizational issues seem pertinent: a) The clearest issue is that the council should be composed of individual editors and other interested persons. Thus the council would speak for itself and not represent the views of any existing societies or associations. b) Editors in the fields of anthropology, economics, political science, and sociology would be the prime candidates for membership, but many persons in geography, history, law, psychology, and statistics share common perspectives. A council should be embracive and not try to draw its boundaries too narrowly. c) American social scientists have been chauvinist in the past, and a council of editors should be expressly international in membership. The council should give high priority to the discussion of problems relating to an international system of information in the social sciences.

We need not go about rediscovering the wheel as we form a council. We have many models of other associations of editors to examine and borrow from (CBE, AESE, ELSE). While the scope of their activities is almost intimidating to a fledgling group like ours, we nevertheless expect to learn from their experiences and to borrow judiciously from their programs.

Conclusions

A council of social science editors is not a panacea for the problems that face publishers of social science journals. Other contingencies such as the availability of funding for research, the pressures of the job market, the credibility afforded to the social sciences by interested publics and the government, and technical innovations in publishing are likely to affect significantly our journals in the social sciences. A council can provide the forum within which these shaping forces are examined, and this might lead to more informed decisions by the large number of groups that publish journals.

The idea of a council is not without problems, however, and as I have noted elsewhere (1976), these might include the following:

> "1. Such a council, like all organizations, would require maintenance, and since editors already complain of being overworked, it might detract from their energies and their priorities.
>
> 2. If it were only a voluntary group of editors, with no no structural linkages to professional associations, it would have relatively little impact on large scale planning for information dissemination in the social sciences. The countervailing threat, of course, is that (a) it might develop into a group of editors who would want to make policy decisions without providing for participation by authors and information users, and (b) sharing of information among editors might lead to uniformity of editorial policies and practices which would stifle imagination and creativity on the part of editors and authors."

Many of us feel that these are not serious problems, at least not compared with the more imminent threat of survival that is facing many journals. Even if journals do disappear in the next couple of decades, the social sciences will need some type of system of disseminating information, and we must plan for that.

We need some bold visions to conceive our apparent handicaps as possible advantages. The scattering of a large number of small journals, for example, could permit a flexibility in publishing new and controversial ideas that is not possible in scientific fields with fewer and more coordinatively managed journals. Although the basis of support for social scientific journals is more narrow, and they are relatively much less subsidized by the government, they could thereby be relatively freer of governmental priorities that shape research.[11] With luck and considerable effort, a council may explore bold visions and not become just another interest group with self-serving priorities.

Coda

In parting, I would urge that we excise the dinosaur as the symbol of journals in the social sciences and substitute instead the House, or English Sparrow (*Passer domesticus domesticus*). It also is small in size, large in number, widely scattered around the countryside and it survives on meager fare. It is not the most beautiful creature, but few dispute its ability to adapt as a species under the most stressful of circumstances. One of its most useful survival mechanisms is the watchfulness of the flock. By keeping our eye on the sparrow we may turn the evolutionary dead-end of journals into an evolutionary beginning of an exciting new system of information in the social sciences.

Table 1. Typical Journal Characteristics
(Life sciences compared with social sciences)

Characteristics	(a) Small Journal	(b) Large Journal	(c) "Average" Social Science Journal
Issues per year	6	12	4
Articles per issue	26	45	11
Subscribers	2,000	5,000	1,500
Mss received per year	206	713	160
Mss published per year	156 (76%)	540 (76%)	29 (18%)
Pages published per year	1,210	4,169	625

Source: Columns a and b from Aspens Systems (n.d.:35). Column c compiled from my questionnaire survey of 43 social science editors (Summer, 1975).

Table 2. Financial, Workload, and Circulation Data about Journals in the Social Sciences

	N	Range	Median	Interquartile Range
Circulation	36	250-19,000	1,500	750-3,250
Number of Manuscripts Received	39	15-780	160	95-350
Percentage of Manuscripts That are Published	39	7.7%-66.7%	18.1%	10.5%-30.0%
Number of Pages Published Per Year	37	70-960	400	220-600
Approximate Annual Budget (20 out of 39 journals reported)	20	\$600-\$120,000	\$17,500	\$5,000-\$31,500

Table 2 (Continued). Financial, Workload, and Circulation Data about Journals in the Social Sciences

	N	Range	Median	Interquartile Range
Sources of Revenue (28 out of 39 editors provided financial information)				
Individual Subscriptions	28	7%-75%	35%	30%-60%
Institutional Subscriptions	25	17%-75%	40%	30%-60%
Other				
Page charges (4)		2%-10%		
Back issues (15)		1%-20%		
Commercial reprints (7)		2%-5%		
Author's reprints (10)		1%-20%		
Advertising (12)		1%-10%		
Mail list (5)		1%-2%		
Institutional subsidy (12)		2%-80% (5 reported 50% or more)		

Source: Compiled from my questionnaire survey of 43 social science editors (Summer, 1975). N's do not total to 43 because some editors did not provide complete information.

Table 3. Percent Distribution of Revenue for Scholarly Journals in Pure Science, Applied Science, and Technology, and Social Sciences: 1969, 1971, 1973*

	Percent								
	Pure Sciences (n=44)			Sceince & Technology (n=50)			Social Sciences (n=65)		
Sources of Revenue	1969	1971	1973	1969	1971	1973	1969	1971	1973
Subscriptions	49	49	51	38	40	41	50	45	46
Society dues	27	28	28	30	30	28	30	39	39
Other publication sales	4	4	4	6	6	6	4	3	3
Advertising	6	4	4	21	18	18	6	6	5
Page charges	11	12	10	2	3	3	-	-	-
Subsidies	3	2	2	3	2	1	4	3	2
Miscellaneous	1	1	-	1	3	3	6	4	4
Total	---------------------------100---------------------------------								

*Reproduced from King, 1976:240.

Source: Fry, Bernard M. and Herbert S. White. Economics and Interaction of the Publisher-Library Relationship in the Production and Use of Scholarly and Research Journals (NSF GN-41398), Indiana University, November 1975.

Table 4. Social Science Editors' Perceptions of Possible Inter-Journal Cooperation*

Type of Cooperation	Possible	Impossible	Don't Know or no answer
Referring manuscripts (forward reviews to other journals, with author's permission)	27	3	0
Sharing printing facilities	14	12	4
Developing a common style format	25	4	1
Exchanging ads (journal announcements)	23	2	5
Soliciting advertising from publishers	20	3	7
Joing mail advertising	22	2	6
Exchanging subscription lists	21	1	8
Publishing a common index	21	5	4
Common distribution center for back issues	23	3	4

*The question was: "What problems or barriers, if any, do you anticipate in having your journal cooperate with other journals in the following ways?"

Source: Compiled from my questionnaire survey of 43 social science editors (Summer, 1975). N's do not total to 43 because some editors did not provide complete information.

FOOTNOTES

1. In a comparative survey of editors, Beyer (forthcoming) found that 70 percent of the editors in the social sciences favored expansion of their journals compared with 27 percent of editors in the physical sciences.

2. In sociology this process of differentiation is evidenced in several ways. Within the American Sociological Association, eight new sections (each with a minimum of 200 members) have been added in the past ten years for a total of 15 sections. The directory of the Association now lists 35 specialties, an increase of ten over the previous directory five years earlier (personal communication from Alice Myers, Administrative Officer, ASA).

3. The American Sociological Association (15,000 members) publishes six journals; the American Economic Association (19,000 members) publishes two journals; the American Political Science Association (11,000 members) publishes two journals; the American Anthropological Association (10,000 members) publishes two journals. Apparently the largest publisher of social science journals is Sage Publications, a commercial publisher with approximately 40 journals.

4. The concept of role, for example, has scores of meanings and referents. If a researcher were to enter an electronic retrieval system with only this one term, he or she would be inundated with information.

5. Some more recent data (King, 1976:139) show that the natural sciences are beginning to fall into the social science pattern of long lag times for publication.

6. Rhoades (1975) data show that about two-thirds of sociology journals have

a generalist orientation.

7. Psychology is an exception to this indictment. The American Psychological Association is now preparing copy for the third revision of its highly successful Publication Manual.

8. In sociology four of the leading generalist journals have different reference and citation formats.

9. Rhoades' (1975) data and mine show that the majority of social science editors serve a term of three years or less.

10. See Freese (1975:2) for an apologia that journal editors simply mirror the work of the profession and do not act as "paradigm enforcers."

11. This is not to suggest that all government priorities affecting social science research are necessarily inimical to the growth of informed knowledge about social phenomena.

REFERENCES

Aspen Systems Corporation
n.d. Editorial Processing Centers: Prepared for the Office of Science Information Service, National Science Foundation. (Aspen Systems Corporation, 11600 Nebel Street, Rockville, Maryland 20852).

Beyer, Janice M.
forth-coming "Editorial policies and practices among leading journals in four scientific fields." Sociological Quarterly.

Freese, Lee
1975 "ASA editors not paradigm enforcers: mirror work of profession." ASA Footnotes 3 (6) (August):2.

Garvey, William D., Nan Lin, and Carnot E. Nelson
1970 "Some comparisons of communication activities in the physical and social sciences." Pp. 61-84 in Carnot E. Nelson and Donald K. Pollock (eds.), Communication Among Scientists and Engineers. Lexington, Massachusetts: Heath Lexington Books.

King Research, Inc.
1976 Statistical Indicators of Scientific & Technical Communication (1960-1980). Vol. II: A Research Report. Prepared for National Science Foundation. U.S. Department of Commerce. National Technical Information Service. PB-254060. May.

McCartney, James L.
1976 "Confronting the journal publication crisis: A proposal for a Council of Social Science Editors." The American Sociologist 11:144-152.

Rhoades, Lawrence J.
n.d. The Author's Guide to Selected Journals. Washington, D.C.: American Sociological Association. (Summary presented to a workshop for sociology editors, San Francisco, August 1975).

Rhodes, Sarah N. and Harold E. Bamford, Jr.
1976 "Editorial Processing Centers: A progress report." The American Sociologist 11:153-159.

Senders, John
1976 "The scientific journal of the future." The American Sociologist 11:160-164.

Senders, John, C.M.B. Anderson, and C.D. Hecht
1975 Scientific Publication Systems. Report submitted to National Science Foundation. Grant No. SIS74-12627 A01. University of Toronto.

Zelditch, Morris, Jr.
1976 "Items." American Sociological Review 41:924, 1094.

Zuckerman, Harriet and Robert K. Merton
1971 "Patterns of evaluation in science: Institutionalisation, structure and functions of the referee system." Minerva 9 (January):66-100.

STABILITY VERSUS CHANGE IN PSYCHOLOGICAL RESEARCH. TOWARD IMPROVED STANDARDS IN PSYCHOLOGICAL JOURNALS

John R. Wettersten
Brookline, Mass.

In order to make the outline of my essay as clear as I can, I will summarize my essay on psychology three times.

ABSTRACT 1

Psychological research has aimed for stability and cumulative advance. Psychologists have attempted to achieve these goals through a combined use of two methodologies -- inductivism and conventionalism. The attempts to achieve stable research and cumulative advance has lead to confusion about the nature and value of experimental research: it has lead to non-testable low level hypotheses with no clear relevance for theory. I propose that if we change our aims from stability and cumulation of whatever truth we can find to criticism and improvement of theory, we can provide a more sensible evaluation of psychological theory - one that is neither condemning nor self-satisfied, but which would provide a clear purpose for empirical research and better standards for both theory and experiment. I will provide proposals for standards for psychological research for reshaping our views about what type of publication is worthwhile and what type is not.

ABSTRACT 2

Psychological journals hope to publish reliable results. This aim is based on the view that science can progress inductively, i.e., by accumulating theoretical and factual knowledge which can be added on to the hard, justified knowledge we already have. The approach is failing however: more and more studies are presented which are not incorporated into any literature. Furthermore, reliability is not being achieved: many studies turn out to be non-replicable; results are not duplicated by others. It is difficult however to show that a result is not replicable because opinions incline toward the published versions. These problems have been noted, but in attempting to solve them the aim of reliability of results has been kept. This has lead to hopes that experiments can be spelled out with more detail so that results reported in journals will be reliable and replicable. This policy naturally leads however to low level hypotheses which are increasingly disconnected from theory.

In Section two, I discuss the method of dealing with the problem. I have not tried much to publish criticism of the sort I am proposing in psychological journals. Yet it seems reasonable to suppose that access to these journals for the discussion of standards is not easy. My reason for believing this is that the standards of the journals, if reasonably adhered to are not easily amenable to criticism: one has to conform to the standards in order to criticize them. So the theoretical problem becomes a practical one as well: how can reasonable standards be formed for criticism of standards?

Miriam Balaban (ed.), Scientific Information Transfer: The Editor's Role, 485-494.

This problem is especially acute since one needs to appeal to the standards one is criticizing in order to present one's criticism. This may sound quixotic but there actually are articles in psychological journals discussing critically the standards of psychological research while attempting to meet, in some superior way these same standards. Problem of writing essay - aim to present clear every time. This leads to the effect that any criticism must also reinforce the standards used: it cannot lead to change but only to increasing the degree to which standards already in existence are spelled out. In particular statistical studies of the use of statistics are presented. It can lead to addition but not easily to removal. Let me turn now to my example of such a problem. In the epilogue I will return to the problem of change of standards in scientific research

In order to solve these problems it is useful to change the aim which journals seek to achieve. If we change the aim from the accumulation of true theories and facts to critical discussion of theories which aims at the improvement and replacement of theories these problems can be solved rather easily. If we change our aim in this way we can publish surveys of theories, problems and methodological appraisals of these theories and criticism of theories. Experiments for the most part should be interesting tests of theories - or in some cases interesting results which perhaps have no explanation but which seem to provide interesting problems. This set of proposals will solve the above problems in two ways. First, since criticism is sought, error can be more readily removed. Secondly, even though theory may be in a bad state, discussion and identification of the problems of theory will enable us to sensibly appraise our situation - we may even find out that things are not so bad after all.

ABSTRACT 3

In this essay I will first discuss the problem, second I will discuss the two traditional methodologies - inductivism and conventionalism - and how they serve to justify the quest for stability in psychological research. Secondly, I will discuss how this approach has failed - or how it has led to the problem discussed in Section One - it has separated experiment from theory and provided bad appraisals of both, i.e., theory is underrated and experiment is overrated. Thirdly, I will explain this failure as a result of the erroneous theory that low level hypotheses can be clearly formulated if we do not know the aim which these hypotheses are supposed to serve. In section four I will provide a new rationale for research, i.e., the aim of research should be to improve theory, and the problems this new rationale poses. In section V, I will explain how these problems can be solved with new standards for psychological research. I will conclude with an explanation of why my proposals are progressive.

Section 1. Inductivism and Conventionalism: The Quest for Stability.

In order to understand the standards for contemporary research in psychology it is useful to explain how these standards have evolved from the application of methodologies. It is needed because if we are to improve standards for research, we need to critically review, from time to time, the rationale for these standards.

Psychologists have traditionally attempted to employ methodologies used, or thought to be used, in the physical sciences.

One of the two most important methodologies used by psychologists, i.e. inductivism -- is the theory that knowledge progresses by the accumulation of facts and/or true (confirmed) theories.

The methodology perhaps began with Bacon - he is at any rate the first methodologist to have his theory widely endorsed by what we now consider the scientific community. His view in various forms has survived to the present. In this essay I can of course only briefly summarize its impact on psychology.

This impact has been felt in two ways. The first is that inductivist methodology and associationist psychology complement each other - perhaps need each other. Since psychologists have often been either one or the other both views are hard to get rid of; in order to remove one, one has to remove both. I will not discuss further however this important influence in this essay. The second major influence is the wide acceptance of two methodological attitudes. The first of these is that the successive gathering of factual knowledge, or at least of very low level, but true hypotheses is a desirable way to increase our knowledge. The rationale for this view is that if we gather facts and correct low level hypotheses we can gradually extend the scope of these hypotheses to explain more and more material. We can thus have a body of psychological theory which is both highly explanatory and secure.

The second methodological attitude, which is due at least in part to the influence of inductivism, is the view that theories which are not confirmed in some very strong sense are unacceptable. It is corollary of the theory that we can discover true theories, even low level ones, by empirical research; that we should rule out of psychological research theories which are not proved or at least highly established by induction.

I will return to the impact of these attitudes on psychology and its research, but before doing so it is useful to discuss a second methodology -- conventionalism -- which has been blended with inductivism in the actual practice of psychology. Conventionalism is at least as old as Copernicus. It is the view that scientific theories are not descriptions of reality but are simply tools which we can use to make predictions. In the late 19th and early 20th century this view has been developed into a sophisticated theory by Duhem, Poincare, and Meyerson. Bridgman, Polanyi and Kuhn have increased its influentale. P.W. Bridgman is not the most important philosopher in this group but he is the most important for my purposes. He has had considerable impact on research in psychology because he recommended and made fashionable a methodological rule which psychologists could use to guarantee the stability of their results. This rule is that one should specify for all terms in a theory operational definitions. The rationale for such a rule is that if we specify the operations which we are referring to in a theory the theory can take no risks - it cannot be overthrown. The reason it cannot be overthrown is that the theory only asserts that known reactions will occur in the future as they have in the past. We can still extend knowledge by increasing our catalogue of observed reactions. We will not on this view attain true theories, but we will attain theories which make true predictions and which are stable: there will be no revolutions.

Psychological methodology is a mixture of inductivist and conventionalist elements. The reason why this mixture occurs is that both views have the aim of creating stable scientific theories -- theories which will not be overthrown but which can be added on them. So psychologists have employed methods in hopes of creating a stable body of theory. The recommendation to use operational definitions seems to provide a method for the achievement of stability. It can be used without making a decision, or without being clear about the final aim: as long as stable true low level hypotheses are established it is not too important whether psychologists are clear about or agree on the use to which these results might ultimately be put. This is especially the case when they all agree a priori that the results can be useful.

The combined influence of inductivism and conventionalism has thus lead to a search for stability in psychological research. And both views have appeared to endorse the formation of low level hypotheses which could be verified and/or formulated in operationally defined terms. This leads directly to standards for psychological research. These standards tend to dictate that one should form reliable low level hypotheses which can be specified by their empirical consequences. This is an overriding aim.

Section 2. The failure of the quest for stability.

The quest for stability in psychology has lead to one remarkable failure: a wide split between theory and experiment. We have on the one hand broad theories of how to do psychology, or how some typed behaviour. On the other hand we have experiments which are often interpreted, or designed to be interpretable by these theories which do not describe or test any description of fact or low level hypothesis deduced from the theories. Or, to make the point another way: empirical research is ineffective in altering and developing theory: it fails to uncover error or provide significant addition to theory. This split and ineffectiveness is evident in a number of ways. Superficially it is evident in that psychological research has broken down into schools. These schools each develop their own theory and experiments, so that the experiments of one school, as well as the discussion of theory, are not relevant to another. This alone would be enough to make my point, within these schools there are difficulties as well. When experiments are put forth as tests, or as relevant to theories they either confirm these theories (or are taken to do so), or theories are modified so as to take these experiments into account. As a result theories are commonly restricted in scope. Extensions, such as they are, are commonly ad hoc. This is of course contrary to the aim of truly increasing our knowledge by extending the predictive ability of theories.

Further difficulties arise for research. According to the model for advance psychologists intend to build on stable results. So, a standard technique in conducting research is to repeat the experiments in the literature in preparation for further research. Such attempts, I am told, at least some of the time block research because appropriate results are not found. If a theory is involved it may be refuted, or it may involve a simple failure to replicate the results of some already accepted study. In either case, the psychologist perhaps properly is inclined to blame himself. But there are no standards, no easily accessible forum, for discussion of these "failures."

One further phenomena is endemic. This is the large numbers of experiments whose theoretical import is unexplained and by no means evident. Experiments which fit a certain form will be published without explanations of their importance. This is likely due to the view that knowledge about particular kinds of phenomena may be collected piecemeal and then used in broad theory constructions. But it does not seem to be working too well.

Expand on conclusion in statistics and relations of statistics to methodology. Elsewhere I have suggested that this type of low level research results in favour of prejudices since we describe events and components of correlation we believe already to exist. "Demonstrating" such correlation is the path to success in psychology.

So the attempt to find stability has failed because it has failed to connect theory and experiment. It has alternately led to high praise and condemnation of theory. Low level hypotheses cannot be established in the way that is desired; this is in part due to the fact that testing of these hypotheses is not a clear-cut matter. And finally the low level hypotheses that are "established" often are not related in any clear-cut manner to theory.

Section 3. The explanation of failure - failure to view statistics as tests.

My explanation of the failure of the quest for stability in psychology is as follows. The rationale for this approach presumes that low level hypotheses can be gathered independently of broader theories. This presumption is false. It is false because in order to have such established low level hypotheses we need to have clear statements of the conditions of prelicability. But this is not possible if we do not specify the aims which we have in formulating a specific hypothesis. Let me explain.

Any low level hypothesis needs to be formulated in such a way that the conditions under which experiments could be run as tests can be formed. It may be that these succeeding experiments would simply be regarded as repetitions of an experiment already performed. In order to meet these conditions we have to know just what desiderata must be duplicated in order to reach the same result. We can of course specify some desiderata quite easily for any hypothesis. But we can always meet these desiderata and achieve a conflicting result by adding some further arbitrary but effective condition. No matter how long a list of conditions we put forth we can always find some effect not mentioned which would change the result. So we need some general standard for ruling out extraordinary circumstances. If however we only have a low level hypothesis this becomes difficult since ruling out interference requires some background knowledge. If this background knowledge is vague and inconsistent; if it involves a mixture of psychological theory, common sense, and statistics we can explain almost any failure away - or, we can simply add another condition to the hypothesis without affecting the purpose of the hypothesis -- primarily since it has none.

The lack of clear-cut desiderata for tests of low level hypotheses cannot be remedied by making them more precise, no degree of precision is sufficient to remove the problem. This difficulty is the same difficulty that caused Bridgman to abandon his theory. It was pointed out to him that to specify the definition of terms operationally it would be necessary to specify mental or paper and pencil observations. He conceded this. He also saw that once this was done we could no longer limit ourselves to statements only about repeatable physical operations since we no longer had a criterion which demarcated those statements which were not simply about physical operations from those which were. The attempt to form operational definitions could not reach its goal of demonstrating these statements only about observed types of events from those which made additional claims.

Let me put this point another way. If we are to guarantee that low level hypotheses specify only repeatable physical operations we cannot introduce mental or paper and pencil operations because these additional operations cannot be demarcated in any appropriate way: we cannot demarcate those mental operations which may introduce risk or ambiguity from those which will not. In the present case we cannot specify completely and adequately the conditions for repetition such that any future case will be decided by the information already given.

This point also explains the difficulty of using experiments to develop and test theory. As I have shown elsewhere it is quite common for psychological theory not to have definite empirical consequences, i.e., they commonly do not specify the conditions under which a test may be run which would refute the theory in question. Because of this psychological experiments have to be evaluated independently of such specification. This allows for leeway. Some condition may always be found which can be so adjusted that any result may be interpreted as confirming the theory. This can occur even if the result initially seems to be a refuting case.

So my explanation of the failure to establish low level hypotheses is that it is not possible to specify conditions of replication without theory. Such theory commonly does not exist or when it does it is not used to specify the conditions or replications. Furthermore, the specifications of low level hypotheses has been confused with the specifications of true low level hypotheses. This confusion is due to operationalism: the theory that we can specify operationally the experimental consequences of hypotheses thereby guaranteeing that we will not be wrong. Since this is not the case - since we cannot guarantee the truth of even low level hypotheses -- the attempt to do so is a mistake. We should seek not to provide low level hypotheses whose results can always be replicated but rather hypotheses of any level which have replicable tests. In statistical terminology more events do not have adequate tests - we cannot show independence and random equally possibly.

The attempt to guarantee the replicability of results explains some of the bad trends in psychological research. It explains both the retrenchment of theory in the face of criticism and it explains the tediousness of psychological research. In each case stability is sought without the means of achieving it.

Section 4. A Rationale for a Remedy - How to test - Statistics Theory?

I will explain.

My remedy for the ineffectiveness of psychologocal experiments by first, proposing an alternative aim, secondly, presenting a theory of how to achieve it and thirdly explaining how it solves the problem of the replicability and usefulness of experiments presented above. In the following section, I will explain how alternative standards for psychological research which follow from the rationale given in section two and which provide a solution to the problem of how this rationale can be implemented.

The new aim proposed in one philosophy of solve is that psychologists seek improvement through change in psychological theory. This may seem trivial and already in practice. But this is not the case. In fact improvement through change in theory has occurred. But this has not been endorsed. As a result change is viewed as bad and retrenchment occurs. The methodological consequences of the proposal I am endorsing are great. Let me explain. The view I endorse is a variant of the theory of Karl Popper. I presume that Popper is right in the general view that the primary problem of the methodology of any science is: how can critical standards be applied to theories to get us closer to the truth? (I happen to disagree with both his theory of criticism -- criticism produces agreement concerning the status of theories - and his standards for criticism - i.e., his theory of high degree testability.) In order to solve this problem we need desiderata for the pursuit of two aims, i.e. how can critical standards be applied to theory to set us closer to the truth? First we need desiderata for determining which theories are worth criticising or which theories we not only can criticize but also learn from the process. Secondly, we need desiderata for good criticism.

The problem of the desiderata for interesting theories is a difficult one. Alternative theories do exist. It seems to me that the most interesting of such theories proposes that theories worth criticism are solutions to deep problems. This seems to me correct but I will not pursue it here since theories of deep problems vary. Let me mention two. Joseph Agassi has proposed that deep and interesting problems are those which enable us to decide between alternative metaphysical theories. In psychology if one applied such a view a deep problem might be one which would enable us to decide between associationist and non-associationist psychology. This theory does have some explanatory power in the history of psychology: it explains the importance of gestalt experiments. A second, different and contradictory theory, though still closely relating is that of Tagdish Hattiangadi. On Hattiangadi's view a problem is deep if it occurs at the intersection of two disciplines, i.e., when a statement in one discipline contradicts that in another.

Let me turn to the second problem: what are desiderata for good criticism? Here we can easily list some inconsistency and experimental refutation (which is, of course, one kind of inconsistency). This is too easy however. We need to add some more since these are often inadequate. One is whether or not a theory solves a problem or any new problems. Even this seems insufficient however. The reason for this is that we do not simply want to know what counts as a good criticism but how to bring such criticism about. I will attempt to provide proposals which solve this problem below in discussing standards for journals.

I wish to turn now however to the philosophical problem of the ineffectiveness of experimental research which I explained above. The problem was that without theory no standards of replicability are possible because we always have possible variations of the set-up not accounted for in the statement of low level hypotheses or experiments. This problem is central because the aim of this essay is to explain how new standards for psychological research may be formed. Much of if not most of this research, in actual quantity, is empirical. Yet this research suffers from the problem of ineffectiveness. I need to explain how this problem may be solved philosophically in order to provide proposals which will solve it practically.

The problem of the ineffectiveness of empirical research due to the failure of standards of replicability may seem especially acute because it may seem to persist even if we do have theory. Let me explain why this is not the case.

A theory provides us with a definite place to start: we have a set of desiderata which, if met, will provide a refutation of a theory. On the basis of this we can deduce low level hypotheses and experiments. In doing so we will specify conditions of the experiment. Without theory such specification is arbitrary.

But suppose something goes wrong. It would appear that any misfire can be explained away whether we have a theory or not. Indeed it can. But if we have a theory we can evaluate how explaining away some refutation changes the theory. If it forces ad hoc changes in our theory, if the explanation is not itself testable, we have grounds for criticism of the attempt to explain the result. Sometimes explaining a negative result away is correct; sometimes it leads to bad theory or deterioration of theory. If we have a theory to begin with we can evaluate these moves as progressive or not. Unfortunately in psychology this has not been done. Any attempt to explain away or modify may be taken as successful. This is partly due to vague theories, but it is also due to vague standards and little discussion of the impact of such moves on theory. The upshot in the situation psychology is what was described above if there is no theory. Any move which is successful in explaining away a failure or modifying a view to block a failure is acceptable: there simply is no way to distinguish an acceptable from an unacceptable move.

Section 5. Standards for Journals.

I have thus far argued that the traditional aim of psychological research is stability. Furthermore I have argued that - due to difficulties in methodologies used to pursue stability this has not been achieved or, when it has been, it is fairly arbitrary. I have then proposed Popper's suggested aim: change theory to make it better approximate the truth. I have discussed methodological problems of this view and argued that it could better solve the problem of the general ineffectiveness of psychological experimentation. I will now explain how this view may be developed and implemented by proposing aims and standards for psychological journals. There is one simple overall aim: improve theory. I do not say this because of the commonly held view that theory in psychology is very bad. The view that theory in psychology is bad is usually adopted because it does not meet unreasonable standards, i.e., it is not proven, unified, and steadily progressing. A more realistic appraisal of theory is needed. I am proposing that whether or not theory in psychology is good or bad the gravest defect in psychological research is the quality of the discussion of theory.

I have five proposals gathered from the literature in philosophy of science. The proposals are that one, surveys of theory be published; two, problems should be developed and published even without solutions; three, theories should include, at least from time to time, appraisals of the possibility of tests and criticism, four, experiments should be introduced with explanations of the relevance for theory, and five, there should be critical discussion of all of the above. In general there is one proposal: discuss the logic of the evaluation of theories in the journals as the discussion of theory proceeds.

Let me explain my proposals. In order to have a chance of having a reasonable evaluation and discussion of theory we need to know which theories are thought to be important and why. We also need to know how these theories do or do not provide alternative solutions to problems. If this is done, we can then begin arguing about which alternative is best. We can end the isolation of schools of psychology from each other. Of course psychologists will vary on how they survey the field. So, to begin we need both surveys and critical discussion of them in order to know what it makes sense to argue about and why.

Secondly, we need to know what problems are important. If we are to progress through change we need to set specific goals which we would like to accomplish. This can be done through discussions of which problems are important. But not only do we need formulation of problems we need critical discussion of the value and interest of problems. This can be accomplished in many cases in surveys: a good survey will explain how theories are related and what problems the survey poses.

Thirdly, theory needs to be encouraged; but not only theory. Theory should also be presented with clear-cut evaluations of the testability, plausibility, and problem solving ability of theories. It seems to me a mistake to require all theory to meet some particular standard of acceptability, e.g., testability. But it is reasonable to have statements of theory include a methodological appraisal of the theory presented.

Fourthly, publication of experiments should include explanation of the relevance or irrelevance to theory of the experiment reported. Experiments which are accompanied by a clear explanation of relevance should be given priority. Some experiments without clear relevance might be published; but if they are, some other explanation of their significance should be provided.

Finally, an important part of the discussion of psychological theory is the discussion of the claims that are made for psychological theories. Essays discussing these claims alone should be published when they can improve our understanding of what theories can and cannot accomplish; of how they can be tested or criticized and how they cannot.

All of these proposals are designed to focus discussion on theories and appraisals of theories. If they were implemented it seems plausible that discussion of theory would improve and that all publications would have a far clearer purpose than what is now common in psychological journals.

Conclusion

The rationale for the proposals I have offered in this essay is that psychology can better improve by taking stock of its theories and how to improve them than by attempting to meet some ideal standard of science whether it be confirmation, testability, or some other standard. Such an attitude can provide for appraisals of psychological theory which need not end in a good - bad classification. They can instead provide problems whose solution might lead to progress. This may not lead to much success; but it should lead to better appraisals. It may even bring more rapid advance.

It is at least clear that the quest for stability through methodological purity which began in the late 19th century is not working very well. This is not to say that no advance has been made. It seems however more in spite of this quest than because of it. By seeking to meet too high standards much research has in practice met too low standards. Of course it is possible to be too liberal as well. But a corollary of too high standards in theory is often too low standards in practice. So the proposals offered here have a reasonable chance of raising the standards for discussion of psychological theory. This is so because they provide tough but still achievable standards. The use of these standards could change the methodological discussion in psychology from the problems of good vs. bad research to the problem: how can we improve what we have, regardless of how good or bad it may be.

Epilogue

In opening my essay I posed a problem. This problem was: how can standards for journals be reasonably changed? In the body of the essay I provided what I take to be an example of the need for wholesale change of standards in a well established discipline. This example then provides a case of a type of criticism which needs to be considered but is not now very effectively dealt with. Philosophers do discuss these matters. And in philosophy there is a wide recognition that to do philosophy of science there should be some contact with the actual practice of science. But philosophers themselves have not resolved the problem of whether they should simply describe, prescribe or do a little of both. I clearly am of the view that a little of both is needed. If however this view is adopted we need a theory of how this can occur. This theory must explain how we can advance by changing our standards, and it must provide for a forum, with standards of its own for the discussion of standards. The lack of such a clear-cut forum leads to both undue influence and undue lack of influence of methodologicasts on standards for scientific research. It leads to undue influence when, as has happened in many of the social sciences, but not only the social sciences, some methodology is taken to be the unquestioned norm for scientific research. It leads to undue lack of influence for the same reason: criticism and alternatives to influential methodologies are not given a serious day in court.

DISCUSSION on "Functional Differentiation of Social Science Journals" by Peter Lengyel

"The Need for a Council of Social Science Editors" by James L. McCartney

"Stability Versus Change in Psychological Research. Toward Improved Standards in Psychological Journals" by John R. Wettersten

JOSEPH BEN-DAVID - I've always admired the courage of editing an international social science journal because unlike in some of the natural science disciplines the difference in the style of writing and standards of publication is so great that I consider writings such as our Latin American colleagues write and compare them with what Amercians write, what Japanese people write. The difference is very considerable and I wonder how you manage, because you do somehow, to select things which really do appeal to more or less every one who reads that journal. I think this will be a problem in this kind of journal.

PETER LENGYEL - It is very kind of you to say that about our journal but in fact I believe that the function of a periodical like the International Social Science Journal precisely to show up these differences, and I do not try to even out the Japanese contributor and the Amercian contributor. On the contrary I try to show the difference between them; between our covers you will find things at very different levels and intentionally so. It is a showcase of differences as well as similarities. One big problem I do have is a stylistic one, in the expression of the ideas, and that is quite a problem, indeed which it takes me a long time to resolve. But I think, you've put your finger on one problem we were discussing with Professor McCartney just before we came into this session; many social science journals are too rigid in their standards and will only accept one type of thing, they expect their contributors, to conform to a paradigm. I don't have a paradigm. I merely commission articles on a theme. I don't even tell people in more detail what they should write on that theme. I want them to think this out. That is the purpose of the roundup. There may be overlaps and repetitions but that is how the game works out.

ARTHUR WILSON - I really shouldn't be speaking here because I am a physicist but I happened to be married to a sociologist and I am wondering whether something could done by training sociologists in the secondary sources that already have available. There is an American publication, Sociological Abstracts, but my wife never has the faintest idea of using it to find something. Her ideas of a source of information is to guess which of her friends is likely to know and ring them up. That is pretty erratic and on occasion I have taken it upon myself to look through Sociological Abstracts myself to find out what she ought to know. It is easy enough if you know the authors but if you want the subject, for a physicist it is pretty difficult.

PETER LENGYEL - I suppose your wife would be called an elite sociologist. These in the more remote corners of the earth see the abstracts and not the originals and are too far away to phone. But it is agreed as I said, quoting an article in my own journal that social scientists are very bad about going back to primary sources, you are talking about secondary sources, abstracts being already secondary sources, but where do they get their data from? Does your wife know where to get the population data for the state of Arkansas or the crime figures for Nicaragua? This is really where the basic material lies and few sociologists know how to get that kind of data. Not that it is easy. It is often very difficult to find where the primary data can be obtained, even in a small, consolidated country.

IRVING AMRON - Also not being a sociologist but an engineer I think that the basic premise that should be taken and start at the very foundation of the problem is the education of the sociologist or the enginee or of the physicist, during the educational processes where it all begins and it seems to me that this is the fault of education more than of anything else. That the chemists during their entire educational career so to say, are trained to use all the facilities that are available. Sociologists and economists, let's say, do not know about Predecast, which exists, which you probably know about, but where you can get graphs on your pages showing you the economic development of certain branches of American industry. So I think the basic premis is the education of all of us to be able to understand it. Because we sitting here and talking among ourselves we ourselves may be able to find it. But I think our problem is going out to the community which we are seving here and seeing that they understand these problems which we are discussing.

PETER LENGYEL - No comment. I entirely agree.

JOSEPH BEN-DAVID - Well if everyone agrees I think we should proceed. Anyway I just wanted to point out an extenuation of the sociologists. If any of you tried to look at such a basic publication as the U.S. Census or any other census and see how many volumes it takes to tabulate even the simplist results by every city and township in a single state and then to consider this material for the whole of Earth, if you ever tried to use this material you would see that there are some real problems which have not been resolved yet. I think that there is also a problem of standardization which is far more difficult than in the natural sciences. So this is not a simple problem but I believe the second, no the third type of your publication, that yould really be a chance for dealing with it and it would be quite useful.

RON SWARTZ - I have a question about the council of editors, but at the intellectual rather than the political level -- at the level of editorial boards when you were talking about the notion that they tried to get a sampling from the different school of thought. So that way it is assumed that your school of thought will get its day in court. What I recommend is to have some kind of discussion of what are the standards for all the schools, rather than seeing the council of editors as some kind of political thing for our profession. Where are we going? I think they could perform a very valuable intellectual function about matters such as what are we trying to do? What are the standards we consider to be disirable for intellectual journal? Why are the standards of my journal different from yours? And the biggest problem I have in dealing with editorial policies is that I very rarely know what those policies are, what the stnadards are. And in relationship to that I just want to say that often it so obvious what the standards are that people don't say what they are. And I think that is more dangerous in a sense than if they had stated them and wouldn't even talk about them.

JAMES McCARTNEY - I don't mean to suggest that I am talking only about the politics of organizing a council of editors and not what a council of editors dould do. I would hope that if a council of editors only dealt with the political issue or how we should divide up the field and matters of this sort I would consider it a waste of my time. I did cut a couple of pages from my paper and the kinds of issues I raise there are essentially intellectual issues. It seems to me that Dr. Lengyel dealt with these types of issues. This can be the vehicle for raising these kinds of issues. Where else can we discuss his ideas? It cannot be an agenda item in any of the professional science associations. I think only among the editors can they begin to raise questions of this sort. For example, are we now bridging the gap between scholars and users of social science information? I take this as a very critical intellectual issue. Now, no one is willing to take the initiative on this. As to the issue of standards. This is highly controversial and I think in the sense, if I understand you correctly the standards for acceptance in journals, I think that issues like this will be discussed and have been discussed in some of the editors groups that have met. The problem is that with the heterodoxy (if you will), the lack of consensus, one must approach this issue of standards, I think, with caution. Now since these fields that I am talking about, sociology, political science and so forth, have rejection rates of eighty percent it means that an article can be rejected on almost any ground. It seems to me that one of the logical conclusions that we have to move toward is that this is a terribly inefficient system, where each journal has to reject eighty percent of what it receives, and since the standards are not that clearly defined, are we really serving ourselves well? I think the editors of journals might well conclude that they are not serving the discipline very well, intellectually.

ARTHUR WILSON - I would just like to ask about that rejection rate. It is slightly rewritten and then given to B and then C, then D, and maybe E accepts it so that if you worked out the rejection rate per paper instead of per journal, the rejection rate would be much lower ultimately?

JAMES McCARTNEY - As Prof. Ben-David knows there is very definitely a pecking order and it is something that sociologists (it is called navel gazing sociology) study the pecking order of journals. The major journals have rejection rates of eighty five percent, and several of the larger ones ninety percent -- ninety two percent was reported I believe for the American Sociological Review which is a very large journal. The data that I am familiar with suggests that something like sixty five percent of articles eventually will be published after four years, and that about thirty five percent somehow fall out of the system.

MARK KELLER - To John Wettersten - I don't have a question, I want to make a comment. These ideas are relevant not only in the psychological and sociological fields but also for the biological field.

AGRICULTURAL AND DEVELOPMENT PUBLICATIONS

Joseph Palti, Chairman

THE NON-CONVENTIONAL LITERATURE OF TROPICAL AGRICULTURAL RESOURCE ASSESSMENT AND DEVELOPMENT: ITS NATURE, SOURCES, QUANTITY, VALUE AND ACCESSIBILITY

W.J. Baulkwill and N.W. Posnett, *Land Resources Division, Ministry of Overseas Development, Tolworth Tower, Surbiton, Surrey KT6 7DY, England*

ABSTRACT

The importance of this literature has been underrated, often because of confusion arising from the application to it of generalisations properly applicable only to the literature of experimentation. An analysis of the Land Resources Division library content (ranging from geomorphology, through soil science and agriculture to economics and sociology) shows that in one annual intake 64% of non-serial acquisitions are non-conventional, the figure rising to 76% for literature directly describing resource assessment and development. It is shown that some criteria applied when judging the value of papers on scientific experimentation (especially the criterion of publication in a journal of repute) have little application to the literature of resources surveys which deal not with experiment but with the inventorisation and classification of data and with the prediction of development possibilities. In practice this literature tends to be non-conventional, semi-confidential, inaccessible and extremely important. An analysis is made of the literature of development at the LRD library and an attempt is made to define it by document type and source. Factors influencing the non-accessibility of the literature are considered and possible actions to increase accessibility are put forward. It is concluded that growing awareness of the value of these documents and positive actions by publishers and client governments on the one hand and by national and international documentation centres on the other will help to increase the accessibility of the literature of tropical agricultural resource assessment and development.

BACKGROUND OF THE INVESTIGATION

LRD has for many years been aware that a high proportion of its library intake consists of informally published or non-conventional literature*. This literature has a specially important role in the field of agricultural development where much of the scientific and technical information on specific regions is to be found in reports of limited distribution issued by agricultural consultants and development aid agencies. LRD is especially aware of this situation because it is both a collector and a publisher of

*Non-conventional literature is variously known as 'informally published', 'unpublished' and 'grey' literature. A portion of it is commonly called 'report literature'.

Miriam Balaban (ed.), Scientific Information Transfer: The Editor's Role, 501-521. *All Rights Reserved.*

such reports. Nevertheless, it has been found in discussion with specialists in some related fields - agricultural research and the secondary literature of agricultural research - that the importance of non-conventional literature is consistently undervalued, because the traditional wisdom of the research world "good research is published in journals of high status" is applied to the totally different world of development surveys in tropical countries. In view of this fairly widespread non-comprehension of the nature of the problem, it was decided to investigate the quantity and nature of LRD's non-conventional holdings and to attempt to define the factors which influence non-conventionality. The main objects of the investigation were to draw attention to the importance of this literature and to encourage publishers and documentation centres to make it more accessible.

PRELIMINARY DEFINITION OF THE NATURE OF NON-CONVENTIONAL LITERATURE

To isolate and study the non-conventional literature held at LRD, a preliminary, working definition was required. It was not expected that this definition would be precise and it was clearly understood that no definition was likely to be valid for all disciplines. For example it was known that the underlying cause of non-conventionality in development literature (reports are sent to a limited clientele) was different from the corresponding cause in the field of extension literature (pamphlets are issued in large numbers to a specific region only). The subjects covered by the LRD library are varied, the important ones being surveys of natural resources, land development for agriculture and forestry, and related investigations in the fields of the physical environment, vegetation, land use, agronomy, forestry and socioeconomics.

Important descriptions of the nature of non-conventional agricultural literature have been made by Wood (1974) and Mei (1975) and for non-conventional scientific literature in general by Holloway et al. (1976). Wood (1974) who was concerned with the signalling of non-conventional documents in Agrindex* described it as "including such material as report literature and dissertations" and emanating "from many different types of organisations, including government departments, universities, industrial firms, independent or sponsored research laboratories and international organisations". He listed other characteristic (but not always diagnostic) features: cheap binding, limited distribution, inaccessibility, lack of sales organisation, speedy publication, presence of detailed research results, non-coverage by established secondary journals and low level of use. Mei (1975) referred specifically, with reference to agriculture, to "economic surveys, feasibility studies, development project reports, enquiries", especially those issued under technical aid to developing countries. Mei also pointed out the absence of powerful distribution networks backing the non-conventional literature and emphasised that distribution was often more related to the "power of the network" than to the document's value to the user. Holloway et al. (1976) noted how governments, a major source of this literature, characteristically listed certain reports but failed to list others of major importance.

Bearing in mind these descriptions, LRD started its enquiry with a broad definition to cover all literature which is non-conventional in the eyes of the librarian acquiring it, regardless of the intentions of the originating authority. For the purposes of our enquiry, non-conventional literature was

*Agrindex is a secondary journal of agricultural references issued by FAO's AGRIS Level 1 service which also issues corresponding material on magnetic tapes.

defined as that part of the literature which would present a non-specialist library, as opposed to a special library, with more than average difficulty in its acquisition. Such literature would not normally occur in booksellers' lists, and booksellers would usually have difficulty in supplying it. This literature includes some of the documents produced by government departments, international organisations, commercial firms, universities, research stations and consultants, conferences and symposia, but not documents issued by recognised publishing houses. These documents are nearly always produced in limited quantities and may be supplied free, at cost price, or on a purely consultant-to-client basis. The special library would tend to apply direct to the producer for this literature rather than to a book-supply agency which would add an appropriate surcharge to compensate for difficulties in acquiring the documents. Users of non-conventional literature are rarely alerted sufficiently to its existence, and they therefore depend heavily on special libraries, processing centres and sometimes clearing houses for news of the literature's existence.

In contradistinction, documents that appear in national bibliographies, on MARC tapes, are known to have been deposited in a copyright library or have an International Standard Serial or Book Number* were, for the purposes of our enquiry, regarded as conventional. Serials and especially annual reports presented a special problem. In making our analysis, serials (other than annual reports) thought to be non-conventional were checked for entry in a book supplier's guide to serials (Ebsco, 1975) - unentered serials were regarded as non-conventional. By our definition a large percentage of annual reports appear to be non-conventional but many are nevertheless extremely well known within professional circles: for this reason our enquiry into the quantities of holdings at LRD excludes annual reports entirely.

It should be noted immediately how the conventionality of a specific type of document may be related to the level of documentation attained in the country of origin. A document with every appearance of an obscure, "informal" origin can easily be 'legitimised' and made much less obscure if it is formally registered and signalled by a well-developed national documentation service. Ideally such a service would not only signal the documents but also be able to make copies of them available on request.

CHARACTERISTICS OF NON-CONVENTIONAL LITERATURE HELD BY THE LRD LIBRARY

LRD-held literature falling within the above definition, whether serial or non-serial (see Appendixes 1 and 2 for examples of sources and titles), but excluding annual reports, was examined and its main characteristics are outlined below.

Types of material

The material which it is possible to call non-conventional in LRD's subject fields is listed below. An attempt is also made to indicate material in our library which is always considered to be non-conventional and that which is sometimes so considered:

---

*The allocation of an ISSN or ISBN was a useful criterion in determining the conventionality of LRD - held books, pamphlets and serials, because such an allocation normally assists in signalling the document's existence. The use of such numbers however does not guarantee that a copy of a document can be made available on request.

Material always considered non-conventional	Material sometimes considered non-conventional
Confidential reports of all kinds	Annual or periodic reports
Consultancy reports	Numbered but irregular series
Internal reports of organisations	Bulletins and pamphlets (e.g. extension material) with a fairly wide distribution within a very limited area
Unprocessed records	
Unpublished survey reports	
Unpublished experimental reports	
Material of limited distribution	

Source

The source of the literature is an important guide to the conventionality of a report, i.e. it commonly indicates whether the material is easily obtainable or not. The typical sources of the non-conventional literature in the LRD library are shown below:

Originating authorities considered to be almost without exception sources of non-conventional literature	Originating authorities often found to be sources of non-conventional literature
Consultants	Government Departments
Commercial development agencies	Quasi-governmental organisations, "statutory bodies"
	International aid-giving agencies, including banks
	Information processing centres
	Universities
	Research Stations
	Conferences

Thus by the very nature of non-conventional literature, the sources of supply are numerous and what each source supplies is a very small percentage of the total: a situation which arouses minimal commercial interest. The publishing houses and the large-scale bookselling agencies act as clearing houses for the conventional literature and so bring it to the attention of their clientele; they can perform no such task for the non-conventional literature, whose existence and sources therefore tend to remain unknown to potential users.

It is not possible to define the sources of non-conventional literature by geographical location, but we nevertheless subscribe to the generally held view that the non-conventional literature forms a higher proportion of the total output in developing countries than it does in developed countries.

Format and binding

Non-conventional literature often has a distinctive appearance related to the short runs printed. Frequently it consists of typed or typed/mimeographed material stapled together with or without soft covers. Such an appearance is not diagnostic however, because the luxuriousness of presentation depends on the purse of the producer (or his client) and his need to please. Some non-conventional material produced by governments and commercial consultants and held by LRD is indeed elaborately presented.

Accessibility

The inaccessibility of non-conventional literature is notorious and is its most significant characteristic.

Accessibility has two aspects. First, the degree to which documents are publicised by their originators or subsequently signalled by information/documentation services, especially the well-known services, determines the extent to which the title becomes known. Second, once a document becomes known, there are difficulties in obtaining it arising from security restrictions, limited number of copies, absence of clearing houses capable of producing hard copy on demand and finally sheer distance of the material from centres of interest: this second aspect, the possibility of obtaining 'hard copy', is called here availability.

Methods of improving the accessibility of the non-conventional literature are discussed at the end of this paper.

Publicity and signalling As Mei (1975) has indicated, initial publicity at the time of issue depends to a large extent on the status, wealth and access to established publication networks of the issuing authority. This is one of the reasons why the proportion of non-conventional literature within developing countries is higher than elsewhere. In addition, the desire to make the material known is often weak, given that the material is semi-confidential, intended for few readers and issued in restricted numbers - publicity would create an unsatisfiable demand. This is LRD's position as a publisher: much of its literature is sent only to client governments and subsequently with their permission to a prescribed mailing list. Often the only "publicity" given to such documents takes the form of lists of publications displayed in the pages of other non-conventional documents! Separate lists of publications are sometimes issued by originators and the LRD library makes considerable use of these. Government lists of publications may be both accessible and comprehensive or at the other extreme sketchy or out of date.

The signalling of documents on non-specialist subjects in the secondary literature follows a fairly regular pattern in many countries - the system of national bibliographies described in Acquisitions in the Third World (Clark et al., 1975).

The signalling of non-conventional literature on specialist subjects depends on a variety of processing and documentation centres and the secondary journals which they produce. This 'system' is unreliable because it is uneven and also because some centres tend to avoid non-conventional literature of unusual languages. The FAO AGRIS system represents an attempt to overcome these weaknesses (G Dubois, 1972; Wood, 1974) and its publication Agrindex has been established with the intention of signalling all agricultural documents, conventional and non-conventional.

Availability In practice some "hard copy" of non-conventional literature becomes available to specialist centres such as LRD through exchange links with kindred centres elsewhere. On the other hand some potential users, for example in developing countries (Ogunsheye, 1969; Edwardes, 1974; Adimorah, 1976), may have difficulty in obtaining the literature because they have no such links, or lack funds, or encounter other obstacles. The AGLINET network of cooperating agricultural libraries has been established (Harada, 1972) with a view to lending or supplying copies of documents which have already been signalled. This network is complementary to the signalling services of Agrindex.

LRD obtains much of its core non-conventional material by exchange and also by direct acquisition in the course of field work, while other similar material is obtained by direct approach to originating authorities and to national and international centres such as the British Library or FAO who are prepared to provide photocopies.

Non-conventional material should not be supposed to be cheap to produce, copy or post. To obtain a free copy of a 1 000-page consultancy report with elaborate maps is a minor triumph for a librarian at a specialist centre.

Security classification

These restrictions make documents inaccessible but they can often be lowered after a passage of time and such measures are described in the last section of this paper.

Speed of publication

A common reason for producing non-conventional literature is the desire to issue results quickly by simple, often in-house printing, rather than by more traditional channels. Rapid publication is thus often a feature of this literature.

Quantity

Attempts to quantify the non-conventional literature have been rare. The AGRIS Study Team of 1971 estimated (Wood, 1974) that about 20 000 non-conventional documents were produced each year on "agriculture and related subjects". LRD, working in the specialised field of agricultural development (in specific countries), collects some 1 000 non-conventional items per year and in our view the study team's estimate is not too high. In this connection it is interesting to note that EEC has commissioned a study of the non-conventional literature of agriculture within EEC which includes its quantity, bibliographic control and accessibility.

Figures from LRD's information store show that, using the definition mentioned earlier, the volume of non-conventional literature represents an important percentage of the whole annual intake; so much so that even if a very limiting and highly specific definition were applied, the resulting figures would still show non-conventional literature to be worth considerable attention. During the period September 1974 to August 1975, the LRD acquisitions lists show that non-conventional literature accounted for 64% of the total library intake of non-serials. Of the soils monographic literature, 50% was non-conventional. In the subject areas of natural resources development studies this figure rose to 76%. These figures apply to monographic material but include single numbers of some irregular series.

Non-conventional literature was not expected to be as common in the field of serials, but even so, under the definition used here, 39% of LRD serial holdings proved to be non-conventional (see Appendix 2 for examples). This is exclusive of annual reports which, depending on precise definition, could be regarded as non-conventional or conventional.

A high proportion of non-conventional literature has also been found in other fields such as applied economics, especially where developing countries are concerned (C P R Dubois, 1973). Littleton (1969) found that 53% of the documents on agricultural economics written in the USA were issued by universities or colleges. It is also evident from Littleton's report that a large proportion of all the documents analysed were non-conventional according to the LRD definition.

On analysing an LRD bibliography on Fiji (Posnett and Reilly, 1973) it was found that approximately 41% of the entries could be classed as non-conventional. This can be compared with a bibliography in a related subject area on Seychelles (Skidmore, 1974) where 70% of the documents could be regarded as non-conventional. These figures exclude annual reports. A further comparison can be made here when it is realised that the Fiji document was compiled from UK secondary sources while the Seychelles bibliography was compiled on site in Seychelles. Having made allowances for different methodologies used in construction and for the incomplete over-lapping of the subject areas, this comparison may be used as an indicator of the potential volume of undetected non-conventional documents in developing countries.

In LRD's final report on its Nepal Project (Berry et al., 1974), taken as representative of the main LRD report series, it was found that the documents referred to could be regarded as 54% non-conventional. In a comparable type of report on New Guinea (Haantjens, 1972) the number of citations to non-conventional literature was estimated to be just under 40%.

This analysis of the quantity and proportion of non-conventional documents in the LRD library emphasises the importance of this literature in our field of study and leads to further questions. How valuable is the non-conventional literature and how much is it used?

Level of use

While LRD makes great use of non-conventional literature in answering enquiries (in nearly half of the enquiries analysed the greater part of the information supplied was from non-conventional literature), it would be wrong to assume a high rate of use per document. Wood (1974) has asserted that the rate of use in the agricultural field is low. For the literature of resources assessment/development planning particularly, use is restricted in the first place to the government and development agencies concerned and is only incidentally extended to other countries where it may be felt that conditions are analogous. Nevertheless, though the rate of use of such literature is low, its importance to the few users is very great.

Value

The value of the non-conventional literature in the LRD collection to its users is very great, as has been illustrated by the high percentage of it which is used in citations and in answering enquiries. In particular, the literature of resource assessment and development planning often represents many years of careful study of the environment and agricultural potential of developing countries. Unlike much research literature which appears in non-conventional forms, the development literature normally never reaches a fully published stage - it is not reported in learned journals or in books except in exceptional circumstances.

Some of the underestimation of the value of the non-conventional literature on development which has occurred in the past results from the experiences of research workers who think (rightly) that the learned journal represents the highest scientific status while to them some of the most familiar non-conventional literature is the ephemeral extension pamphlet. Some confusion between the words 'extension' and 'development' has heightened this confusion.

An attempt is made below to set out clearly the nature of development literature, to distinguish it from research literature and to show why the standards often used in assessing the non-conventional literature of research cannot be applied to the literature of development.

Differences between research and development literature Confusion sometimes arises when the criteria used in evaluating agricultural research literature are applied to agricultural development literature. The same criteria cannot be applied because of the fundamental difference between the two types of literature.

Whereas the literature of agricultural research concerns the validation of hypothesis by experiment, the literature of agricultural development contains the elements listed below:

1. Inventorisation of environmental and socioeconomic resources
2. Classification of those resources in terms of land capability or suitability
3. Prediction of desirable development
4. Detailed feasibility studies of areas chosen for development
5. Planning for development
6. Description of actual implementation of development plans

The two most commonly applied criteria for evaluation of research literature are: statistical validity, and publication in a journal of standing. Neither can easily be applied to development literature as described above. Development literature often appears in the form of non-conventional consultancy reports sometimes of nondescript appearance! In particular it is unwise to discard resource inventories, however superficial they may be. An incomplete resource inventory may still serve as a useful 'first approximation' for more thorough investigations which may be made later. Development literature has a long life.

Misconceptions about the non-conventional development literature Some common misconceptions about this literature are recorded and corrected below:

	Misconception	Corrected view
1.	It is physically slight, abbreviated	The reverse is often true of resources inventories - a 400 page report is commonplace
2.	It is scientifically sub-standard	Although sometimes true of research literature this is quite untrue of non-conventional development literature which is commonly written by known experts in their fields and is often the result of prolonged study

	Misconception	Corrected view
3.	It is cheap	Initially non-conventional literature may be supplied free of charge to clients (although one report may represent an input of thousands of dollars). Subsequently the material is obtained with great difficulty, tends to be bulky and expensive to transport, reproduce or translate
4.	It represents a small proportion of the whole literature	Untrue of development, although true of research
5.	It is always repeated somewhere else in published form	Untrue of development, although often true of research. Consultancy reports, for example, are rarely re-issued, repeated or reviewed
6.	It is recorded (somewhere) by the secondary services	Only partially true. Commercial secondary services do not always abstract or refer to hardcore non-conventional literature. They do however commonly abstract some material which is non-conventional by wider definition, e.g. annual reports and irregular series of better known research stations and development agencies
7.	It can most easily be obtained in the developing country to which it refers	Not always true. Often more easily obtainable at an agency (not in the country concerned) responsible for production of the literature
8.	Consultancy literature intended for a client government cannot be obtained by information services	Often true. Nevertheless such material can often be released by permission of the client government. LRD always seeks such permission to distribute its major reports to the LRD mailing list.

Summary of characteristics

The non-conventional literature in the LRD library, principally of agricultural development and related topics, was examined. It was found that the initial definition was substantially correct: non-conventional literature is that which is not easily obtainable through the commercial book agencies, since the originating authorities do not participate in powerful distribution networks.

A closer view of the non-conventional literature was obtained. It proved to be mostly of high scientific value, playing a dominant role in our information service. There was a sharp distinction between the literatures of research, development and extension, each of these having its characteristic non-conventional sector. Non-conventional literature was especially important in the field of resource assessment and development surveys, partly because of its scientific content and partly because unlike non-conventional research literature it very rarely reaches the stage of being published in a journal or book.

TABLE 1 The spectrum of conventionality in the literature of tropical agriculture

Characteristic	1 Confidential documents, research/survey	2 Semi-confidential documents, research/survey	3 Weakly distributed documents a research/survey	3 Weakly distributed documents b extension	4 Published documents, research
Degree of obscurity	 Unknown outside originating authority	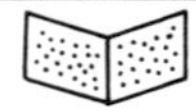 Known to few organisations	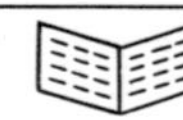 Little known	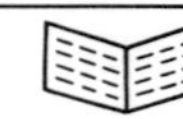 Little known	 Commonly known
Content Research	Research reports, intermediate stage	Research reports, intermediate stage	Research reports, intermediate stage	-	Research papers, FINAL stage
Survey	Survey reports, intermediate stage	Survey reports, intermediate stage	Survey reports, FINAL stage //	-	-
Extension	-	-	-	Extension documents, FINAL stage //	-
Typical security class	Confidential *	In confidence /	Limited distribution	Free distribution in limited region	None
Typical appearance	Typed sheets	Stapled mimeographed document	Various. Often bound copies reproduced from typed pages	Printed pamphlets	Fully bound printed journals and books
Typical speed of reproduction	1 week	1 month	2-9 months	2-9 months	4-12 months
Typical no. copies	1-10	Less than 100	100-1000	More than 1000 (regional public)	More than 2000 (world public)
Typically available from	-	Originators and closely related organisations	1. Originators 2. Related organisations 3. Specialist processing centres 4. Clearing houses	Authorities in the extension region	Book agencies
Typically signalled by	-	1. Originators, via lists of documents issued 2. Library accession lists	As for (2) plus: 1. Specialist processing centres. 2. Occasionally commercial secondary journals	Occasionally higher - level material by commercial secondary journals	1. Publishers 2. Book agencies 3. Secondary journals

* Implies an imperative restriction of use in the interests of secrecy
/ Implies that the reader is in honour bound not to pass on the information to "outsiders"
// Typically survey and extension documents are weakly distributed, the latter having a regional distribution. Nevertheless, certain series of extension documents have

An attempt is made in Table 1 to set out the salient features of the non-conventional literature of tropical agriculture in the form of a spectrum of document types ranging from "unknown" literature for internal consumption only, through moderately distributed documents with a few hundred readers, to fully published book or journal articles.

The table is intended as a guide to the actual situation and is not intended to place the non-conventional literature in a straitjacket. On the contrary there is every indication that a great deal can be done to lift literature out of the "obscure" into the "commonly known" category and that this depends, as is explained later, on greater publicity by issuing authorities and on developments in documentation at national, regional and world levels.

SOLUTIONS TO THE PROBLEMS OF INACCESSIBILITY

In the field of tropical agricultural resource assessment and development, the non-conventional literature is mostly of high scientific and technological value and it is therefore important to devise means whereby this can be made more accessible to potential users.

The inaccessibility of non-conventional literature is sometimes inevitable for reasons of security, but is more often the more or less accidental consequence of a certain typical mode of publication characterised by the publisher's desire to distribute quickly to a small selected number of users; other, often associated,factors are: limited funds available for publishing, small initial print run, absence of a powerful distribution network and failure on the part of the originator to seek registration in the secondary literature or in an institutional, national or international documentation centre. This initial obscurity, which may be laid at the door of the issuing authority, is aggravated by a widespread ignorance of the value of non-conventional literature and by certain weaknesses in existing documentation/information services.

One purpose of this paper has been to underline the value of the non-conventional development literature: a more general understanding of the situation will help to ensure that more non-conventional literature reaches the hands of potential users - the users themselves will perfect means of obtaining it.

The remaining solutions all involve actions by issuing authorities or by a variety of documentation/information institutions. It is true to say that whereas many steps have been taken in recent years to improve the bibliographic control of the non-conventional literature by establishing documentation/information institutions, originating authorities have continued to aggravate the problems of these institutions by a steady multiplication of the obscure literature.

Action by originators

First, originators should attempt to free their documents from unnecessary or outdated restrictions imposed by confidentiality. These restrictions, on examination, can often be revised when original conditions of security no longer apply. In addition,restrictive conditions which are thought to be imposed by clients on consultancy agencies can sometimes be removed with the client's consent. It is often a question of making suitable enquiries.

Second, originators should be as generous as possible at the initial printing stage and keep a large mailing list. Thus even within the confines of the traditional modes of producing non-conventional literature the number of users can be increased. With the same object in view, production, distribution and storage in microform should also be considered.

Third, originators should wherever possible give their documents a certain identity and authenticity by allotting to them an International Standard Book Number and/or an International Standard Serial Number. This authenticity will commonly mean that the document will be considered for entry into a variety of national and international information networks. According to the definition used in this investigation at LRD,a document so numbered* is conventional.

Fourth, originators should do everything in their power to bring their publications to the notice of the appropriate bodies who in one way or another will signal the existance of the publications to the world: national copyright libraries, national documentation centres, specialist clearing houses, and those institutional, national and international information services who produce secondary literature (abstracts, references, bibliographies, reviews etc) and tape services.

Fifth, the assignment to documents of fully descriptive titles and abstracts by the originators as recommended by the International Council of Scientific Unions Abstracting Board and by editing associations such as EDITERRA and ELSE could clearly assist the passage of the documents into the secondary literature.

Sixth, originators may decide, when circumstances are favourable, to make their publications entirely conventional. In the case of non-conventional periodicals this has often been achieved by placing them in the hands of commercial publishers. This is a logical solution but one which often means that users can no longer obtain "free copies" by exchange or other means. It seems therefore that there is a price to be paid for true conventionality.

Action in the field of documentation/information

Once a document has been non-conventionally issued, or even conventionally published, its future recognition and use depends upon the extent to which a variety of documentation/information processing centres receive or discover it and then signal its existence to the world. Whether such centres seek, find and process the non-conventional literature depends on a number of factors, notably the obstacles to be surmounted in finding it, its reputation for quality and its availability to users. Most centres, for example, have in the past catered primarily for published material and have often adjured the non-conventional literature on the grounds of its inaccessibility and sometimes, wrongly, on the grounds of its supposed lack of quality: often however such centres have simply failed to obtain these documents because of the diversity and obscurity of the sources.

While there are a number of important document recording and processing centres covering the whole or parts of the field of tropical agriculture (AGRIS, the Commonwealth Agricultural Bureaux, the Royal Tropical Institute, Amsterdam and several others), and while these centres have, to varying degrees, sought to record non-conventional literature, there are still large quantities of non-conventional documents which remain unsignalled by any of

*The use of such a number, as explained in a previous footnote, is not a guarantee that material can be made available on request. It is nevertheless a valuable signalling device.

these major services because they have gone unrecorded in their country of origin or because some countries do not yet have the capacity to complete the necessary input into major central services such as AGRIS. A service such as AGRIS depends largely for its coverage of non-conventional literature on inputs or information originating from national documentation centres responsible for the agriculture literature. It therefore follows that the complete cover of the non-conventional literature in our fields of interest will never be complete until every country has an efficiently functioning national documentation centre or equivalent institutions covering the literature of agriculture. It is to complete this infrastructure of national documentation centres for knowledge in general that NATIS (National Information Systems) has been founded as a UNESCO programme (Zaher, 1975; Unesco, 1976).

The encouragement of the formation of national documentation centres in developing countries directly benefits the bibliographic control and wider distribution of the literature of agriculture including the non-conventional literature of resource assessment and development. UNDP/FAO in the past decade have made a notable contribution to the formation of national information centres in Morocco, Algeria, Brazil and Zaire, these centres using a basically FAO style of published indexes (computer-based) and replication of documents via microfiches. Further efforts to establish such centres are desirable.

A system of national libraries, national archives and national bibliographies has long been in existence in many countries, including developing countries, and these are described in Acquisitions in the Third World (Clarke, 1975). It should not be supposed however that these older institutions have always coped with the scientific literature explosion or with the accumulations of consultancy documents which have accompanied aid programmes. Clarke (1975) also gives other helpful information to collectors of non-conventional literature from the developing countries - for example, lists printed by Government Printers and the addresses of publishers - but information officers will know that such guides are only a partial help in answering the question "What non-conventional literature on our subject has been produced in or for specific developing countries?".

There are a number of regional documentation centres in the field of agriculture capable of collecting and registering the existence of non-conventional literature arising from individual countries. Such regional centres have the function of making documents (conventional and non-conventional) more widely known and available within regions and have the additional purpose of bringing the regional knowledge into the international information networks. Again, encouragement of such centres will tend to facilitate the recognition and use of the non-conventional literature and a number of steps have already been taken in this direction. UNDP/FAO have contributed to the establishment of regional documentation centres, e.g. those of the Organisation pour la Mise en Valeur du Fleuve Senegal (for states bordering on the Senegal River) and the West Africa Rice Development Association, Liberia. UNDP/UNESCO established the Centre de Documentation de la Commission du Bassin du Fleuve Niger for states bordering on the River Niger. Two other notable examples of existing centres are AIBA (Agricultural Information Bank for Asia) (Cooney and Viloria, 1975) in South-East Asia and IICA/CIDIA covering Latin America and the Caribbean. AIBA in the Philippines provides a bibliographic service for Asia based on documents produced within the region and is also the AGRIS input centre for the region. IICA/CIDIA, in Costa Rica, has a similar function for Latin America and the Caribbean.

Turning to international documentation centres for agriculture, the AGRIS system (Wood, 1974) offers considerable hope of covering the non-conventional literature of agriculture, both in the form of Level 1 (East, 1976), which aims to cover the literature of agriculture and related subjects completely, and Level 2, still at a preliminary stage, which is "envisioned as a network of specialised information centres, data banks and data analysis centres with responsibility in depth for particular subject fields" (Caponio and Moran, 1975). AGRIS Level 1, currently recording (1976) some 77 000 titles per annum plays an important part in signalling the non-conventional literature: the role of Agrindex in reporting the literature from developing countries is especially significant because the input is world-wide and based on national and regional contributions. Nevertheless the cover achieved by Agrindex at present is far from complete. Clearly the encouragement of national and regional documentation centres will accelerate this movement towards complete cover of the non-conventional literature. AGRIS Level 2, with its proposed networks of specialised information centres, seems in principle well placed to assist in bringing the non-conventional literature of agricultural resource assessment and development (FAO, 1976) to the notice of potential users.

The development of multiple tape services in recent years, giving rapid access to CAB, AGRICOLA, BIOSIS, AGRIS and other relevant stores of information on agricultural literature, has also raised the chances of "detecting" the difficult-to-locate non-conventional literature of agriculture.

Apart from the national, regional and international information systems mentioned above, covering both conventional and non-conventional material, there are clearing houses, notably NTIS (National Technical Information Service of the US Department of Commerce) which specialise in the collection and distribution of the report literature. In the field of agriculture it is possible that there is scope for more of such clearing houses to be set up nationally or regionally for specialist purposes, sometimes in connection with national or regional research and development associations. Biotrop is a regional clearing house for tropical biology with a base in Indonesia (Sahertian-Bakhoven, 1975). The West African Rice Development Association, established by fourteen West African countries with UN assistance is an example of an institution which as a secondary function acts as a clearing house for the collection and distribution of literature, much of it non-conventional: it publishes a Current Bibliography (WARDA, 1976). DEVSIS (Wild and Woolston, 1975) is a project which intends to operate in the whole field of development, industrial and agricultural.

The publication of registers of research and development projects, whether national, regional or international, is a further means of making known the progress of investigations and thus an indirect means of publicising the non-conventional literature in which it is reported. CARIS, organised by FAO, is an example of such a register, restricted at present to West Africa.

As new organisations are founded, either primarily or secondarily to provide information, with orientations which may be based on subject, region or document-type, there is a prospect of considerable overlap between institutions. Although the emergence of these overlapping institutions may be in many ways uneconomic, it does indicate that the hitherto fugitive non-conventional literature will, by one means or another, eventually be brought to the notice of the relevant specialist readers.

Action to ensure availability

When references to documents have been seen and abstracts read, it is still necessary to obtain "hard copy" of the most relevant documents, which, when non-conventional, are particularly difficult to obtain. It is therefore desirable to ensure that information services have a capacity to reproduce documents in the form of photocopies, microfiches or reproductions from microfiches, or equivalent arrangements. The present procedure for obtaining the non-conventional literature recorded in Agrindex (FAO, 1976) is that it 'should be obtainable through the organisation which is indicated by an asterisk (at the end of the reference)'. When such an organisation is an official input centre with microfiche-making equipment, it is probable that copies can easily be obtained. Again the importance for our purposes of having national documentation centres, capable both of signalling and reproducing the national non-conventional literature, is apparent.

Cooperative action between libraries is another means of providing hard copy once a document has been tracked down. As already mentioned, AGLINET is an example of a network of libraries which aims to complement the signalling service of Agrindex.

It should be noted however that even when 'hard copy' is made available by information services there are still barriers which prevent potential users, particularly in developing countries, from obtaining the documents. These barriers may be related to currency problems, customs regulations, lack of funds in libraries, poor telephonic or postal communications and a number of other causes*. It is hoped that a determined effort to surmount these barriers may be made in future.

CONCLUSIONS

In the section on solutions we have discussed numerous ways in which originators of documents can make their material better known and have indicated developments in the field of documentation/information which will tend, if encouraged, to make the non-conventional literature, especially the literature produced in and for developing countries, more accessible. Our plea is that those who encourage the further development of services for agricultural literature should give thought to the detection, reporting, analysis and reproduction of the non-conventional literature, especially that which deals with land resource assessment and development in the tropics.

ACKNOWLEDGEMENTS

This paper is published with the permission of the Director of the Land Resources Division. Acknowledgement is made to Mrs S A Hitchin and Mr P M Reilly of the Land Resources Division for their assistance in reading the early versions of this paper and to Mr H East, Chief of AGRIS Coordinating Centre, and Mr F C Hirst, Librarian of the Ministry of Agriculture, Fisheries and Food, United Kingdom, for reading the manuscript.

The view expressed by the authors are their own and do not necessarily represent an official view of the Land Resources Division.

---

*The International Development Research Centre of Canada (Lebowitz, personal communication, 1976) is currently sponsoring a study of these problems and their solution with special reference to inter-library loan and photocopying services available to developing countries in South-East Asia.

REFERENCES AND RELEVANT WORKS

ADIMORAH E N O (1976) Problems of scientific information work in developing countries. Information Scientist, volume 10, no. 4, 139-148.

BAULKWILL W J (1972) The Land Resources Division of the Overseas Development Administration. Tropical Science, volume 14, no. 4, 305-322.

BERRY M J, LAURENCE J F, MAKIN M J & WADDAMS A E (1974) Development potential of the Nawalparasi area of Nepal. Land Resource Study no. 17.

CLARKE D A ed. (1975) Acquisitions from the Third World. Papers of the Lique des Bibliotheques Européenes de Recherche seminar, University of Sussex, Brighton, UK, 17-19 September, 1973. London: Mansell.

CAPONIO J F & MORAN L (1975) AGRIS. Science, volume 187, no. 4173, 233-236.

COONEY S & VILORIA R V (1975) Agricultural Information Bank for Asia. Paper presented at the Fifth World Congress of IAALD, Mexico, 14-18 April 1975.

DUBOIS C P R (1973) INSCOPE. A co-ordinate index in the field of applied economics. Aslib Proceedings, volume 25, no. 9, 336-339.

DUBOIS G (1972) Un systême internationale d'information pour les sciences et la technologie agricoles (AGRIS). Quarterly Bulletin of IAALD, volume 17, no. 2, 55-64.

EAST H (1976) AGRIS level one: progress in the first operational year. Quarterly Bulletin of IAALD, volume 22, no. 2, 43-56.

EBSCO SUBSCRIPTION SERVICES (1975) Librarians' handbook, 1975-1976. Birmingham, Alabama: EBSCO Industries Incorporated.

EDWARDES P M J (1974) The IAR library. Quarterly Bulletin of IAALD, volume 19, no. 3/4, 209-220.

FOOD AND AGRICULTURE ORGANISATION (1976) Agrindex, volume 2, no. 10, i-x.

HARADA K (1972) The world-wide network of agricultural libraries (AGLINET). Quarterly Bulletin of IAALD, volume 17, no. 2, 68-77.

HAANTJENS H A comp. (1972) Lands of the Aitape - Ambunti area, Papua New Guinea. CSIRO Land Research Series no. 30.

HOLLOWAY A H, RIDLER E H, FUCCILLO D A, WILSON M E & YATES B (1976) Information work with unpublished reports. London: Deutsch.

LEBOWITZ A (1976) Personal communication.

LILLEY G (1975a) Characteristics of the production of scientific and technical agricultural literature in more developed countries and its relation to information networks. Paper presented at the Fifth World Congress of IAALD, Mexico, 14-18 April 1975.

LILLEY G (1975b) The Fifth World Congress of the International Association of Agricultural Librarians and Documentalists. UNESCO Bulletin for Libraries, volume 29, no. 6, 336-340.

LITTLETON I T (1969) The literature of agricultural economics. North Carolina Agricultural Experimental Station. Technical Bulletin 191.

MEI A (1975) La littérature non-conventionelle sur le développement dans le cadro de la documentation agricole mondiale. Rivista di Agricoltura subtropicale e tropicale, volume 69, no. (1-3/4-6), 99-102.

OGUNSHEYE F A (1969) Problem of bibliographic services in Nigeria. Nigerian Libraries, volume 5, no. 2, 62-68.

POSNETT N W (1973) The Land Resources Division and its information services. African Research Documentation, no. 2, 16-18.

POSNETT N W (1975) The Land Resources Division and its feature card system. Quarterly Bulletin of IAALD, volume 20, no. 1, 23-33.

POSNETT N W & REILLY P M comp. (1973) Fiji. Land Resources Bibliography no. 4. Tolworth, Surrey, UK: Land Resources Division, Ministry of Overseas Development.

SAHERTIAN-BAKHOVEN P (1975) The potential supporting role of Biotrop - SEAMEO Regional Centre for Tropical Biology Clearing House and Information Centre in enhancing the effectiveness of national information networks. Paper. Fifth World Congress of the International Association of Agricultural Librarians and Documentalists, Mexico, 14-18 April 1975.

SKIDMORE G L comp. (1974) Annotated bibliography of scientific and technical information compiled from work and records of the Department of Agriculture, Victoria, Mahé, Seychelles to August 1972. London: Eyre and Spottiswood Limited.

UNESCO (1976) NATIS news. UNESCO Bulletin for Libraries, 30(6), 306-310.

WEST AFRICA RICE DEVELOPMENT ASSOCIATION (1976) Current Bibliography. Monrovia: West Africa Rice Development Association.

WILD I L & WOOLSTON J E (1975) Social and economic development. How DEVSIS would link into the emerging network of international information systems. Paper presented at the Fifth World Congress of the International Association of Agricultural Librarians and Documentalists, Mexico, 14-18 April 1975.

WOOD D N (1974) Access to primary documents in the fields of agriculture, food and related subjects. FAO Report no. AGRIS 16.

ZAHER C R (1975) National information systems (NATIS), a new UNESCO programme for improving access to information. Paper presented at the Fifth World Congress of the International Association of Agricultural Librarians and Documentalists, Mexico, 14-18 April 1975.

APPENDIX 1 SOME ORGANISATIONS WHICH ARE CHARACTERISTIC ORIGINATORS OF NON-CONVENTIONAL LITERATURE OF TROPICAL AGRICULTURAL RESOURCE ASSESSMENT AND DEVELOPMENT

Agricultural Information Bank for Asia
South East Asia Regional Centre for Graduate Study and Research in Agriculture College
Laguna 3720
Phillipines

Arid Zone Research Association of India
c/o Central Arid Zone Research Institute
Jodhpur
Rajasthan
India

Bureau of Resource Assessment and Land Use Planning
University of Dar es Salaam
PO Box 35097
Dar es Salaam
Tanzania

Commonwealth Scientific and Industrial Research Organisation
314 Albert Street
East Melbourne
Victoria 3002
Australia

Commonwealth Development Corporation
33 Hill Street
London W1A 3AR
England

United Nations Economic Commission for Africa
PO Box 3001
Addis Ababa
Ethiopia

Institute for Scientific Cooperation
7400 Tubingen
Landhausstrasse 18
Federal Republic of Germany

Interafrican Committee for Hydraulic Studies
BP 369
Ouagadougou
Upper Volta

Office de la Recherche Scientifique et Technique Outre-Mer
70/74 Route d'Aulnay
93140, Bondy
France

Royal Tropical Institute
63 Mauritskade
Amsterdam - 5
The Netherlands

US Agency for International Development
Department of State
2201C Street NW
Washington DC 20523
USA

University of the West Indies
St Augustine
Trinidad
West Indies

APPENDIX 2 EXAMPLES OF NON-CONVENTIONAL LITERATURE OF TROPICAL AGRICULTURAL RESOURCE ASSESSMENT AND DEVELOPMENT HELD IN THE LRD LIBRARY

Secondary Periodicals

Africa Index	United Nations Economic Commission for Africa PO Box 3001 Addis Ababa, Ethiopia
Arid Lands Abstracts	Office of Arid Land Studies, University of Arizona, Tucson, Arizona 85721, USA
Current Bibliography	West Africa Rice Development Association PO Box 1019 Monrovia, Liberia
Index of Biology, Agriculture and Agroeconomy	Bibliotheca Bogorensis, JI. Ir. H Juanda no. 20, Bogor, Indonesia
Nedo Abstracts	Royal Tropical Institute, 63 Mauritskade Amsterdam - 5 The Netherlands

Primary Periodicals

Actuel Développement	Actuel Developpement, 110 rue de l'Université, Paris-7e, France
Bulletin de Liaison	Association Sénégalaise pour l'Etude du Quaternaire Africain 227-56 Dakar-Fann, Sénégal
Bulletin of the Regional Office of Science and Technology for Africa	UNESCO, PO Box 30592 Nairobi Kenya
Liaison Bulletin	Interafrican Committee for Hydraulic Studies BP 369, Ouagadaugou, Upper Volta
Natural Resources and Development	Institute for Scientific Cooperation Landhausstrasse 18, 7400 Tubingen Federal Republic of Germany
Proceedings, Agricultural Society of Nigeria	The Editor, Dept. of Agric. Economics and Extension, University of Ibadan Ibadan, Nigeria

Monographs (LRD Library catalogue cards)

ANDRIESS J P
SIM E S
SARAWAK, DEPARTMENT OF AGRICULTURE, RESEARCH BRANCH

Report on soil investigations in the proposed Sungai Sarawak padi scheme area - 1st division

Report No. 125
Vol. 1 text, Vol. 2 maps

Sarawak: Department of Agriculture, Research Branch 1968

Book 38 pp 10 maps

CONWAY G
ROMM J
THE FORD FOUNDATION

Ecology and resource development in South-east Asia.

Bangkok, Thailand: The Ford Foundation Office for South-east Asia 1973

Book

BRASIL, MINISTERIO DE AGRICULTURA, CENTRO NACIONAL DE ENSINO E PESQUISAS AGRONOMICAS

Levantamento de reconhecimento dos solos da regiao sob influencia de reservatorio de Furnas.
(Reconnaissance soil survey of the region of the State of Minas Gerais influenced by the Furnas Dam)
Portuguese text, English summary.

No. 13

Rio de Janeiro: Centro Nacional de Ensino e Pesquisas Agronomicas 1962

Book 462 pp bibl.

FOOD AND AGRICULTURE ORGANIZATION OF THE UNITED NATIONS (FAO)
UNITED NATIONS DEVELOPMENT PROGRAMME (UNDP)

A land capability appraisal - Indonesia. Interim report.

AGL/INS/72/011

Rome: FAO 1974

Book iv + 47 pp bibl. 5 maps

GERHART J
MEXICO, CENTRO INTERNACIONAL DE MEJORAMIENTO DE MAIZ Y TRIGO (CIMMYT)

The diffusion of hybrid maize in western Kenya. Abridged by CIMMYT.

Mexico City: CIMMYT 1975

Book vi + 57 pp

PRADHAN M L et al.
NEPAL, DEPARTMENT OF AGRICULTURAL EDUCATION AND RESEARCH, SOIL SCIENCE SECTION

Soil survey of Sunsari District, Kosi Zone

Report No. 8

Nepal: Ministry of Land Reform, Food and Agriculture, Department of Agriculture Education & Research. 1966

Book

HERNANDEZ S G
AGUILERA H N
UNIVERSIDAD NACIONAL AUTONOMA DE MEXICO

Ecologia de suelos copreros y su potencialidad economica en la Costa Grande, Estado de Guerrero. (Ecology of coconut soils and their economic potential in Costa Grande, Estado de Guerrero.) Spanish text.

Mexico: Universidad Nacional Autonoma de Mexico 1971

Book 41 pp bibl. 1 map

INSTITUT D'ELEVAGE ET DE MEDECINE VETERINAIRE DES PAYS TROPICAUX (IEMVT)
GASTON A et al.

Project Assale-Serbewel - étude agrostologique des paturages. (Agrostological study of pastures: Assale-Serbewel project).

Notice d'utilisation de la carte des paturages du projet Assale-Serbewel. (Note on the use of the pasture maps.)

Projet Assale-Serbewel - Cartographie
IEMVT - Etude Agrostologique No. 041
France: IEMVT 1974 Book

UNITED STATES, NATIONAL ACADEMY OF SCIENCES

CONFERENCE ON AGRICULTURAL RESEARCH PRIORITIES FOR ECONOMIC DEVELOPMENT IN AFRICA
ABIDJAN, Ivory Coast, April 5-12 1968

Vol. I and Vol. II

Washington DC: National Academy of Sciences. 1968

Books 612 pp

SAKIYAMA T
NEPAL, MINISTRY OF FOOD AND AGRICULTURE,
ECONOMIC ANALYSIS AND PLANNING DIVISION

International assistance to Nepalese agriculture. Subsectoral and project analysis (1951-70).

EAPD Staff Paper No. 4

Kathmandu: Ministry of Food and Agriculture. 1971

Book

WHITESELL C D
UNITED STATES, DEPARTMENT OF AGRICULTURE (USDA), FOREST SERVICE

Growth of young Saligna eucalyptus in Hawaii: 6 years after thinning.

USDA Forest Service Research Note PSW-299, 1975

Berkeley, California: US Pacific South-west Forest and Range Experiment Station

Pamphlet 3 pp

UNITED STATES AGENCY FOR INTERNATIONAL DEVELOPMENT (AID)

Improving farm production in tropical and sub-tropical regions of limited rainfall.

Agriculture Technology for Developing Countries Technical Series Bulletin No. 4

Washington DC: AID Technical Assistance Bureau 1971

Book 25 pp bibl.

A RATIONAL SYSTEM FOR EDITING AND PUBLISHING THE LITERATURE OF TROPICAL AGRICULTURAL RESOURCE ASSESSMENT AND DEVELOPMENT PLANNING

W.J. Baulkwill, *Land Resources Division, Ministry of Overseas Development, Tolworth Tower, Surbiton, Surrey KT6 7DY, England*

ABSTRACT

The literature of tropical agricultural resource assessment and development planning is characteristically written by government departments or by teams of consultants and is non-conventional, bulky, fragmented and difficult to retrieve. Since the early 1960s, the Land Resources Division has publicised its main findings but has encountered difficulties in meeting conflicting needs: in particular the client government's need to receive recommendations for development quickly and in confidence, the writer's desire to meet the most exacting demands of his fellow scientists, the requirements of the developing world at large and the needs of information services to retrieve the documents. There have also been complaints of delays in presenting results and of the excessive bulk and technical detail of the numerous reports which emerge from an assessment project. To improve this situation LRD has analysed the contents of its six different series of project reports and has related these to specific categories of readers, e.g. client government executives and scientific advisers ("participants") and non-client scientists and aid agencies ("spectators"). A flexible system has consequently been designed at LRD whereby the documents produced by a project team are visualised as outputs of a production line to be prepared individually or assembled in combinations to meet the needs of specific reader categories. Each document so produced is coded to relate it to the project and in addition provided with a document control sheet to facilitate recognition and retrieval. The main types of documents under the new system are Project Records (mainly supportive data of prime interest to scientific advisers), Project Reports containing recommendations for action and Land Resource Studies (printed volumes, derived from the Project Reports and supported by microfiches of selected Project Records for the permanent use of all reader groups). The editorial section of LRD has thus sought a creative role in structuring, condensing and distributing the results of tropical resource assessment. To fulfil this role, editors must work in the field to assist team managers in processing materials for the benefit of specific readers. Finally it is shown that the lessons learnt by LRD are applicable in any situation where complex projects of resource assessment and agricultural development are reported.

Miriam Balaban (ed.), Scientific Information Transfer: The Editor's Role, 523-542.

1. THE WORK AND PUBLICATIONS OF THE LAND RESOURCES DIVISION

The Land Resources Division is a scientific unit of the Ministry of Overseas Development charged with evaluation of land resources and development of land and with associated consultancy and training work. All the work is done in response to requests from developing countries.

The Division has about 60 scientists covering earth sciences, ecology, agriculture, forestry, sociology and economics.

LRD adopts an integrated approach to resource assessment and land development projects, that is, it makes for each project a collaborative study, starting with air photo interpretation, of all the relevant factors - geology, geomorphology, climate, water resources, soils, vegetation, land use and economic and social conditions - leading to a land evaluation which normally includes selection of certain areas for specific development possibilities. It is common for such projects to include second and subsequent phases in which the feasibility of proposed developments is tested and precise plans for actual development are prepared.

Since 1965, the Land Resources Division has undertaken some 30 major projects (each lasting at least 12 months) and a similar number of minor projects, the major projects being described in printed Land Resource Studies. In addition to its printed Land Resource Studies, LRD has issued several hundreds of unpublished reports, mostly in confidence and with very restricted circulation, covering technical aspects of the projects.

2. THE PUBLICATIONS PROBLEM

In this section the defects of the editing/publishing system up to 1976 are explained and it is shown how the most persistent problems were only overcome after an analysis of the main constituents of project reports and their relevance to different classes of readers. LRD's problem-solving efforts are introduced by a description of the nature of the written products of a project and of the way in which these were initially handled by the "traditional" methods of research institutes - categorising the reports as they were placed before the editor into "series" according to content.

2.1 Content of LRD reports: the three main streams

In the past there have been numerous categories of LRD project reports which are described below. Close examination of these documents shows that there are three main streams of written material which enter to varying degrees into all project reporting. Much of this paper is concerned with the problem of finding suitable outlets for these three streams.

1. Survey data, most of which are best retained in archives but some of which in condensed form, are used in technical reports to show the validity of the conclusions and to be used in further investigations. Examples are landscape analysis descriptions, soil profile descriptions, soil chemical analyses, descriptions of soil classification methods, meteorological records, sociological survey data. This material, when appearing in technical reports, is generally described in this paper as supportive material.

2. Interpretation There is an interpretative element in every survey; that is, the extraction from the data of resource classes whose distribution can often be shown on maps (e.g. soil association or forest type maps) or of trends or possibilities which will effect eventual development. At a later stage these interpretations are integrated to give an evaluation of the land's suitability for a variety of development possibilities.

3. Recommendations These arise directly from interpretation and are of prime interest to administrators and their scientific collaborators.

   Figure 1 shows schematically how, during a land resource assessment project, there is a continual process of condensing the data while at the same time there is a growth of the interpretative element leading to recommendations.

2.2 The previous system

From 1967-1973 there was a severe staff shortage and the attention of the editor was concentrated on the most prestigious publications, the printed Land Resource Studies but, as the complete list below shows, there were other series, nearly all confidential, and of very limited distribution.

1. Appraisal/Proposal Report: report on LRD's initial appreciation of the terrain with proposals for a resource assessment/land development project.

2. Land Resource Report: supplying client governments with technical reports on individual subjects such as physiography, climate, vegetation and soils. These included some interpretation of the data and were heavily loaded with supportive material.

3. Mid-Term Review: review of progress with some recommendations to the client government.

4. Draft Land Resource Study: the key publication, a report to the client government containing interpretations of the data and recommendations for development; it also contained much descriptive and supportive matter which was in reality composed of late-arriving, technical material of the Land Resource Report type. The Draft Land Resource Studies were always required urgently, were quickly printed from camera-ready typed copy and were usually accompanied by black and white maps.

5. Land Resource Study: a much-edited, and revised printed version of (4) provided with high-quality (often coloured) maps, the printing process normally taking nine months. It was sent to the client government in quantity and also to a world-wide mailing list.

6. Supplementary Report: containing 'useful data' such as meteorological tables - essentially a recent device to deflect supportive material from the main printed report.

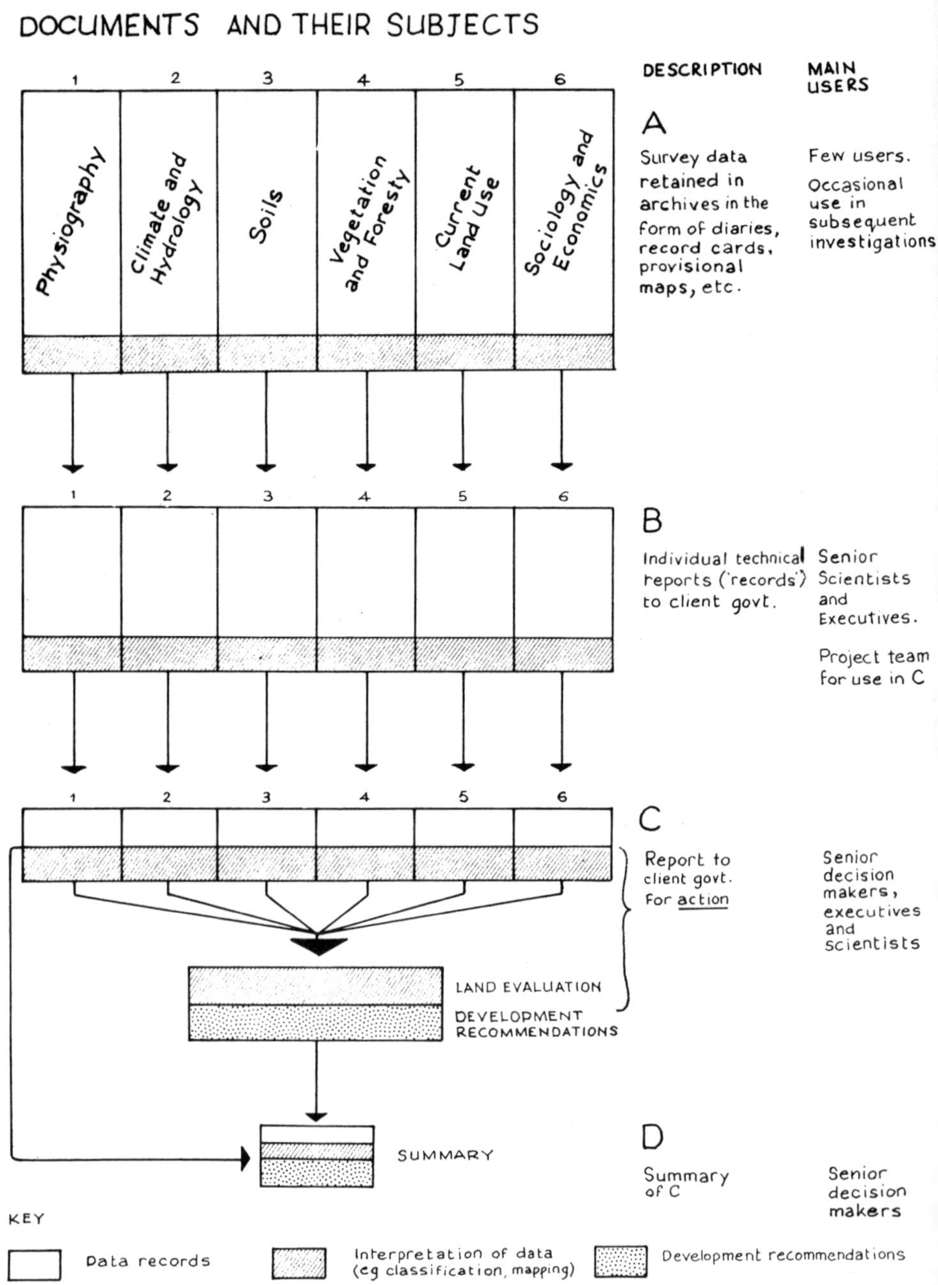

FIGURE 1 - Reporting of a land resource and development project. A schematic display of the successive processes of interpretation and condensation of data in relation to broad classes of users. A system of routine summarising of stage B documents for use in Stage C is assumed here.

This sixfold array of LRD series successfully categorised the total output arising from projects in general, but failed to unite the various outputs of any one project. Thus it was easy to assemble all LRD Project Proposals/Appraisals but relatively difficult to assemble all reports on, for example, the North-East Nigeria project.

2.3 Advantages and defects of the previous system

Most development agencies comparable with LRD act in confidence and make no attempt to publish their results, which become the private property of the client government; when such results are published, we see only the tip of the iceberg, the final printed product, minus its supportive material. It is therefore impossible for us to compare the success of LRD's publishing system with that of other similar organisations and our judgements can only be made in terms of LRD objectives.

On the credit side, LRD published, with the client government's permission, the Land Resource Study, a polished final report of every major project, well equipped with maps and available to a world-wide mailing list of some 300 ministries, universities, research centres and aid agencies, the majority of addresses being in the developing countries. The remaining documentary output of LRD projects was carefully categorised and serialised and although of very limited distribution was not impossible to find.

The key document giving vital information to the client government, the Draft Land Resource Study, although bulky and sometimes slowly written, was edited, typed and reproduced at high speed.

Also on the credit side there were comprehensive Instructions for Typists and Authors, including both house style and final report structure.

On the debit side the documents were undoubtedly overladen with data from beginning to end of a project: an example was the sometimes unnecessary inclusion of massive soil analysis data.

There was tendency to interlace the three streams, supportive data, interpretation and recommendations, which constituted a major fault, led to confusion in the minds of the less scientific readers, and sometimes obscured the need for action. At the same time the dead weight of the supportive material caused delays in writing, editing, printing and expense in dispatch. This confounding of the three streams proved to be the hard core of our problems to which we return later.

A related defect was the excessive elaboration of accounts of certain studies due to the author's desire to present a good academic report, using a degree of refinement not demanded by the terms of reference. The result tended to be a lengthy section making the whole report unbalanced.

At the stage when the Draft Land Resource Study was presented to senior decision makers, there were complaints of lack of impact, despite the fact that all such reports contained a well-signposted summary of recommendations. It was made clear that such senior readers required a separate volume for their own purposes, containing not merely summarised recommendations but a synopsis of the whole report, presented in an appropriate style.

Another defect, already mentioned, was that there was no coding system to link all the documents emanating from one project. The obstacles thus caused to the retrievability of a complete set of project documents are obvious.

Finally there was insufficient staff before 1973 to guide the production of reports and to edit texts. The result was shallow editing of all but final reports and inability to overcome defects in the system.

2.4 Problem identification

Awareness of the problems of editing and publishing project reports grew during the period 1971-74, which also coincided with a heavy increase in output of documents.

Briefly reviewed, the problems were

1. Insufficient staff to control and edit material
2. Excessive bulk due to presence of unnecessary supportive data
3. Imbalance and excessive bulk caused by overelaboration of some scientific elements
4. Slowness of production (writing, editing, printing) of some reports and their maps, resulting in part from (1), (2) and (3)
5. Confusion caused by juxtaposition of supportive data, interpretation and recommendations and consequent lack of impact of reports on senior decision-makers
6. Failure to provide a coding system to permit easy assembly of all documents emanating from a single project.

3. THE SOLUTION

The problems were attacked in two successive waves, the first a partially successful attempt to solve the problems by better staffing and administration (1971-74) and the second a more radical approach to the remaining, "hard core" problems in 1975-76.

3.1 Initial improvements (1971-74)

The problem of staffing was met by the addition of two staff giving a 3-tier system of senior editor, editor and sub-editor.

In addition a system of 'technical appraisal' (internal refereeing) was introduced whereby individual elements in reports such as soil science and hydrology were scrutinised by appropriate senior scientists. Finally the editing staff was buttressed, not always with success, by the use of graduate contract editors equivalent to an additional one half of an editor.

The administrative system was improved by the use of a plan chart showing planned and actual achievement in editing, map preparation, all the printing stages and dispatch. At the time of writing there are some 35 major volumes on the chart.

An attack was made on the problem of excessive bulk, and, given the several types of report already described, the problem was seen as one of directing the materials into the appropriate types of report and in particular diverting bulky data-rich supportive material out of the final printed report by early planning and consultation with team managers. It was at this stage that the Supplementary Report (e.g. of meteorological data) separated from the printed Land Resource Study was invented.

The "first-wave" improvements were successful in improving the depth of editing, imposing more uniformity and more rigorous technical appraisal, reducing the length of reports by a fraction and giving better control of printing.

Nevertheless, a hard core of persistent problems remained, mostly centred on the bulk and imbalance already mentioned, for despite all our attempts at early planning, team leaders still tended to bring to the editor's desk a key Draft Land Resource Study which was too long, with unhealthy growths in certain subject areas, and still weighed down with appendixes of data for which no other suitable place had been found. Since the Draft Land Resource Studies always had to be issued with great speed it is not surprising that the task of revising them for a final printed version proved to be a time-consuming one!

3.2 Solution of persistent problems (1975-76)

The persistent problems were now well known but not yet clearly understood. It was evident that editorial control of the documentary output at the early (field) stages was too lax, allowing too many data to creep into documents, or into the wrong documents. It was evident too that the sixfold array of LRD Serials, each a 'category' with a defined 'content', might be an evolutionary failure, causing material to appear too late, to go to the wrong reader or to lose the desired impact.

It was therefore decided to devise a new system, putting more emphasis on the needs of the project, with the overriding objective of bringing the necessary information to appropriate users in the client country without delay. If necessary the existing serial categories would be abandoned entirely. In other words we sought a goal-directed system.

To meet these newly stated objectives, the desired new system was redefined as:

1. Directing specific elements of the total documentary output of a project to the appropriate classes of readers, with maximum impact on the readership.

2. Relating especially to the needs of the prime users (decision makers, administrators and scientists in the client country).

3. Catering for the needs of interested observers (development agencies, universities, etc.) outside and within the client country, but in no way interfering with (2).

4. Restricting supportive data in reports to what is necessary to justify conclusions and provide tools for action.

5. Enabling the editor in consultation with the project manager to retain control of the documentary output throughout the project.

6. Ensuring that the total written output from one project would constitute for LRD, the client and other users a recognisable corpus of documents.

With these aims in mind we now set out to define the common elements in the documents of major projects, to define typical readership classes and their needs in relation to these elements.

3.2.1 The relationship between document content and readership

3.2.1.1 Analysis of document content

An abbreviated analysis is set out in Table 1 of the documents emanating from a typical major land resource assessment and development project. It will be seen that the various types of document content are directly related to a sequence of project activities, these activities being broadly speaking common to all our integrated projects, regardless of scientific discipline.

3.2.1.2 Analysis of readership

The various classes of reader and their needs were defined in Table 2 after intensive consultations with staff responsible for project implementation and project scheduling and with experienced project leaders.

Note that the first five groups are all participants in that they are directly committed to utilise the reports. The sixth group may be called spectators in that they may use the matter for secondary purposes - education, etc. But the spectators are not regarded as "useless": they may become participants at a later stage and from LRD's viewpoint they also contribute considerably to the Division's store of knowledge under document exchange arrangements.

In general there was little difficulty in defining the classes, but senior development executives and senior scientists, originally defined as two separate readership groups, were eventually merged to form Class 2 because of practical difficulties in separating their roles and reading needs.

3.2.1.3 The document content/reader matrix; or who needs to read what

In Table 3 the types of document outlined in Table 1 are related to the six classes of reader.

The object of issuing supportive material (dsm in Table 3) is twofold: to validate the conclusions and to provide detailed information which scientists may use for a number of purposes. It will be seen immediately that three classes of reader, Class 2 (Senior development executives and senior scientists), Class 3 (Officers in charge of local research) and Class 6 (Educationalists/foreign observers (the 'spectators')), require the supportive material and that Class 2 must have it in order to judge and implement the proposals made in the reports.

It is also clear that the senior and junior reader classes in the direct line of plan-implementation, Class 1 (Senior decision makers) and Class 4 (Local development executives) will not normally need supportive reports e.g. technical reports on environmental and other data, and certainly do not need to have these reports repeated (as was often done in the past) in the printed Land Resource Study.

TABLE 1 Analysis of the content of the total *corpus* of documents emanating from a typical land resource assessment and development project.

Types of document content associated with project activities	Classification*	Explanatory comment with examples from actual reports
Initial problem identification with proposals for establishing a project	ra	Known at LRD as 'Project Appraisals/ Proposals'
Method (including classification methods)	dsm	Example: soil analysis methods
Collation from previously existing records of data	dsm	Previous surveys used as sources of information
Recording of data	dsm	Examples: climatic records, soil description, record of forest species, sociological and economic data
Classification and interpretation of data	i	Typically classification of forests, etc., and geographical distribution of classes
Mapping of interpretation	i	Typically soil associations, forest types, etc.
Integration of interpretations of various kinds of data	i	Example: combination of landscape + soils + vegetation + current land use + economics to give land suitability evaluation
Solution of problems and recommendations	ra	Typically indicates possible areas for specified kinds of development

*A rough classification according to the leading element in the subject matter:

dsm = data and other supportive matter;
i = interpretation of the data;
ra = recommendations for action

Class 5 (Extension workers) do not normally need any project report material because it has no direct application to extension work. They have important special needs which are considered below.

3.2.1.4 Special needs of certain users

It has been recognised for some time that senior decision makers (Class 1 users) need special treatment, i.e. that they should receive a special summary, possible a separate volume, summarisi g all elements of the report. This has already been done in recent LRD reports.

TABLE 2 Analysis of the readership of LRD project reports

Organisational level	Class	Type of reader	Function in relation to report	Reader's requirement
National or State	1	Senior decision makers Political leaders Senior administrators Financiers	Weigh arguments for options. Take advice from various sources including Class 2 readers Make decisions in relation to political and financial possibilities	Condensed but substantial report <u>excluding</u> bulk of supportive material but containing: Interpretations Cost-benefit summary Recommendations <u>Readers with limited time need a separately bound summary</u>
National or State	2	Senior development executives* and senior scientists Heads of departments Scientific advisers Senior officers of research organisations	Advise Class 1 readers on action, explaining technical implications Prepare plan and supervise its execution Arrange related extension work	As for Class 1, plus supportive material as follows: Descriptions of methods Supportive (e.g. environmental) data which can be examined in detail.
Local	3	Officers in charge of local research	Conduct research as indicated in plan Use report as research tool	As for Class 2
Local	4	Local development executives*, including district agricultural officers	Execute plan as indicated by Class 2 readers	As for Class 1. May also receive annotated digests of supportive data from Class 2 readers
Local	5	Extension workers	Help farmers to develop technical skills as indicated in plan	Do <u>not</u> require project reports Require special extension material possibly derived from the reports by Class 2 readers
Not applicable: international clientele	6	"Spectators" Educationalists and students in client country and abroad Research workers and planners outside client country	Use knowledge in report with view to <u>future</u> development or for application to <u>other</u> countries with analogous conditions	As for Class 2: including supportive material, but do <u>not</u> require reports urgently

*The term "development executive" has been coined to designate officers responsible for executing development plans. Many have scientific/ technical qualifications.

TABLE 3 Who needs to read what?

Type of document content	Classification*	Classes‡ of readers requiring	
		Summary only	Full details
Problem indentification/statement of objectives	ra		1,2,3,4,6
Details of methods of all kinds, including classification methods	dsm	1,4	2,3,6
Collation from previously recorded data	dsm	1,4	2,3,6
Recording of data	dsm	1,4	2,3,6
Classification and interpretation	i		1,2,3,4,6
Mapping of interpretation	i		1,2,3,4,6
Integration of interpretations	i		1,2,3,4,6
Solution of problems/recommendations	ra		1,2,3,4,6

*dsm = data and other supportive material; i = interpretation of the data; ra = recommendations for action

‡Readership classes: 1. Senior decision makers, 2. Senior development executives and senior scientists, 3. Officers in charge of local research, 4. Local development executives, 5. Extension workers, 6. 'Spectators'.

The field extension staff (Class 5) of the client country are a class who have a need for specially prepared material showing how the field demonstrator has to act when helping to implement the plan - in other words, extension manuals, extension leaflets, audio-visual material etc. Such material is not normally produced by a development project, although the terms of reference could provide for the preparation of extension aids if these were requested by the client government.

Since the dispatch of heavy printed volumes to addresses on a world-wide mailing list has been an expensive operation, the needs of the 'spectators' (Class 6 users) have to be reviewed critically, although there is a quantifiable return to LRD from this mailing list in the form of monographs and journals related to development. As already explained, these spectators have comprehensive needs (full supportive data as well as interpretation and recommendations) whose satisfaction is essentially of very long-term benefit to developing countries. A reasonable compromise solution seems to supply the spectators with interpretative material and recommendations in printed form plus a selection of the supportive material on microfiche to reduce the cost of printing and transport.

3.2.2 Reorganisation of the output of project documents

Following the analysis of the readers and their requirements, and bearing in mind the persistent problems identified in Section 3.1 (excessive length, confounding of supportive data, interpretation and recommendations, need of a coding system to assemble documents from one project), it was decided to reorganise the documentary output to meet the following description:

1. In future all reports from a project would be regarded as a sequence of outputs from the project, each one being a contribution to the whole - the solution of problems for the benefit of the client government.

2. It would be extremely unusual to repeat any individual output - it being assumed that the client government would collect and assemble the parts which make the whole.

3. The documents of the project would fall into two main categories: supportive material (to be called Project Records) and reports recommending and explaining action (Project Reports). The difference in the readership of the two categories would be kept in mind.

4. Unlike the end-of-project Draft Land Resource Studies of the old system, Project Reports would appear as and when dictated by the progress of the project.

5. The printed Land Resource Study, intended both for bulk supply to the client government and individual supply to 'spectators', would continue, but would be derived from one or more Project Reports, not as before from a single magnum opus (the Draft Land Resource Study). Essentially it would be a more condensed work than before, shorn of distracting supportive material. In this way it would best serve the needs of the client government who would have already received the supportive data in the form of Project Records. Nevertheless a concession would be made to the needs of 'spectators' by supplying them with microfiches of selected Project Records.

There would thus be far fewer categories of report than previously, as shown in Figure 2 where the equivalence between the old and new categories is shown.

Figure 3 compares the old and new systems and shows the clear distinction now made between reports recommending action and reports supplying scientific support. It also shows how the previous galaxy of supportive reports, supplements and appendixes now all have the same status as Project Records and are issued as soon as the authors can produce them. Finally, it shows how the principle of continual condensation (already shown in Figure 1) works out in practice, with summaries of Project Records moving via the Project Manager* into the Project Reports which are themselves summarised to provide an administrator's synopsis for busy decision makers.

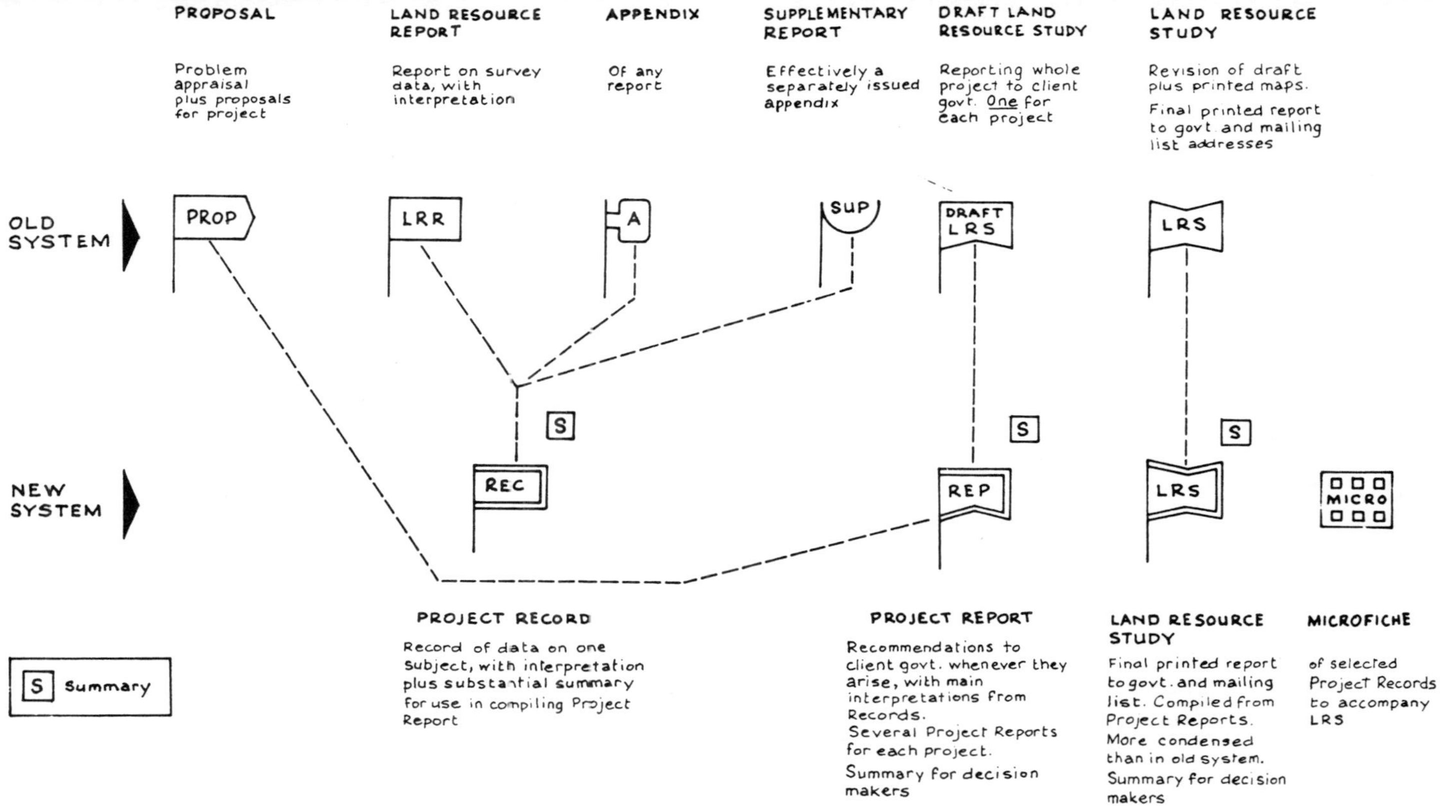

FIGURE 2 - Representation of various types of report in the old and new systems of editing. The dotted lines show equivalence between reports in the old and new systems. The new system includes routinely produced summaries of Project Records and separately bound summaries of Project Reports.

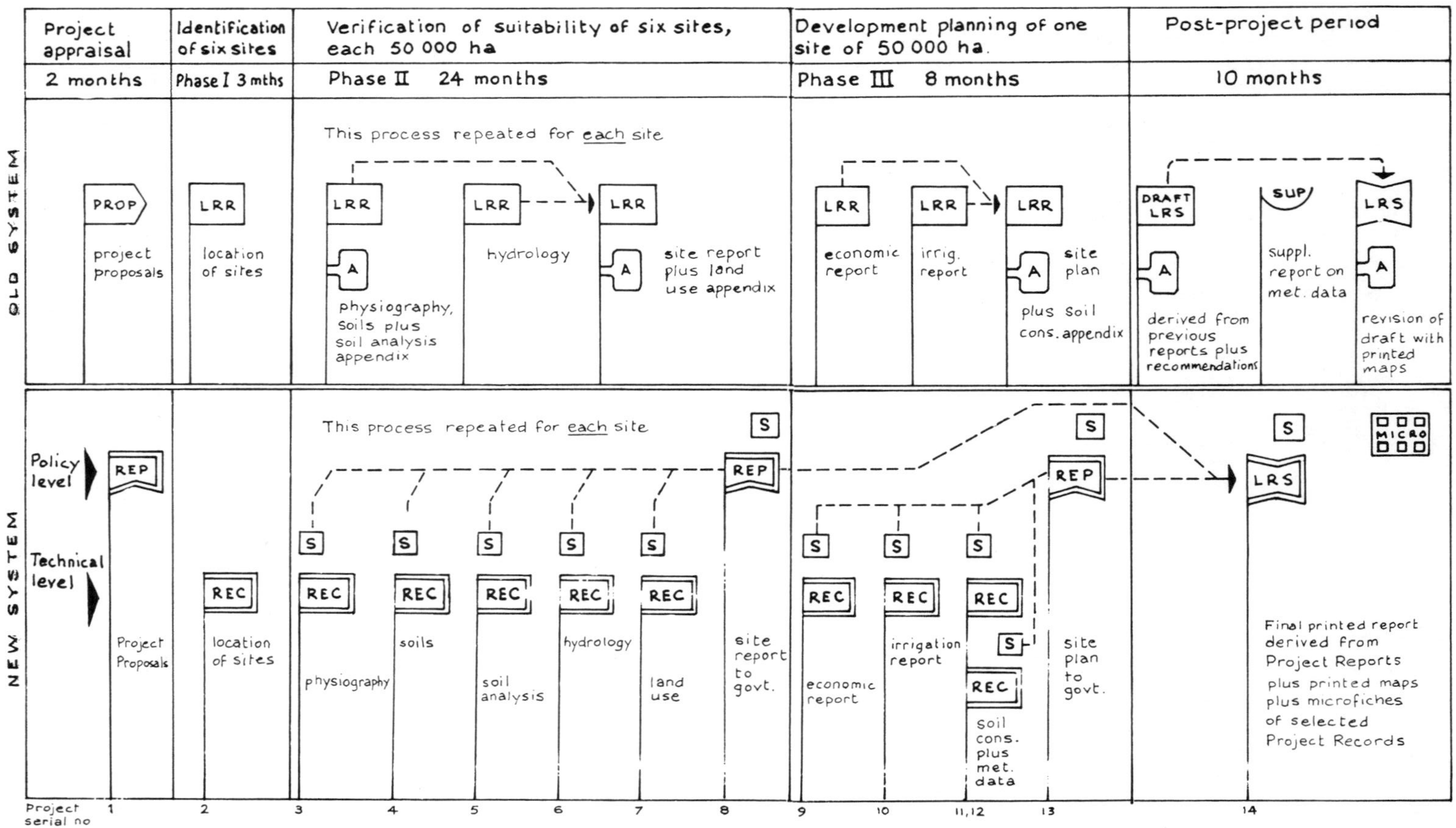

FIGURE 3 - The old and new systems of reporting a three-phase project, showing the distinction in the new system between reporting at "policy" and technical levels. Note the use of Project Record summaries in the production of

For easy assembly of the whole corpus of reports from one project it was decided that within-project serial numbers would be allotted as shown in Figure 4. Additionally for the benefit of librarians and collectors generally, LRD serial numbers (Land Resource Study No. 29 etc.,) would continue to be allotted. The use of within-project serial numbering is of course not a novel idea: it has long been used by FAO for example.

To make quite clear to all recipients the reasoning behind this double coding, it was decided to introduce into all reports a tear-out document control sheet as illustrated in Figure 5. This control sheet has another purpose - to remove some of the mystery from non-conventional literature by clarifying the nature, origin and security classification (if any) of each document.

3.2.4 Advice to writers on the nature of the readership

Once the new simplified reporting system had been defined it was possible to explain for the benefit of LRD authors the readership they should have in mind (and not have in mind) when writing particular kinds of report: this "by-product" of the analysis used in formulating the new system is considered to be of prime importance in concentrating the attention of writers on specific readership classes.

Table 4 shows how in the LRD instructions for authors the readers are defined for one type of report (the Project Report to client governments) and how the analysis of the readership also assists the definition of the subject matter. There are similar instructions for all other types of LRD reports.

4 CONCLUSION : APPROACH TO A STANDARD METHOD OF REPORTING LAND RESOURCE ASSESSMENT AND DEVELOPMENT PLANNING PROSPECTS

From the methods devised at LRD it may be possible to derive lessons of general application to resource studies of the developing world.

The first lesson is that methods commonly applied by research institutes - channelling of written output into a number of categories of serials, Technical Bulletins, Technical Monographs, Memoirs, Advisory Bulletins, etc., etc. (as well as into scientific journals) can be dangerous if applied to projects. In the end the nature of the specialised serial begins to dominate the content and the link with the aims of the project is lost. LRD's Land Resource Study for example in its early days acquired a reputation as an inventory of resources, became progressively more compendious and commonly contained more information than was required by the terms of reference of the project.

---

*This movement is an idealised state of affairs. In practice the Project Manager has to make his own synthesis of numerous Project Records. In this work the existence of good summaries of the Project Records is extremely helpful.

Project Serial No.	AC1*	AC2 to 6	AC7	AC8	AC9
Category and content of report	Project Report (proposal for project)	Project Records (five, on physiography soils, hydrology, forestry, land use)	Project Report (mid-term review of progress; some recommendations)	Project Report (final recommendations)	Land Resource Study (printed report for general disbribution, with microfiches of AB2 to 6)
LRD Series No.	Rep. 25	Rec. 60 to 64	Rep. 30	Rep. 33	LRS 29

* Simplified code shown here. In practice a more elaborate code is used catering for several projects in one country, as in Figure 5.

FIGURE 4 - The flow of documents from an integrated project in Acadia and the new system of double coding according to position in the project and in LRD series. The numbers of the various LRD series are not continuous because there are other projects in progress at the same time.

BIBLIOGRAPHIC DETAILS		DETAILS OF REFERENCE CODE	
Security class	In confidence	Country (site of project)	Acadia
Distribution	To Acadia Govt. only	Country code	ACA
Issuing authority	Land Resources Division	Project code no	01
		Within - project serial no of report	4
Report title	The land resources of northern Acadia and their development	LRD report series	Project Report
LRD Author(s)	R G Green and E Brown; Non-LRD authors and their sponsoring organisations	No. in LRD series	55
No. of pages	105; No. of refs 31; No. of separate maps 5	Date of issue	1976
REFERENCE CODE	quote in references and when ordering →	ACA - 01 - 4/REP-55/76	

Additional information	Material from this report will be incorporated in a printed Land Resource Study
Abstract(s)	A two-phase resource assessment and development project in northern Acadia is described. The first phase was an integrated evaluation of resources including physiographic, hydrological, meteorological, soils, current land use, sociological and economic surveys leading to a land suitability evaluation. In the second phase development options were defined for two specific areas.
Keywords	Land resources, agricultural development, landscape analysis, hydrology, meteorology, soil survey, current land use, maize, cotton, groundnuts, irrigation, economics, marketing, sociology, population studies, land suitability evaluation.

FIGURE 5 - A sample document control sheet included as a tear-out page in LRD reports. (Issued by the Land Resources Div., Ministry of Overseas Development, Tolworth Tower, Surbiton, Surrey, England KT6 7DY)

TABLE 4 Sample from LRD's instructions to authors, specifying purpose, readership and subject matter of each type of report. The instructions for a *Project Report* akd its separately bound summary are shown here.

Title of report series	Purpose and timing	Readers		Subject matter	
		Included	Excluded	Included	Excluded
Project Report	To convey advice and information to client government (as shown in Table 4a) and in addition to convey all the relevant supportive data which in Table 4a is consigned to Project Records The only report on a short-term project, to be issued immediately after its completion	Senior decision makers Senior development executives and Senior scientists attached to client government Local development executives Officers i/c local research	"Spectators"* Extension workers**	All the material shown in Table 4a (recommendations and interpretative matter) plus material as in Table 4c (supportive data)	Field data not needed for validation of conclusions or to meet terms of reference. Such data, after appropriate condensation, to be consigned to the archives
Project Report Executive Summary	To provide busy senior decision makers with a separately bound shortened version of the Project Report. Remarks: not longer than twenty pages; separately bound	Senior decision makers (Other Project Report readers will also receive the summary)	"Spectators"* Extension workers**	Synopsis concentrating on recommendations, but also making clear the background, objectives and method of the project	Terms and concepts incomprehensible to the non-scientific reader

*"Spectators" comprise (a) educationalists and students in the client country and abroad and (b) research workers and planners outside the client country

**Extension workers require specially prepared material

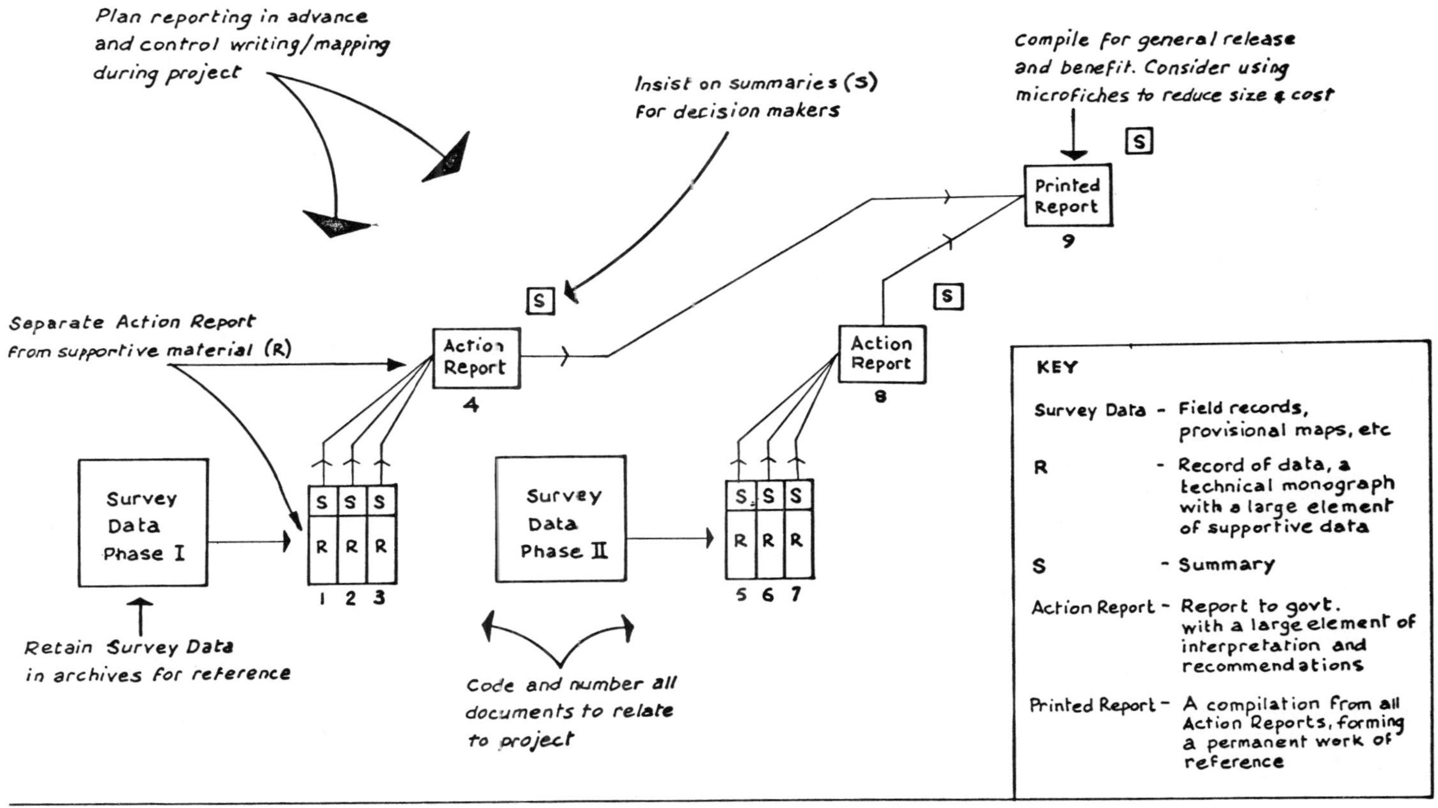

FIGURE 6 - A basic approach to the reporting of a resource assessment and development planning project. A two-phase project is illustrated. The thin lines indicate derivation of one document from another.

It is necessary therefore to devise a system of reporting which is specifically geared to the objectives of projects and to the typical nature of the project output. It is also necessary to make a close study of the readership of the project documents to ensure that the various types of written output go to the appropriate readers.

Figure 6 is a generalised scheme derived from LRD procedures and shows how the written outputs of land resource assessment and development projects can be handled systematically to produce archival matter, technical monographs (LRD's Project Records) primarily for scientists in the client country administration, action reports (LRD's Project Reports) giving options for development and finally a printed report for worldwide circulation.

Such a system concentrates attention on the action report and its speedy delivery to the client government, freed both of the technical detail which is not required by all readers and of the need to address a worldwide audience. The worldwide audience is nevertheless felt to be important and its needs can be met by compiling a special printed report out of one or more action reports and adding to it microfiches of selected technical monographs. This printed report is of course also supplied to the client government as a permanent work of reference. Finally to facilitate assembly of a full set of project documents each document is given a consecutive project number.

Such a system can only be operated under close editorial supervision and requires collaboration between the editorial staff and the project leader in the field, where the editor helps the leader to adhere to the schedule of document and map production (a cartographer in the field is often essential too), to separate action reports from supportive material and to prevent the excessive growth of technical monographs whose authors are over-zealous in this direction. Finally the important task of compiling a printed report for worldwide readership will fall more heavily on the editor than on the project leader who may well require to begin the preparation of a new project at this stage.

It will be seen that the editor's role in the preparation of project reports has special aspects which are rarely seen in the editing of the literature of research. These special aspects are the editor's early involvement with the authors, even before writing begins, his continued collaborative action at the centre of scientific activity, and finally his compilation of several documents to produce a work of reference. The degree of interaction between editor and authors is a reflection of the diversity of sciences involved and the continual process of condensation and integration of the written output which is necessary to achieve the objectives of a development project.

ACKNOWLEDGEMENTS

The author thanks the Director, Deputy Directors and senior staff of the Land Resources Division for their assistance in preparing the revised system of project reporting which is recorded in this paper. Special thanks are due to Mr D M Lang, Dr G Murdoch, Mr C A Robertson, Mr R N Jenkin and Mr J R F Hansell for their comments on the analysis of the subject content and readership of LRD reports. Thanks are also due to Mr C S Griffin and Mr N W Posnett of the Publications, Library and Information Section for their ad on document coding and control and to the former for his constant assistance in the preparation of the revised system.

SCIENTIFIC EDITING AND PUBLISHING FOR THE THIRD WORLD: A REVIEW FROM THE TROPICAL PRODUCTS INSTITUTE

Shirley M. Blatchford, *Tropical Products Institute, 56/62 Gray's Inn Road, London WC1X 8LU, England*

ABSTRACT

The paper reviews the publications of the Institute, all of which are primarily published for readers in the Third World as part of the UK Aid Programme. The Institute's role, as a scientific unit of the Ministry of Overseas Development, is to co-operate with the Third World in deriving greater benefit from its plant and animal resources after harvest. The ways in which these publications reflect this role are outlined, eg in the subjects covered (most having been investigated in the Institute) and the readers served (mainly scientists, technologists and economists, but also administrators and extension officers). Attention is drawn to the editorial features of the different types of publications (ie journal, reports, digests, advisory bulletins and extension literature) in relation to the needs of their main readers. Publications aimed primarily at different and specific groups of scientists and technologists range from Tropical Science (the quarterly journal of the Institute), through two series of reports (including those dealing with marketing and industrial economics), to advisory-type publications such as the Crop and Product Digests and Tropical Stored Products Information. Finally, at the other end of the spectrum to Tropical Science, is the Rural Technology Guide series which the Institute is just about to start publishing for extension officers. This new venture is an endeavour to meet the current policy of the Aid Programme which is to emphasize the provision of aid for the poorest countries and for the rural sectors within these countries. Experience in launching this series is described and Tropical Science is also highlighted.

INTRODUCTION

This paper reviews the publications of the Tropical Products Institute, which are primarily published for readers in the Third World as part of the UK Aid Programme.

The Institute is a scientific unit of the UK Ministry of Overseas Development and exists to co-operate with Third World countries in deriving greater benefit from their plant and animal resources after harvest. It specializes in the post-harvest sector and so is concerned with such matters as: crop handling; drying and other forms of preservation; storage; transport; processing; quality control and safety; standardization; the development of

Miriam Balaban (ed.), Scientific Information Transfer: The Editor's Role, 543-551.

new products, processes and equipment; national and international markets and marketing; and the problems of existing or new industries based on plant and animal products (particularly rural and agro-industries). All these subjects are dealt with in the Institute's publications but in different ways for different types of readers.

The paper is divided into two sections: editorial procedures and publishing procedures.

EDITORIAL PROCEDURES

1. Tropical Science

This is the quarterly journal of the Institute and is intended primarily for scientists, technologists and economists. It is concerned with the plant and animal products of tropical and sub-tropical countries, particularly the scientific, technological, economic and social problems arising in their handling, processing, preservation, storage, quality control, marketing and utilization, and related agricultural, forestry and fisheries topics.

Papers (by Institute staff and by outside authors) in any one of the four following categories are accepted, after refereeing: development (experimental) papers; review papers; case studies; and short communications. The features of these categories are described in the 'Guide for Authors' which is published in each issue of the journal. We look particularly for original papers which deal with problems being experienced by the developing countries and which offer some solutions to these. Critical book reviews are also published.

Referees comment on the papers by using a specially prepared form: they answer a series of questions posed by the editors; and they complete the form so that a portion can be sent, anonymously, to the authors.

Authors are asked to write their papers in keeping with the 'Guide for Authors'. An abstract in English, French and Spanish - the principal languages of those developing countries with which the UK is in frequent touch - appears at the start of all papers, except short communications. The abstracts are presented in as clear a style as possible to aid translation. The conventions used in the text are similar to those used by many scientific journals: they include specialized terms, abbreviations (in keeping with international standards), and SI units. The Harvard system of referencing is used, and the terminal list of references is given in as much detail as possible to assist readers in the developing countries to locate the original documents.

Tables of data are frequently included. Illustrations are used only when they are essential for clarifying the text, line drawings (especially graphs) being more frequent than half-tone photographs. On rare occasions colour photographs are published where these are essential to illustrate an important point.

A great deal of editorial input goes into ensuring that the text can be clearly understood by a wide range of readers - the English is often considerably altered.

The journal is designed to a 24 x 15 cm layout, well spaced, set in Times Roman (mainly 10 on 12 point) and ranged left with the minimum use of capitals. The line drawings are professionally drawn.

2. Papers in non-Institute journals

A wide variety of papers and articles by Institute staff is published in external journals, mainly special subject ones; this enables staff to communicate with their subject peers throughout the world. These papers are more specialized and detailed than those published in Tropical Science and are not primarily aimed at Third World readers, but at subject specialists.

3. Economic and technical reports

The Institute publishes two series of reports: a G (General) series and an L (Laboratory) series. These are written by Institute staff and represent the results of major project work conducted in the Institute.

Many of the G reports are economic studies of the markets for a wide range of tropical and sub-tropical products, and on the feasibility of setting up small-scale industries in the developing countries. These are aimed at economists, administrators and planners in Government departments in the developing countries and at the trade in both developing and developed countries, but they are also of interest to some scientists and technologists. The rest of the G series comprise technical reports and bibliographies which are considered to be inappropriate for publication in a journal since they are either too specialized or too lengthy. They are aimed at scientists, technologists and some administrators who have a close interest in the subject of the report.

The L series of reports presents the findings of laboratory or field work conducted by the Institute, eg many of the reports are of pulp and paper-making trials carried out by the Pulp and Paper Section of the Institute. These are useful to forestry departments in many developing countries in offering guidance concerning the best varieties of trees to plant for pulping purposes.

All these reports are preceded by a summary in English, French and Spanish. The style of the text is similar to that used in Tropical Science and, likewise, tables and graphs are frequently used, particularly in the economic and pulp and paper-making reports.

The report series are designed to an A4 layout and have a clear, well spaced typeface (Univers - mainly 10 point) with professionally drawn illustrations.

4. Biennial report

This reviews the work of the Institute - at home and abroad - over each two-year period, the information being arranged primarily on a commodity basis and supported by a brief general report and relevant appendices.

It is intended for use by scientists, technologists, economists and administrators throughout the world who are concerned with the subjects dealt with by the Institute. It is also of value for handing to the many visitors to the Institute and sending to interested enquirers.

The initial contributions are written by the specialist staff (according to guidelines) but these are edited centrally to a common and simple style, and illustrated mainly with photographs.

5. Conference proceedings

These are published after each biennial TPI conference. The purpose of these conferences is to bring together people from the developing and developed parts of the world who have a common interest in a subject of practical importance to the developing countries. A careful selection is made of the subjects suitable for conferences, eg the conference last held in 1976 dealt with the 'Handling, processing and marketing of tropical fish' - a subject of major importance to the Third World.

The Proceedings are of value to scientists, technologists, economists and administrators throughout the world who are concerned with the subjects of the various conferences. They are also of considerable interest to the trade and commerce.

The papers presented to a conference are edited by the relevant specialist section in the Institute and are published in the Proceedings, considerable care being taken to edit the papers into a simple, readable style. English, French and Spanish abstracts of each paper are published. An A4 double column, well-spaced layout is used.

6. Advisory publications

The next three publications - the Crop and Product Digests, Tropical Stored Products Information and Oil Palm News - are what can be termed advisory publications: the information they present has already been sifted and reviewed rather than the reader being presented with original research findings.

Crop and Product Digests. This is a series of occasional numbered publications each dealing with one group of commodities. Two digests have been published so far, 'Oils and Oilseeds' and 'Root Crops', and the third, 'Grain Legumes', is scheduled for publication in 1978.

Each digest gives, in quick reference form under standard headings, basic data concerning the more important economic plants of the world with special emphasis on the tropics and sub-tropics. These data include growth requirements, planting and harvesting procedures, yields, products and their uses, processing techniques, and trends in production and trade. The data on each crop within a digest are supported by short bibliographies to the key literature, and each digest concludes with indexes of botanical and common names.

One of the original aims of the digests was to provide information which would be helpful in selecting new crops for cultivation in particular geographic and economic circumstances. However, they have also proved extremely useful for general reference purposes, particularly to organizations and workers in countries with limited resources of technical literature. It is sometimes difficult to realize that many workers in the developing countries do not have easy access to good library resources. The digests can thus be regarded as a useful starting point for specialists and non-specialists requiring information on a particular crop and, with their selective bibliographies, indicate the more important papers that have appeared in recent years in the literature received by the TPI library.

One of the most valuable aspects of the digests is that information in the literature has been sifted and assessed and is presented in a concise way. The information used comes mainly from that stored in the Institute's Technical Index and Library. It is now clear that in order to produce a digest which contains up-to-date information, a thorough study is required of the ever-increasing literature on the various crops covered by a digest. In addition to standard reference books and recognized scientific and agricultural journals, the information scattered in reports from research and experimental stations, conference and symposium proceedings must also be studied and evaluated.

After evaluation and collection of the data for a digest, a considerable amount of editing is required particularly to standardize the very numerous units of measurement, abbreviations, and references.

Tropical Stored Products Information. This is a biannual advisory bulletin which is edited by the Storage Department of the Institute. It deals with practical problems of the storage of durable agricultural products (particularly cereals) in the tropics and sub-tropics and is intended to keep stored products workers in touch with the results of research and developments in the Third World.

It publishes review articles, articles on overseas storage organizations, accounts of techniques, annotated bibliographies, book reviews, and regular news items. Because practical problems are dealt with, basic research results are not published.

Most of the papers are by Institute staff though, occasionally, papers by outside authors are also published - all papers are refereed.

Authors are asked to write their papers in keeping with a 'Guide for Authors' which appears at the back of each issue. All papers are published with an English, French and Spanish abstract. Illustrations, usually in the form of half-tone photographs or line drawings, are frequently used.

Because losses of stored cereals are particularly high at the farm level in the developing countries, this bulletin is of special use to field workers, including those at the extension level. Thus, it is important that the information is presented concisely and in as readable a form as possible so that it can be quickly applied. As far as we know this is the only advisory publication on this subject - the Journal of Stored Products Research being for scientific research workers.

Oil Palm News. This, like Tropical Stored Products Information, is a biannual advisory bulletin, and does not publish basic research findings. Its purpose is to collect and disseminate current news concerning all aspects of oil palm growth and the utilization of its products throughout the world. It is intended for specialists and non-specialists, mainly in tropical countries, who are concerned with the growing of oil palm and the processing and use of palm oil. Once again, it is a bulletin which is intended for the field worker rather than the research scientist, and is, therefore, of value at the extension level.

It publishes short papers, extracts from other much longer publications, notes, and three regular features - an editorial, an 'oil palm market survey' and abstracts of recent publications. Papers by Institute staff and outside authors are accepted after refereeing.

The papers are edited by the Oilseeds and Edible Nuts Section of the Institute into a simple, clear style, the references being given in considerable detail. The abstracts are carefully selected and written under the supervision of the specialist section.

The bulletin has recently been designed into an A4, double column layout. The abstracts are printed so that they can be cut out and attached to 13 x 8 cm index cards for storage in personal index systems.

Two interesting collaborative arrangements with other organizations exist with this bulletin. The French journal Oléagineux receives typescript copy for translation into French and publication in Oléagineux when the editors consider a paper particularly useful. Also, the Commonwealth Agricultural Bureaux have recently started publishing Tropical Oilseeds Abstracts (monthly); this contains abstracts of papers on oil palm, some of which are taken from Oil Palm News under an arrangement with the Institute.

7. Secondary publications

Three secondary publications are produced by the Institute: Tropical Storage Abstracts; Bibliography of Insecticide Materials of Vegetable Origin; and the Library Accessions Bulletin.

Tropical Storage Abstracts. This is published every two months and is a current awareness service presenting abstracts of recently published information on the storage of durable agricultural produce in the tropics and sub-tropics. It is intended primarily for field workers in the Third World, but is of use to some scientists, technologists and administrators.

It is compiled and edited by the Storage Department, as is Tropical Stored Products Information. It comprises about 30 abstracts per issue, arranged alphabetically by author, with addresses of authors and/or organizations for readers to obtain further information.

The type of abstracts published is directly related to the needs of users which were expressed in a user survey conducted in 1970, prior to the commencement of publication.

The abstracts are longer than most other published abstracts because the aim is to be as informative as possible for readers who lack good library services.

The publication is designed to an A5 layout, the abstracts being printed on one side of the page to enable them to be cut out and attached to 13 x 8 cm index cards for storage in personal index systems.

A collaborative arrangement exists with the Commonwealth Agricultural Bureaux by which they use some of the abstracts from Tropical Storage Abstracts in their monthly Sorghum and Millets Abstracts, Tropical Oilseeds Abstracts and Cotton and Tropical Fibres Abstracts.

Bibliography of Insecticide Materials of Vegetable Origin. This is a biannual publication which presents selected references to publications relevant to the manufacture and use of insecticide materials of vegetable origin. It is intended for scientists and technologists throughout the world who are working on these subjects.

The Technical Index selects and collates the references from literature received in the Library, and edits them. The technical accuracy of the titles, particularly the nomenclature used, is checked by the Pesticide Analysis Section. The references are arranged alphabetically by author under a few broad subject headings. The Technical Index has recently increased the coverage by use of computer-based services such as BIOSIS, AGRICOLA, CHEMCON and MEDLINE.

Library Accessions Bulletin. This is a current awareness service which is published at approximately three-weekly intervals. It comprises references to publications recently acquired by the TPI Library dealing with plant and animal products of tropical and sub-tropical countries, and related topics. It includes annual reports and publications in series, books and pamphlets, periodical articles arranged by subjects and then by author, and the contents of recent reviews. It is intended primarily for librarians and documentalists throughout the world.

A common feature of these three secondary publications is that they are printed and published in as economical and in as speedy a way as possible, camera-ready copy for two of them being prepared within the Institute.

8. Rural Technology Guides

This is a new series of publications which is being launched next month (May 1977) by the Institute and reflects our parent Ministry of Overseas Development's policy to emphasize its co-operation in rural development work in the Third World.

The purpose of the series is to provide working details of devices or techniques which are suitable for use in rural communities. The devices and techniques chosen for inclusion in the series are those which can be used in villages where capital is scarce, labour is seasonally plentiful, and fossil fuels are expensive. The first two guides (No. 1 'A wooden hand-held maize sheller' and No. 2 'Pad-batch dyeing of plant fibres') are by our own staff, the ideas having been developed and tested by the Institute. We are not, however, precluding the publication of guides by outside authors.

The guides are intended principally for extension workers, instructors and other field staff concerned with rural development in the Third World. Because of this, each guide is being fully illustrated and written in simple, practical language rather than in more complex scientific terms. It is then hoped that the reader will, if he wishes, be able to make the device or carry out the technique himself.

Since the series is being published in English, each guide will only be useful in areas where the English language is widely used. For other areas, we are suggesting that the text might be modified to take local conditions into account, translated into the local language and reproduced locally. We are also suggesting that visual aids - such as wall charts and slide sets - might be prepared from the illustrations in the guides, for use by instructors and other field staff. In some cases, the Institute plans to produce visual aids itself.

Experience so far in editing this series shows that a considerable editorial input is needed to convert the writing of scientists into a form which is suitable for use at the extension level. In particular, a simplistic style of writing is essential, with short sentences, a minimum of technical terms, abbreviations and units, the avoidance of jargon and cross referencing, and the presentation of one idea at a time - in the text and in each illustration. The text and illustrations need to be as free as possible from in-built cultural barriers so that they are quickly understood by as wide a readership as possible. The balance between the text and illustrations has also proved to be of great importance, both being needed but in the right proportions. The illustrations need to be as realistic as possible - for example, one needs to avoid the use of symbols and sectional drawings, and the close cropping of photographs.

We have had assistance in editing and pre-testing Guide No. 2 on pad-batch dyeing from the Agricultural Extension and Rural Development Centre of Reading University, a concern of which is overcoming communication problems at the extension level in the developing countries. By publishing this series the Institute is moving into a new publishing area: in the past we have concentrated on communicating with fellow scientists, economists and technologists, but now we are increasing our expertise in communicating with extension staff.

A graphic designer has played a large part in the final preparation of the two published guides. Although a basic layout for the guides has been agreed, it is clear that within this framework each guide will need to be designed individually. In fact with Guide No. 2 (pad-batch dyeing) it was found - as doubtless others have discovered - that it was desirable to adopt a team approach involving regular meetings with the authors, the editor, the graphic designer, the photographer, and the communications expert all present; these meetings ensured that all concerned understood the goal.

Although the foregoing is not directly concerned with scientific editing, it illustrates the problems which face scientific editors when they need to communicate the results of research and development work to a non-scientific readership.

9. Publicity literature

The Institute publishes a Newsletter three times a year. It is intended as a useful bridge between the Institute and a wide ranging readership of administrators, planners, diplomats, scientists, technologists, economists, extension workers and, indeed, anyone with an interest in the post-harvest sector of development in tropical and sub-tropical countries. The purpose of the Newsletter is to make information about the activities of the Institute more speedily and more widely available than is possible through TPI's other publications. The Newsletter gives information about assignments and visits abroad by TPI staff and about trainees and visiting workers in the TPI departments. Special activities, such as TPI conferences, are referred to and notes on particularly interesting progress in research and development projects are given. Summaries of recently published reports and scientific papers are included.

Although the editing required for the Newsletter is not truly scientific, the editor needs to be able to translate scientific information into a form which can be widely understood.

PUBLISHING PROCEDURES

Some brief reference to printing, distribution and charging matters may be of interest.

The printing of our publications is arranged by Her Majesty's Stationery Office (HMSO), except for a few which are printed by our parent Ministry's print unit. Offset litho printing is used except for Tropical Science which is printed letterpress.

The Institute organizes the distribution of publications except for Tropical Science which is distributed by HMSO. We have access to an HMSO addressing system in which the master address plates are held. There are about 5,000 names on this publications mailing list, about two-thirds of this total being recipients in developing countries. The serial publications are mainly distributed automatically on sub-lists of this total and Table 1 gives details. However, since everyone on the mailing list receives a copy of the Newsletter, some people choose to order new publications as they are announced in the Newsletter.

Regarding charging, a price is printed on all our publications but 'public bodies in countries eligible for British Aid' are able to receive free copies as part of the Aid Programme. This applies to all the Institute's publications except for Tropical Science which is either on subscription or on exchange for publications required by the Library.

IN CONCLUSION, this has been a brief résumé of the publications produced by a scientific organization mainly for developing countries. More details can be given to anyone who feels these would be useful.

Table 1

TPI publications mailing list April 1970 to April 1977 : number of names with number of reports in brackets

Publication	1970	1971*	1972	1973	1974	1975	1976	1977
TPI Newsletter**and total mailing list	1550	1700	2160	3200	3800	4400	4800	5034
TPI reports:								
G (General)	500 (8)	600 (16)	924 (11)	1050 (8)	1160 (7)	1380 (4)	1540 (16)	1650 (3)
L (Laboratory)	300 (7)	400 (2)	470 (6)	576 (6)	590 (5)	630 (5)	653 (3)	702 (1)
Periodicals:								
Oil Palm News	400	500	792	1039	1060	1268	1450	1640
Trop. Stored Prod. Inf.	800	900	1187	1538	1725	1920	2174	2477
Trop. Abstr. †	-	-	-	1538	1725	1920	2174	2477
Bib. Insect. Matls Veg. Origin	455	500	665	719	760	800	939	995

* Mailing list expansion exercise carried out

** Commenced December 1974

† Commenced February 1973

THE INTERNATIONAL DEVELOPMENT RESEARCH CENTRE: ITS UNIQUE ROLE IN WORLD DEVELOPMENT

Reginald MacIntyre, *Associate Director,*
Publication Division, IDRC, Box 8500, Ottawa, Canada K1G 3H9

ABSTRACT

Canada has a unique organization concerned with helping developing countries: the International Development Research Centre seeks to initiate, encourage, and support research into problems of such countries, and into the means for applying and adapting scientific and other knowledge to the economic and social advancement of those regions. IDRC places great emphasis on developing regions building up their own research capabilities and innovative skills, and stresses coordination of research across national boundaries and between regions. Founded in 1970 the Centre gave grants of about $73 million during its first six years of operation. The main focus for assistance has been to improve the well-being of rural people, in five main areas: agriculture, food and nutrition sciences; information sciences; population and health sciences; social sciences and human resources; and publications.

The Canadian Parliament established the International Development Research Centre in 1970, as a public corporation, to stimulate and support research in, by, and for the benefit of developing countries.

The research is directed at improving the well-being of people in less-developed countries by adapting and applying science and technology to their needs. The Centre also supports programs that will help developing regions to build both human and institutional capacity for bringing the methods of scientific inquiry to bear on the solution of their own problems.

The Centre supports research in areas that directly affect the day-to-day lives of people -- farming, nutrition, education, health, the impact of technological change on traditional societies, population, communicable diseases, water supplies, etc.

And because rural areas of developing countries are usually the last to benefit from the advantages of science and technology, research that can improve rural life is emphasized.

NEW APPROACH

IDRC has many unique features: it has sought to place in perspective the role of the scientist in international development, and has encouraged developing countries to tap the talent of their own scientific communities; it obtains its funds from the Parliament of Canada, and reports annually to Parliament, but its operations are guided by an international and autonomous

Miriam Balaban (ed.), Scientific Information Transfer: The Editor's Role, 553-554.

Board of Governors; it was the first aid organization set up specifically to support research projects that are identified, designed, conducted, and managed by developing country researchers in their own countries, in terms of their own priorities; it has helped to create research networks through which developing countries can share common experiences, conduct studies with a common design in areas of mutual concern, and learn from each other as they work toward common goals; and it has fostered cooperation between developing country researchers and their counterparts in Canada or in other industrialized countries.

Some examples of IDRC-supported development research:

- A milling system developed in Maiduguri, Nigeria, has overcome many drawbacks - such as limited services, unsatisfactory quality control, and unstable prices - associated with existing neighbourhood grain mills. The Maiduguri Mill project is a practical demonstration of a systems approach to postharvest handling of grains.

- The adoption of policies to utilize paramedical or nonmedical personnel in health systems has stimulated the need for teaching, supervisory, and management skills. The development of curricula including these skills is indicated and applied research projects will be supported in teaching and management systems adapted to the health field.

- Project Impact, an innovative attempt to provide universal primary education in Southeast Asia without crippling costs, has produced promising results in the Philippines and Indonesia, and the experiment is to be extended to Malaysia. Self-instruction, mutual help, and flexibility are key features of this system.

- With IDRC assistance for training, equipment, and initial staffing, two major regional centres - in Latin America and Southeast Asia - are playing important roles in collecting and distributing agricultural information within the global information system AGRIS, operated by FAO.

PUBLICATIONS

The Publications Division acts primarily as a service group to the other program divisions of the Centre, though we also support projects that are aimed at devising media suited to link scientists and rural people, and that help improve science writing and science reporting in developing countries. The division disseminated the results of IDRC-supported research through the publication of monographs, technical studies, booklets for policymakers, publications for the general reader, and audiovisual material. The staff also write and supply articles on science and technology to the media in developing countries, as well as in Canada.

DISCUSSION on "The Non-Conventional Literature of Tropical Agricultural Resource Assessment and Development: Its Nature, Sources, Quantity, Value and Accessibility"
by W.J. Baulkwill and N.W. Posnett

"A Rational System for Editing and Publishing the Literature of Tropical Agricultural Resource Assessment and Development Planning"
by W.J. Baulkwill

"Scientific Editing and Publishing for the Third World: A Review from the Tropical Products Institute"
by Shirley M. Blatchford

"The International Development Research Center: Its Unique Role in World Development"
by Reginald MacIntyre

BASIL WALBY - I got the impression that your publications are in English. If they are, what steps are being taken to have them translated into the vernacular?

SHIRLEY BLATCHFORD - They are mainly in English. We have summaries in French and Spanish in our journal, *Tropical Science,* in all our economic and technical reports, and in our storage bulletin. We have a few cover-to-cover translations of reports. We are aware of the need to do more along these lines and will probably do so. However, there is a problem with our *Rural Technology Guides* since so many local languages are involved at the rural level: we are encouraging the extension services to organize the translation of local editions.

ELI COHN - Can they be bought only on subscription and are they copyright?

SHIRLEY BLATCHFORD - They are free to public bodies in countries eligible for British aid. The rest are charged for. They are all Crown copyright, but we are willing to consider their reprinting, etc. With the *Rural Technology Guides,* which are also Crown copyright, we offer to help with their modification and the production of local versions.

JOHN BAULKWILL - Of the projects which you have abroad, do you distribute them to the public in that sense?

REGINALD MacINTYRE - No, not at all. Our scientific texts are made available free to any bona fide research worker in a developing country or to an organization in a developing country, much like TPI's policy. Some of our publications of course have global interest, and for North American and some European requests we do demand payment. We put a price tag on almost everything we publish and we try to make say, a North American pay for the copy that we give to someone free in a developing country. This works quite well and our publications are growing in demand. We sell a few thousand dollars worth of books each year.

EDITOR-SECONDARY SERVICES INTERFACE

Developments in Secondary Information Services and Information Transfer

Udo Schützsack, Chairman

DEVELOPMENTS IN INFORMATION TRANSFER

E.J. Mann
Commonwealth Bureau of Dairy Science and Technology,
Shinfield, Reading, Berks. England

ABSTRACT

The author identifies 3 distinct phases in the history of information transfer. The first phase held until the beginning of this century when there were relatively few scientists and relatively few scientific journals and new information was transmitted from scientist to scientist directly via these primary journals.
The scientific literature explosion, which has continued more or less unabated during the last 25-50 years, led to the second phase, the advent of Abstracting and other Bibliographic services, acting as a sifting mechanism for scientists and overcoming both the volume and language barriers for them.
The primary literature explosion was followed by a secondary literature explosion which is leading us into the third phase, brought about by the advent of the computer, enabling large secondary data base producers to store their data bases in machine-retrievable form for purposes of current awareness and retrospective searching for specific scientific information.
An attempt is made to assess the likely impact of these developments on secondary services and on the form they might take in the future.
The author also refers to certain aspects of a recent study attempting to improve the information flow between primary journals and secondary services in the biological field, a study which has underlined the need for greater collaboration between primary and secondary journal editors.

We are coming to the closing stages of an interesting and challenging Conference and our task this morning is to examine various aspects of secondary information services and information transfer. In setting the scene for this particular session, I want to delve a bit into the history of scientific information transfer so that we may gain a better understanding of the reasons for the general turmoil that exists in scientific publishing today.

On a brighter note, I will tell you something about the efforts of various organizations to introduce a degree of order into the general chaos and finally, being slightly tarred with the prophets brush of this great land, I will risk my neck by doing some crystal ball gazing into the future.

Miriam Balaban (ed.), Scientific Information Transfer: The Editor's Role, 559-562. All Rights Reserved.

There is little doubt that in the beginnings of science, the spoken word was the supreme medium of information transfer. It is recorded from the early days of Greek science that learned scholars thought nothing of travelling from Greece to Asia Minor, the Black Sea, Alexandria, Italy and North Africa as a matter of course.

One suspects that, unlike the jet-set scientists of today, who sometimes 'do' Laboratories in 3 different countries in one day, the scholars of the Greek era took many months over their journeys.

The first great scientific revelations made in print were in the De Revolutionibus of Copernicus in 1543. However, it appears to have been the Royal Society of Britain, whose foundation in the middle of the 17th Century marked the beginning of a new stage in scientific communication, in the form of the scientific periodical as well as with the inception of the modern scientific society. The Transactions of the Royal Society first published in 1665, have been a channel in scientific communication for well over 300 years.

During the first part of the 18th Century, numerous Scientific Academies sprang up in different parts of Europe, notably the Academy of Science of St. Petersburg (1725), now the Academy of Sciences of the USSR, as well as the Swedish and Danish Academies. One of the earliest scientific journals of note was the Verhandlungen der Königlichen Akademie der Wissenschaften zu Göttingen in Germany in 1752. Other prominent journals of this period were the Transactions of the American Philosophical Society (1771) in the USA and the Annales de Chemie et de Physique (1789) in France.

The 19th Century saw an enormous growth in science and with it an equal growth in scientific literature, the specialized scientific journal, such as Liebig's Annalen der Pharmazie (1832), making their first appearance.

One suspects that in those days, scientists specializing in any subject, could still keep more or less up-to-date by perusing 10-20 scientific journals in their discipline.

However, we were now on the verge of the scientific literature explosion which gathered momentum during the second half of the 19th Century and has continued more or less uninterupted ever since.

What we have witnessed is frightening in many ways. It has been estimated that the number of scientific and technical journals published has doubled every 10-15 years since 1900 and that the number of scientific papers published increases by 150,000 each year. The same scientist of the 19th Century who could keep up-to-date in his field by reading for only about 30 minutes per day, would, today, have to read for 13 hours a day for each day of the year for 12 years in order to keep up-to-date.

Fortunately, and this brings me to the second phase of scientific information transfer, the advent of the secondary information services has, to a large extent, saved the scientist from this ordeal.

Here, the general and specialized Abstracting Services have played a leading role. By predigesting and condensing vast quantities of primary scientific information into a quickly absorbable form as well as by breaking through the language barrier in scientific communication, these services have played an invaluable role by helping scientists in keeping abreast of developments in their subject any where in the world. They have performed a dual function in providing both a current awareness service as well as a retrospective archival service in the various branches of science.

Unfortunately, secondary services have proliferated at almost the same rate as primary journals, and in a somewhat disorganized manner at that, resulting in a considerable amount of duplication and, as a consequence, confusion for the poor scientist, who may have to turn to as many as 10 Abstract journals to make sure that he has got all the information on the specific subject that he requires.

It is indeed a sobering thought that since the publication of the Pharmaceutisches Zentralblatt in 1830, Abstracting and Indexing Services have proliferated to such a degree that today, according to a recent survey of the World's agricultural documentation services, that is to say in a relative small area of the whole field of science and technology, there are no less than 700 Abstracting and Documentation Services in existence. It is only fair to point out at this stage that Abstract journals are by no means the only printed form of secondary information available to scientists today. There are a whole range of services available, such as titles only journals, indexes, Current Contents journals and review journals, all performing valuable functions in their own right. We shall, of course, be hearing in greater detail about a number of these services later this morning.

If it was the advent of the secondary journal that helped to solve the problems created by the primary literature explosion, it is probably equally true to say that it is the computer that is helping to solve the problems created by the secondary literature proliferation.

This does, in fact, bring me to the third phase of information transfer ushered in with the arrival of the computer on the information scene. The computer has a multiple role in information transfer ranging from computer-aided printing and index production to the retrieval of highly specific information from one or several data bases.

The computer has, indeed, added a new mobility, or given a new lease of life to valuable items of scientific information, stored away in abstract journals and sometimes not retrievable on account of poor indexing. It also facilitates the linkage of items of information from one data base with relevant items from another data base, thus helping the user to obtain a more complete picture. Most of the world's leading data bases are now available in machine-readable form and there are a number of Information Retrieval Services with highly sophisticated software and hardware who process large numbers of the world's leading data bases and make them available 'on-line' - that is to say directly via International telephone networks - to the end users. All the user, in fact, requires is an information receiving terminal in the form of a printing device and/or a visual display unit, plus a certain amount of know-how.

The information transfer scene of tomorrow, and to some extent of today, is summarized admirably in the following picture.

We certainly seem to have travelled a long way since the early Greek scholars went on their long journeys to discuss their findings with colleagues in distant lands.

However, the road ahead in the field of scientific information transfer is far from clear. Will primary journals as we know them survive? Will the newly created Synopsis journals, designed to replace the 'old-style' primary journals, survive? Do Abstract journals have a future? Will there ultimately be no printed journals at all and will we be slotting into one world data base via our television screens?

I promised you a little bit of crystal ball gazing at the beginning and for what it is worth, my view is that there will be changes in the directions just indicated but that these changes will be far slower to materialize than some people think. In other words, some primary journals will disappear, some synopsis journals will appear (and perhaps disappear again), some Abstract journals will disappear, but by and large these changes will be very gradual. What is certain is that we are living in a time of change and that this demands close cooperation and frequent dialogue between all people and organizations associated with the transfer of scientific information. That is why we are here to share our experiences for the first time on a world-wide basis. That is why we have been working towards closer collaboration among Editors of both primary and secondary journals through organizations such as ELSE and EDITERRA. That is why we have attempted to improve the quality of presentation in scientific journals through the publication of Manuals like "Scientific writing in English". That is why we are attempting to improve the flow of information between the primary journals and secondary services through in-depth studies of the problems involved, about which you will be hearing later this morning.

I have taken you on a whirlwind tour through the history of scientific information transfer stretching from the days of ancient Greece to the uncertain times of today and the unpredictable situations of tomorrow. I have had to leave out a great deal but I hope that I have succeeded to some extent in setting the scene for this Session.

INFORMATION FLOW BETWEEN PRIMARY JOURNALS AND SECONDARY SERVICES: REPORT ON A STUDY

R.B. Weiss, *Information Officer*
Centre for Agricultural Strategy, University of Reading, Earley Gate, Reading RG 2AT, England

ABSTRACT

An internationally sponsored project was initiated by the European Association of Editors of Biological Periodicals to study primary journal abstracts (their presentation, quality and suitability for use by the secondary services) and to develop standard guidelines for author abstracts.

User panels (totalling 35 scientists and 5 editors of secondary journals from 13 countries) evaluated abstracts of 400 papers from 100 periodicals; subsequent panel discussions were crucial in developing the guidelines.

The proposed guidelines consist of two distinct parts:
i) BRIEF, STANDARD INSTRUCTIONS for inclusion in the 'Instructions to Authors' normally provided in primary journals.
ii)DETAILED BACK-UP INFORMATION of standard layout but with provision for details and examples to be adapted according to the needs of individual journals. The back-up information should be available to an author on request or sent to him by an editor. It stresses the importance of title and abstract and includes instructions on: presentation and style; a check-list of points for possible inclusion in the abstract; examples relevant to the journal.

Proposals are made for practical evaluation of the guidelines so that they can be refined and tempered in the heat of real-life publishing.

Miriam Balaban (ed.), Scientific Information Transfer: The Editor's Role, 563-566. *All Rights Reserved.*

Introduction

The development of guidelines for author abstracts was part of a project[1] to study the flow of information from primary journals to the secondary services in the biological sciences. The entire project had an international flavour - proposed by ELSE (European Life Science Editors), funded jointly by the British Library and the West German Institut für Dokumentationswesen, aided by the work of scientists and editors from thirteen countries. This paper discusses the role of author abstracts, describes the development of guidelines for author abstracts and outlines a proposal for practical evaluation of the guidelines.

The importance of author abstracts

The basic function of an information system is to aid the flow of ideas from scientist to scientist. Transmission of ideas by direct contact between scientists is necessarily limited. Much information is transmitted through primary journals - in this case the title and abstract of a paper are helpful in arresting a reader's attention and in evaluating the relevance of a paper to his interests. Faced with ever more literature, the scientist relies increasingly on the secondary services to locate relevant information - in this case abstract and/or title are crucial in deciding whether or not to obtain the original paper.

It would be logical for primary journal abstracts to be used by the secondary service. In practice, this is rare. In many cases, primary journal abstracts have to be adapted to suit the subject emphasis and length of abstract required for a particular secondary journal; such changes are relatively easy if the author's paper and abstract are well written. Unfortunately, secondary services often have to write or re-write abstracts because primary journal abstracts are absent or inadequate, yet this is a time-consuming and expensive process.

Clearly, a good author abstract is of benefit to all. The author benefits because his work can be disseminated rapidly by the secondary service; the primary journal benefits because improved abstracts enhance the quality of a journal; the secondary journal benefits because less time is spent on re-writing abstracts.

As regards the quality of author abstracts it has been suggested that 'the scientist', as a reader, seems to know precisely what he wants from an abstract yet 'the scientist', as a writer, often fails to produce it! Some reasons for inadequate abstracts resulted from a survey of instructions to authors and from editors' replies to a questionnaire about editorial policy:
(i) Lack of clear instructions. Less than a third of the 240 journals in our sample gave the author any detailed guidance on the required content and style of abstract.
(ii) Lack of awareness of importance of abstract. It is essential that authors are aware that for every person who reads a paper, hundreds read the abstract, thousands scan the title. Only an author who recognises the vital importance of title and abstract for retrieval of his work is likely to take the time and trouble to produce a good abstract.

[1] Landau, N and Weiss, R.B. Information flow between primary journals and secondary services in the biological field. The Library, University of Reading, 1976.

(iii) Lack of writing ability. It is hard to accept the suggestion put forward by one editor that 'scientific ability and writing ability are incompatible'. More widespread training in scientific writing would be useful. Meanwhile, improvement could result from better motivation, clearer instructions and editors' insistence on a satisfactory abstract.

(iv) Lack of willingness to read or follow instructions. It is understandable that authors have little patience to wade through lengthy instructions when writing up their work. It is therefore proposed that guidelines in 'Instructions to Authors' be brief but that fuller information should be available to the author on request or sent by the editor when requesting revision of an abstract. Even an author normally reluctant to read instructions is likely to do so if acceptance of his paper depends on it.

(v) Lack of objectivity. This is an inherent difficulty for the author. The aid of colleagues and the vigilance of referees and editors is essential.

In order to assess the possible value of clearer guidelines primary journal editors were asked whether they consider that guidelines might be of value in improving author abstracts; the majority (about 60%) were optimistic.

Development of guidelines

A pilot study (preceding the main project) had put forward tentative guidelines for abstracts of different types of biological papers and proposed that the project team should write or revise abstracts of papers in representative primary journals in accordance with these guidelines. The abstracts were to be evaluated by User Panels of scientists and secondary journal editors. This procedure was carried out using 400 abstracts of papers from 100 journals.

The work had two main outcomes. First, use of the pilot project guidelines was discarded as impractical, due to the difficulty of classifying different types of papers. The diversity of papers in the biological sciences seems to preclude categoric recommendations for the essential content of an abstract. Second, the evaluation of abstracts by the User Panels and subsequent Panel discussions were crucial in developing the proposed guidelines. Our confidence in the basic structure of the proposed guidelines stems from this 'grass-roots' approach. Instructions are not imposed by an editor: they have developed from the experience and views of working scientists.

The main features of the proposed guidelines are:

(i) They consist of two distinct parts - BRIEF GUIDELINES for inclusion in the primary journals 'Instructions to Authors'. BACK-UP INFORMATION available to the author on request or to be sent to him by the editor.

This division was prompted by the fact that more than 80% of primary journal 'Instructions to Authors' are of one page or less, so incorporation of lengthy guidance on abstract writing would not be acceptable.

An advantage of this division is that the back-up information can be sent to the author by the editor when requesting revision of an unsatisfactory abstract, i.e., timed to have maximum impact on the author.

(ii) The standard BRIEF GUIDELINES consist of only about 120 words but include basic instructions on presentation (length, location and language of abstract); importance of title and abstract; need for abstract to be intelligible without reference to the main text; brief guide to contents - and reference to the back-up information.

(iii) The BACK-UP INFORMATION stresses the importance of title and abstract and gives a check-list of possible points for inclusion in abstracts of both research studies and reviews or discussions. Some information and examples are given concerning style of writing.

In compiling back-up information efforts have been made (i) to avoid what the scientists termed 'information science jargon' (such terms as 'indicative abstract' and 'informative abstracts' were particularly disliked); (ii) to spell out the reasons for instructions, for instance, to point out that the requested 200 word maximum for an abstract is not an arbitrary limit but represents the 1,000 character maximum demanded by the automatic recording system used by many abstracting services; (iii) to provide examples of abstracts relevant to the journal.

The back-up information is seen as a basic outline; editors will be encouraged to adopt it (especially as regards check-list of contents and selection of examples of abstracts) according to their journal's needs and interests.

Evaluation of proposed guidelines

Guidelines will be used in practice in a follow-up trial due to start soon. The trial will be funded by the British Library and carried out under the auspices of the Commonwealth Agricultural Bureaux. It is proposed that 30 primary journals will test the guidelines over a period of eighteen months. The aim is to refine and temper the guidelines in the heat of real-life publishing so they can be confidently published as definitive instructions. In addition, if guidelines can be shown to work successfully for some editors they will be more readily adopted by others.

THE PAST AND FUTURE OF THE CONVENTIONAL ABSTRACTS JOURNAL

H. Edward Kennedy, *Executive Director*
BioSciences Information Service
2100 Arch Street, Philadelphia, Pennsylvania 19103, U.S.A.

ABSTRACT

Over the past several decades, the conventional abstract journal has changed much less than how the journal is created. Many abstract journals are remarkably similar to their predecessors of three centuries ago. The computer has vastly changed how we produce publications but until now the journal itself has changed much less than is predicted for the future.

The impact of modern technology on information transfer suggests indicators for the future, particularly the abstract journal. The changing role of the abstract journal in an atmosphere of on-line bibliographic and abstract services, electronically stored primary information, and synoptic journals is discussed. The opportunity provided by technology for resource sharing by primary and secondary publishers and its impact on the secondary journal is explored, as well as potential impact of data handling by secondary information services.

The abstracts journal was invented in 1665 when, on January 5 of that year, Denis de Sallo issued the first number of the first published abstracts journal and called it Journal des Scravans (1). The record shows that de Sallo quickly offended the authors whose works he abstracted and after only 13 weekly issues, a mere 3 months, his permit to continue his journal was revoked because he refused the compromise of accepting a censor. However, his permit was transferred to a less controversial editor and his Journal des Scravans survived uninterrupted for 130 years.

The account implies that de Sallo wrote "critical" abstracts and so suggests that the first abstracts journal might not have been a conventional one in today's terms. Almost 20 years later it was followed by the Nouvelles de la Republique from Amsterdam and three years after that the Histoire des Ouvrages was published in Rotterdam. All of these preceded the famous Monatsextracte, the first German abstracts journal, which was issued in 1703. Often credited with being the first abstracts journal, it was followed by several other German abstracts publications which were introduced early in the 18th century. By the middle of the 18th century Central European abstracts journals were fairly numerous, although the age of the subject-oriented abstracts journal did not arrive until the latter part of the 18th century, with the introduction of Chemisches Journal fuer die Freunde der Naturlehre in 1778. The Chemisches Journal was the first in chemistry, and of course, the abstracts journal with emphasis on science and technology survived from then until today.

The journal, Neues Jahrbuch fuer Mineralogie, Geologie, und Palaontologie

Miriam Balaban (ed.), Scientific Information Transfer: The Editor's Role, 567-572.

was the first new abstracts journal of the 19th century. By this time the Chemisches Zentralblatt was about to be born and was followed in the 19th century by abstracts publications in medicine and law.

Although not characteristic of the conventional abstracts journal as we know it today, the index to the New York Times from 1851 had all the characteristics of one -- detail, comprehensiveness, conciseness, and impartiality. Engineering Index was first published in 1884 and the last decade of the 19th century saw several important abstracts and index journals introduced, including Physics Abstracts. Very early after the turn of the century the American Chemical Society's Chemical Abstracts was introduced.

Thus, the general framework of today's abstracting and indexing services was established in a short 250 years -- from the mid 17th century to about the turn of the 20th century.

Since the late 1900's the conventional abstracts journal has changed very little. It has been said that "there is precious little new under the sun (2)," and perhaps that is so; the abstracts journal was introduced more than 300 years ago and the scientific abstracts journal, much as we know it today, has been on the scene almost 100 years. A look at excerpts from abstracts journals over the last 75 years supports that statement. An abstract in the first issue of Chemical Abstracts looks totally familiar, with all the usúal elements of information -- author, primary journal reference, and a very typical seeming abstract. Some twenty years later, an abstract in the first issue of Biological Abstracts (BA) looked essentially the same as one in other abstracts journals of the times, only the subject content was materially different. And in 1977, in the year of Biological Abstracts' 50th anniversary, an abstract in BA looks essentially the same. All outward appearances would indicate that after almost 100 years the abstracts journal is basically the same as it was in the beginning.

What is an abstracts journal? A reasonable definition of the abstracts journal, "a concentration in a single source of knowledge that is recorded in a multitude of sources" has remained unchanged over the centuries. And a reasonable statement of the purpose of the abstracts journal, "to identify quickly and accurately the basic content of a document and to decide whether or not to consult the original publication" has remained unchanged, certainly for the history of the abstracts journal as we know it today.

Functions served by abstracts have been described as: alerting, awareness, scanning, screening,browsing, retrieval, and sometimes as a provisional substitute for the original (3). However, the aggregate volume of primary literature, which reportedly doubles on the average every 15 years (4), has reached a level that is causing a shift in the emphasis of these functions, i.e., more and more scientists need the abstracts journal to enable them to select from the mass or to eliminate the non-essential, rather than to make them aware of the existence of a piece of work.

Why has this shift of function taken place? The National Science Foundation (NSF) (5) has published statistics that indicate a steadily progressive relationship between the number of scientific and technical articles and the numbers of scientists and engineers, for the period 1960 to 1980. The period 1974 to 1980 was projected. The growth in numbers of scientific and technical articles published through 1974 and projected through 1980 indicate that the steady increase in growth rate over the past decade is not projected to level off during the remainder of this decade. The National Science Foundation also published statistics on the number of world and United States scientific and technical journals published; the growth rate for journals published worldwide is essentially the same as for journal articles. Those statistics suggest that Price's (6) data of more than a decade ago are still valid. He not only showed the growth in numbers of scientific journals since the beginning, but also a parallel growth in the number of abstracts journals from about the middle of the 19th to the middle of the 20th centuries. A World Inventory of Abstracting and Indexing Services, to be published by the National Federation of Abstracting and Indexing Services and the International Federation of Documentation later this year, will list 2500 such services, an increase from a total of 1855 when the "inventory" was last published ten years ago.

Recognizing, then, that the abstracts journal has changed little in appearance and substance over the past century, and faced with the ever growing body of literature and proliferation of new abstracts journals, what would the future of the conventional abstracts journal seem to be? Another century of the same? I, and many others, hardly think so.

We should remember that the current status of the abstracts journal, conventional as it might seem, is largely predicated on the application of computer technology; otherwise the large, comprehensive services would have crumpled under the load, unable to cope with the volume, long before now. Although perhaps at a lesser rate than would be the case if more manual processing were still used, and electronics notwithstanding, costs have continued to rise. Macklup (7) has suggested a strangulation hypothesis which states that we may have only 5-10 years left to come up with economically viable alternatives to today's publication practices. Nevertheless, the forces of economics and electronics, I believe, more than demand by users for different or better services, will dictate change -- will dictate that the abstracts journal of the future be radically different from that of the past.

The secondary services that publish the abstracts journals will attempt to provide services to accomplish the purpose of the abstracts journal, but increasingly via products of the electronic age, and aimed for a user community that is attuned to electronic innovations. The flexibility of today's information storage, retrieval and dissemination techniques is having far reaching effects on the methodologies of abstracting and indexing services. The possible forms of output from these highly sophisticated systems is seemingly limitless. They create the machine manipulatable file from which not only the abstracts and index journals are derived, but the machine readable file is also the source of magnetic tape products, on-line services and microforms. With these non-print alternatives at hand, it is impossible to predict the future of the conventional abstracts journal.

Chacko (8) reported at the recent meeting of the American Association for the Advancement of Science that the rising cost of publishing scientific journals and the rapid proliferation of articles a scientist must survey to find information he needs may lead to a revolution in scientific communications. The alternative to paper journals he suggests would be an electronic system to deliver the results of current research. Also, a University of Toronto study team (9) says the favorable outlook for computer memory developments and telecommunications availability suggests the practical possibility of an electronic journal, accessible through remote terminals. They estimate that a cost crossover point would be reached in about 20 years, when the electronic publication scheme would cost less than paper publication.

EDUCOM (10), a cooperative organization of 200 colleges and universities operated out of Princeton, New Jersey, recently was quoted as seeing electronics supersede printing in the future. A spokesman from EDUCOM said recently that the inter-university communications networks will make significant inroads on the print medium for the delivery of information to users. EDUCOM has the purpose to "promote effective collaboration among its members in the development, operation, and use of technology-related systems in higher education and research." Students educated in the atmosphere of EDUCOM are tomorrow's scientists, technologists, and information users. Will they settle for the unchanged, conventional abstracts journal? I think not.

Use of the electronic medium for documents is not a new idea. Already many of the abstracting services are in part participating indirectly in the electronic publication concept through such systems as Lockheed and Systems Development Corporation (SDC), which produce printouts of references in on-line searches. Their systems are examples of growing number of information centers that provide on-line machine searches of secondary information services' files (in essence the abstracts journal on tape). They print on-demand citations and abstract text, if after the user is alerted to the existence of the reference he indicates the desire for hard copy.

In 1972 Baker, Parkins and Poyen (11) wrote "on-line access to information, which persists as a dream of most information users, is today a reality for a favored few and for a limited scope of material --- a correct vision for many, with the only unresolved question being 'when'." Today, only five years later, on-line searching of bibliographic and abstract files is a rapidly growing reality. The list of users is growing daily, and as facilities for high capacity storage and transmission become less costly through further development and as a result of competition between the on-line centers, the number of users will grow even faster.

The abstracting and indexing, library, information center, and user communities are anxiously watching for signs of the impact of on-line services on use of the abstracts journal. It is too early yet to have clear signs of its impact. However, the impact when it becomes known, could already have had profound influence on the future of the conventional abstracts journal.

The synopsis journal (12), should it become a common replacement for the conventional primary journal, could have a significant effect on the future of the abstracts journal. However, the concept of the synopsis journal is not new, and its acceptance seems to have been remarkably slow. Should it become the way of the future for primary publication, or even if it should become commonplace in an interim mode prior to the electronic journal, the secondary journal must reckon with it.

The technology used today in generating secondary services, and especially the machine derived and photocomposed abstracts journal, sets the stage for greater cooperation then ever before between primary and secondary information processors for mutual benefit and survival. It presents opportunity for direct input of bibliographic headings and abstract text to the secondary service from magnetic tapes prepared for deriving the primary journal. Input may occur before or simultaneously with the appearance of the primary journal. Tradeoffs such as preparation of indexes for the primary journals from a single processing step by the secondary processors could become commonplace. That sort of cooperation is already happening. The potential for the sharing of input, and possibly even hardware and software for the purpose of generating primary and secondary services will increase. These services will see themselves more as partners in the information transfer system, as they recognize these opportunities and accept them as necessary for ensuring, or at least increasing the likelihood of their respective economic viabilities.

For years, some have predicted the demise of the printed page, and indirectly therefore the demise of the abstracts journal. I think both the printed page and the journal are here to stay. But I also think that alternate forms will continue to erode the monopoly held by the printed page and the conventional journal, whether primary or secondary. The conventional abstract could be the first component to go -- substituted by elements of information that could be intellectually assimilated by the user without full verbage of a printed or machine readable, in toto abstract. The "structured" abstract, i.e. an "abstract" derived by filling in blanks about a document, would serve a number of purposes in the electronic system. It would replace the conventional, often wordy abstract, provide index words for automatic indexing, and lend itself to standardization and inclusion of "data" in the secondary information system.

In recent years, secondary information processors have felt a growing demand from users for not just references to where the data were published, but for the actual data. Additionally, if data out of context of the full document are to have meaning they must be "evaluated" data, which to me, suggests another opportunity for cooperation between the primary and secondary publishers. The advantages for data handling in the electronic journal system are obvious, and the prospects for secondary accessing systems to deal with data are encouraging.

In summary, today's abstract journal, while not outwardly different from earlier generations, is not so conventional anymore. It is the product of a system that has added many opportunities to the process of information dissemination. The future of the printed journal is bright, and while it will be with us for a long time, new transfer media and mechanisms derived from the same systems that are essential to enable economic survival of the conventional journal will compete for its livelihood. Those systems and the products and services generated therefrom would seem to offer promise for mutually rewarding cooperative efforts involving the primary and secondary publishers and the user community.

Bibliography

1. Klemper, Irving M. The early history of abstracts. In: Collison, Robert L.; ed. Abstracts and abstracting services, ABC-CLIO, Inc., Santa Barbara, CA, 1971, 59-64.
2. Smith, J. Ronald. The future of publications and services. In: Primary publications and secondary services - partners in information flow. ICSU AB, Paris, 1974, 132-139.
3. Kennedy, H. Edward. Abstracts in secondary services. In: Primary publications and secondary services - partners in information flow. ICSU AB, Paris, 1974, 168-177.
4. Price, Derek J. de Solla. Little science, big science. Columbia University Press, New York, 1963.
5. King, D.W., McDonald, D.D., Roderer, N.K., Wood, B.L. Statistical indicators of scientific and technical communication 1960-1980. Vol. 1. A summary report for: National Science Foundation. 1976.
6. Cummings, Martin M. The biomedical communications problem. In: de Reuck, Anthony and Knight, Julie, ed. Communications in science: documentation and automation, J. & A. Churchill Ltd., London, 1967, 110-133.
7. Macklup, Fritz. The production and distribution of knowledge in the United States. Princeton University Press, Princeton, N.J. 1962.
8. Chacko, George K. Beyond Gutenberg: communication without paper? American Association for the Advancement of Science, Annual Meeting, Denver, Col., February 21-25, 1977.
9. Central electronic journal conceptualized. Information Retrieval and Library Automation, 11:6, 1976, 1-2.
10. EDUCOM sees electronics supersede printing. Information Retrieval and Library Automation, 12:1, 1976, 2.

11. Baker, D.B., Parkins, P.V., and Poyen, J. The future of access (abstracting and indexing) services. In: Chernyi, A.I., ed. Problems of information science, FID 478, Moscow 1972, 142-166.
12. Synopsis journal idea catching on in Europe. Chemical and Engineering News, 53:15(April 14), 1975, 14-15.

USER REACTIONS TO SECONDARY JOURNALS

Knut Faegri, *Botanical Museum, Bergen University, Bergen N5014 Norway*
Past President, International Union of Biological Sciences

I am very glad that I was asked to speak about abstract journals. In this case we can by reference to well known quotation consider something that is usually printed on usually rectangular pieces of paper. However, when it comes to contents I think that what I am going to say will not be particularly different between a journal and these other types of information transfer. It was very heartening to hear Dr. Kennedy's belief in the printed journal. To me the idea of the electronic journal or whatever you like to call it is based on a terrible fallacy, namely that man is a rational animal and that the subspecies of homo sapiens called the scientist shares this ethological characteristic. He does'nt. We all know that if we think on the way we work. In the old textbook of logic which was used at the University of Oslo in my time (I think it's obsolete known), there was one sentence which all of us remember. This is the only sentence we ever remembered later. And my rather crude translation runs: life is manifested, and thought likewise. The word follows thought on its winding paths. I think this is a very wise sentence. The scientific thought is manifested too, and the scientific word has to follow. You can live in an enormous row or block of flats. You can exist there, but most of us to live will have to get out into the country and perhaps live in a primitive cottage. The cottage is primitive. It hasn't got any of the amenities of your flat, but after all you feel a certain fulfilment. And this is very similar to the difference between just recording the one thing you are asking for and to be able through browsing - and I was very glad to see that notorious word was still there - browsing through the material and getting your inspiration not from the answer to your question but from the answer to a question which you would never have asked if it hadn't been presented as an answer.

Now this is very important to my conception of what a journal - a secondary journal - should give you. But we'll have to distinguish quite clearly between the two main functions of the secondary journal: the alerting function and retrospective search. The alerting function is comparatively simple. Current Contents - in some way you are alerted to the material which is produced by your colleagues most of whom you know or have heard of the labs in which they work. And just a catchword will be enough, and you'll understand the rest of it. You don't need much, hardly much more than the title in an alerting service. The demands on the contents of the abstract in an alerting service are very small indeed. Sometimes of course, we all know, the title is so obscure that it does not convey any idea of what the paper contains or perhaps even the opposite idea, in which cases a small (we could almost call it a correcting abstract that will be a new type of abstract in addition to the informative and so on) will be necessary, to alert the right persons to the right subject matter. But as I said, on the whole, this is a very simple function which can be carried out by a few key words perhaps, in addition to the title. Again I would emphasize that the browsing is perhaps at this stage even more important than at the other stage. Because you'll find out that other labs working in the same field have perhaps a completely

Miriam Balaban (ed.), Scientific Information Transfer: The Editor's Role, 573-574. *All Rights Reserved.*

different angle which you never thought of and which you would never have found if you went straight for your object. But approaching it about in the same manner as an insect approaches a source of odour by going zigzagging and all the time passing through the main stream will in the end come up to the source of the odour. It will never go in a straight line, but it will cover the whole area and in this way I think zigzagging through the alerting services is a terribly important way of learning what is happening on the borders of your own field.

When it comes to the retrospective search, the demands are much more stringent. One - it's easier to form a profile, because you are looking for information on a thing you are working on, you are looking at the information that exists already - not the information in status nascendi. It's all much easier. But you'll have to define your profile and instead of coming in comparatively small amounts of data you have this immense amount of data covered by decades of accumulated wisdom in the journals. There is great difference there in the various types of scientific work. In experimental science, you are not particularly interested in papers beyond say 10 years of age. The half year life of the experimental papers in biology is about 7 years which means that after 7 years, only half the papers are still more or less alive. That means that after 15 years, very little is really of interest. So you can stop your retrospective search rather early. And also all the papers which are of interest will be referred to in your recent documents. In descriptive biology, (I'll stick to biology - that's the only thing I know anything about) the half life of a paper is something between 15 and 16 years (according to the little statistics I have made), and that means that you will have to scan at least 30 years to be fairly certain to get what you need. As a matter of fact, the thing is even worse because the curve flattens after about 25 years so that it does'nt go as rapidly down as the other one. So it's an enormous amount (even 10 or 20 years, it's very much to scan) of material that has to be scanned and you'll have to define your profile quite accurately. And it's no use defining your profile if the abstract isn't equally well structured. This is an important thing.

The other thing is that whereas an indicative abstract just telling what the paper is about is in most cases perfectly suitable for an alerting service, it is completely useless for retrospective search because in retrospective search, the material is so enormous and because much of it is unobtainable (or obtainable only with great difficulty and time lag). It is no use to know that so and so has produced some data about this in, say the Ukranian Journal of Agriculture. Point one is that it doesn't exist anywhere in North. Europe as far as we have been able to find out, and point two is that I don't read Ukranian. Of course you can find it at the end at the BLL, and you can find somebody who knows Ukranian. By that time, half a year has passed and you have lost interest in the thing or it isn't useful to you any more. There you really need the concrete data in the abstract.

There is no use to know that it has been done. You must know what has been done. There are papers which cannot be abstracted that way. I've been abstracting papers for Biological Abstracts for almost 40 years, and many times I've been obliged to send an indicative abstract because it was simply impossible to derive any general principles out of the paper. In many cases it is an extremely important archival description, but you can't summarize a paper that consists of, let's say, phyto-sociological tables with no general conclusions. This material is very important but you can't get anything but a description into an indicative abstract. But apart from these cases where the material itself doesn't lend itself to an informative abstract, the user's first and primary demand is that the abstract be informative - that it give not only a technical description, but that it gives me the information I need.

DISCUSSION on "Developments in Information Transfer," by Ernest J. Mann

"Information Flow Between Primary Journals and Secondary Services: Report on a Study," by Ruth B. Weiss

"The Past and Future of the Conventional Abstracts Journal," by H. Edward Kennedy

"User Reactions to Secondary Journals," by Knut Faegri

MICHEL GIBB - Mrs. Weiss explained how the ultimate objective is to educate the author into writing correct, usable abstracts. I was quite fascinated to hear Dr. Kennedy saying that there was a possibility of link-ups between primary journals and secondary journals. I always thought this was Utopian because you could never get the material in machine readable form from the primary journals in a proper manner. Now I understand that this is possible. I wonder if Dr. Kennedy could elaborate on this because it seems quite important.

H. EDWARD KENNEDY - In that part of my talk I was attempting to be a bit provocative and suggest some of the things we see as being possible in the future. I did say that that type of cooperation is taking place to a limited extent already. We are not at present taking direct input from primary journal publishers; however, we are talking with primary journal publishers now who are photocomposing their journals from machine readable records. We are discussing with them the possibility of having sufficiently standard character sets etc., so that perhaps this exchange may be accomplished in the future. We already provide indexes for several primary publishers from the processing steps that we go through for the secondary journals, i.e. the subject indexes, based on keywords, are available to these primary publishers from a master file that we have built. So to that extent this type of cooperation is effective already, and we see promise for the future, although I don't think progress will be rapid as there are an awful lot of problems to be worked out. It depends on how much people want it, how much the primary publishers want it. We all recognize that while there are many problems, they are not insurmountable.

CARL KEREN - I would like to make a plea not only for informative abstracts but also for more informative titles. We find that for browsing and current awareness the first stage of a search is often done via the titles and one has only to look at the phenomenal success of *Current Contents* in order to see that this procedure is probably adopted by most users of scientific material. We believe that title formulation and contents is much more neglected by many editors of primary journals than abstracts formulation. I recommend that a good informative title should be insisted upon by every editor. I would like also to question the usefulness of guidelines for authors. Our experience tends to confirm that most authors do not look at such guidelines, unless they are very strictly enforced. We could perhaps call on the universities to ensure that each of the potential future authors write their first papers, theses, already in the proper form. If we could convey to the universities and to the educators some of the importance which form and format of the presentation of the student's work have on his future activities in his field, this could possibly be the most effective means in order to create a future generation of authors and would make life very much easier for the editor.

MYRA KAYE - In this remarkably interesting session I found two fundamental questions which were not posed and I would be interested to hear the comments of any of the speakers. Using an analogy, if you have a cardboard box factory, you *know* why you are making cardboard boxes. But what are we transferring information *for*, why are we doing it? And are we transferring *information* or are we transferring *data* and if the latter is true, is that the primary need of the scientist? The scientist sometimes wants to know whether it was a pH of 7.7 or 7.1 at which the experiment was done, but I do not believe it is his *primary* need. I think his *primary* need is to be stimulated to think, to have new ideas, to find a system in which he can show an effect, to find an effect which will throw some light on a current problem. Where does he get these ideas? He gets them usually in the speculative part of papers, precisely what is abstracted*out* of the literature by the secondary media, which then feeds the computer. How *does* a scientist get stimulus? Have we ever posed this question and seen how it relates to information on transfer? Archimedes apparently got it in his bath. Leibnitz, I understand, derived an equation of which he said he had "known" the derivation twenty years before it suddenly came to him in a form in which he could write it down explicitly. How did it come to him? Are there experiments on this? There are, there is the famous experiment with kittens which shows that kittens learn better if they are taught by their mothers than by surrogate mothers, or other means. Now I'm not suggesting that scientists take their mothers with them to the laboratory, but I am suggesting that we do have to think about the learning process, about problem recognition, storage, short term memory, long term memory and transfer from the one to the other, to try to find out what *we* can provide that will stimulate the scientist to creativity, rather than place the emphasis on acquiring, storing and selecting data. I don't want to throw the baby out with the bathwater. I'm not suggesting we do not provide means of pinpointing data, rather, that we should place more emphasis on providing the requisites for creativity. I would like to see our next Conference perhaps turn to these questions, why are we transferring information in the form and way we do, what is it for, and consider whether we should not be doing something else as well.

M.R. PRICE - I'd like to reinforce the comment on inadequate preparation of author's abstracts. My main function is preparing *Desalination Abstracts*. I suffer chronically from all the faults that have been mentioned on inadequate preparation of abstracts, but I feel that the onus for insuring better abstracts rests with the editors of the primary journals. Then where I can envisage ensuring better abstracts is to return papers and insist on getting a better abstract. Following Mr. Keren's suggestion of teaching at university or some earlier level this seems to be to me analogous to trying to improve the standard of driving by improving driving tests. You can pass a driving test and then go off and still massacre people on the road if there aren't enough policemen to enforce the tests. I feel something like this is necessary on the part of editors to insure that adequate abstracts are presented with the papers.

PETER LEA - It has been suggested recently that *Chemical Abstracts* will not be able to print in conventional print on paper format mainly because of the expense within about five years time. Does Dr. Kennedy have any comment on this? In relation to *Biological Abstracts* does he forsee the possibility of his particular service not being available in print on paper?

H. EDWARD KENNEDY - I had not heard that prediction concerning *Chemical Abstracts*. As far as *Biological Abstracts* is concerned, we have not predicted the demise of our services on printed paper in the next five years. It is true that costs are rising all the time and we are constantly looking for ways to hold costs down as much as we can, but we have no such prediction.

JANINA BARD - I would like to refer to a comment of Dr. Kennedy's concerning the eventual replacement of the traditional abstract by a new format with less written text. In fact at IIIC we started already producing this type of journal. Our format is the result of my personal frustration over a twelve year period with abstracts. In abstracts one cannot give any evaluation or assessment. You inevitably and quite correctly, have to give an absolutely equal weight to a publication that is mediocre and one that is good. If one happens to be fortunate enough to have a team of highly intellectual abstracters one can instruct them to indirectly insinuate an evaluation, but this is no way of working. When I was given the opportunity of setting up a completely new service I tried to work out a system by which we could give some evaluation of the publication, but without offending the author.

Briefly we give the bibliographical details as usual. We also give a short indicative abstract which states quite clearly what the paper is about. But in addition we add something which we call The Scope. For The Scope we selected very carefully some basic facets of the subject of irrigation and we consistently indicated the presence and the absence of these facets. An intelligent reader using our Scope should be able to assess the breadth and depth of a particular paper and say: Does this paper contain the information that I would expect it to contain?

One byproduct of our new format, which as I have said incorporates a mechanism for indicating absence of information will, I hope be to educate the author. Ours is an irrigation journal, and yet the number of authors who do not bother to put down what irrigation system they used is impressive. I hope that authors will now notice which basic information they left out, and it may inspire them to improve the quality of their articles.

A.J.C. WILSON - I should like to comment on a couple of points that arise from Dr. Kennedy's paper. The first is his suggestion of structured abstracts that could be produced simply by filling in a form. For one part, Short Structural Papers, which amounts to about one-quarter of Section B of *Acta Crystallographica*, the abstract is practically that already. I haven't a copy of the instructions with me - 100 of them were delivered here, but they have all disappeared - I hope into useful places. The instructions are roughly like this:
The *Abstract* will consist of the crystal system, space group, unit-cell dimensions with an indication of accuracy (normally the standard deviation in units of the last quoted decimal place enclosed in parentheses), formula, cell content, measured and calculated density, source of the material, and such other information (especially structural) as can be conveyed in approximately fifty further words.
(The rough summary given at the conference has been replaced by the exact instructions given to authors). That is at least half way towards a structured abstract.

The other point was Dr. Kennedy's suggestion that one might be able to go straight from the tapes used for controlling photocomposition of the Journal to the input to the secondary services. The International Union of Crystallography has done just a little experimenting about trying to go direct from the tapes used to print Section B of *Acta* to the tapes of the Cambridge Crystallographic Data Centre. The experiemnt has been very preliminary, but technically it seems promising. There may, however, be copyright difficulties in implementing it on a large scale. Does the tape, and does the information on the tape, belong to the printer or to the International Union of Crystallography?

YEHUDAH WERNER- By this time what I want to say may be redundant, but is perhaps unpleasant enough so that I would like to ease it in by an example. Some twenty years ago we founded the Society for the Protection of Nature in Israel. A founding assembly was called in by an organizing committee. Two or three of us told the organizers in the assembly that you cannot get people to conserve nature if they do not love nature, and you cannot get them to love nature if they do not know nature and therefore the newly founded society has to teach natural history. The organizers said, "No, no, this is not our goal, we cannot compete with the school system. We will simply instruct people to protect nature, and help guide the government authority which will enforce it."

After a year or two they changed their opinion, they started a large operation of instruction in nature which has since become a success story, leading to efficient nature conservation. Wild flowers which had almost disappeared, cover the countryside again, as some of you, I know, noticed already. The lesson for us this. The state of the art of writing abstracts is really worse than we could have understood from what Mrs. Weiss told us, even where according to statistics there were usable author abstracts, probably in half the cases this was so only after heavy handed guidance by the editor, not to say rewriting. Why should this be so? After all an intelligent, logical, considerate author, should be able, from this term, "abstract", to conclude the basic aims and requirements of an abstract and compose a useful abstract without any further guidance, tools and enforcement. And in fact these do not work, as has been commented here, even guidelines do not help. We in our journal give very specific, detailed and comprehensive guidance to writing an abstract, and that is to simply tell the authors we want an abstract as for *Biological Abstracts* and I cannot think of any more detailed explanation than that. And it does not work. We have to more or less rewrite half the abstracts. I think it has already been hinted here, that we are trying to improve the examinations system when really we should press for an improvement in the teaching itself, and at a rather lower and younger level than the author stage.

RUTH WEISS - I will take up the points in order. First, the importance of title. In our study the evaluation of titles was on the same basis as evaluation of abstracts. Of the titles we evaluated, seventy three percent were considered satisfactory, about six percent were "more or less" satisfactory, more than twenty one percent were considered unsatisfactory." These are figures for published material, i.e. after editing. Twenty one percent of titles unsatisfactory is, to say the least, no reason for complacency. Second, the importance of teaching. Yes, I agree this is where any improvement should start. This does not necessarily minimize the importance of useful guidance to the author (if the author requests it) - and this brings us to the third question again of whether authors actually read guidelines. It seems they generally don't unless either they are Australian, or unless acceptance of a paper depends upon re-writing the abstract in accordance with the guidelines.

At this critical stage the guidelines will be read. There is a temptation for the editor to say "in view of the postage and the time and the backwards and the forwards I'll forget about being educational and rewrite the thing myself." Whether this pays in the long run is a question I leave with you.

ANDY JESSE - We have heard a lot about the flow of information from primary sources to secondary sources - but what about the flow from secondary sources to primary sources? Of course you can answer that when an author writes a paper he "taps" the secondary sources, thus providing circular information. But this, ladies and gentlemen, to my mind is a vicious circle, because these authors do not "tap" the secondary sources correctly. They must be guided, for example by a scientific information officer who helps them to trace relevant publications and who feeds new sources back into the information circuit. I have made an experiment to this end in an interdisciplinary field, in automatic microscopy which is used in medicine, biology and materials science, by providing a bibliography and contacting authors asking them to prepare a contribution.

We published several review papers on these individual fields and we had a common bibliography which was our resource of all previously published papers. However, when it came to list them in abstracting journals, Physics Abstracts, Computer Abstracts and the like, they included some papers only and gave the bibliographic remark "no references", and they even omitted the bibliography. Thereby the information flow was incomplete.

I should like to take up the comments that the speakers before me made: We must train authors to use abstracting journals. Thus when people undertake diploma work, why doesn't the professor ask them to prepare an abstract of his or his assistant's paper, just for the exercise. Then they get some training. So let's teach them and let's try to get a good information flow.

JOHN ZIMAN - From Ed Kennedy's remarks there was a feeling somewhat of a disappointment on the part of the computer people. Somewhow they have not made all the progress in taking over the field of abstracting and information services - secondary services, that they had been promising themselves and us over the last decade, to my knowledge, and that they are promising us for the future. In other words he said it is not going as fast as they like to say. Whether he was for or against I was not quite sure. I think that the reason for this is that the analogy with takeover by the computer of other aspects of life, does not necessarily hold in the information service business.

Where the computer has taken over is precisely in the field where there is a lot of number punching to be done or the equivalent of number punching, that is to say where there can be within the computer enormous and very rapid transformation of the input into a completely different form of output.

There are things that the computer can do, for example in weather forecasting that are completely impossible by any other means. That is to say taking innumerable bits and pieces of data from all over the world and turning them over and over millions of times and then coming up with a weather chart for the next day. You simply cannot do that in any other way. But the characteristic of so much of the material that goes into the abstract and information services we are talking about is such that it cannot be transformed very far away from its original form. Indeed what you want mostly to do is to retrieve it in its original form as a human readable text. This means that the great power of the computer, its logical transformation power and mathematical transformation power is not used, is not usable in a way that makes it economic and valuable in other ways.

It seems to me that almost all the things the computer is said to be or is going to be used for in this field can already be done. They are not technically impossible. They are in fact being done, at what will be a slightly higher cost but not at, in the case of the analogies I have given, the numerical work, a million times the cost. They will be done at twice the cost. And this is why I think that the penetration of the computer into all the processing techniques in this field will be very much slower than the optimists of computer technology have been saying. They have been disappointed in the past and I think they will go on being disappointed because they fail to realize this fundamental aspect of their whole activity.

ERNEST MANN - I think if I understood the last two speakers correctly, particularly Dr. Jesse, they were referring to a phenomenon which is worrying us in the information business. I think it is matron science, and sophisticated computer technology can be a block. It has almost become a science of its own and I would say it has become a self-perpetuating exercise, processing information, and I think we have forgotten that our main job is to transfer information from the producers of information to the users of information. We have created a block in the middle called information science and sophisticated computer technology. And I really believe that this can create problems for the user and I think this is what the speaker meant by the vicious circle.

Now one problem with computer technology in some cases is that it can produce a high error rate. There have been figures published recently that show that computer produced journals have a higher error rate in them than manually produced journals. Well errors in photographic data are very serious for information uses for scientists, and that is one aspect I think we have to solve.

Mrs. Kaye - I always enjoy listening to her - raised really such basic issues which you could argue about for the whole conference, and maybe there should be a conference on "What Makes Scientists Tick?" I remember that when my wife was a scientist many years ago, she worked very seriously. She went at night to the laboratory and injected her rats. Today it is all done mechanically, and when I went to a laboratory in America recently and saw a chemist sitting and reading the New York Times while the William Stein amino-acid analyser was putting out results, I wonder what is a scientist today. I don't have the answer.

URI BLOCH - I think there is some misunderstanding where the block really is. You see, there are some things the computer can do, apart from crunching numbers. He is, for example, a first class civil servant. You have probably not met the Israeli civil servant, otherwise you would understand why I prefer the computer. It doesn't come late, it doesn't get sick leave, it does not go away because it is pregnant and it does not drink tea. A computer can do most of the simple processing that a clerk does and I think you can expect it to do it better and more faithfully and these mistakes need not be in them. But you must know what you are doing. Somebody mentioned before two computers which could not read each other, but this is not the computer's fault. These are the mistakes of system-analysts who did not think beyond their noses, who had not heard of such a thing as standardization and who were not aware of the existence of ASCII codes, MARC tapes or even ISO standards.

We must realize that, apart from software and hardware, there is something which for lack of a better term I'll call humanware. We must realize that the most important of the three resources is the humanware and we should begin to think how we can minimize expenditure on it. It hurts my heart to hear that secondary services rewrite everything that exists already in the primary sources, simply because there has not been enough cooperation, because this would be an excellent way to lower prices so that librarians could afford to buy more journals. I am sorry that this cooperation is not coming rapidly, and I don't understand why Dr. Kennedy cannot give it a push. I know that there are technical difficulties, but I am quite sure that they are solvable.

Of course there are problems with computers too. I began by saying that computers can fill any clerical job, but editors and abstracters do not only clerical work, they have a highly intellectual job to perform and it is here that the computer falls completely down. There are people who are working on the problem of automatic information processing which is one of the more interesting problems. Unfortunately, here in the Levant we do not always know what is going on in the world, so I am not really aware of how far things have progressed, but the problem of artificial intelligence and specifically the problem of processing natural language text in a meaningful way is one goal which would truly serve the scientific community and incidentally also save a lot of money. It would be much easier to process such texts automatically, but the text would have to be in a suitable language. I am thinking especially of abstracts where we are using the most convoluted sentences to save three words, which makes it very difficult for the computer to handle. In that field, the computer still has a long way to go and I probably won't be around when we get somewhere. But from that statement to come to the conclusion that the computer cannot be of tremendous aid to us this very day I will not accept.

ON THE INFORMATION TRANSFER FROM PRIMARY TO SECONDARY SOURCES AND FROM SECONDARY TO PRIMARY SOURCES - AN EDITING EXPERIMENT IN AN INTERDISCIPLINARY FIELD (Automatic Image Analysis)

Andreas Jesse
Avenue Wendt 22, CH-1203 Geneva, Switzerland

Abstract

A model of co-ordinated information transfer between primary and secondary sources is presented. Its application in the production of state-of-the-art papers on Automatic Image Analysis (quantitative microscopy) is described: an extensive bibliography was made available to scientists of selected discilines who amended it and used it to prepare papers which were read at an International Conference; these contributions together with the updated, common bibliography were subsequently published. The procedure and results of this co-ordination between editors, documentalists, conference organizers and authors are discussed.

1. Information transfer between primary and secondary sources

In scientific communication, one oftens distinguishes between primary sources (such as articles, monographs, theses, conference proceedings) and secondary sources, which indicate the existence of these primary publications in the form of indexing and abstracting journals, bibliographies, computer files, etc. When preparing a scientific paper, an author is expected to trace relevant literature by means of secondary sources (which include here the references or bibliography at the end of a primary publication). However, this is not (and cannot be) done in a very efficient way.

The problem is as follows (Fig. 1a): many relevant primary contributions are not covered by the secondary sources for various reasons (1); some of the secondary sources may remain "isolated" and thus un-used (2); new contributions are often produced in a sort of duplicated way (3). These factors are the more pronounced the more diverse the field of research is, e.g. interdisciplinary fields.

In order to make best use of existing literature and of existing know-how, a co-ordinated information transfer is suggested (Fig. 1b): the primary contributions should be covered as completely as possible by a limited number

Miriam Balaban (ed.), Scientific Information Transfer: The Editor's Role, 581-586.

of secondary sources (1); when preparing new primary publications, these reference sources and the initial primary publications should be made apparent (2); the new primary publications in turn should be completely covered by the new secondary sources (3); etc.

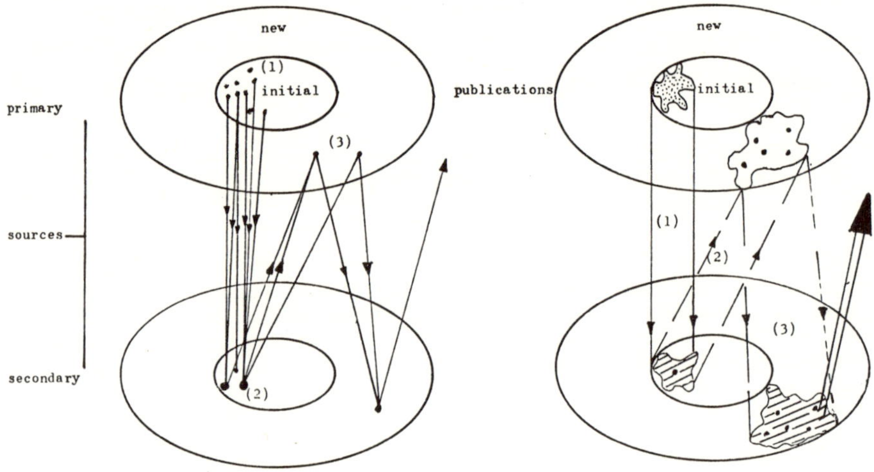

a) not co-ordinated b) co-ordinated

Fig. 1 Information transfer between primary and secondary sources (see text for details)

This paper describes the procedure and results of such a co-ordinated information transfer in an interdisciplinary field. The salient features consist of the steps (1) and (2), i.e. the elaboration of a comprehensive bibliography and the urging of authors to make best possible use of it.

2. The interdisciplinary field Automatic Image Analysis

Automatic Image Analysis deals with the quantitative investigation of images, which, although normally two-dimensional, are taken as representative for three-dimensional structure. Microscopical methods are usually necessary to obtain these image, whereas stereology is employed to deduce the quantitities and relations required for describing the structure of the object. The degree of automation of the instruments used varies from simple measuring devices with logic circuits to sophisticated pattern recognition equipment with data processing facilities.[9]

Automatic Image Analysis is now commenly employed in materials science, in the medical and biological sciences, and others. Its relations to these various fields of science is depicted in Fig. 2.[17] Clearly, the advancement

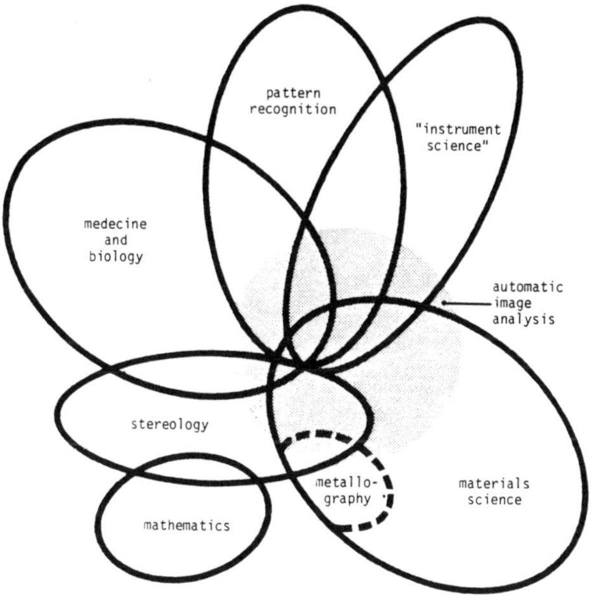

Fig. 2 Automatic Image Analysis and its relation with other fields of science [17]

of Automatic Image Analysis depends on the quality of scientific communication within and across the borders of these disciplines. Whilst the coverage of scientific literature within the individual fields mentioned above may be considered adequate, this is not the case for Automatic Image Analysis. An earlier evaluation of the information sources revealed that 70 % of the publications appear in periodicals, 12 % in proceedings, 8 % in or as books, 3 % as reports, and the remaining 7 % as dissertations, prospecti and others; of the roughly 300 papers in periodicals, within three years, half appeared in the issues of four journals - the other half being scattered among 90 different scientific journals! This makes tracing of relevant papers quite difficult, especially since secondary sources do not and cannot cover all literature of such a diverse field.

3. An editing experiment in Automatic Image Analysis

In 1971, the first reviews on Automatic Image Analysis were published in the January issue of The Microscope. It became clear that in this interdisciplinary field, "complete" coverage of recent advancement could only be achieved by combined efforts. Therefore, for the subsequent review issues of The Microscope (January 1974 and 1976), the papers were co-ordinated by the present author as follows:

- a draft version of a structured bibliography containing recent and relevant references was sent in the Spring to a number of scientists working in Automatic Image Analysis;
- they ammended it and used it as a basis for their papers presented at the INTER/MICRO-Conferences (Cambridge, England; Summer 1973, 1975);
- all collaborators were kept constantly informed about any changes of the bibliography (additions, numbering, re-arrangements);
- the scientists' contributions were submitted for publication in the the Autumn, togeth r with the updated, common bibliography.

It is claimed that by this co-ordination, reviews and bibliographies could be obtained which otherwise in such an interdisciplinary field are extremely difficult or even impossible to be produced by an individual. Other scientists were urged to send their papers to the editor of the project, in order to continue the bibliography.

The advantages of this procedure are obvious:

- scientists normally cite only those papers which they happen to come across; here they do not have to spend much time for literature searching, and the references were put into a standard bibliographic form by the documentalist which also saved considerable time on the authors' side;
- references which an author omits, because they do not fit into his review, will not be lost — in other words: whilst a review paper may be considered as an author's <u>subjective</u> description of the progress made in the field, the bibliography may be considered to form an <u>objective</u> collection of relevant elements of information; reviews and bibliography are thus supplementary;
- the flow of information from primary sources to secondary sources is normally much more pronounced than the other way round (cf. Fig. 1a); with the present editing experiment, the bibliography (in its function as a secondary source) is directly linked to the individual papers and assures an optimum information transfer (cf. Fig. 1b).

4. Results of the editing experiment

The conclusions that may be drawn along the steps (1)-(3) in Fig. 1b are as follows:

(1) The information transfer from primary sources to the bibliography (secondary source) was considerably improved: many authors, conference organizers, instrument manufacturers kept the editor constantly informed about new papers, courses/conferences, and equipment — but more could be done.

(2) The existence of a comprehensive bibliography made it easy to "persuade" authors to prepare reviews on the topics as indicated in Fig. 2. These state-of-the-art articles together with the common bibliography [1-8, 9-16, 17-24] are considered to present a unique source of information. However, it was noted that some authors did not fully exploit the bibliography — they continued their citation habits although they were given clear instructions what they were expected to do.

(3) On the side of the "new" secondary sources some discouraging occurrences were observed: one leading abstracting jounal could not include the papers because the copyright question had to clarified; some abstracting journals [25] did include most of the review papers (together with the remark "no refs."), obmitting to mention the common bibliography (!); once, only the bibliography was mentioned;[26] it was also abserved that the coverage of one and the same abstracting jounal was different from issue to issue of The Microscope.

5. Outlook

The usefulness of this "co-ordinated information transfer" in the field of Automatic Image Analysis has been welcomed and acknowledged at several occasions — but it has been impossible to find a successor of the project as the present editor cannot continue it. However, the procedure as such may be of interest to other people wishing to improve the information transfer in their field. It is strongly recommended not only to improve the information transfer from primary to secondary sources but also from secondary to primary sources. These "publications on command" would meet the requirement of user-oriented information.

References

Microscope 19 (1971):

1) C. Fisher, The new Quantimet 720 (p. 1-20)
2) A. Jesse, Quantitative image analysis in microscopy — a review (p. 21-30)
3) H.P. Hougardy and A. Rose, Quantitative image analysis in the field of commercial metallic materials (p. 31-39)
4) M. Mertz, Quantitative image analysis in medicine and biology (p. 41-63)
5) A. Jesse, Quantitative image analysis in the field of nuclear materials (p. 65-75)
6) P.J.C. Dinner and B.H. Kaye, Some pattern recognition concepts for automatic image analysis in fine particle technology (p. 77-86)
7) M. Cole, Instrument errors in quantitative image analysis (p. 87-103)
8) A. Jesse, Bibliography (p. 104-112)

Microscope 22 (1974):

9) A. Jesse, A review of Automatic Image Analysis (p. 1-4)
10) H.P. Hougardy, Instrumentation in Automatic Image Analysis (p. 5-26)
11) I.A. Cruttwell, Pattern recognition by Automatic Image Analysis (p. 27-37)

12) S. Nazaré and G. Ondracek, Automatic Image Analysis in materials science (p. 39-58)
13) S. Eins, Automatic Image Analysis in the medical and biological sciences (p. 59-68)
14) E.E. Underwood, Stereology in Automatic Image Analysis (p. 69-80)
15) A. Jesse, Information sources in Automatic Image Analysis (p. 81-88)
16) A. Jesse, Bibliography on Automatic Image Analysis (p. 89-115)

Microscope 24 (1976):

17) A. Jesse, Automatic Image Analysis — 1976 (p. 1-6)
18) H.P. Hougardy, Recent progress in Automatic Image Analysis Instrumentation (p. 7-23)
19) J. Slater and B. Ralph, The status of Automatic Image Analysis in Materials Science and Technology (p. 25-28)
20) S. Eins and E. Wilhelms, Assessment of preparative volume changes in central nervous tissue using Automatic Image Analysis (p. 29-38)
21) H. Aus, U. Gunzer and V. ter Meulen, A note on the usefulness of multi-color scanning and image processing in cell biology (p. 39-44)
22) E.E. Underwood, Trends in Stereology (p.45-48)
23) E.E. Underwood, Quantitative Shape Parameters for microstructural features (p. 49-64)
24) A. Jesse, Bibliography on Automatic Image Analysis (1973-1975) (p. 65-102)

25) Physics Abstracts (Aug. 1974), abstracts 51044-51049
Computer & Control Abstracts (Aug. 1974), abstracts 17727, 17910, 17911, 1795
Physikalische Berichte (Aug. 1976), abstracts 2792-2795

26) Computer & Control Abstracts (June 1976), abstract 12851

HOW SERVICES FROM THE INSTITUTE FOR SCIENTIFIC INFORMATION[R] (ISI[R]) AID JOURNAL EDITORS AND PUBLISHERS

Eugene Garfield, *President*
Institute for Scientific Information
325 Chestnut Street, Philadelphia, Pennsylvania 19106 U.S.A.

ABSTRACT

The ways by which scientific editors and publishers can utilize ISI[R] services to advantage are described. Seven key editorial practices that improve the retrievability of scientific articles through secondary information services are also identified. Also discussed are the impact of secondary information services on the economics of publishing scientific journals and how publishers can use citation analyses for market research.

As the head of an organization that produces secondary information services, I consider ISI to be in an active and mutually beneficial partnership with journal editors and publishers.

While I'm sure each of you, especially the publishers, readily understands how ISI benefits from the existence of your journals, past experience tells me that a good number of you are not quite so sure how you benefit from the existence of ISI and its services. The purpose of this talk is to present the basis for my belief of mutual advantage.

According to F. Peter Woodford, who used to conduct an excellent training program for scientific editors at Rockefeller University, the editor is in a position of "some power, considerable prestige and great responsibility."[1] The first responsibility Woodford identifies for the editor is to assure the scientific worth of the articles published in his or her journal. This involves evaluating the magnitude of each paper's scholarly contribution to an important topic, its adequacy of experimental design and methods, and the validity of its inferences and conclusions. The second responsibility is to assure the comprehensibility of each article, that is, the ease and accuracy with which a reader can grasp the author's thoughts. And the third responsiblity is to assure the retrievability of each article. This includes the certainty and rapidity with which an article will reach the attention of another scientist interested in the subject, now and in the future.

Let's take the editor's first responsibility, to assure the quality of the papers accepted for publication, and see how ISI services can help.

In the reviewing process an editor frequently needs background information to support or refute referees' and authors' criticisms and claims. While most editors are attuned to the use of traditional reference tools for such purposes, many do not take full advantage of the recent innovations in information retrieval. For example, how many editors here subscribe to an

Miriam Balaban (ed.), Scientific Information Transfer: The Editor's Role, 587-595.

SDI (selective dissemination of information) service?

By creating a profile that defines the subject scope of his or her journal, an editor can receive an SDI report at regular intervals indicating relevant articles published in other journals. In this way, the selection of manuscripts and other editorial decisions are made with more confidence.

SDI services, such as ISI's ASCA[R] and ASCATOPICS[R], are relatively inexpensive for what they provide ASCA (its full name is Automatic Subject Citation Alert) can cover a subject area on as broad or as narrow a basis as you think necessary . ASCA is available with weekly reports for about $100 per year, and with monthly reports for only $50 per year. Its reports are computer-generated, and they bring you lists of relevant articles published within the past few weeks. ASCATOPICS costs less than $100 per year and gives you broad coverage of a subject area through standard profiles.

Of course, you may argue that you can more than adequately satisfy your current awareness needs by using Current Contents[R]. And many editors do! But I think the focus of the ASCA services is important.

ASCA can also be used to provide an innovative service for authors that makes it more attractive for them to publish in your journal rather than with a competitor. Through ASCA, you can list all the current articles that cite any article ever published in your journal. Or you could limit the listings to citations of articles published during the past few years. This is especially useful for new journals where there is a strong chance that other journals will be the main source of citations.

Besides the constant general awareness an editor requires, specific questions must frequently be researched. Or, you may need to decide on the originality of a submitted manuscript or the completeness of the bibliography it contains. In these cases, the use of online search services can be quite helpful. Online services would also make it easier to provide referees with supplementary bibliographical information. If editors used such quick-response services more often they might be able to speed up decisions and thereby earn the respect of authors who justifiably resent the time lags between submission of their articles and final publication.[2] In actuality, however, the use of online services by editors is astonishingly low.

Once again, the cost is minimal. With ISI's online services, SCISEARCH[R] and SOCIAL SCISEARCH[R], once you acquire your terminal, you pay only for the exact amount of time you are on-line and searching. There is no minimum charge for access to these data bases.

ISI can also help you assure the worth of the articles you accept in other , unique ways.

In producing our information services we have created a massive, permanent record of who has published what and where in the journal literature. What's more, for about 3,500 journals, we have also recorded the references to other published work contained in each article processed. These records permit a wide range of quantitative, qualitative, and other descriptive analyses of scientific communication through journals. While some applications of these analyses, like measuring an individual's research performance, have created controversy,[3-9] other uses, such as defining the history of a scientific development and measuring the activity and interaction of scientific specialities, seem to be more palatable.

Citation analysis can be especially helpful to editors of new journals. By using it to identify the important people in a specific field, editors can obtain some objective input in establishing editorial advisory boards and in soliciting potential authors to solicit for review articles.

Through citation analysis one can determine those topics which are currently more active.[10] And for those articles you have already published, citation analysis can help determine whether your selection criteria are corroborated by the research community.[11] Indeed, we did a study for one journal that helped evaluate the papers they had rejected.

In today's interdisciplinary world, if you are having trouble determining just where your journal fits in, citation analysis can help resolve your identity crisis. In other words, if you know which journals cite yours and which journals yours cites, you can get a clearer picture of which journals are most closely related to your field. Undoubtedly, the dozens of journal citation studies I have published in Current Contents[R] have produced many surprises.[12-13]

While ISI services can help you improve and measure the scientific worth of your journal, there isn't much we can offer that directly assists you in improving comprehension. I've spoken out repeatedly on the need for better and clearer style in scientific writing.[14] I've even cajoled scientific authors to use a little humor now and then.[15] But, basically, you editors will have to fight this battle on your own. In this respect, the work of Woodford[16] and DeBakey[17] and others is more relevant.

ISI can, however, provide some important help in the third area of responsibility--making your articles retrievable. Remember that in this context retrieval includes dissemination--making certain that all potential readers see pertinent articles.

In my experience, this is the area where journal editors have the most problems. Many of you are still using editorial practices appropriate to another era. In the old days the typical scientist personally received all the journals he or she considered relevant and scanned over them cover to cover. Today's researchers generally use Current Contents or other information services to maintain their current awareness and to do retrospective searches.[18] These services either supplement or even replace scanning the journals. Thus, a completely worthwhile and well presented article might be overlooked by the research community if the journal in which it appears is not covered by the various secondary information services. Even if the journal is covered, its articles can still be overlooked if certain editorial practices prevent them from being effectively abstracted and indexed.

To get a clearer picture of what I'm talking about, consider these numbers. There are nearly 30,000 subscribers to our Current Contents publications. Our subscriber studies indicate that over 250,000 researchers read Current Contents every week. This readership is matched only be a few journals like Nature and Science while the circulation of most other journals is a few thousand or even less. Obviously, then, being listed in Current Contents is important to the visibility of small journals. I do not know the readership figures for other services like Chemical Abstracts or Biological Abstracts, but certainly they too have significant impact on the use of small journals.

Later on I'll talk about the economic impact of having your journal covered by information services. What I would like to touch on now is the fact that even after your journal is covered, there remain many things you can do to improve retrievability. I recognize that the personalities of editors and publishers give journals much of their distinct character. But the checklist of editorial practices I am about to suggest does not infringe on your liberty or expressiveness and will help your journal immensely.

1. Use clear titles which include informative key words. A title that does not adequately identify the subject of the work reported is inexcusable.

2. Include an abstract or summary for each article that can be understood by scientists outside the particular area of specialization involved. I am delighted to note, after years of prodding, that even Science has now agreed to include such summaries for all of its lead articles. This is as important to those who actually scan the journal itself as to those who read the articles in the form of reprints.

3. The title of each article should be displayed on the table-of-contents in a bold face ahead of the author's name. Most names will not be known to the average reader, but if you insist on catering to the vanity of authors, put their names in a separate column.

4. Prominently display your volume and issue number on the contents page. Too often this vital bibliographic information is too small or located in an obscure location.

5. Provide English titles on your contents page--even though the articles are written in another language. You can also supply the original title in parentheses, but the problem of bilinguality is probably best solved by having contents pages in separate languages. I offer this approach because the suggestion to use English exclusively may be offensive to some of you. I recently published an article in La Recherche in which I suggested that French scientists publish in English.[19] The reaction in France was explosive.[20]

   I can only say that my stance is not one of "linguistic imperialism," as one of my critics asserted. In fact, my position is based on citation analysis of French journals which showed that articles published in French journals are infrequently cited not only by scientists outside France but also by French authors. My conclusion is that people are not browsing through these articles because of the language barrier. Even mediocre French scientists would get more attention if their articles were published in English.[21] As I stated in the article, "Throughout the world, practically all prospective readers--even those who might prefer some other language--can immediately understand a scientific article published in English."

Now to move on to a less controversial suggestion:

6. On the first page of all articles, provide the complete address (including departmental affiliation and postal code) of each author. Ideally, the address should immediately follow the author's name.[22] You may think that this is a minor point, but at ISI we process nearly a million addresses each year, and it is incredible how often addresses are incomplete. And it often is completely impossible to tell which address goes with which author when there are three or more authors. Since ISI publishes Who Is Publishing In Science® each year, our own editors know all too well the difficulties of dealing with the inconsistencies of author addresses from one journal to another as well as within individual journals. One wonders if editors and publishers are fully conscious of the importance of these addresses in promoting the sale and distribution of reprints. Current Contents and ASCA, alone,

produce about 10 million reprint requests each year.

7. Finally, some advice on reference citations. Naturally, ISI has a special interest in the way you handle reference citations since we are the only major information service currently using citation indexing. But what I suggest will reduce printing costs for your journal and avoid frustrating readers. So, stop using archaic citation phraseology like *op.cit*. and *ibid*. And when was there ever a footnote important enough to publish that could not be included in the text? Or, if some force compels you to put footnotes at the bottom of your text pages, at least separate them from the reference citations which should be gathered together at the end of an article rather than dispersed through its pages. Naturally, the numbered or alphabetic arrangement makes it easier for ISI to process your articles. But it really is an important service to your readers, too. One of the most common things researchers do when they find an article of interest is to use its bibliography to obtain additional reading material. This is frequently done by supplying an assistant with a photocopy of the reference list so he or she can obtain the checked items. Gathering the references at the end of an article can sometimes make the difference between 20 pages of copying and just one. I will also mention, although I'm sure it will be for naught, the disservice done to readers by *Science* and other journals which omit the titles of cited articles. I have published my objections on this before, but I assure you it makes little difference to ISI in its work.[23-24]

Following these seven suggestions will help your journal's articles attain maximum exposure to the research community. You may then wish to see how well your journal is doing in terms of its relative importance. One way you can monitor this is through our *Journal Citation Reports*®, which is available as an individual publication and as part of the *Science Citation Index*®. In *JCR*™ you can observe how your journal compares with other journals in total citations, impact, immediacy, and in other measures of importance.[25-27] If you have a favorable citation analysis, you will be more likely to attract better authors. Today's more sophisticated researchers want to know in which journals their articles will be read and responded to most quickly. Most important of all, you can observe whether your journal's impact is increasing or decreasing from year to year.[28]

Until now I've dealt primarily with the concerns of journal editors. But ISI is also very much aware of the publisher's role in scientific communication.

Probably the one question most asked of me by publishers is, "Won't having my journal covered by a secondary information service reduce my total subscription level?" Such publishers are especially concerned about ISI services since they are backed up by a tear sheet service, called OATS®, which provides copies of articles from the journals we cover. Many publishers fear that users of our services will continuously order tear sheets without ever asking for the actual journal.

Realistically, economic considerations work against this happening unless you have a *very* poor journal which only infrequently publishes articles of interest. Here's why. Having your journal covered by an information service should cause an increase in the number of requests for its articles. If a library does not subscribe to your journal, copies will have to be ordered through inter-library loan. This is always slow and is now becoming quite expensive as the larger libraries increase their fees to discourage heavy use of their resources by others. Frequent purchases of articles through OATS at three dollars per 10-page article can quickly run up some big bills, too. So what usually happens is that librarians do some basic arithmetic and realize that it's cheaper to subscribe to a journal than to continue to buy its articles one by one.

This works on an individual basis, too. Researchers who are regularly ordering individual articles from a specific journal are going to get tired of the hassle, delay, and expense and want their own copy. This will be true whether they have been buying tear sheets or writing authors for reprints. Even a free reprint has a cost associated with it that is not trivial.

Even if a library already subscribes to your journal, having it covered by an information service may cause a second or third copy to be purchased or influence the annual decision to renew. As more requests for articles from your journal are received, librarians are forced to do more in-house photocopying. These days nearly everyone realizes the true costs of this procedure. Furthermore, in the United States, new copyright legislation seems likely to limit the number of photocopies a librarian may legally make. The upshot of this is that most librarians will realize the economy and wisdom of buying a second copy of your journal.[29]

Of course, if your journal is really bad or of limited interest, it isn't likely that it would be covered by ISI services in the first place. In some cases this lack of exposure may well be fatal. This is not necessarily a bad thing, either, since I believe in euthanasia for poor quality journals.[30]

But why has ISI persisted in operating its OATS service for over 10 years in the face of publishers' strong concern that it is depriving them of subscription revenue? Basically, we do it because we want subscribers to ISI services to know that when traditional channels fail them they can get the article they want from us. I believe that even if a subscriber comes to ISI for an article just once a year, he or she must be serviced without a delay of weeks.

We also persist with OATS because we feel we have not done badly at all when it comes to protecting publishers' rights. Since its inception, OATS has been a royalty-paying service. Each OATS transaction is recorded and, at the end of each year, publishers receive royalty checks based on volume. Until recently that royalty was equivalent to five cents per page. Now, this amount has been doubled and we have established a minimum payment of 50 cents per article. And we are prepared to go higher if publishers insist on it. It is their privilege and ISI supports the principle of copyright protection even when it causes ill-will with the librarians and researchers who are our customers. But if publishers insist on <u>exorbitant</u> royalty fees--one publisher expects to receive \$3.75 per article--they will limit the effectiveness of voluntary arrangements such as the one set up by the Association of American Publishers. What they will get instead will be a compulsory licensing system similar to the one that exists for music. Under such an arrangement, publishers will be likely to receive <u>less</u> than 10 cents per page.

But despite the fears of some publishers about having their journals covered by ISI services, others are quite eager to participate. And just as an editor faces the wrath of an author when it is necessary to reject a manuscript, so it is when ISI turns down a publisher's offer to cover a new journal.

We even have some problems with editors and publishers who, once their journals are covered, try to tell us in which edition of <u>Current Contents</u> they should appear. I have in mind the editor of a botanical journal who insists that it should be in the *Life Sciences* edition instead of the *Agriculture, Biology, and Environmental Sciences* edition. We know from careful analysis, however, that the journal is of primary interest to botanists and agricultural workers. And since we can only cover a few of the *most* prestigious botanical journals in *Life Sciences*--because of our absolutely comprehensive coverage of areas like biochemistry and molecular biology--our decision had to stand. I only wish that it were not necessary to make such decisions.

I now want to offer a few final suggestions on how ISI services can help publishers. For one thing, we can show you in what areas new journals are needed. A citation analysis can provide evidence that articles related to a specific subject appear in a broad range of journals. Frequently, this scattering indicates that a new journal in this emerging specialty will be well-received. Back in 1973, I published a citation analysis on pathology journals and stated that there seemed to be a field of "applied virology" developing.[31] The recent publication of the *Journal of Medical Virology* seems to have confirmed the correctness of this analysis. ISI citation analyses can also help publishers identify prospects in given specialties for mailing lists for book and journal promotions.

As I said earlier, I see ISI in a partnership with editors and publishers. I hope my talk has illuminated some of the areas in which we can work together for our mutual benefit, and I would like to continue this dialogue. For this reason, I am thinking about conducting a series of workshops at ISI that will offer more complete instruction to scientific editors and publishers on how to interface with information services. I would appreciate any comments you may have on this idea.

REFERENCES

1. Woodford, F. Peter, "Training Professional Editors for Scientific Journals," Scholarly Publishing, 2, 41-46, 1970.

2. Rodman, H., "The Moral Responsibility of Journal Editors and Referees," The American Sociologist, 5, 351-57, 1970.

3. Wade, N., "Citation Analysis: A New Tool for Science Administrators," Science, 188, 429-32, 1975.

4. Carter, G. M., "Peer Review, Citations, and Biomedical Research Policy: NIH Grants to Medical School Faculty," Rand Report R-1583-HEW (Santa Monica, California: Rand Corporation, 1974), 90 pp.

5. Johnson, A. A. and Davis, R. B., "The Research Productivity of Academic Materials Scientists," Journal of Metals, 27, 28-29, 1975.

6. Gustafson, T., "The Controversy over Peer Review," Science, 190, 1060-66, 1975.

7. Bayer, A. E. and Folger, J., "Some Correlates of a Citation Measure of Productivity in Science," Sociology of Education, 39, 381-90, 1966.

8. Bernier, C. L., Gill, W. N., and Hunt, R. G., "Measures of Excellence of Engineering and Science Departments: A Chemical Engineering Example," Chemical Engineering Education, 9, 194-97, 1975.

9. Roy, R., "Comments on Citation Study of Materials Science Departments," Journal of Metals, 28, 29-30, 1976.

10. Garfield, E., "Citation Indexing for Studying Science," Nature, 227, 669-71, 1970.

11. Margolis, J., "Citation Indexing and Evaluation of Scientific Papers," Science, 155, 1213-19, 1967.

12. Small, H. and Griffith, B. C., "The Structure of Scientific Literature I: Identifying and Graphing Specialties," Science Studies, 4, 17-40, 1974.

13. Griffith, B., Small, H., Stonhill, J. A., and Dey, S., "The Structure of Scientific Literature II: Towards a Macro- and Micro-Structure for Science," Science Studies, 4, 339-65, 1974.

14. Garfield, E., "On Style in Scientific Writing," Current Contents, 10 January 1977, No. 2, p. 5-14.

15. Garfield, E., "Humor in Scientific Journals and Journals of Scientific Humor," Current Contents, 20 December 1976, No. 51, p. 664-71.

16. Woodford, F. Peter, Scientific Writing for Graduate Students, The Rockefeller University Press, 1968 (New York).

17. Debakey, L. and Debakey, S., "Muddy Medical Writing--is the Culprit Bad Grammar, Technologic Terminology, Committee Authorship, or Undisciplined Reasoning?," Southern Medical Journal, 69, 1253-54, 1976.

REFERENCES (cont.)

18. Garfield, E., "Primary Journals, Current Contents and the Modern System of Scientific Communication," Interdisciplinary Science Reviews (In Press).

19. Garfield, E., "La Science Francaise est-elle Trop Provinciale?" ("Is French Science Too Provincial?"), La Recherche, 7, 757-60, 1976.

20. Garfield, E., "Le Nouveau Défi Américain," ("The New American Challenge"), Current Contents, 18 April 1977, No. 16, p. 5-12.

21. Garfield, E., "Let's Erect a New Tower of Babel," Current Contents, 6 November 1974, No. 45, p. 507-09

22. Garfield, E., "An Address on Addresses," Current Contents, 14 July 1975, No. 28, p. 5.

23. Garfield, E., "The Value of Article Titles in Bibliographical Citations," Current Contents, 8 November 1968, No. 45, p. 7-8.

24. Garfield, E., "Citations in Popular and Interpretive Science Writing (letter to the editor of) Science, 141, 392, 1963.

25. Virgo, J. A., "A Statistical Procedure for Evaluating the Importance of Scientific Papers," Library Quarterly (In Press). Dissertation, Graduate Library School of the University of Chicago, Chicago, Illinois, December 1974; 105 pp.

26. Garfield, E., "Citation Statistics May Help Scientists Choose Journals in Which to Publish," Current Contents, 16 February 1972, No. 7, p. 5-6.

27. Koshy, G. P., "The Citeability of a Scientific Paper," Proceedings of Northeast Regional Conference of American Institute for Decision Sciences, Philadelphia, Pa., April-May, 1976, p. 224-27.

28. Garfield, E., "Significant Journals of Science," Nature, 264, 609-15, 1976.

29. Garfield, E., "Citation Studies Indicate that Two Copies May be Cheaper than One!," Current Contents, 7 June 1972, No. 23, p. 5-6.

30. Garfield, E., "Is There a Future for the Scientific Journal?," Sci/Tech News, 29, 42-44, 1976.

31. Garfield, E., "Citation Analysis of Pathology Journals Reveals Need for a Journal of Applied Virology," Current Contents, 17 January 1973, No. 3, p. 5-8.

THE DIFFICULTIES OF PREPARING THE INTERNATIONAL SERIALS CATALOGUE. HOW EDITORS COULD HELP.

Jeanne Poyen
Secretary General, ICSU AB, 17 rue Mirabeau, Paris, France

ABSTRACT

ICSU AB has prepared the list of the 30 000 periodicals covered by 19 of its Member Services. This paper underlines some of the difficulties which were met for this project and arise all along the information transfer chain to identify journal titles uniquely. It explains how editors and publishers could help simplifying the problems for a better dissemination of scientific and technical information.

1.- INTRODUCTION

First let me introduce ICSU AB to those of you who might not be familiar with this organisation.

As to the International Serials Catalogue, I shall try to explain :
-How it originated,
-What it is,
-The difficulties we met to prepare, and where these difficulties came from.
How we solved these difficulties is another story which I would be glad to talk about upon another occasion. I shall also try to show how similar difficulties are met all along the information chain, from the scientists and engineers who produce the information, to the scientists and engineers who use it.

2.- ICSU AB

2.1. History and purpose

Established in 1952 as an outgrowth of a joint UNESCO and International Council of Scientific Unions commission organised to study the problems of abstracting the literature of science and technology, the International Council of Scientific Unions Abstracting Board (ICSU AB) is a non governmental, not for profit organisation.

The purpose of the Board was to improve the flow of public information between the world's scientists and engineers by establishing an international forum for the scientific and technical abstracting and indexing services. The proliferation of the literature, the ongoing development of technical storage and transfer devices, and the evolution of media have kept this forum and its purposes as vital today as it was in the fifties.
The ICSU AB is the only international organisation that is devoted exclusively to the interaction of the secondary information services with the primary publishers, the library and documentation communities, and the dissemination and data centres in the interest of the scientific and technical users.

2.2. Membership

Membership in ICSU AB includes countries (National Members),

Miriam Balaban (ed.), Scientific Information Transfer: The Editor's Role, 597-601.

scientific unions, abstracting and indexing services and Associate Members. It is sponsored by ICSU.
The Member Services of ICSU AB comprising the leading abstracting and indexing services of the world in the natural and physical sciences, mathematics and technology, in all of the major languages, produce over three million indexed abstracts each year.
The ICSU and the Scientific Union Members find an opportunity within the Board to contribute to the information transfer process reflected by the abstracting and indexing journals and products that serve their members.
For the National Members, the ICSU AB is the forum in which they represent those national sectors of the information community (the libraries, dissemination centres, primary publishers, etc.) that are not members of the Board, but whose standard practices are affected by Board decisions.
The class of Associate Members, has been established for organisations that do not qualify for full membership but who want to remain aware of the activity of the international abstracting and indexing community.

2.3. Activities

ICSU AB conducts a number of ongoing and special purpose activities. These include : Annual Meetings, Symposia, Working Groups and Task Forces, Special Projects, Publications.

3.- THE INTERNATIONAL SERIALS CATALOGUE

3.1. Origin

In her opening speech on Monday, Mrs Balaban stated that "we will have to find ways to publish less or to absorb more".
Publish less is the problem which publishers and editors of primary literature have to face for the future. Absorb more is the very problem which abstracting and indexing services have to face at present : thev have to find ways of coping with the inflation of primary literature which they have to cover. And because their role is to provide organised routes of access to this primary literature, it is essential that the more this literature is abundant, the better they are able to fulfil this role.
Abstracting and indexing services do not have many solutions to solve this problem.
One is to increase their budget, and therefore the cost of their services at a rate not only equal but superior to the growth rate of the primary literature, since bigger and bigger data bases become, more and more complex they are to operate their search.
The other solution is to try to reduce their costs. This is the solution which abstracting and indexing services, members of ICSU AB have agreed to explore when, in 1971, they decided to develop a plan for cooperation among them for the coverage of primary literature. This plan known as the "Input Plan" aims at reducing the unnecessary overlap at the level of serials coverage and scrutiny (selection of papers).

A first feasibility study for this plan showed that no comprehensive list of the serials covered by abstracting and indexing services existed and that therefore, ICSU AB did not have either the list of serials covered by its member services. It was evident that such a list was an essential pre-requisite for the development of the Input Plan.
This is why, we embarked in the preparation of the International Serials Catalogue which is the Union List of Periodicals covered by ICSU AB member services. This list is now in its final stage of pre-

paration and will be available in printed form in 1978.

3.2. Content

The International Serials Catalogue contains all the serials currently scrutinized by 19 member services of the Board, which amounts to a total of approximately 30 000 titles.
For each serial, it gives the following information :

- the full title,
- the abbreviated title,
- the codes of the serial (ISSN and CODEN),
- the name of the abstracting and indexing services which cover the serial.

All titles are given in Roman alphabet which means that titles in non Roman alphabets have been transliterated or transcribed; it contains also indexes giving concordance between ISSN and CODEN and vice-versa.
It seems simple. It is simple. It seems also very simple to prepare until you try to find out what the title of a journal is. I am now trying to explain some of the difficulties we met, by showing you some of them.

3.3. Difficulties met

3.3.1. Multiple titles

Let us take for example an hypothetical journal with 3 titles on the cover :

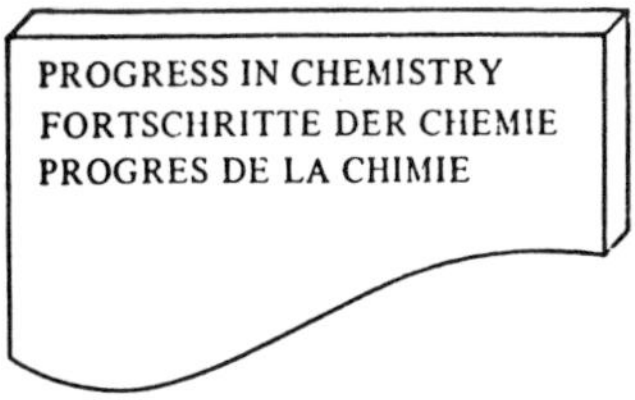

which title is to be selected for the serials catalogue ? Most probably, our member services whose data base is in English language would take the first title, those whose data base is in German would take the second title, and those whose data base is in French would take the third title.
Thus the same journal could be cited by 3 different ways both in their serials list and in their bibliographic descriptions.
Let us abandon for one moment the preparation of the serials catalogue and turn to the scientist and engineer. With the help of the secondary services products (data base on tape of printed journals) he found the references of a paper which seems of high interest to him and turns to a library to get the journal. Why should the library he goes to has to select the same title as the secondary service out of these three ? If it has not, and if the journal is not a well known one, the scientist or the engineer will not get his paper.
It would have been so much simpler if the journal either carried only one title or if at least one title would have been given prominence.

3.3.2. Title changes

Another series of problems we met was caused by title changes. There are many valid social, economic and technical reasons why editors change the title of their journals from time to time and the information community has learned to live with this situation. However, a very annoying type of title changes is when the word order is changed :

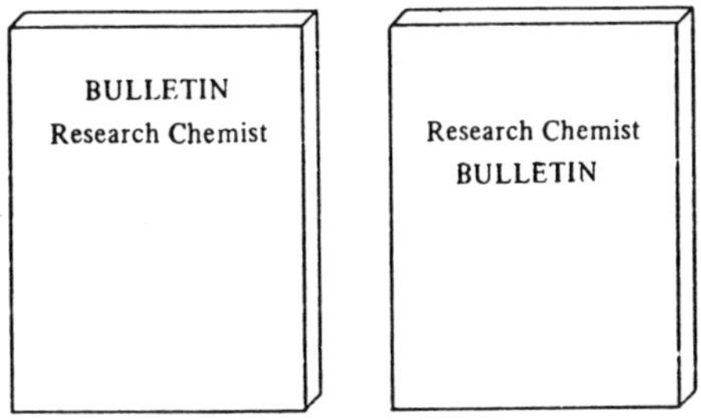

Some abstracting and indexing services and libraries will recatalogue the publication, some others will not. And again this might cause problems to the end user : scientists and engineers.

3.3.3. Acronyms

Journal titles that consist of or contain acronyms were also a source of problems for the preparation of the serials catalogue and are a source of trouble for the scientist :

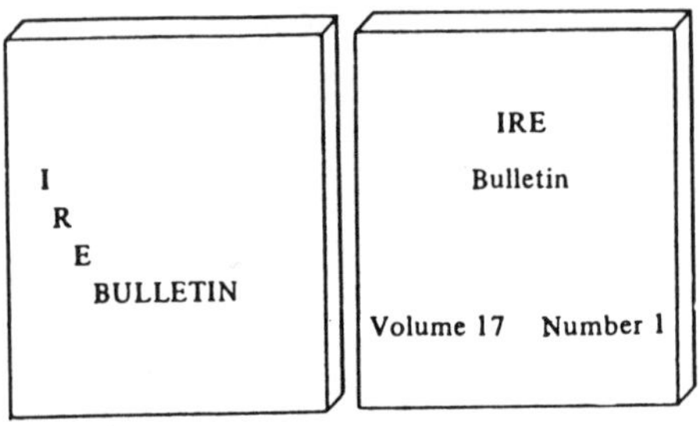

(Institute of Radium Extraction) (International Radio Engineers)

3.3.4. Other problems

Now, there are even worse examples. There are journals for which it is almost impossible to find out what the title is. One title appears on the cover, a different one on the spine, and a third one on the content page. We had some problems with these journals too.
There are also a significant number of different periodicals which carry exactly the same title, in which case, additional information such as the place of publication, for instance, must be added to each title to distinguish among them. Also journals which are published in several parts or series cause difficulties if the part of serie does not appear as an integrant part of the title on the cover, front page, spine, etc.
Another series of difficulties was caused by journal titles in non Roman alphabets. Different secondary services use different transliteration or transcription schemes and it is sometimes difficult to recognize that a title transliterated in different ways pertains to the same journal :

Щедрин

Shchedrin
Szczedrin
Chtchedrin
Sjtsjedrin

These problems I have just briefly identified are not hypothetical ones. If you look at the periodicals which are exhibited on the shelves in the lobby, you will find several examples of each case.
In the preparation of the International Serials Catalogue, approximately 10 000 titles out of 30 000 caused problems. Although the catalogue lists only the different forms of a title used by ICSU AB member services, there are approximately 5 000 titles for which 2 forms are given.

4.- CONCLUSION AND RECOMMENDATIONS

I would like to conclude by advocating the eminent editors who are present to-day, and others to follow some very simple rules which would save a considerable amount of time and money to abstracting and indexing services, libraries, documentation centres, and above all to scientists and engineers.
As far as the titles of journals are concerned, these rules are :
1.- Give prominence to one title and print it on the cover, front page, masthead, spine, so that it is easily recognizable as "the" title of the journal.

2. If you know of any journal which has the same title as yours, add an information element to your title to enable to distinguish between the 2 journals, for example the place of publication :

NATURE (London)
NATURE (Paris)

3. If your journal is published in several parts or series, always make the part or serie integrant part of the title : "Acta Crystallographica, section A" "Acta Crystallographica, section B".

4. Don't make unnecessary title changes. If you do have to change your title :
(a) please give advanced notice of such changes by indicating the succeeding title on the cover, front page, etc,. before the change is made.

(b) after the change is made, indicate the formal title (or titles) of the journal in a pre-eminent place.
You can see on journals in the lobby that this practice is in use.

5. Avoid acronyms in the title, instead, spell the full name.

6. If your title is in non Roman alphabet, please give a transcription in Roman alphabet, preferably without diacritics.

And also, in any case, get an ISSN and/or a CODEN for your journal, print them on the journal cover and be sure to check whether they need to be changed when you change your journal title or when you split your journal into several parts.
These simple rules will, as I believe and explained, save time at each stage of information transfer chain.

As simple time and cost saving rules as these ones can be listed for other items such as volume and issue designation, page numbering, etc., but this is out of the scope of my presentation to-day.

They can be found in a paper which Dr J.L. Wood delivered at the Conference "Primary Publications and Secondary Services, Partners in Information Flow" organised by ICSU AB with the support of the Royal Society some years ago, and from which some of the examples I gave you, have been extracted.

I shall only ask you to remember that it depends mainly upon primary journals to make the information they publish easily retrievable.

.-.-.-.-.

DISCUSSION on "How Services from the Institute for Scientific Information[R] (ISI[R]) Aid Journal Editors and Publishers"
by Eugene Garfield

"The Difficulties of Preparing the International Serials Catalogue. How Editors Could Help"
by Jeanne Poyen

ITAMAR PERATH - My question is to Dr. Garfield. Do you usually give ratings in citation analysis to readers' letters? And do you consider the reader's letter a useful means for an author to publicize an opinion?

EUGENE GARFIELD - There are two questions here. Certainly we consider... and we encourage them to write us. It's amazing how few do, considering the large number of readers. The second part, do you mean letters to the editor?

ITAMAR PERATH - No, my question was whether a reader's letter in a journal is rated as a citation, like any other article?

EUGENE GARFIELD - Do you mean that if it cites some other article, would it be counted as a citation? Yes.

ITAMAR PERATH - Would the letter itself rate as a kind of article?

EUGENE GARFIELD - Well we code it as a letter. In the *Journal Citation Reports* we eliminated certain codes including letters and that's why some journals changed in impact as in the case of *Nature*.

Mr. Chairman I think I might point out one of the main sources of miscitation. In fact it is the author's own reprints. These are published and distributed quite widely. I think proper studies would show that reprints are a source used by scientists more often than libraries. Most of those reprints are printed without the proper citations. People who receive them have to go to a great deal of trouble to obtain the missing citations. As a result they tend to cite from memory. But now that the copyright law has changed and publishers are going to print citations on the first page along with other statements about payments, we may see some improvements.

URI BLOCH - What are full citations?

EUGENE GARFIELD - Well at least on the first page one ought to include the authors, volume, pagination and year as well as the name of the journal. This condensed citation should appear as one line.

DIMITRI VIZA - I should be in agreement with Dr. Garfield about the use of French and the use of English, but as a matter of fact I am not. I have changed my mind by listening to him. If I understood him correctly he said that all the rubbish which is published in French, for instance, could read much better if it was published in English. I would say "no!; let the rubbish be published in parochial languages." The less the data are important the more the language should be prestigious. Let's have the non-important data published in prestigious languages like French, Latin, Ancient Greek or even Sanskrit. This will unclog the information inflation we have these days. So I think we should encourage people to publish in national languages, and leave the English for important data.

I come now to the second point and a question I'd like to ask Dr. Garfield is whether he has considered his role as a referee of scientific journals. The Institute for Scientific Information has become very important and provides an essential service to the scientist. But it is also a kind of test and a matter of survival for a new journal if it is covered or not by *Current Contents* for instance. I suppose no new international journals would survive for very long if they were not covered, providing of course they don't receive any grants from national organizations or governments. Now, has he considered his duty and the role he may play in the scientific community, apart from his duty to his shareholders? I was trying to emphasize in this conference that the scientific editors have a duty to the scientific community by the impact they can have on scientific policies. In this respect Dr. Garfield has a duty, I think, in eliminating irrelevant journals. I believe that there are irrelevant journals created every day to serve what I call the "flogging dying horses." And I know that in my own discipline new journals have been created just to accommodate the data which cannot be accommodated by the existing journals. Now, supposing that you exclude these new journals from your coverage system, these journals would not survive and science as well as circulation of information would be healthier.

EUGENE GARFIELD - Well I thought at first you were getting ready to take me out to Calvary. I recognize that any large scale service has some effect on the future of certain journals. It is a very unfortunate situation that one possesses such awesome power. I was speaking about that in the rest of my paper. At ISI we know we can make the difference between life and death for certain kinds of journals. But unfortunately for us, there are also many journals that do not worry about ISI. They go their own merry way, and somehow seem to survive. We only cover five thousand journals in all fields of natural, physical and social sciences. But there are many more -- 25,000 presumably that we could pick up. We do not attempt to be absolutely comprehensive for the entire world. We keep pruning the journals that we cover. We have dropped dozens of journals in past years. We are going to continue to do that. We do our best to prevent including even new journals that seem, as you say to be, irrelevant. But this is, as I said, an awesome responsibility. After all, if a group of scientists has done everything necessary put together a refereed journal why shouldn't it have a chance, for a few years, to make its impact felt. Journals published by big societies or publishers will make it or not with or without our help. But an extra margin of profit or success, whichever you prefer, may make the difference. So we have given many small journals a chance for exposure. Then after a few years we dropped them because of little reader interest.

URI BLOCH - I would very much like to commend Jeanne Poyen's paper and I would call it "the potentially greatest contribution of this conference." I hope she will write it up in the proceedings, including her advice about pagination, authors, titles and so on, but what I would really like to happen is that the editors and the publishers will truly adhere to these principles! If we achieve this then we can safely say that this conference has contributed to the transfer of scientific information. So I hope there will be at least a slight nudge in the right direction.

M. JEUKEN - I want to support the remark on footnotes. In my opinion the footnote is the invention of the devil. When they are important, they should be inserted in the text, and when they are not, they are only a display of more knowledge, not of the value of the paper. I would suggest the congress support to withdraw footnotes as much as possible, especially in papers of authors' disciples.

MARTIN CREMER - It seems to me that there is another very important competition to printed information services. From our experience, I think more and more people concerned with research are interested to get immediate information from data banks. And there is another kind of user who needs practical information such as in the field of planning transport or road construction. The need for data collection is growing as an important medium of dissemination. A second remark. This is a conference of scientific editors about scientific information transfer. We discussed mainly users in research - in universities and big research institutions. But there are many users in industry who have quite another approach to the use of scientific information. And there is a different approach in big industry and another in small industry. We have more and more challenge from users connected with administration and with societal problems connected with planning modern society. So the scientific editor is responsible not only to the user in research but to the broader community of people interested in scientific information but are not scientific and research people themselves.

HISTORY OF ASCE INDEXING AND ABSTRACTING

Irving Amron, M.ASCE, M.ASIS
Editor of Information Services, Publication Services, American Society of Civil Engineers, 345 East 47th St., New York, N.Y. 10017, U.S.A.

ABSTRACT

Development of abstracts, as well as composition from manual cards, through Listomatic cards, keypunching for computer-aided composition, OCR entry and CRT-cassette units. Financial data for computer-aided composition, as well as future prospects of setting up a civil engineering information center for members of ASCE. Reduction of error rate with in-house use of CRT-cassette unit, in comparison with previously used methods. Use of abstracts at least four times results in financial viability of used method. Description of coding method of entries, with separation of fields for smallest type of entry: first, last name, ASCE membership, job title, place of work, Society name, individual bibliographic entries, as well as coding of key words matched with Thesauri of EJC, Burec, etc.

Introduction

The ASCE was formed in 1852 and produced its first formal publication in 1867. Since that time the Society has been producing both subject and author indexes. The first indexes were very detailed (Fig. 1). Whenever a name was mentioned in the text, it was included in the author index with a reference to the considered subject.

It should be noted that Engineering Index, the forerunner of all other indexes in engineering fields, was created by Professor John Butler of Washington University and the University of Wisconsin, a member of ASCE. Professor Johnson was the originator and manager of the "Index to Current Engineering Literature" from October 1884 (the date of its inception) to December 1895, for 11 years. The memoir issued upon the death of Professor Johnson was prepared by a past president of the ASCE, Robert Moore. Professor Johnson was a very distinguished member of our Society in addition to creating Engineering Index, and being the first secretary of the American Society for Engineering Education.

Until the 1960's ASCE published its indexes by the traditional method, using 3 x 5 in. cards. In the early 60's the ASCE began using Listomatic cards, which were photographed for the annual indexes and cummulative 1960-69 index (Fig. 2).

Because of our indexing tradition, it was easy for the ASCE to adopt in the early 60's the proposal of the EJC for source indexing and abstracting. This is very important for the production of abstracts and indexes because we now receive 99% of our abstracts from authors. A page from bimonthly ASCE Publications Abstracts (Fig. 3) is produced on the basis of the mentioned

Miriam Balaban (ed.), Scientific Information Transfer: The Editor's Role, 605-618.

POWER.
See **COMPRESSED AIR: ELECTRIC POWER: STEAM: STEAM POWER: WATER POWER.**

POWER BRAKES.
"— for Freight Trains." William P. Shinn. XIV, 405 (1885).

POWER HOUSES.
See **POWER PLANTS.**

POWER PLANTS.
"A High-Voltage Power Transmission." M. H. Gerry, Jr. (With Discussion.) L, 212 (1903).
"Bibliography on Valuation of Public Utilities." LXXVI, 2133 (1913).
"Chemi-Hydrometry and Its Application to the Precise Testing of Hydro-Electric Generators." Benjamin F. Groat. (With Discussion.) LXXX, 951 (1916).
"Construction of the Power-House of the Rochester Power Company, adjacent to Genesee Falls, Rochester, N. Y." Robert Cartwright. XXVIII, 19 (1893).
"Electrical Power: Generating Stations and Transmission." L. B. Stillwell. (With Discussion.) (Inter. Eng. Cong. 1904.) LIV, Part D, 357 (1905).
"Ice Diversion, Hydraulic Models, and Hydraulic Similarity." Benjamin F. Groat. (With Discussion.) LXXXII, 1138 (1918).
Power house, Astoria, Oregon, Water-Works. XXXVI, 35 (1896).
Power house equipment at South Boston Terminal. XLIII, 150 (1900).
— at Ontario, California. LV, 173 (1905).
—, Pennsylvania Railroad Terminal, New York. LXIX, 284, 333 (1910).
Power stations of the New York Central & Hudson River R. R. LXI, 78 (1908).
"The Design of Hydro-Electric —." J. D. Galloway. LXXIX, 1000 (1915). Discussion: William P. Creager, E. Newman, H. Homberger, and Arnold Pfau, LXXIX, 1036.
"The Electric Station of the Citizens' Light and Power Company of Rochester, N. Y." Robert Cartwright. (With Discussion.) XXXI, 335 (1894).
"The Guadalajara Electric-Light Installation, Utilizing the Famous Juanacatlán Water-Falls 28 Km. Distant from Guadalajara." Rafael M. de Arozarena. XXIX, 689 (1893).
"The Hydraulic Plant of the Puget Sound Power Company." Edwin H. Warner. (With Discussion.) LV, 228 (1905).
"The Hydro-Electric Development and Transmission Lines of the Canadian Niagara Power Company." A. H. Van Cleve. (With Discussion.) LXII, 199 (1909).
"The Lock 12 Development of the Alabama Power Company, Coosa River, Alabama." E. L. Sayers and A. C. Polk. (With Discussion.) LXXVIII, 1409 (1915).
"The Necaxa Plant of the Mexican Light and Power Company." F. S. Pearson and F. O. Blackwell. (With Discussion.) LVIII, 37 (1907).
"The Ninety-Sixth Street Power Station of the Metropolitan Street Railway Company, of New York City." L. G. Montony. (With Discussion.) XLIV, 119 (1900).

POWER PLANTS—*(Continued)*.
"The Power Plant, Pipe Line and Dam of the Pioneer Electric Power Company at Ogden, Utah." Henry Goldmark. (With Discussion.) XXXVIII, 246 (1897).
"The Pumping Plant of the Morenci Water Company." W. L. Du Moulin. (With Discussion.) LXXIX, 1268 (1915).
"The Twenty-Eighth Street Central Station of the United Electric Light and Power Company." H. W. York. (With Discussion.) XXXV, 429 (1896).

POWER STATIONS.
See **POWER PLANTS.**

POWER TRANSMISSION.
"A High-Voltage —." M. H. Gerry, Jr., L, 212 (1903). Discussion: F. O. Blackwell, P. H. Thomas, W. B. Potter, and J. R. McKee. L, 245.
"A Problem in Continuous Rope Driving." Spencer Miller. (With Discussion.) XXXIX, 165 (1898).
"A Simple Method of Computing Deflections of a Cable Span Carrying Multiple Loads Evenly Spaced." F. C. Carstarphen. (With Discussion.) LXXXIII, 1383 (1919-20).
"A Water Power and Compressed Air Transmission Plant for the North Star Mining Company, Grass Valley, Cal." Arthur De Wint Foote. (With Discussion.) XXXVI, 171 (1896).
Comparison of electric and compressed air transmission of power. XXXVI, 192 (1896).
"Electrical Power: Generating Stations and Transmission." L. B. Stillwell. (With Discussion.) (Inter. Eng. Cong. 1904.) LIV, Part D, 357 (1905).
"Electrical Transmission from Niagara." Benjamin Rhodes. XIV, 205 (1885).
"Overhead Construction for High-Tension Electric Traction or Transmission." R. D. Coombs. LX, 505 (1908). Discussion: Joseph Mayer, W. K. Archbold, Charles Rufus Harte, Farley Osgood, and W. S. Murray, LX, 539.
"Reinforced Concrete Towers." D. W. Krellwitz. (With Discussion.) LX, 160 (1908).
"The Hydro-Electric Development and Transmission Lines of the Canadian Niagara Power Company." A. H. Van Cleve. (With Discussion.) LXII, 199 (1909).
"The Lock 12 Development of the Alabama Power Company, Coosa River, Alabama." E. L. Sayers and A. C. Polk. (With Discussion.) LXXVIII, 1409 (1915).
"The Necaxa Plant of the Mexican Light and Power Company." F. S. Pearson and F. O. Blackwell. LVIII, 37 (1907.) Discussion: Edwin H. Warner, H. F. Labelle, F. G. Baum, and J. D. Galloway, LVIII, 51.
"Transmission of Power in Operating Cable Railways." Robert Gillham. XXIX, 543 (1893). Discussion, XXX, 548 (1893).

PRECIPITATION.
See **RAINFALL.**

PRESERVATION OF TIMBER.
"A Proposed Method for the —." F. A. Kummer. (With Discussion.) XLIV, 181 (1900).
Data for vulcanizing. XXXVII, 320 (1897).

Fig. 1 1920 Index Page

SUBJECT INDEX 311

PORTS
Proceedings of Tenth Conference on Coastal Engineering--1966, New York, N.Y. 1967, Vol. I and II, 1546 pp.
TR:133:688
Systems Approach to Petroleum Port Site Selection, William S. Gaither and John Patrick Sides, WW3:Aug 69:395:6750

PORTUGAL
Arch Dams (Symposium), M. Rocha, J. Laginha Serafim and A.F. da Silveira; Carlo Semenza; Dino Tonini
TR:125I:909:3048
Rockfill Dams: Foundation Treatment of Paradela Dam, Walter J. Weyermann, PO4: Aug 58:1748
TR:125II:419:3077
Rockfill Dams: Paradela Concrete Face Dam, Luis Henrique Gomes Fernandes, Edgard de Oliveira, and Nuno de Vasconcelos Porto, PO4:Aug 58:1747
TR:125II:365:3076

POST TENSIONING
Prestressed Pipe for Power Site in New Zealand, Robert H. Weight, A. Max Jensen, and A.M. McConnell, PO2:Nov 68:143:6229

POTENTIAL FLOW
Free Steamline Analysis of Two Dimensional Jet, Anat Arbhabhirama, HY4:July 69:1139: 6658

POWER
Additional Aspects of the Enrico Fermi Atomic Power Plant, N.L. Scott and R.F. Mantey, PO1:Feb 60:39:2375
Aesthetic Design of Electric Distribution Lines, Thomas A. Bettersworth, UP1:Jul 65:31: 4400
TR:132:68
Arch Dam Analysis with an Electric Analog Computer, Richard H. MacNeal, EM4:Aug 60: 127:2578
Arch Dam Layout Facilitated by Use of Computer, Alfred L. Parme, PO2:Oct 67:69:5523
Arch Dams (Symposium), M. Rocha, J. Laginha Serafim and A.F. da Silveira; Carlo Semenza; Dino Tonini
TR:125I:909:3048
Atomic Power, A Failure in Engineering Responsibility, Adolph J. Ackerman, PP2: Oct 61:43:2975
TR:128V:17:3497
Bibliography on Civil Engineering Features of Thermal Plants: Progress Report of the Task Committee on Compilation of Bibliography on Civil Engineering Features of Thermal Plants of the Committee on Power Project Planning and Design of the Power Division, M.G. Salzman, Chmn, PO1:May 62:1:3118
Boiler Column Straightening Under Load Conditions, James F. O'Sullivan and Dallas I. Downs, PO2:Apr 66:115:4751
TR:132:560
California--California Water Project, Power Plans and Demands, Alfred R. Golzé, CE: Feb 65:35
TR:131:100
Civil Engineering for CVNPA Nuclear Power Plant, W.L. Klehm, PO1:Sept 63:45:3645
TR:129:452
Compact Analysis for Tower Structures, Alfred H. Palmer, PO1:Mar 67:1:5127
TR:133:273
Comprehensive Planning in the Southeast, George E. Tomlinson, HY5:Sep 65:81:4480
TR:132:91
Conception, Investigation, and Design of Arrow Dam, Keith A. Henry and P. Nigel Grant, PO1:May 68:41:5940
Concrete Reactor Shield, Harvey L. Meyer, PO2:Jul 62:143:3198
TR:128III:266:3472
Concrete Volume Change for Dworshak Dam, Houghton, Donald L., PO2:Oct 69:153:6825
Construction of Rocky Reach Hydroelectric Project, W.N. Evans and J.H. Boyd, CO3: Nov 60:45:2662
TR:126II:516:3202
Containment Sphere for Dresden Station, Leonard P. Zick, James T. Dunn, and James B. Maher, PO2:Apr 58:1:1601
TR:125I:536:3033
Containment Studies for an Atomic Power Plant, William McGuire and Gordon P. Fisher, PO3:Jun 60:27:2508
Control of Cracking in TVA Concrete Gravity Dams, Walter F. Emmons, Olav Lavik, and Poul L. Hornby, PO1:Feb 60:11:2372
Cooling Pond Design in the Southwest, T.J. Cotter and A.W. Lotz, PO2:Jul 61:85:2861
Cooling Water for Steam Electric Stations on Tidewater, R.W. Spencer and John Bruce, PO3:Jun 60:1:2503
TR:126III:280:3151
Cooling Water Studies at Electric Power Station, Gray, R. Fred, and Stephenson, J. Ben, PO2:Oct 69:293:6853
Cost and Economics of Maintenance of Turbine Runner Cavitation, Harold G. Hunt, PO1:Jan 66: 107:4634
TR:132:95
Current Developments in Hydro-Electric Plant Design, William H. Wolf, PO3:Nov 61: 55:2995
Dam Safety--An Objective of Federal Licensing, F. Stewart Brown, PO1:Mar 67:67:5155
Design and Construction Features: Boundary Project, H.V. Strandberg, PO2:Apr 66:157:4753
TR:133.000
Design and Performance of Mammoth Pool Power Tunnel, Bruce R. Laverty and Klaus R. Ludwig, PO1:Sept 63:9:3642
TR:130:416
Design and Selection of Hyperbolic Cooling Towers, R.F. Rish and T.F. Steel, PO5: Oct 59:89:2227
TR:126III:399:3193
Design Features of Glen Canyon Dam, Ernest R. Schultz, PO2:Jul 62:113:3197
TR:128III:236:3471
Design of Arch Dams by Trial Load Method of Analysis, Merlin D. Copen, PO4:Aug 60:13: 2569
TR:127III:236:3340
Design of Karadj Arch Dam, Jan A. Veltrop and Roman P. Wengler, PO1:Mar 64:1:3827
TR:131:213
Design of Karadj Hydroelectric Project, Richard D. Harza and Robert F. Edbrooke, PO4:Aug 60:31:2579
Development of Arch Action in Arch Dams, K.V. Swaminathan, PO1:May 65:39:4329
TR:131:135
Digital Computers for Trial-Load Analysis of Arch Dams, L.R. Scrivner, PO4:Aug 60:1:2568
TR:127III:116:3302

Fig. 2 1969 cumulative index obtained from Listomatic Cards

CALIFORNIA'S ANSWER TO 1971 EARTHQUAKE

6818

KEY WORDS: **Buildings (codes); Dams;** Dam stability; **Earthquakes; Geology; Legislation;** Regulation; **Urban planning**

ABSTRACT: California's recent earthquake laws are breaking new ground in public policy matters. In turn, these changes in law are also having significant impact on the approaches to civil and structural engineering designs for specific projects as well as for other aspects of engineering practice. For one example of the intent of the new laws, new hospitals must not only remain safe, but must remain functional after the event insofar as practicable. Secondly, although dams are to be constructed to the state-of-the-art of earthquake resistive design, they are not necessarily held to be 'perfectly' safe since downstream inundation maps are required and thereby an acceptable level of risk is implied for downstream construction. Lastly, land use restrictions are having increasing influence with respect to new construction in earthquake geologically hazards areas such as fault zones.

REFERENCE: Steinbrugge, Karl V., and Degenkolb, Henry J., "Meeting the Earthquake Challenge: California's New Laws," CIVIL ENGINEERING—ASCE, February, 1975, pp. 44-47

RECYCLING AND ZERO WASTEWATER DISCHARGE

6819

KEY WORDS: **Florida;** Irrigation; Sewage disposal; **Sewage effluents; Sewage treatment; Viruses; Wastewater disposal**

ABSTRACT: St. Petersburg, Florida, committed itself to total recycling of its wastewater with zero discharge to its surrounding bays. They plan to accomplish that goal through land disposal techniques. The techniques are not entirely new, however, the findings of viral studies have indicated that commonly used effluent irrigation and landspreading systems are inadequate. Increased treatment is required prior to disposal to obtain virus-free conditions. This new system will incorporate the suggested changes. As an alternative to advanced treatment facilities, Spray irrigation and landspreading of effluent and sludge, respective[illegible] still a viable alternative.

REFERENC[illegible] Dove, Lloyd A., "Recycling Allows Zero Wastewater Discharge," CIVIL ENGINEERING—ASCE, February, 1975, pp. 48-50

JOHN STEVENS AND PANAMA CANAL

6820

KEY WORDS: **ASCE (Presidents); Canal construction;** Construction; Construction procedure; **History; Panama Canal**

ABSTRACT: John Frank Stevens was appointed by President Theodore Roosevelt in 1903 to take over the flagging Panama Canal project when John Wallace quit. Stevens had been an outstanding railroad engineer and he reluctantly accepted the job. He applied railroad techniques to the job, convinced the Congress to build a lock canal rather than a sea level canal, built up the morale of the men and backed Dr. Gorgas in his methods of ridding the Isthmus of disease. All in all, he is the man who should have been credited with making the canal possible. He left unexpectedly in 1906 and Goethals, his successor, is instead generally given credit for Stevens's accomplishments. Stevens later became a president of ASCE.

REFERENCE: Fairweather, Virginia, "The Forgotten Engineer: John Stevens and the Panama Canal," CIVIL ENGINEERING—ASCE, February, 1975, pp. 54-57

Fig. 3 Abstracts from ASCE Publications Abstracts

abstracts which were previously included in sixteen Journals of the Society and in Civil Engineering--ASCE.

We have to consider that civil engineers have different needs than other engineers; they work in small offices. Basically there are two or three engineers in each office and they have different needs than those working for larger corporations in electronics, automobile manufacturing, etc.

ASCE Publications Abstracts have removable cards. At the end of each book is a bingo card where the number of the Proceedings Papers are circled to receive at low cost reprints of all our Proceedings Papers. We have in our ASCE Publications Abstracts, a permuted key word index (Fig. 4). This is helpful because of the small volume of publications. All the key words are permuted. The title, author, and abstract number are also shown. Turning to the abstract according to the abstract number it is possible for readers to order the Proceedings Papers.

Computer-Assisted Composition

In the early 1970's the computer entered our indexing efforts and we compared several organizations. We finally selected John Vitelli who set up our information system which we are using for almost seven years. Within two months after we had decided going with John Vitelli we were entering data for the 1970 Annual Index and within four months after starting, we were preparing abstracts for our current journals. The worksheets that were developed at that time are still being used (Fig. 5). The coded key words are from the thesauri of the EJC, Bureau of Reclamation, Highway Safety, and some terms added by the ASCE. We have an accession number, title, notation of content, data on authors, including first, second, last name, membership in the ASCE, job title, and place of work.

There may be up to 13 authors for each article. We also show the name of the Society, figuring that we could also include in the ASCE system abstracts from other societies just by changing the entry in this field or by changing the accession number. The code for each ASCE journal or miscellaneous publication, volume, issue, month, year, first and last pages are entered. These fields are used for other indexes that are produced and they can be seen in Fig. 5a.

For our National Conventions for example, we produce a Preprint Index including the session number and date of the convention. These Preprint Indexes are handed out free to all those attending ASCE National Conventions.

The abstract can theoretically consist of 26 + 8 = 34 paragraphs. The coding of each paragraph begins with T2 and can use all the letters of the alphabet and from 2 to 9. This can be used for larger abstracts for our miscellaneous publications. I consider this a very important facet of what we are using because there were few changes in our system and I've seen several different systems that are being used and it seems to me a detailed breakdown such as the one that we use here can be used by larger abstracting organizations.

We receive computer printouts after we send out our worksheets. When we started working with computers the worksheets were sent for key-punching. Later we used different methods for input, such as optical character recognition. Fig. 6 shows the sheets used until October 1975.

We do not consider it necessary for the ASCE to be involved directly in the programming and processing of our data. This is a problem that should be considered only by computer specialists. We are also not concerned with the printing or the production of paper. What we have to do is the intel-

Environmental engineering; Filtration; Sewage effluents; Zeta potential
Physicochemical Approach to Water and Wastewater Filtration, Mriganka M. Ghosh, Thomas A. Jordan and Ronald L. Porter, JF75:6735

Environmental engineering; Flow regulators; Hydrology; Overflows; Runoff; Sewage treatment; Storage tanks; Storm sewers; Surface water runoff; Water pollution; Bacteria; Combined sewers; Environmental effects
Urban Runoff Pollution Control—State-of-the-Art, Richard Field and John A. Lager, JF75:6738

Environmental engineering; Impact; Inspection; Refuse disposal; Solid wastes
Evaluation of Environmental Impact of Landfills, Patrick Caffrey, Martin David and Robert K. Ham, JF75:6734

Environmental engineering; Limestone; Models; Wastes; Acid treatment
Use of Crushed Limestone to Neutralize Acid Wastes, Frank H. Pearson and Archie J. McDonnell, JF75:6740

Environmental engineering; Ponds; Thermal pollution; Thermal power plants; Cooling water
Makeup Requirements for Cooling Ponds, John C. Sonnichsen, Jr., JF75:6731

Environmental engineering; Sanitary landfill; Solid wastes; Waste disposal; Costs
Future of Sanitary Landfill, Bruce Thompson and Iraj Zandi, JF75:6733

Environmental engineering; Sewage; Sludge; Sludge drying
Chemical Sludge Dewatering on Sand Beds, John T. Novak and Gail E. Montgomery, JF75:6730

Environmental engineering; Sewage treatment; Statistical analysis; Biochemical oxygen demand
Modeling Sewage Treatment Plant Input BOD Data, Paul M. Berthouex, William G. Hunter, Lars C. Pallesen and Chun-Yen Shih, JF75:6739

Environmental engineering; Water pollution; Automation; Combined sewers; Computers; Drainage
Automatic Control of Large-Scale Combined Sewer Systems, John W. Labadie, Neil S. Grigg and Bruce M. Bradford, JF75:6732

Environmental engineering; Water pollution; Water treatment; Air pollution; California; Cost effectiveness
Clean Air Equals Dirty Water, Donald C. Tillman and Robert C. Horii, JF75:6736

Environmental quality; Human resources; Hydraulics; Project feasibility; Water quality
Environmental Values and Water Quality Planning, Peter Richerson and Robert Johnston, JF75:6765

Erosion; Exchange reactions; Flocculation; Geotechnical engineering; Pore water pressure
Pore and Eroding Fluid Influences on Surface Erosion on Soil, Kandiah Arulanandan, Paripurnanandan Loganathan and Ray B. Krone, JF75:6744

Erosion; Gullies; Hydraulics; Sedimentation; Sediment transport
Soil Erosion and Sediment Transport from Gullies, Robert F. Piest, Joe M. Bradford and George M. Wyatt, JF75:6755

Estuaries; Harbors; Oregon; Environmental effects
Strategic Approach to Estuarine Environmental Management, David A. Bella, JF75:6809

Ethics; Professional activities; Professional practice; Social needs; Bids; Civil engineers
Public Versus Client Interests - An Ethical Dilemma for the Engineer, Terry L. Turnock, JF75:6723

Evaporation; Interbasin water transfers; Union of Soviet Socialist Republics; Water resources
Russia Redistributes River Flow, Norman Precoda, JF75:6825

Exchange reactions; Flocculation; Geotechnical engineering; Pore water pressure; Erosion
Pore and Eroding Fluid Influences on Surface Erosion on Soil, Kandiah Arulanandan, Paripurnanandan Loganathan and Ray B. Krone, JF75:6744

Failure; Geotechnical engineering; Photoelasticity; Soil mechanics; Cofferdams
Photoelastic Analysis of a Cofferdam, Naveed K. Burki and Rowland Richards, Jr., JF75:6747

Failure; Probability theory; Reliability; Structural engineering; Deterioration
Reliability Theory of Deteriorating Structures, Hiroyuke Kameda and Takeshi Koike, JF75:6788

Fatigue (materials); Fractions; Low temperature; Service life; Structural engineering; Supports; Bridges (structures)
Low-Temperature Tests of Simulated Bridge Members, Charles G. Schilling, Karl H. Klippstein, John M. Barsom, Stephan R. Novak and Gilbert T. Blake, JF75:6772

Fig. 4 Permuted subject index page from ASCE Publications Abstracts

Col 1 MO DA YR	Col 7 ACCESSION NO	Col 2Ø	Col 3Ø	Col 4Ø	Col 5Ø	Col 6Ø
		HØ1 ?KA E279280	?KB W032300	?KC E519280	?KD E217060	?KE A683800
2/12/75	0001123200	HØ2 ?KF	?KG	?KH	?KI	?KJ
		HØ3 ?KK	?KL	?KM		
		HØ4 ?LA E559210	?LB E139720	?LC E041560	?LD E686770	?LE
		HØ5 ?LF	?LG	?LH	?LI	?LJ
		HØ6 ?LK	?LL	?LM		

KEY KEYWORDS (Underlined) and Keywords:

ASCE COMPUTER WORKSHEET

ACC. NO. 0001123200

(TØ) TITLE Wastewater Reclamation and Recharge, Bay Park, N.Y.

RØ TR VOL

R1 TR PG

(NØ) MISC.

AUTHORS' NAMES

(A1) First John	(A2) Middle	(A3) Last Vecchioli	(A4) Jr./Sr.	(A5) Chmn.
(A6) Membership	(A7) Job Title Hydro.	(A8) Place of Work U. S. Geological Survey, Mineola, N.Y.		
(B1) James	(B2) A.	(B3) Oliva	(B4)	(B5)
(B6)	(B7) San. Engr.	(B8) Dept. of Public Works, Nassau County, Mineola, N.Y.		
(C1) Stephen	(C2) E.	(C3) Ragone	(C4)	(C5)
(C6)	(C7)	(C8) U.S. Geological Survey, Mineola, N.Y.		
(Ø1) Henry	(Ø2) F.H.	(Ø3) Ku	(Ø4)	(Ø5)
(Ø6)	(Ø7) Hydro.	(Ø8) U.S. Geological Survey, Mineola, N.Y.		
(11)	(12)	(13)	(14)	(15)
(16)	(17)	(18)		
(21)	(22)	(23)	(24)	(25)
(26)	(27)	(28)		

MØ MEMOIR	M1 TECH NOTE	SØ SOCIETY ASCE	P4 PPofDSC&TN	P5 NO of PP	P6 PREPRINT NO	SN SESSION
JØ JOURNAL EE	J1 VOLUME 101	J2 ISSUE 2	J3 MONTH 04	J4 YEAR 75	J5 FIRST PG	J6 LAST PG

(T2) ABSTRACT (attach here)

Fig. 5 Indexing/abstracting worksheet

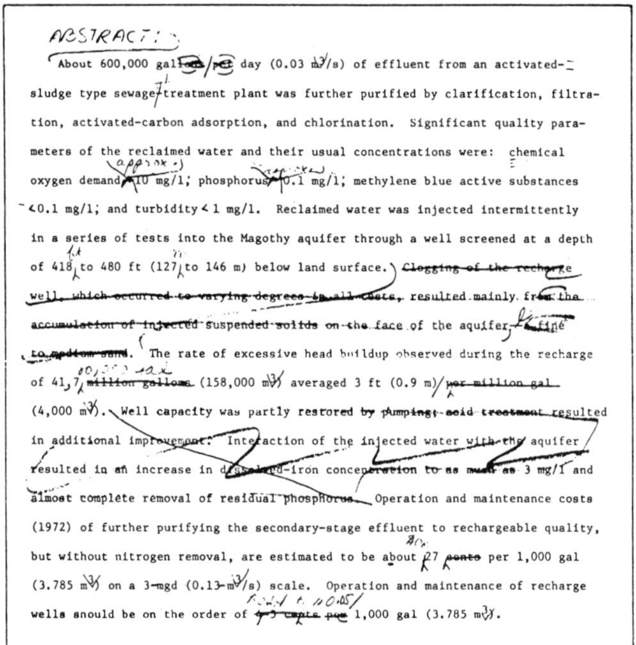

ABSTRACT:

About 600,000 gallons per day (0.03 m3/s) of effluent from an activated-sludge type sewage-treatment plant was further purified by clarification, filtration, activated-carbon adsorption, and chlorination. Significant quality parameters of the reclaimed water and their usual concentrations were: chemical oxygen demand, approx. 10 mg/l; phosphorus, approx. 0.1 mg/l; methylene blue active substances <0.1 mg/l; and turbidity <1 mg/l. Reclaimed water was injected intermittently in a series of tests into the Magothy aquifer through a well screened at a depth of 418 ft to 480 ft (127 m to 146 m) below land surface. Clogging of the recharge well, which occurred to varying degrees in all cases, resulted mainly from the accumulation of injected suspended solids on the face of the aquifer, of fine to medium sand. The rate of excessive head buildup observed during the recharge of 41,700,000 gal million gallons (158,000 m3) averaged 3 ft (0.9 m) per million gal (4,000 m3). Well capacity was partly restored by pumping; acid treatment resulted in additional improvement. Interaction of the injected water with the aquifer resulted in an increase in dissolved-iron concentration to as much as 3 mg/l and almost complete removal of residual phosphorus. Operation and maintenance costs (1972) of further purifying the secondary-stage effluent to rechargeable quality, but without nitrogen removal, are estimated to be about $0.27 cents per 1,000 gal (3.785 m3) on a 3-mgd (0.13-m3/s) scale. Operation and maintenance of recharge wells should be on the order of 4-5 cents per 1,000 gal (3.785 m3).

Fig. 5a. Figure 5(continued)

lectual work that goes into the input of our abstracts.

It should also be noted that when we were doing key-punching, the delay between the time of preparation of abstracts until we reached the camera-ready copy varied from 3 to 4 weeks. When we went to OCR it dropped to 2 weeks to 10 days. Theoretically, we can now, sending our output by telephone to the computer, receive camera-ready copy between 2 to 8 days, and if there is anything urgent we can add an abstract and have it included the next day in the camera-ready copy.

In October 1975 we introduced a CRT/cassette unit. The CRT and dual cassette are connected by modem with the computer at Thomas National.

All input for abstracts is prepared on the CRT/cassette unit, with transmission to Thomas National. This could be either an abstract, or the data for a miscellaneous publication (conference proceedings). All are entered on cassettes and transmitted to the computers at a rate of 300 baud.

Next we obtain an edit listing. The edit listings (Fig. 7) can be obtained on the cassette, with further editing and transfer back to Thomas National. The coding in the upper part of Fig. 7 comes back with the key words written out for checking input. After the corrections are sent back we receive the next printout of the abstract in the way it will appear in the Journal (Fig. 8) with the notation of content, key words, abstracts and reference. The pages are omitted as there is an interval between entry of abstracts and composition of the pages by the printer. Upon receipt of the page numbers, the next day we receive our camera-ready copy to be sent to the printer for inclusion in our Journals.

During training of those working on the CRT/cassette unit, we tell the trainee that they are working on a simple tape recorder. We type on the CRT,

Job ______ No. ______ Keyboard ______ Page 320 Lines Per Inch ______

ALIGN FIRST CHARACTER UNDER THIS ARROW

INSTRUCTIONS (NOTES)

Proofread by

Corrected by

Scanned by

Printout by

PP232

//0001123200=C=KAE279280 =KBW032300 =KCE519280 =KDE217060 =[redacted]E683800
=LAE559210 =LBE139720 =LCE041560 =LDE686770 =TOWastewater Reclamation and
Recharge, Bay Park, N.Y. =WASTEWATER RECLAMATION, BAY PARK, N.Y. =A1John
=A3Vecchioli =A7Hydro. =A8U.S. Geological Survey, Mineola, N.Y. =B1James
=B2A. =B3Oliva =B7San. Engr. =B8Dept. of Public Works, Nassau County, Mineola,
N.Y. =C1Stephen =C2E. =C3Ragone =C7Hydro. =C8U.S. Geological Survey, Mineola,
N..Y. =O1Henry =O2F.H. =O3Ku =O7Hydro. =O8U.S. Geological Survey, Mineola,
N.Y. =SOASCE =JOEE =J1101 =J22 =J304 =J475 =T2About 600,000gal/day¤¤600,[redacted]
gal/day (0.03 m/un3/unt/s) of effluent from an activated-sludge type
sewage treatment plant was further purified by clarification, filtration,
activated-carbon adsorption, and chlorination. Sigini¤¤Si[redacted]ficant quality
parameters of the reclaimed water and their usual concentrations were:
chemical oxygen demand approx. 10 mg/l; methylene blue active substances
/lt0.1¤¤//lt0.1 mg/l: and turbidity /lt1 mg/l. Reclaimed water was injected
intermittently in a series of tests into the Magothy aquifer through a well
screened at a depth of 418 ft to 480 ft (127 m to 146 m) below land surface.
The rate of excessive head buildup observed during the recharge of 41,700,000
gal (158,00 m/un3/unt) averaged 3 ft (0.9 m) 1,000,000 gal(4,¤¤(4,000
m/un3/unt). Operation and maintenance costs (1972) of further purifying
the secondary stage effluent to rechargeable quality, but without nitrogen
removal, are estimated to be about $0.27 per 1,000 gal(3.785 m/un3/unt)
on a 3-mgd (0.13-m/un3/unt/s) scale. Operation and maintenance of recharge
wells should be on the order of $0.04 to $0.05/1,000 gal (3.785 M/un3/unt).==

DO NOT TYPE BELOW THIS LINE

Fig. 6 OCR Indexing/abstracting typing by ASCE staff

```
07/26/76  0001244700  C  *KA*E317890Industrial wastes      *KB*W293650Ion exchange          *KC*E384310Metals
                         *KD*E477670Pollution              *KE*W259350Heavy metals          *KF*E686530Water pollution
                         *LA*E105460Cementing              *LB*E486190Precipitation (chemistry)
                         *LC*W128950Costs                  *LD*W382050Neutralization        *LE*E582970Sorption
                         *LF*W579200Sulfides               *LG*E684610Wastes                *LH*E684430Waste disposal
                         *T0*Removal of Heavy Metals from Industrial Effluents              *T1*REMOVAL OF HEAVY METALS FROM E
                  6      FFLUENTS                          *A1*Jim                          *A2*V.
                         *A3*Rouse                         *A7*Physical Sci. Specialist     *A8*U.S. Environmental Protection
                  3      Agency, National Enforcement Investigations Center, Denver, Colo.  *S0*ASCE
                         *J0*EE                            *J1*102                          *J2*5
                         *J3*10                            *J4*76                           *T2*Increasingly stringent effluen
                  3      t discharge requirements have spurred an interest in more effective and economical methods of removing
                 17       heavy metals from industrial wastes. Increasing scarcity and prices of metals also hasten efforts to
                 32      achieve metal removal and recovery. Neutralization and chemical precipitation have been utilized for y
                 46      ears. Operating data indicate metal removals are possible beyond those indicated by metal hydroxide so
                 60      lubilities, apparently as a result of coprecipitation. Reported costs of neutralization are given as a
                 74       function of waste character. Metal sulfides exhibit lower solubilities than do metal hydroxides. This
                 88       fact is used to advantage in sulfide-precipitation systems, utilizing a variety of sources of sulfide
                103       ion. Other treatment systems, such as cementation, ion exchange, sorption, or reverse osmosis are uti
                118      lized in special cases, and are examined.
```

Fig. 7 Edit listing of abstract

```
P
223
A 12447 REMOVAL OF HEAVY METALS FROM EFFLUENTS
201
CKEY WORDS: Cementing; Costs;  FYHeavy metals;  TF FYIndustrial wastes;  TF FYIon exchange;  TF FYMetals;  TFNeutralization;  FYPoll
ution;  TFPrecipitation (chemistry); Sorption; Sulfides; Waste disposal; Wastes;  FYWater pollution  TF
201
DABSTRACT: Increasingly stringent effluent discharge requirements have spurred an interest in more effective and economical methods
of removing heavy metals from industrial wastes. Increasing scarcity and prices of metals also hasten efforts to achieve metal remov
al and recovery. Neutralization and chemical precipitation have been utilized for years. Operating data indicate metal removals are
possible beyond those indicated by metal hydroxide solubilities, apparently as a result of coprecipitation. Reported costs of neutra
lization are given as a function of waste character. Metal sulfides exhibit lower solubilities than do metal hydroxides. This fact i
s used to advantage in sulfide-precipitation systems, utilizing a variety of sources of sulfide ion. Other treatment systems, such a
s cementation, ion exchange, sorption, or reverse osmosis are utilized in special cases, and are examined.
201
EREFERENCE: Rouse, Jim V.,  Removal of Heavy Metals from Industrial Effluents,"  MOJournal of the Environmental Engineering Division
,  TMASCE, Vol. 102, No. EE5,  FZProc. Paper 12447 TF, October, 1976, p.
223
```

Fig. 8 Computer printout of composed abstract

ıave our input on the tape and the unit has all advantages and disadvantages of a tape-recorder. It is very difficult, for example, to correct the lines. Using the dual cassettes it is possible to make corrections.

After the abstracts are entered on the master tape they are "massaged" and you can see the way the abstract looks when it goes into the Journal of the Hydraulics Division and the way the abstract looks when it comes out in ASCE Transactions (Fig. 9).

ASCE Transactions have their own history. Previously, complete papers were put into transactions, with all discussions and the closure. Afterwards digests were prepared by the author. In other words, there were too many papers and digests were used. As the number of digests increased we changed to abstracts (author-prepared abstracts). The same title, the authors, those who have discussed the paper are included in ASCE Transactions, as well as the closure of the paper and the abstract. If one reads the permuted key word index at the end of ASCE Transactions and is interested in the paper he can very easily go to the source. We use the same key word index for our conference proceedings. The one on Nuclear Power Plants is shown in Fig. 10 again with the permuted key word index, title, authors, and page. This is prepared in a very simple way. We receive the contents and the first page of all articles, and, if required, the entire paper is scanned and key words are selected. This index can be prepared within a week after receiving the first pages with camera-ready copy of the indexes ready to go to the printer using the same CRT/cassette unit. The average length of a permuted subject and author index is about 10 pages for a 40 paper book.

Costs

Let us now go over to the basic problem of costs. At a meeting of NFAIS in Philadelphis about 2 or 3 years ago I was trying to get somebody to give me figures on costs, and nobody had them. Let us consider ASCE costs for composition for the past five fiscal years (Fig. 11).

In the production of our indexes, there was an average expenditure of about $25,000 a year for the production of about 5,000 entries per year. However, these entries are used several times. In addition, we also produced the conference proceedings, ASCE Transactions, Directory and the 1970-74 Cumulative Index, which was a larger expense.

During the last two years our total composition costs have dropped. We have gone from $30,000 when we had a lot of expenses on programming of ASCE Transactions, during the second year of computerized composition. It should be noted that the reduction from $30,000 to $26,000 and even lower during the current year was caused by the introduction of in-house entry of our original documents; the trend for 1975-1976 is similar because of the use of the CRT/ cassette unit.

All errors are picked up before we transfer the data to the computer and this leads to much lower expenditure at the computer, with much lower expenditure for composition. We are only considering composition of our information services.

The cost per page would be about $3 for ASCE Transactions, because all we spend is for the camera-ready copy. It can go up to $7 and $10 for the composition of the Directory, for example, but that is the maximum -- $10, from $3 to $10 and only for composition. Salaries are not included in these costs. Only computer processing data is considered.

Our monthly cost as I mentioned, is about $2,000 a month. It should also be mentioned that we have eliminated messenger service. Previously,

9298 UNIVERSITY OF CINCINNATI URBAN RUNOFF MODEL

KEY WORDS: **Hydraulics; Hydrographs; Hydrology;** Infiltration; **Mathematical models;** Overland flow; **Runoff;** Sewer pipes; Storage; **Urban areas; Watersheds**

ABSTRACT: The purpose of this model is to provide a reliable method for obtaining the hydrograph of runoff at any selected point in a sewer system for any given rainfall. This is accomplished by simulating separately the processes involved during a rainfall: infiltration, depression storage, overland flow, gutter flow, and routing through sewers. The model has been developed by using independent hydraulic and hydrologic principles and equations, and by combining the processes sequentially. It is believed that the output of this mathematical model may be helpful in the management of existing sewer systems. It also should be useful in the design of new sewer systems. The model is intended for an urban watershed of at least several hundred acres or more in area; however, it may be applied to any size by appropriate scaling. The main input information required for the simulation of the rate of hydrologic runoff includes a rainfall hyetograph and data from topographic maps or aerial photographs, or both.

REFERENCE: Papadakis, Constantine, and Preul, Herbert C., "University of Cincinnati Urban Runoff Model," *Journal of the Hydraulics Division,* ASCE, Vol. 98, No. HY10, **Proc. Paper 9298**, October, 1972, pp. 1789-1804

University of Cincinnati Urban Runoff Model

Constantine N. Papadakis, A.M. ASCE, Teaching Fellow, Dept. of Civ. Engrg., Univ. of Michigan, Ann Arbor, Mich., and Herbert C. Preul, M. ASCE, Assoc. Prof. of Civ. & Environmental Engrg., Univ. of Cincinnati, Cincinnati, Ohio

With discussion by **Vijay P. Singh,** Grad. Student, Hydr. and Water Resources Program Dept. of Civ. Engrg., Colorado State Univ., Fort Collins, Colo.

REFERENCE: Paper 9298, *Journal of the Hydraulics Division,* Proceedings of ASCE, Vol. 98, No. HY10, October, 1972, pp. 1789-1804. Discussion: Vijay P. Singh, HY7:July 73:1194:9823. Closure: HY4:Apr 74:608:10441.

The purpose of this model is to provide a reliable method for obtaining the hydrograph of runoff at any selected point in a sewer system for any given rainfall. This is accomplished by simulating separately the processes involved during a rainfall: infiltration, depression storage, overland flow, gutter flow, and routing through sewers. The model has been developed by using independent hydraulic and hydrologic principles and equations, and by combining the processes sequentially. It is believed that the output of this mathematical model may be helpful in the management of existing sewer systems. It also should be useful in the design of new sewer systems. The model is intended for an urban watershed of at least several hundred acres or more in area; however, it may be applied to any size by appropriate scaling. The main input information required for the simulation of the rate of hydrologic runoff includes a rainfall hyetograph and data from topographic maps or aerial photographs, or both.

> >< <

Fig. 9 Abstract in card format and in Transactions

446 NUCLEAR POWER PLANTS

Earthquakes; Industrial plants; Nicaragua; Powerplants; Damage; Earthquake damage; Earthquake resistant structures
Industrial and Power Plant Damage from the Managua, Nicaragua Earthquake of December 23, 1972, Peter I. Yanev and Roland O. Marsh, 397

Excitation; Nuclear power plants; Pressurized water reactors; Seismic design; Dynamic structural analysis
Dynamic Analysis of Structural Systems Excited at Multiple Support Locations, R. P. Kassawara and D. A. Peck, 73

Fatigue (materials); Nuclear power plants; Piping systems; Pressurized water reactors; Rupturing; Stress concentration; Coolants; Cooling systems
Determination of Location and Orientation of Pipe Breaks in the Westinghouse Primary Coolant Loop, Kiran M. Vashi, Donald F. Miller and Henry J. Thailer, 173

Floors; Machine foundations; Nuclear power plants; Responses; Seismic design; Vibration; Computer applications
Generation of Seismic Floor Response Spectra for Equipment Design, Kanwar K. Kapur and Lawrence C. Shao, 29

Fluid flow; Impingement; Nuclear power plants; Piping systems; Predictions; Pressurized water reactors; Rupturing
Fluid Reaction and Impingement Loads, F. J. Moody, 219

Foundations; Mats; Nuclear power plants; Shear strength; Containment vessels
Design of Containment Base Mats, Anil K. Kar, 323

Geophysical prospecting; Geophysical surveys; Geotechnical Engineering; Nuclear power plants; Ohio; Seismic design; Bedrock
Bedrock Verification Program for Davis-Besse Nuclear Power Station, Richard A. Millet and Douglas C. Moorhouse, 89

Geophysical surveys; Geotechnical Engineering; Nuclear power plants; Ohio; Seismic design; Bedrock; Geophysical prospecting
Bedrock Verification Program for Davis-Besse Nuclear Power Station, Richard A. Millet and Douglas C. Moorhouse, 89

Geotechnical Engineering; Nuclear power plants; Ohio; Seismic design; Bedrock; Geophysical prospecting; Geophysical surveys
Bedrock Verification Program for Davis-Besse Nuclear Power Station, Richard A. Millet and Douglas C. Moorhouse, 89

Impingement; Nuclear power plants; Piping systems; Predictions; Pressurized water reactors; Rupturing; Fluid flow
Fluid Reaction and Impingement Loads, F. J. Moody, 219

Industrial plants; Nicaragua; Powerplants; Damage; Earthquake damage; Earthquake resistant structures; Earthquakes
Industrial and Power Plant Damage from the Managua, Nicaragua Earthquake of December 23, 1972, Peter I. Yanev and Roland O. Marsh, 397

Internal pressure; Nuclear power plants; Codes; Concrete; Concrete (reinforced); Containment vessels
Analysis of Internal-Pressure-Relieving R C Structures by the Proposed ACI(359)-ASME Code, O. G. De Pineres, 375

Jets; Loads (forces); Nuclear power plants; Seismic design; Shields; Steel construction; Containment vessels; Earthquake resistant structures
Analysis and Design of Sacrificial Shield Wall, V. K. Datta, 291

Loads (forces); Nuclear power plants; Probability theory; Quality control; Risks; Safety; Tornadoes; Aircraft
Probabilistic Treatment of Problems in Nuclear Power Plant Design, M. P. Singh, A. Morcos and S. L. Chu, 263

Loads (forces); Nuclear power plants; Seismic design; Shields; Steel construction; Containment vessels; Earthquake resistant structures; Jets
Analysis and Design of Sacrificial Shield Wall, V. K. Datta, 291

Loads (forces); Piping systems; Rupturing; Transient flow; Dynamic structural analysis; Dynamite
Structural Evaluation of Nuclear Plant Facilities for Pipe Rupture, R. Broman, 151

Machine foundations; Nuclear power plants; Responses; Seismic design; Vibration; Computer applications; Floors
Generation of Seismic Floor Response Spectra for Equipment Design, Kanwar K. Kapur and Lawrence C. Shao, 29

Mats; Nuclear power plants; Shear strength; Containment vessels; Foundations
Design of Containment Base Mats, Anil K. Kar, 323

Models; Nuclear power plants; Nuclear reactors; Seismic design; Structural analysis; Containment vessels; Earthquake resistant structures
Structural Modeling and Analysis of Containment Internal Structures, Lazar I. Skundric, 343

Nicaragua; Powerplants; Damage; Earthquake damage; Earthquake resistant structures; Earthquakes; Industrial plants
Industrial and Power Plant Damage from the Managua, Nicaragua Earthquake of December 23, 1972, Peter I. Yanev and Roland O. Marsh, 397

Nuclear reactors; Seismic design; Structural analysis; Containment vessels; Earthquake resistant structures; Models; Nuclear power plants
Structural Modeling and Analysis of Containment Internal Structures, Lazar I. Skundric, 343

Ohio; Seismic design; Bedrock; Geophysical prospecting; Geophysical surveys; Geotechnical Engineering; Nuclear power plants
Bedrock Verification Program for Davis-Besse Nuclear Power Station, Richard A. Millet and Douglas C. Moorhouse, 89

Piping systems; Predictions; Pressurized water reactors; Rupturing; Fluid flow; Impingement; Nuclear power plants
Fluid Reaction and Impingement Loads, F. J. Moody, 219

Fig. 10 Permuted subject index for a conference proceedings

we were always in a rush to get our printouts back and everything was sent by messenger. Now we have a much longer lag time that we can use and our processor sends back the printouts by mail, absorbing the postal costs.

Our approach seems to have been good because even though we did not attempt to have any interaction with any other services, Engineering Index has used our tapes for the production of their 2,000,000 abstract, and we are also sending our tapes monthly to the Institution of Civil Engineers in Great Britain for their production of an abstract publication.

We are producing abstracts at a reasonable cost and we are selling about 1,000 abstracts a year to about 3,000 subscribers at $8 a year for the regular edition.

Our basic approach in indexing and abstracting has been to utilize existing facilities with the help of our contractors and we seem to have been very lucky. First , we worked with John Vitelli who developed our system, and then with Thomas National which has improved to a large extent the initial work started in February, 1971. And this is basically what we have been doing in computer assisted composition.

We check each key word against the thesaurus -- manually, and it is the only way that I see it can be done. Apparently with the development of the systems that are existing now at Lockheed and SDC it is possible to go directly to the texts of the abstract; it may be that in a year of two we may drop our current approach. But for type-setting, for composition, for the publication of our abstracts, for the use of our individual engineers working in small offices, this may still remain the method that we will have to use for civil engineers.

TABLE 1. COMPARISON OF ANNUAL EXPENSES FOR COMPOSITION

Budget year	Miscel- laneous publica- tions	ASCE Trans- actions	Directory	Cumulative index	Subtotal (-)	Total (+)	
1971-2	4,504.20				4,504.20	29,272.14	24,767.94
1972-3	7,254.95	6,118.57			13,373.52	40,623.44	27,249.92
1973-4	3,121.68	8,457.89	3,038.27	4,570.33	14,617.74	49,696.25	30,508.18
1974-5	4,685.49	3,827.89			8,513.38	37,461.21	26,947.83

It should be noted that the reduction in costs during the 1974-5 budget year was caused by the introduction of-house processing of original documents. The trend for 1975-6 is similar, with the use of the CRT/cassette unit. All errors are picked up before transfer to the computer.

A PRACTICAL SEMINAR-WORKSHOP COURSE IN SCIENTIFIC WRITING

Myra Kaye, Consultant, Israel Institute for Biological Research
Arnold Colodner, Teud, Technical Publication Services, Israel Aircraft Industries

The history of our Science-writing Seminar Workshop perhaps has its earliest antecedents in 1667, when the first formal recording of experiment - the famous experiment involving a beetle and powdered unicorn's horn - was made by the sages of England's Royal Society.

Its more immediate history began in 1971, when Prof. Aharon Nir, then Director of the Soreq Nuclear Energy Establishment in Israel, asked an author (M.K.) if there was anything she could do, as a professionally-trained science editor, to improve the standards of reporting at the Soreq laboratories.

She did what any scientist would do in like circumstances - consulted a fellow-professional, Mr. Arnold Colodner of Teud. Together we combed the literature on scientific writing and, specifically, outlines of courses that had been given.

We were looking for a course that we could adapt to our special needs, which were - a brief course, suitable for working scientists; a course amenable to the kind of time restrictions such participants have, that is, one that could be broken up into self-sufficient units which could be presented at virtually any time intervals, to fit into work schedules; a course at a level intellectually suitable for Department Heads as for graduate students; a seminar-workshop style course, in which the students could interact with us as individuals with individual problems; and, lastly we wanted a course that would be oriented toward the needs of scientists writing in a language, English, that was not necessarily their mother tongue.

We found a variety of courses - but some exhibiting the very faults they were supposedly trying to correct - pompous, complex language structure, verbosity, and the use of complicated terms and constructions

Miriam Balaban (ed.), Scientific Information Transfer: The Editor's Role, 619-632. *All Rights Reserved.*

where simpler ones were just as accurate and apt.

We found a number of textbooks to some of which we are greatly indebted. A list of the principal sources we used and consulted is given in Table I.

What seemed to be generally missing from the literature was any analysis of what scientific writing actually was and in what way it differed from other kinds of writing, for example a novel or a letter.

The attributes of the medium were described - it had to be precise, simple, concise, factual - but this is rather like describing a dress in terms of its colour, shape, size, material, style, without saying what it is and what it is for; for example, in what way does it differ from a pair of trousers, and why?

It seemed to us that to interest scientists in the tedious process of scientific writing, and to present the task as an intellectual challenge, such a pragmatic approach would not be sufficient. We would have to derive the basic principles, analysing the medium and the necessities which shaped it to its specific form, just as if we were teaching any other scientific technique, such as, for example, gas chromatography. It would be unthinkable simply to demonstrate how the chromatograph can be operated to a scientist - he would want to understand the underlying principles of the technique - and this would have to apply to our teaching of scientific writing also.

With the required basis of intellectual comprehension, one could expect the scientist to approach the techniques of scientific writing with understanding and respect. The tools of the technique would be used more intelligently, when there was a real understanding of why scientific writing differed from other writing forms, and what were its precise aims, its range and constrictions. (To emphasize our analytical approach, our first course had an accompanying lecture series. See First Course Announcement, Table II.)

We therefore decided to spend one whole session in analysing the medium, comparing it with the free unrestricted medium of literary writing, and seeing how the different aims of the two forms of written communication determine restrictions and constraints, shape attributes and dictate the tools they may employ.

In the course of this analysis, we emphasized three vital differences, leading to literary writing as a free "anything goes" medium using language tools, and to scientific writing as a highly constrained medium, with compensatory powerful 'mechanical' tools.

An example of one of these tools is given in Figure I. As told in literary style, the story of Shula, who had never felt happier than when her young man, Shmuel, returned home safe from war, might occupy many pages and give us valuable insights. To convey only the facts, however, we can utilise the powerful graphic tool of scientific writing, and convey the data with economy, clarity and precision. We planned a ten session course, consisting of 1½ hour seminar-workshops. Subsequent evaluation by students of several courses indicated that a larger course, perhaps 12-15 sessions, would be preferable. We welcomed the enthusiasm and tend to agree with the conclusion. We constantly review the course and will try to prepare an extended course should this prove practical.

A year of planning preceded the course in its basic form. It was subsequently adapted to meet the needs of engineers, then physicians, the adaptations consisting of a quite radical change in the material presented for the engineers, and less radical changes, mainly in the student exercises, for the doctors.

Of our 10 sessions, the first is devoted to orientation and analysis of what we propose to teach. To repeat, we emphasize that scientific writing differs from other writing forms, explore why, and show how its special audience and requirements determine its characteristics.

In Sessions II and III, we are concerned with steps before the writing process *per se*, i.e. the decision to publish, selection and sorting of material, outlining, titling and abstracts.

Sessions IV and V are devoted to style and use of language. No attempt is made to teach English grammar, but rather choice of appropriate forms and selection of writing modes to fulfil style and language requirements for our special medium discussed.

Sessions VI and VII cover the various forms of presentation of data and results - tabular, graphic, photographic, and what considerations govern choice.

'Good' forms, from which data can easily be read, and 'bad' forms, where the reader must interpret, are contrasted.

Session VIII examines language at the word level, and text revision to identify faulty constructions in phrasing.

Session IX is concerned with responding to the editor as well as aspects of text revision not concerned with language. It is a summarising session.

Session X covers a number of topics which are optional; they can include special writing assignments like theses, 'popular' articles, reports, research grant proposals and oral presentation at seminars or conferences.

The course for engineers includes material on specifications, presentation of engineering data, and reports.

Mr. Colodner presents sessions II, III, VI, VII and IX and the others are presented by Mrs. Kaye. We do not always agree on specific points, and at times present different preferences. We consider this healthy, as we are dealing with an art as well as a science, in which there is room for differing viewpoints.

Our agreement on the basic principles is complete, and I would recommend to anyone planning a Science Writing Course to make sure that they are working with colleagues with whom they can achieve the amity we have found - if constructing the course is fun, the course will be fun for the participants.

If it does not provide both an intellectual challenge and a goodly dose of sparkle, it will be tedious. One has to remember that writing papers is not the scientist's favorite occupation. The Course has to provide stimulus as well as knowledge.

The Course Handbook is a loose-leaf compendium of reference material as well as material presented at the Course. Much of its contents was culled from the literature. Some of it is original. A list of contents is presented in Table III. Not included are some pages handed out at the Course Sessions, on 3 hole-punched paper, ready for insertion in the Handbook; these are mainly student exercises.

Since 1972 to date, the Course has been presented some 12 times, at Research Institutes and Universities in Israel. It has been given in part within the framework of a credit course at the School of Pharmacy of the Hebrew University for several years. We have given the Course over a week-end as well as spread over 5 weeks.

The Course will make its first foreign debut for participants who are, or will be, science writing teachers at the Nordic Universities in the summer of 1978. We are eagerly looking forward to the opportunity of meeting with our colleagues and express our admiration of the initiative of NOP and the Nordic Universities in pioneering the teaching of science writing as part of the University curriculum.

Blaker, A. Photography for Scientific Publication. San Francisco: Freeman, 1965.

Bernstein, T. The Careful Writer. New York: Atheneum, 1968.

Douglass, P. Communication Through Reports. Englewood Cliffs, New Jersey: Prentice-Hall, 1957.

Gilman, W. The Language of Science. New York: Harcourt, Brace and World, 1961.

Gowers, Sir Ernest. The Complete Plain Words. London: Her Majesty's Stationery Office, 1954.

Graves, H. and Hoffman, L. Report Writing. Englewood Cliffs, New Jersey: Prentice-Hall, 1965.

Halverson, J. and Cooley, M. Principles of Writing. New York: Macmillan, 1965.

Kapp, R. The Presentation of Technical Information. London: Constable, 1957.

Menzel, D., Jones, H. and Boyd, L. Writing a Technical Paper. New York: McGraw Hill, 1961.

Mitchell, J. Writing for Technical and Professional Journals. New York: Wiley, 1968.

Morris, J. Principles of Scientific and Technical Writing. New York: McGraw Hill, 1966.

O Rourke, L. J. Self-Aids in English Usage. Florida: Psychological Institute, 1964.

Perrin, P. Writer's Guide and Index to English. Chicago: Scott-Foresman, 1965.

Peterson, M. Scientific Thinking and Scientific Writing. New York: Reinhold, 1961.

Quiller-Couch, A. On the Art of Writing. New York: Putnam, 1916.

Rathbone, R. Communicating Technical Information. London: Addison-Wesley, 1966.

Richards I. and Richards P. Proper Words in Proper Places. Boston: Christopher, 1965.

Table I. Some Sources Screened in Planning the Course

Sherman, T. Modern Technical Writing. Englewood Cliffs, New Jersey: Prentice-Hall, 1966.

Strunk, W. and White, E. The Elements of Style. New York: Macmillan, 1959.

Tichy, H. Effective Writing. New York: Wiley, 1967.

Trelease, S. How to Write Scientific and Technical Papers. Cambridge, Mass: MIT, 1958.

Ward, R. Practical Technical Writing. New York: Knopf, 1968.

Woodford, F. Scientific Writing for Graduate Students. New York: Rockefeller University, 1968.

Table I Some Sources Screened in Planning the Course (continued)

The Israel Chapter of the Society of Technical Writers and Publishers (STWP) will conduct a special course in Scientific Writing. The purpose of the course is to promote an understanding of good writing and to help the scientist select proven techniques to improve his writing.

The course features lectures (open to the public) followed by seminar-workshops. Groups will be limited in size to allow "on-the-job" assistance in writing papers and articles for journals, as well as reports.

SCHEDULE

The course will be held at the Feinberg Graduate School of the Weizmann Institute of Science, beginning 1 February 1970. It will consist of ten sessions, each with a lecture from 1530 to 1630 and a seminar-workshop from 1630 to 1800. The exact days and schedule will be announced.

REGISTRATION

Please register with the secretary of your department, or call:
Mrs. L. Binstock at 952075, ext 285, between 0800 and 1600, Sunday - Thursday.

Only a limited number of participants can be accepted. Registration closes 1 January 1970.

A Practical Course in
SCIENTIFIC WRITING

Organized and Conducted by the
Society of Technical Writers and Publishers
Israel Chapter

Sponsored by the
Soreq Nuclear Research Center

Lectures

1. The Needs of Scientific Writing: Prof. I. Berenblum, Weizmann Institute
2. To be announced.
3. Pitfalls in the English Language: Dr. R. Venezky, University of Wisconsin
4. Obscurantism in Scientific Writing: Dr. A. Kohn, Israel Biological Inst.
5. Historical Review of Disseminating Scientific Information: Mrs. M. Kaye, Soreq Nuclear Research Center
6. To be announced.
7. Library Search Techniques: Mrs. R. Shalev, Tadiran Ltd.
8. Requirements for Presenting Scientific Results: Mr. A. Colodner, STWP
9. View from the Editor-Publisher: Mrs. M. Balaban, Science Services Ltd.
10. Philosophy of the Scientific Method: Dr. B. Barnett, Soreq Nuclear Research Center

Seminar-Workshops

1. The Writing Process: purpose and organization of the course; overview of technical writing
2. Analyzing the Writing Requirement: review of journals; publishing (when, why, what); titles and abstracts
3. Organizing the Content: purpose and scope of journal; instructions to authors; outlining
4. Beginning the Draft: introduction; references; scientific style
5. Materials and Methods: chronological presentation; grammar
6. Results: constructing graphs, curves, tables and charts; photography; layout
7. Results: order of presenting data; logical flow; combining data
8. Discussion: editing for readability
9. Final Steps: conclusions; final title and abstract; typing; review
10. Other Assignments: reports; proposals; theses; reviews

Table II. Announcement of First Course.

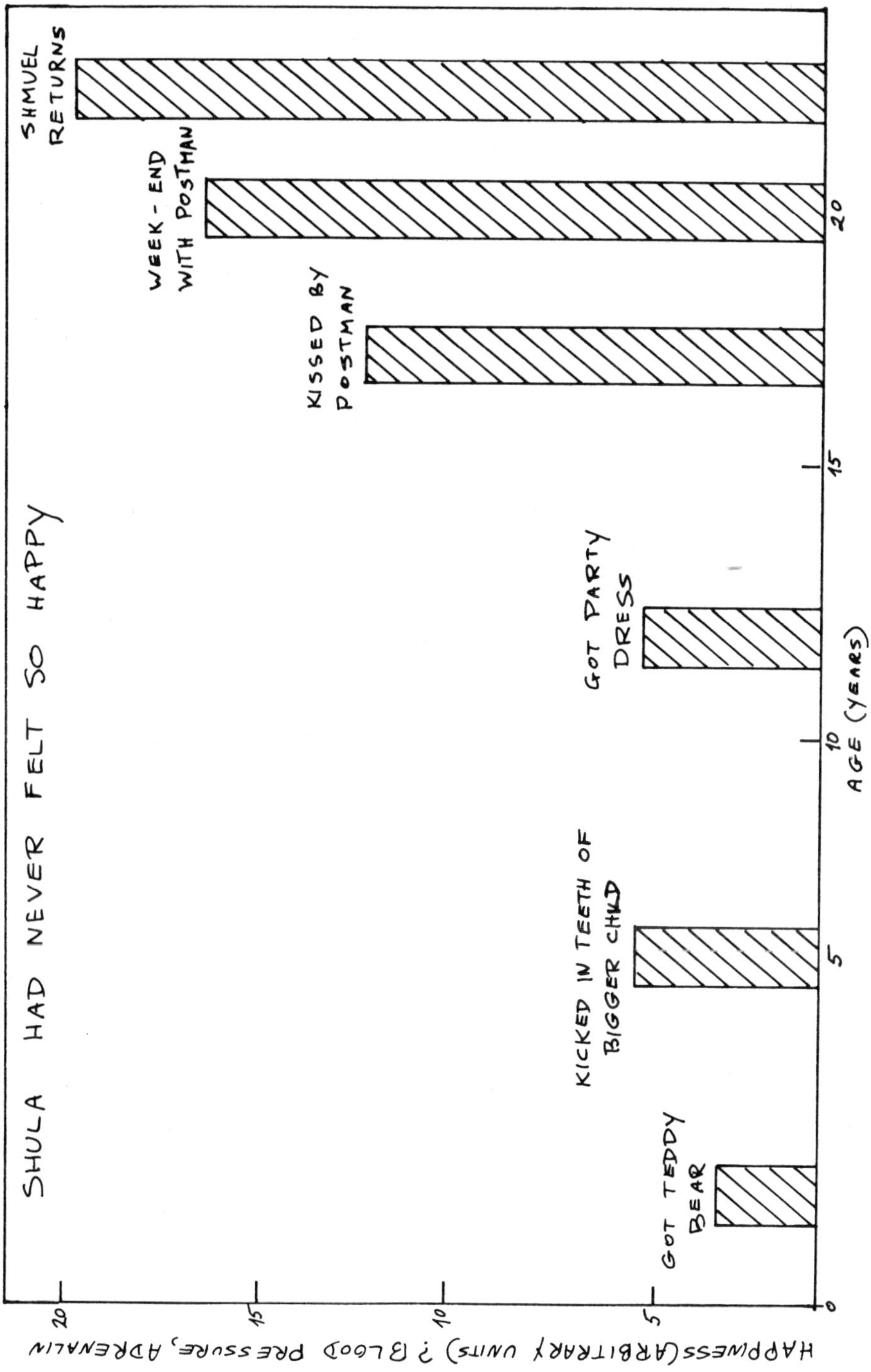

Figure I. Shula's Happiness

TABLE OF CONTENTS

Table III. Handbook: Table of Contents

Table III. Handbook: Table of Contents (continued)

16. Byline

17. Greek Alphabet and Recommended Prefixes

18. Numbering

19. International System of Units

    a. Rules for Use

    b. Physical Constants and Conversion Factors

20. Classified List of Abstracting and Indexing Services

21. Dictionary of Roots, Prefixes and Suffixes

22. "Deadwood": Words and Phrases

23. Choice of Tense

24. Curves and Charts

    a. Scales

    b. Changes in Scale

    c. Histograms

    d. Cumulative Frequency

Table III. Handbook: Table of Contents (continued)

Table III. Handbook: Table of Contents (continued)

Table III. Handbook: Table of Contents (continued)

CLOSING SESSION: Summary, Recommendations, Resolutions and Future

Miriam Balaban, Chairman

IFSEA

TOWARDS AN INTERNATIONAL FEDERATION OF SCIENTIFIC EDITORS' ASSOCIATIONS

Anders Martinsson *Editor, Lethaia*
Department of Palaeobiology, University of Uppsala
Box 564, S-751 22 Uppsala, Sweden

ABSTRACT
Librarians (IFLA), documentalists (FID), archivists (ICA) and publishers (*stm*, etc.) are represented by fully international organizations, contributing vigorously to international development in their fields. There are no reasons but historical ones why editors should not be organized, recognized and active in a similar way. Until too recently, scientific editorship remained at a non-professional level, as an accessory occupation for active scientists, without much methodological involvement. During the last decade a considerable number of mainly continent-wide and subject-specialized editors' associations has been created, and there is enough activity and coverage for the creation of an International Federation of Scientific Editors' Associations (IFSEA), as has been the repeatedly declared intention of Unesco. — It is still evident that editors and their associations have difficulties in identifying their tasks, largely concentrating their activities to the publishers' and printers' fields. Standardization in editing has mainly been dictated by librarians and documentalists. A widely available publication for the development of ideas, methods and standards in scientific editing is still lacking, which means slow development and the duplication of efforts.

Librarians, documentalists, archivists and publishers are organized in well-developed and fully international non-governmental bodies. I am referring to the International Federation of Library Associations (IFLA), the International Federation for Documentation (FID), the International Council on Archives, and the International Publishers' Association (IPA). The latter organization sponsors, since 1968, an affiliated group of special interest in the present context, namely the International Group of Scientific, Technical and Medical Publishers (*stm*).

All these organizations have widely differentiated activities and have received most authoritative recognition, by intergovernmental organizations, national institutions and other non-governmental organizations. This enables them to act forcefully in the interest of their profession and make their members' proposals, work and ideas effective in the development of scientific communication. With great differences between the organizations their recognition is reflected materially in, e.g., national and intergovernmental support for interesting projects.

The scientific editors do not have such an organization. They have recently emerged from an old-fashioned state of things under which they held honorary positions, highly respected scientifically but without much professional involvement, letting an ill-defined person called the "printer" do

Miriam Balaban (ed.), Scientific Information Transfer: The Editor's Role, 635-638.

the real editing for them and paying for this from the allocations for publication in a way that has never been properly understood or questioned.

When the editors started to become organized about ten years ago, it was deemed impossible to assemble them on a global basis. The organizations that were created, in most cases under Unesco sponsorship, were made continent-wide and given a limited disciplinary range, with the express but not strongly emphasized aim of eventually becoming united in a global federation. The associations have done a lot of good, and the economical and technical development in the world has contributed to giving the scientifical editors a professional identity. However, the editors' associations have not yet had the international and national impact that some of the global organizations in the field of scientific communication have had. It is, for example, still impossible or abnormally difficult for a scientific editor to obtain funds to take part in a meeting, if his status is merely that of an editor without the backing of a specialized institution. Editing is not yet in the serene funder's mind integrated with the methodology of science.

Standardization, which has occupied much of our attention at this conference, provides a good example of the situation. Several of the editors' associations have devoted most of their efforts during the last decade to working out editors' handbooks or guides of good practice for editors and authors within their disciplines — but the standards on which these handbooks and guides are largely based have been worked out by librarians and documentalists!

In the International Organization for Standardization (ISO), questions on editing are accommodated within a Technical Committee designated TC 46 (Documentation) and a Subcommittee designated SC 7 (Presentation of Publications). When TC 46 last year established a Steering Committee, representatives of IFLA and FID were immediately included besides the national representatives who, incidentally, all are librarians. ICSU and Unesco are also represented.

This symptomatic example does not mean that the scientific editors have remained entirely undiscovered by the international organizations that standardize or coordinate scientific communication. Some of them have been employed by their original sponsor, Unesco, in different parts of the UNISIST programme. For a number of years ICSU Abstracting Board had a Committee for the Cooperation between Editors of Primary Publications and the Secondary Services, subsequently transformed into a new body that has not yet started its activities. ISO TC 46 SC 7 (preceded by a body designated WG 7) has indeed very actively involved those few individual editors with an interest in standards who could be tracked down. However, the number of editors now involved (an SC consists of national delegates, a WG of individual specialists) seems to be only two, and the situation in national work and for the publishers seems to be about the same.

Towards a federation

Evidently, then, the scientific editors need a global federation of wide disciplinary range in order to be identified and recognized. It also seems to be evident that the editors' associations, although some of them are new or weak from economical and organizational points of view, provide a sufficient basis for such a federation. The earth sciences are now served by three such associations, Editerra (Europe and adjoining areas), AESE (North America) and Alegeo (Latin America). A fourth association, Editeast (for the countries bordering the West Pacific) has been established for science in general but is for historical reasons still biased towards the earth sciences. Two associations have been formed within the life sciences, namely ELSE (Europe) and CBE (North America). There is a fully international Committee of Editors of Biochemical Journals (CEBJ) and a corresponding body within clinical chemistry (CEJCCC). Also fully international but specialized with regard to discipline are the Association of Philosophy Journal Editors and the *ad hoc*

Council of Social Science Editors. Editors in European chemistry (EdEuChem) and engineering (AEEP) organized themselves in 1970 but seem to have been less successful in getting their associations off the ground. The European physicists decided not to create an editors' association.

However, a federation of editors' associations would be much strengthened if the structure of constituent international bodies were supplemented by a structure for national adherence, through which the results of international work could also be implemented at a national level. Active national associations of editors for the entire range of science would then be the preferable type of adhering organization, although a certain flexibility might have to be allowed to accommodate countries with different social systems. If associations of this kind exist, they must be exceedingly rare.

The International Federation of Scientific Editors' Associations (IFSEA) would according to this model consist of continent-wide or global unidisciplinary associations that would probably increase considerably in number and multidisciplinary national bodies that would have to be created at the initiative of the federation. The need for such national bodies is by no means an artificial construction — that they do not exist is due to the fact that the editorial professionalism that has penetrated the earth and life sciences and some of the exact sciences has only started to affect the structure of editing in other fields.

Identifying the editor's tasks

Without trying to enumerate and evaluate all the editor's tasks, I would like to recommend a certain amount of organizational self-criticism in this field before a federation goes ahead with a programme. In my previous example we have seen that the scientific editors have found it to be a dominant natural task for them to enforce the use of standards, namely by issuing handbooks and guides, but that there has been less interest in creating the standards.

A fashionable field, connected with power, is the economy of publishing. However, the economy discussed by the editors is mostly typically the publisher's economy or the printer's economy. I do not question at all that a deep insight into these fields is essential for the editor, but editorial economy can also be singled out. The editor is still in the position of being able to reduce the costs of publication of an average journal by most considerable amounts and without information loss just by applying professional skill in the planning of text and illustrations. The entire complex of concealed costs in the form of rewriting and repeated keyboarding of texts and readaptation of figures are within the editor's reach but outside the interest spheres of the publisher and the printer.

A fashionable special field with economical implications is that of copyright. The great efforts in these fields have clearly been in the publisher's interest, not in the editor's. In taking the function of the author's arm in scientific communication, the editor might of course have been expected to involve himself in the author's (scientist's) copyright, but this has not been the main attitude. The efforts of the editors' associations have not been different from what would be expected by the IPA or the *stm*. Of course it is excellent and necessary sometimes to join forces. For our identity it would be good to know, however, whether we are the publishers' editors or the scientists' editors and when we even have to safeguard the interests of our own profession.

A field where much more can be done in the latter respect is active education. The scientific editor should not take on the responsibilities and work of the author, nor of the publisher, nor of the printer. His relations to the author are particularly complex in this respect, and the author often tends to leave all his methodological problems in publishing, by ignorance or ruthlessness, to his fellow scientist, the editor. Guides for authors are of course intended to counteract this, but they are passive instruments, and authors keep turning up for endlessly repetitive individual teaching of the same elementary things. Reporting on results in publications is part of the normal methodology of science, and this methodology is amenable to normal teaching in groups. In order to save his own time and labour the editor has to see to it that this teaching is organized. He is himself the natural teacher.

Last but not least the scientific editor has to realize his own professionalism and recognize his task as being as integral a part of his academic position as for example teaching or curating a museum. He also has to convince his employer and academic environment of this fact.

Such things have to be done in groups, on a broad front and in all countries. The area in which the reformative measures are to be taken is one of utmost conservatism and much strength is needed. The status of a labour that academic people have always depreciated has to be changed.

When I submitted the abstract of this paper, I intended to stress the need for a publication in which questions on scientific editing could be treated in a universally available, scientific form, with due respect and reference to previous work. Hitherto advancement has taken place in an uncoordinated way, with many parallel and unnecessarily repeated efforts, the results of which have only been available in the form of unpublished reports in restricted circles. General journals are still lacking in the field — technology is served by the *Journal of Technical Writing and Communication*, and editorial matters have appeared in *Scholarly Publishing* which is expressly a journal for authors and publishers. A few months ago it was decided that *Earth & Life Science Editing* will be available for this kind of articles, and it is likely that the *Journal of Research Communication Studies*, whose introduction was announced by Elsevier Scientific Publishing Company only a few weeks ago, will become the general medium that we have been waiting for, not always passively. In general, however, scientific editors have been much better at publishing for others than publishing in their own professional interest.

In summary, an International Federation of Scientific Editors should work for the integration of scientific writing with the methodology of science and for incorporating scientific editing in academic activities at the same level as teaching, supervision of research, or curating. It should further the methodology of scientific publication in a creative way and contribute to both the creation and the enforcement of standards. It should work towards the introduction and maintenance of competent teaching in the methodology of scientific publication, in all countries at the authors' level and in national or international courses at the editors' level. It should provide its members with a forum for competent discussion of these questions and for formulating plans and strategies.

CLOSING SESSION

MIRIAM BALABAN - First of all I would like to thank you all very much for coming and for such lively participation in all the sessions. Maybe we've had too many lectures, but I hope that for the beginning this gave everyone a chance to lay the ground for stimulating thought and exchanging ideas towards implementation of some of the things we can concur on.

This is the closing session, but I hope that this is the opening to future activity. Now we have heard a proposal for the future by Anders Martinsson, and Ernest has more to add.

ERNEST MANN - Although I agree with quite a lot of what Anders has said, I have some reservations on matters of emphasis and timing of such a federation. ELSE which is a very beautiful lady - European Life Science Editors - with which I've been associated with for many years has been active in the fields of scientific writing in English, and also in the area of information flow between primary and secondary journals which you heard about this morning. Now both these areas cover the whole stratum of science, and we became aware through that work that what we were doing really wasn't applicable only to life science editors but crossed the whole area of science. So it wasn't unnatural that that beautiful lady ELSE should contemplate some form of marriage with somebody else, and who better than that very down to earth gentleman, Editerra. We have made a start in two ways. We have got a joint newsletter now and we have a joint secretary. That's a very recent innovation. So that is quite a serious start in collaboration between two quite important editors' associations.

Now having listened to the various viewpoints expressed at this conference, there is certainly some conflict in my own mind over where we go from here. The sort of questions I've been asking myself are: Do we want an IFSEA, the international federation that Anders mentioned? Do we want to limit it to scientific editors or do we want a broader based body including publishers, librarians, documentationalists, archivists and others, all of whom form an important link in the flow of information from the producer to the receiver?

And the second question, do we want an international association to grow naturally out of a number of trial marriages such as the one we just mentioned, or do we want to superimpose it over a somewhat already disorganized array of existing organizations? Frankly I don't know the answers at this stage. Perhaps you do, but I believe that the best compromise lies in the following resolution, which is broad based and allows for a working party to consider your views and the various alternatives available in some depth.

1. The conference recognizes the excellent work being done by the many bodies concerned in the transfer of scientific information and encourages the establishment of new groups as necessary.
2. This conference affirms the need for continuing and sustaining discussions to identify matters of common interest and to provide a forum for the exchange of ideas.
3. The conference resolves that to further these aims:
   a. A working party should be set up to consider how best the activities of the various parties can be coordinated.
   b. A second international conference of scientific editors and others involved in scientific communication should be held in approximately three years.

It is our hope that more concrete proposals could then be put to another conference in about two or three years time. Let us not kill what looks like a good idea by forcing it through prematurely without sufficient thought.

ANDERS MARTINSSON - I don't think that there are any considerable differences in opinion here, with regard to the timing in particular. The documentalists among you will notice that there is a noise word in the beginning of the title which is "Towards an International Federation...", and I think that starting the work by trying to find the right form and scope for this federation is something that we should do. It is just that we shouldn't work endlessly on that. The only thing I would like to stress that could imply a very slight initial difference is that we really have to care about the editor's identity here, because the other organizations already exist and are comparatively stronger.

LARRY LESTER - At this stage I don't think it is terribly important that we have a roof organization or combine several organizations. What is important is that we know who the working group is composed of and what it is doing. There are several excellent newsletters being published, especially ELSE-Editerra newsletter edited by Mrs. Morris which I highly recommend. If the working group would periodically provide information that would be published in these newsletters it would be a great service to us so if there is a conference in three years we will know what is going on.

JEFFREY WATSON - I think it is a pity we didn't have a prewarning of this. I find it very difficult to comment on something just out of the blue like that, particularly since I have an ingrained skepticism towards international organizations. I wonder whether we may be creating a dinosaur when we already have a considerable number of highly efficient sectorial groups which could possibly get together initially and work quite well. We haven't really yet worked out an effective cooperation between North American biology editors groups and the European ones. Some of the problems I heard listed here seem to be more unique to Europe than to North America in terms of the standardization and representation of professional editors on standard boards.

The other thing that I wanted to find out about particularly is to what kind of ideas have already been hatched in terms of composition of the working group. It would be rather nice to know whether any ideas have been put out as to the most efficient size of the working group.

MIRIAM BALABAN - Well actually forewarning to this session exists in its very presence as the program for the closing session. We are here to summarize and think about the future. It is true that an effective cooperation between American and European groups only in biology has not yet been effected. In fact it was exactly this that planted the kernel which germinated into this conference.

We are here considering what cooperation is desirable among our organizations and in which way - international or national and regional, interdisciplinary or disciplinary, or in the wider context of scientific information transfer. We have opened the floor to discuss this and see whether and how we should continue this forum.

NANCY MORRIS - I think it would be a great pity to be precipitous and form an organization that is going to founder. I think the next thing to do is to hold another conference, and as Mr. Mann suggested, have a sort of bridging committee with an address. But when you form an international organization somebody has got to pay for it. You may get grants to start with but you've got to have a continuing income, and are we all going to be asked to join this federation. Speaking for Editerra, we want people to join us, we don't want to have to pay to join something else and if you have individuals belonging to the thing they don't want to have to join more than one thing. They want to join Editerra, ELSE or one of these, but not an international thing as well, so I think finance does come into it. Also what exactly is it going to do? It requires a great deal of thought as to what its terms of reference would be and what its work would be and where its secretariat would be. Who is going to run it, and is this gathering sufficiently large to create such an important organization? But I think it is a splendid idea to have another conference, because obviously this has been awfully useful. If we can help earth and life science editing and go on as a sort of continuing thing to tell people what is going on, we will certainly do our best with that.

MIRIAM BALABAN - Thank you Nancy. The intention would not be a place where individuals would have to apply twice, that is to their own organization and to this one, but rather it would provide a link between the editors' organizations so that the individual would not be troubled twice. Of course if he didn't belong to any organization he could apply directly, and that would answer the question of financing too. In turn the editors' federation could link up with other parts of the information cycle.

KNUT FAEGRI - I completely agree, but this is an extremely sensitive and very difficult matter. In spite of a good start it took ELSE five years I would say to get off the ground, and you have to remember, it was in a well-defined area and it was a comparatively small geographic area. Now we are discussing a thing which is practically unlimited both ways, and there are several orders of magnitude of difficulty inbetween. It needs a very careful evaluation and there is one thing I think we must all agree upon. This meeting should not fizzle out today. We must go on - that is very important - and we must find a way to go on. I think the idea to have a working group with a very specific mandate is very important. I quite agree that let us not create dinosaurs. If we are going to have something, let it hatch out of an egg. But there is another point where I am terribly disappointed to disagree with my American colleagues and that is from necessity. Working within the pleasant confines of a continent of the same culture, speaking the same language and so on, it is very easy to say that we don't need it. If you are working outside of that, whether you are working in a small country in the so-called developed part of the world, and even more so if you are working in the developing part of the world you feel this as a desperate need. So let us go on.

JEANNE POYEN - Speaking from the point of view of the A and I services there is nothing that we need more than a body to talk with about mutual problems - like the ones I have just underlined this morning. And I think, as Knut said, that this conference must not end, but must continue. My only worry is that for the 25,000 scientific and technical periodicals covered by the secondary services at present, how many editors are represented in the existing organization? And what are these editors? They are the editors of the best journals, people who already know how to edit a journal and what international cooperation is. So whether it is through an existing organization or whether it is through the federation, what we need really is to have a means to get in touch with the editors of these much less famous journals, who do not follow the rules like the editors of good journals. That is one point I would like to see discussed.

ANDY JESSE - I consider the proceedings as an important means of information for going on in this work, so I should like to make a plea. Do not include only the papers and discussions, but include all the complete addresses of these associations for further contact. Please include a minibibliography of the styles and standards presented here, and this could form a basis for going on. Now as just was mentioned, to reach the many editors it seems to me that we have to get more than 25,000 copies printed. See to it that many buy them so that at least half of the editors can see them, and then one would perhaps have a basis to go on.

DAVID SHEPHARD - First thank you Miriam for putting on this conference. I certainly wouldn't have come to Jerusalem otherwise. But coming to international conferences is not an easy matter, and I think I agree certainly that there should be another international conference. But I hope that in the meantime there will be a clear delineation of first of all whether the group should consist entirely of editors or others. Second, how to define editors. I think this is something that could be done through working groups, but in order to have the working groups work effectively, I think it is very important to have a liaison between groups such as ELSE and CBE. Here I would like to support what Dr. Watson said about working through such groups as CBE.

Now the other point I'd like to make is that, as I said, coming to an international conference is interesting, it takes an effort on one's own part, but it certainly has been rewarded in my own case here. It has given me an opportunity to meet different types of editors, but I think at another conference the emphasis should be on working groups - fewer papers with more discussants perhaps. I think that we must perhaps prevent the diffusion of interest. And for the next conference I hope that the working parties will emphasize the way the second conference can achieve definite objects.

MIRIAM BALABAN - Well it seems as if there is one point on which everyone seems to concur on, and that is that a second conference of scientific editors should be held within approximately three years.

DIMITRI VIZA - Well I'd like really to say that I would agree with Professor Faegri in that it takes five years to create something, and take it off the ground. Now if we don't make a decision today to create something, we have to wait three years until the next conference to make the decision and then wait another three years to see something get off the ground. I'm against haste, and I think we should consider it, but we should maybe make a decision on principle on whether we would like to create something or not, and then we have maybe three years to discuss all the details - the byelaws the statutes, who is going to finance it, and who is going to adhere or not to this organization. I think that we should decide on the principle today.

WOLFGANG LÖHNER - First of all I would like to recommend that we make a very clear distinction between having a second conference and creating a working party on whatever subject. I would be in favour of having a second conference, and I would propose that we adopt this resolution which is proposed here. But I would like to see it separate from the working party which is also proposed.

My second point I would like to make concerns the establishment of an international federation of scientific editors or of editors' associations. Here we have also a distinction to make concerning the editors or the persons involved in such an association or federation, and the associations which might be involved in such a federation. As you know, about ten years ago UNESCO has started to develop a program for editors as one part of the program. We deal of course not only with editors; we deal with authors, librarians, archivists, secondary services, and even with the users. So I think that, in the light of this program, and in the light of what has been said for instance by Anders Martinsson, I would strongly recommend to consider at least for the moment, the inclusion only of scientific editors in such a federation. I think it would be difficult and dangerous to address yourselves to persons or to scientists or to information handlers outside the editorial process - at least for the moment.

Anders Martinsson has said something about the role or the degree of organization of the other components, or the other elements, of the information transfer chain. I do very much support this idea. I can tell you that the librarians, the archivists, the secondary services, information processing people are very well organized on a national, regional and international level with a very strong impact on the international and intergovernmental organizations with the effect that by this kind of organization they have access even to financial sources.

I would say that only authors and editors are still living in a certain isolation, and when UNESCO started to implement its program we started on a discipline-oriented basis. We created regional associations, we initiated them, but these associations had to find their own style of life. Because always the final target of UNESCO was one day to create an international federation of editors, and in fact I am very pleased today to have this proposal on the table. I don't know whether this proposal is already ripe, I don't know whether the time has come to start or to create this international federation. I would like to encourage you to consider this very carefully. Many speakers have talked careful consideration that should be given to these questions and I would say that it would be against the mandate of UNESCO and against the interest of scientific editors if UNESCO would impose international structure on you.

But I can assure you that UNESCO is open to any constructive proposals you might wish to make, and I think that if you can provide certain proof for a long life of such a federation, if you can assure us that there is a need, and if you can tell us that there is a will to continue this activity, UNESCO would definitely be willing to lend its morale support of course, but also financial.

MIRIAM BALABAN - Thank you very much. That's a concrete offer.

PETER LENGYEL - Just to insure that you do not think that we are a monolithic organization, I'd like to speak after my colleague Wolfgang. I am editor and I happen to be employed be an international organization, but my main function and my main sympathies are those of an editor with other editors, cosufferers in this process. That is the angle in which I see it irrespective of my employer, which happens to be UNESCO, because I have the problem of putting out a quarterly journal. In this perspective I must agree with several of the suggestions that have been made.

First of all I like the idea of using the proceedings here as a kind of continuing instrument. Socondly and again, as I said, by coincidence, simply out of my own experience, I wouls support Mr. Lohner's idea that editors should be treated separately. We are of course a link in the chain, but we are a peculiar link and one which I think up until quite recently has had the rather a poor status. Many of us were in fact amateurs. I'm talking now mainly of the social sciences where I am more familiar with the situation than in the natural sciences, but I think it also exists in the natural sciences.

It was a sideline, it was amateurishly done, there were no particular professional standards to keep, and so anything went. We've heard several talks here about saml1 journals in the way they have recouped from their amateur status, even on the matter of advertising and things of that nature. So I would like to support the idea of editors being separately treated as a segment of the chain, and if there is going to be a next meeting, then it should concentrate on the problems of the editor as that particualr point of the chain.

May I make another plea however. An international conference must give the opportunity to people who do not speak English, to express themselves in their own languages, and if you are going to widen the geographical representation, you must have some form of interpretation, because there are many of our colleagues who of course do not speak English as well as some of us do.

ANDERS MARTINSSON - Madam Chairman, as a professional paleontologist I would of course have nothing against recreating a dinosaur, and then spend the rest of my life studying how dinosaurs became extinct. But I think that that is not the attitude we should take. Since I asked for the floor I think very much of what I intended to say has been said here, and I would like just to stress two points of view that have been presented here. It was Nancy Morris who said that we would certainly have to have another conference specifically for the purpose of creating a federation so that it has been announced properly that this is the main purpose of the conference.

The other point of view is that of Jeanne Poyen who stressed the existence of the rest of all these 25,000 journals - plus several thousand.

I think that one of the very important areas for the work of such a federation would be to go down to the national level and that is something that we should not wait with too long. We should try in different countries, or in this instance, groups of countries, to reach and organize these people. Many of them will certainly not be present at the forthcoming meetings of the federation, many of them will drop out of the business because it becomes too laborious to be professional enough to be an editor. We will get a lot of better editors, but there is a fair chance that we will have good representation at the inauguration meetings - national representatives who then can carry the message downwards in their hierarchy and give a solid and sound structure to the federation and to the international cooperation that is needed.

UDO SCHÜTZSACK - Ladies and Gentlemen I must tell you I am getting lost now. I'll tell you why. I am not against a federation of editors. We have federations of documentalists, of librarians, of archivists, and so on and so on. And why should we not have, additionally, an association of editors. However this conference was under the auspices of information transfer. I have learnt in the five days I have been here, that there are quite a lot of problems between authors, editors, secondary services and so on. And I am afraid that if you are only going to create an association of editors, then you are not going to take care these very pressing problems of information transfer and of cooperation and coordination between the different services.

KLAUS HELBIG - I would like to say two things. First, lest Nancy Morris mentioning of finance and her being afraid that a federation might be a competitor to the existing associations be taken as the reluctance of Editerra in this field, (I cannot speak of course for Editerra now because this was not announced and we did not have a council meeting on this) I can speak as an individual editor but I can also speak as a private individual who happens to be the President of Editerra. I think would support fully such a federation and I do not think that finances are important. It is important to think about it, but history shows that the formation of international federations or international unions from local bodies might be difficult, but the finances which are necessary have always shown up.

It happened for instance when the Geophysical Union was founded. There was no money from the local organizations, but when the union was founded with a very small levy on the individual organizations, the money was in the end there.

I would also like to comment on what Dr. Schützsack said. It is true that our present conference had the motto of "scientific information transfer". but it had also the motto, "the editor's role", and we have been here essentially as editors. We had a few sections of the conference which dealt with problems that were not properly editor's problems. In the first day we heard more about the production of journals than of editing of journals, but I cannot imagine that we have a federation of editors or a federation at all which deals with authors and publishers and archivists and documetalists and editors.

These are problems which have to be dealt with between these groups, not in one organization, but in a dialogue between the organizations. That is the way I see it. Of course, this is a private opinion, but I think we should try to establish a federation of editors, and editors only. There are enough scientists - authors who happen to be editors - as most of us editors are authors on the side too. There are many documentalists, many archivists who happen to be editors on the side. There are even publishers who are editors and the other way around, so I think there is enough linkage between the different fields by personal union, but I would not like to have this union cemented, established and institutionalized. I don't think that would help at all.

WOLFGANG LÖHNER - I would like to agree with what Prof. Helbig said. I think that creation of a federation on information transfer would just be a still-born child. There are so many people who feel responsible for information transfer, and I think that it would be an overestimation of the editors themselves to call themselves the federation of information transfer. I'm sorry, but you are just one segment, one element, of this whole chain and you are just one profession. Please do not try to embrace areas which are not really your concern. I think that the librarians, the archivists or the secondary services would not at all like such an idea. What they look for is a professional counterpart in the area of editing, and I think here you have a chance and a very good chance to succeed.

URI BLOCH - I'm afraid I'm rather siding with Dr. Schützsack. It may be that there are quite a number of people who are responsible for the transfer of information. I tend to claim that they do not discharge their responsibility, and that is what we have heard during all these days. We have pinpointed a few problems that I feel one could handle and could really come up with solutions. And therefore instead of building a dinosaur - incidentally when I heard the word dinosaur I thought of something quite different because the latest space shuttle is built on the principle of a dinosaur written "soar" at the end - I don't see why a dinosaur of that kind couldn't be built and that we shouldn't soar a little bit above lowly politics and see things slightly higher up. We shouldn't shelve sitting down to brass tacks.

We have found a few subjects for which we feel that we could find a solution. I for one would certainly side with those who feel that we should get a few working parties going in the fields for which we can hope that a reasonable proposal can be found. I think that most of these problems are problems of interfaces and we would do the scientific community as a whole and the editors in particular a greater service dealing with interfaces than in creating another international federation. I can very well understand that the editors feel the need for their own organization and I am the last one to tell them what to do. My personal inclination would be that since we have so many groups, try and merge between them. Whatever you decide, I hope you decide on working parties that get things solved so that in three years at the next conference, we can suggest certain ways that editors can behave.

MIRIAM BALABAN - Uri, I think you missed the point. Fortunately there are no politics involved here.* So both those who are defending creation of a federation of editors' associations and those, like you, who feel we should at once move to the higher level of creating a superstructure for all links in the information chain are genuinely concerned about getting down to the brass tacks you are talking about. It is a matter of how, not whether.

I personally think that if editors do get together in one forum they might be able to cooperate where cooperation is urgently begging and then in turn interface with other areas of information transfer which are already represented by strong bodies such as ICSU, the scientific societies themselves, IFLA, FID and others.

---

*It is interesting to note here Sir Robert Robinson's words at the 1948 Society Scientific Information Conference: "I notice that writers of certain notices in the newspapers have allowed themselves to dwell, with evident relish on the prospect of a clash of ideologies and the probable conflict between the planners and those who don't want to be planned. They are mistaken and will be disappointed. No delegate to this conference has any aim other than the development of scientific information, but there may be differences of opinion about the methods."[1] (Ed. note).

ERNEST MANN - I have a suggestion to make. We have heard the views of a lot of the delegates here. I think the organizing committee together with Prof. Martinsson should draft a resolution which would be submitted to you in writing as soon as possible for your consideration.

MARTIN CREMER - Well Madame chairman, I would like to propose a resolution. I think that everything is in what we have to do. I do not see any opportunity to solve the problem of whether we need an international federation of editors' associations, because some of the members of the conference including me, are not concerned with editors' problems alone, I'm concerned with the interdependence of documentation, information, information transfer, and editors, so I cannot see whether editors need an international federation or not.

Personally I am against it because I think you have an overdevelopment of international management, and you have more and more secretariats and journals, but that is not my problem so if the editors need something, I would strongly recommend to adopt your excellent recommendations and I think that if we add to the working party some people interested in the creation of an international federation of editors, then it will be all right.

STEPHAN KERTES - I think that as far as the next conference is concerned, it is pretty obvious that it is all settled. I don't see the audience here voting itself out of existence. I think we will all vote to meet again in three or four years.

What I am more concerned with is the international federation. My Canadian colleague tried in half a sentence to define what "scientific editor" means. I do not think that scientific editing is a profession. Career scientists serve for a period of two, three or perhaps five years as editors of a journal, and they are replaced by colleagues. It is not a permanent position. On the other hand, an information scientist - that is a profession, it is a job, a life long job. Now I hate to contradict Dr. Schützsack by saying that I do not think that an international organization, having scientific editors, which are to my definition career scientists whose main profession is to generate new scientific information, are scientific editors for a limited period of time that do not intermix, are not interchangeable, with a profession of information scientists or librarians, or the majority of the audience here. I am sorry for speaking for a minority. I tried to estimate what the number of career scientists in this audience is, and I came up with the number of fifteen. I asked a colleague of mine to make that estimate and he came up with almost thirty. It is still a tremendous minority in the audience. We, career scientists who contribute to primary literature source, have been lectured on how to write. I had the impression that we are infantile, that we cannot write, that we do not know science, that we do not know writing, we do not know English. And I think that the working group which I am in favour of establishing should have 85% of career scientists on it if we are dealing with establishing an international association.

EUGENE GARFIELD - I would support any move for the establishment of an international organization. I am not going to speak about any specifics. I know that ISI and a number of other secondary services, which have to deal with dozens of different organizations in order to make any headway along the lines I tried to talk about in my paper, would find it vastly more efficient to deal, if possible, with one international organization.

I would like to turn to the more practical side of this. I realize that perhaps this particular conference is a little more slanted towards the interface between secondary services and primary journals than other conferences might be in the future. From the point of view of the secondary services, I would like to see the various financing groups, including ISI, organize international workshops for the working scientist-editors, not the people already converted. I am prepared to say right here that ISI would spend money for such development, and if necessary finance part of their travel.

I am prepared right now to set up a workshop in Philadelphia next year where each one of these editors is put on trial, and we are going to go down a checklist and say: are you doing this, and are you doing that, why aren't you doing that, why aren't you doing this, and we will get the kind of results we need. In that way I think that we can really help to teach these people. I really agree with the previous speaker: the 85% are the kinds of problem editors that are most difficult to reach and I never see most of those people at these conferences. So I think UNESCO can help, and others with travel grants. I myself would be prepared to participate in a workshop like that. Some speakers talk of national ones, I will go to Norway, anywhere, to get fifty of those editors and preach the gospel to them. If you want practical results you must have that sort of an approach. There we won't have a long list of papers like we had here. They are very interesting, and I will read every word when I get the proceedings - but I would much rather have sat down, as somebody said - nuts and bolts.

MIRIAM BALABAN - I think we should come to some kind of summary, certainly not a conclusion, but a summary of the beginning. We have these proposals before us. Let me take them issue by issue. One - a second international conference for scientific editors and others involved in scientific communication should be held in approximately three years. Do we accept this as a resolution? No opposition.

A working party should be set up to consider how best the activities of the various parties can be coordinated.

KLAUS HELBIG - I don't think it is quite clear whether the working group is to prepare the next conference or is to work out the byelaws of the federation.

MIRIAM BALABAN - You are quite right. Actually there are two two things here. One is the conference, and the other one is the federation. We should make it clear we are talking about both.

ERNEST MANN - I support Dr. Cremer and propose we should adopt the resolution as it stands - a broad based resolution. If you can accept it that's fine. If you can't, then I suggest that the organizing committee coopting Prof. Martinsson and an American representative reformulate the proposal taking into account the various views expressed.

JEANNE POYEN - I agree entirely with Wolfgang Löhner. I would like the working group to devote itself only to problems of scientific editors and not to concern itself with all the problems of the information transfer chain.

PETER LENGYEL - I second it.

KLAUS HELBIG - What would the composition of the working group be? If the objective is to get these organizations together, I don't think we can designate names because the organizations themselves will designate representatives.

DIMITRI VIZA - Who will coordinate?

JAMES McCARTNEY - By what mechanism can the constituency react to the organizing committee and also receive materials from it?

MIRIAM BALABAN - The organizing committee together with Prof. Martinsson, an American representative and Dr. Löhner will prepare proposals to be presented to the organizations who will designate representatives.

The meeting is adjourned!

Thank you everyone!

SEQUEL AND PREVIEW

FIRST INTERNATIONAL CONFERENCE OF SCIENTIFIC EDITORS

To all participants of the Conference
and to those who didn't make it but were with us in spirit

Dear Friends,

This is to thank you for your share in making our conference a success. We look forward to any suggestions towards the second which we hope will take place in 1980, most probably in Amsterdam.

I am pleased to inform you that we are proceeding with the suggestion to form an International Federation of Scientific Editors' Associations (IFSEA) and we will keep you informed of the activities of the organizing committee.

The following draft proposal for the statutes of IFSEA will be remolded and formulated for acceptance at the founding meeting at UNESCO in Paris on June 5-6, 1978.

We hope then to create a roof for collaboration of our associations, disciplines and regions to advance the role of the editor for more efficient scientific information transfer for better service to all.

With best wishes,

Miriam Balaban

Miriam Balaban

Jerusalem
February 1978

IFSEA

INTERNATIONAL FEDERATION OF SCIENTIFIC EDITORS ASSOCIATIONS (in formation)

FOUNDING OF IFSEA AND PROPOSED STATUTES

Organizing Committee

Miriam Balaban*	Knut Faegri	Anders Martinsson
Michael Bowen	H. Edward Kennedy	Jacques Michel
Leonard Cohen	Wolfgang Löhner	Gillian Page
Martin Cremer	Ernest Mann	Udo Schützsack

As recommended, the organizing committee met on June 8 and August 10 at UNESCO in Paris and on November 24, 1977 at the Commonwealth Institute of Entomology in London to work out a proposal for founding IFSEA and its statutes.

These have been circulated to representatives of editors' associations and others interested in editor's affairs listed in the following pages who have been invited to the founding meeting of IFSEA scheduled for June 5-6, 1978 at UNESCO in Paris.

At the same time that we were furthering our organizational structure, with the initiative and cooperation of Maeve O'Connor and the Ciba Foundation, we got down to brass tacks appealed for at the Jerusalem Conference by holding a workshop on the Consolidation and Adoption of a Rational System for Bibliographic References. In the following pages are the resulting suggestions.

We will welcome comments, suggestions and information on parties interested in IFSEA and its aims.

---

* Correspondence to POB 4059, Jerusalem, Israel; Tel: (02)34405
Telex: RECO IL 25488; Cables: RECO Jerusalem

1978-06-09

INTERNATIONAL FEDERATION OF SCIENTIFIC EDITORS' ASSOCIATIONS (IFSEA)

The idea of IFSEA was presented at the First International Conference of Scientific Editors held in Jerusalem, April 24-29, 1977, and the organization was constituted at a meeting in the Unesco house in Paris, June 5-6, 1978. These statutes, adopted *ad interim* at the Paris meeting, are to be ratified or revised at the First General Assembly of IFSEA.

IFSEA INTERIM STATUTES

Adopted at the constituent meeting of IFSEA, Paris, June 5-6, 1978.

ARTICLE 1 - TITLE

The name of the association is: International Federation of Scientific Editors' Association designated by the acronym IFSEA, hereinafter referred to as "The Federation".

ARTICLE 2 - AIMS

The Federation is an independent international non-governmental association, without a profit motive, whose purposes are:

(a) to coordinate and complement but not duplicate the activities of its members in the different disciplines and regions, and to assist as necessary in the formation of further scientific editors' associations;

(b) to represent scientific editing in relation to other organizations particularly those concerned with the transfer of scientific information;

(c) to contribute to the formulation of standards and good practices in editing and to promote their international implementation;

(d) to encourage editors to take an interest in the planning, production and distribution of publications and in the development of information systems;

(e) to ease author-editor-publisher relationships and to optimize the scientific information process;

(f) to provide a forum for exchange of information and ideas on matters of interest or concern to editors by (i) promoting the participation of its members in meetings organized by individual associations, and (ii) organizing regular international interdisciplinary meetings on broad or specific topics, not covered by other meetings;

(g) to initiate and encourage international research on the most effective methods of achieving the aforementioned purposes.

ARTICLE 3 - MEMBERSHIP

(a) Full membership can be accorded to editors' associations in all scholarly and technical fields;

(b) Associate membership can be accorded to:

(i) other institutional organizations concerned with scientific editing;

(ii) editorial representatives of scholarly and technical publications in geographical areas not covered by editors' associations.

Members may be admitted by the Board and are subject to ratification by the Assembly.

ARTICLE 4 - RIGHTS AND OBLIGATIONS

4.1 Membership of the Federation shall constitute an agreement:

- to pay such dues as may be fixed and approved by the Assembly;
- to actively contribute to the objectives of the Federation.

ARTICLE 5 - ASSEMBLY

5.1 The Assembly shall consist of all members of the Federation.

5.2 The Assembly shall meet every third year and on such other occasions as the circumstances warrant.

5.3 Meetings of the Assembly may be called either by the Board, or by the President at the written request of one third of the members. At least 90 days' notice shall be given.

5.4 The duties of the Assembly shall include:

a) Election of Members of the Board (see article 7);

b) Approval of fees, budgets and accounts;

c) Approval of the triannual reports and proposals on future programmes;

d) Constitution of commissions of the Federation;

e) Ratification of membership of the Federation;

f) Approval of the rules of procedures and changes.

ARTICLE 6 - VOTING RIGHTS

Each full member shall have one vote.

ARTICLE 7 - THE BOARD

7.1 The Board shall consist of a President, a Vice-President, a Secretary-General, a Treasurer, and two or more ordinary members. The Board shall be elected for a period of three years. The President and Vice-President may be re-elected once to the same office. The Board should consist of no less than six members.

7.2 The duties of the Board shall include:

a) Supervision of activities of the Federation between the meetings of the Assembly;

b) Admission of members to the Federation;

c) Preparation of budgets and accounts;

d) Preparation of proposals for programmes and their supervision.

7.3 The President shall preside over the meetings of the Assembly and the Board and shall represent the Federation in its relations with other organizations. The President may appoint those Committees of the Federation that he deems necessary.

7.4 The Secretary General shall be the executive officer of the Federation, keep its minutes and correspondence and prepare annual reports that shall be submitted to the Board and presented to the Assembly at its triannual meetings.

7.5 The Treasurer shall keep the accounts, collect the fees and disburse the monies of the Federation.

ARTICLE 8 - FINANCE

The Federation shall derive its finances from:

8.1 Members' dues. Membership dues are to be fixed and revised periodically by the Assembly.

Special consideration may be given to non-governmental organizations who during an initial period or owing to unforeseen circumstances are unable to pay the fees set for their category.

8.2 Gifts, grants, legacies and other resources, subject to a decision of the Board.

8.3 Income from services, such as the sale of publications, studies and others.

ARTICLE 9 - COMMISSIONS AND COMMITTEES

Commissions shall be permanent working bodies of the Federation and may be instituted by the Assembly.

Committees are working bodies of the Federation which may be instituted by the President to solve specific problems within a limited period of time.

Commissions shall submit annual reports on their activities to the Board.

ARTICLE 10 - INTERNATIONAL CONFERENCES

International conferences shall take place every three years, normal in connection with the meetings of the Assembly. Attendance will not be restricted to members of the Federation. The organization of the conferences may be entrusted to a local organizing committee or a Committee of the Federation. Each conference shall have its own independent budget.

ARTICLE 11 - RULES OF PROCEDURES

Details concerning the internal operation of the Federation shall be dealt with in Rules of procedures established by the Assembly on the recommendation of the Board.

ARTICLE 12 - GENERAL PROVISION

12.1 Amendments to these statutes may be made at a meeting of the Assembly by a majority of two-thirds of the Members in good standing.

12.2 The Federation may be dissolved by a majority of two-thirds of the members in good standing.

12.3 In the case of conflict arising between the Federation and one of its regular members, an arbitration committee should be established if one of the parties requires this. Each party appoin one member, and they appoint a third member. The decisions of the arbitration committee are advisory.

Correspondence: Miriam Balaban, Secretary General, POB 4059, Jerusalem
Tel. 02-34405 Telex 25488

BIBLIOGRAPHICAL REFERENCES IN SCIENTIFIC PUBLICATIONS: SUGGESTIONS FROM A CIBA FOUNDATION WORKSHOP, LONDON, 25 NOVEMBER 1977

New suggestions from an international and interdisciplinary group of editors

---

Authors, editors, typists and typesetters often devote more time to references than to any other part of a typescript on its way into print. In November 1977 an international group of editors representing the earth and life sciences, chemistry, physics and engineering met at the Ciba Foundation to discuss this problem. The following document summarizes suggestions made at and after the November workshop.

SUMMARY

A unified system for bibliographical references is described. The suggestions, which are in the main compatible with ISO DIS 690, allow the name/date (Harvard), the numeric-alphabetic or the sequential-numeric system to be used for references in scientific books and serials; an optional dual-access system is also outlined. The 'master typescript' system described allows numbers to be substituted for names/dates *in the text* when a paper is submitted to a publication that uses a numeric system. In *reference lists*, depending on what the author's target publication requires, references must be arranged either alphabetically or sequentially, in the order of their citation in the text. In all lists, whether alphabetic or sequential, references should contain the same elements, arranged in the same order, as follows:

Abel B C, Cain A D, Adam E, Eve A 1976 Bark thickness in apple trees. Tree J 12:90-99

Extra punctuation marks can be added if required. Authors may leave further capitalization and italicization to be marked by editors/copy editors. Titles of articles should be included; they can be easily deleted if the publication does not require them. Journal titles should be given *either* in full *or* abbreviated in accordance with the standards named in this document.

INTRODUCTION

These suggestions deal with the citation and arrangement of bibliographic references in scientific books and serials. They are addressed to authors, typists, editors and copy editors. The *'master typescript'* system explained here can be easily adapted for publications using either the name/date or one of the two main numeric systems for references (sequential-numeric and numeric-alphabetic).

The aim of citations (in the text) and references (in the list) is the identification and retrieval of documents (usually an article or a book) referred to in a scientific text.

If there are differences between the suggestions in this document and an editor's 'Instructions to authors', the latter must take precedence in the final manuscript submitted.

The connection between citation and reference is made either by repetition of the author's surname and the *year* of the publication cited, or by repetition of a reference *number*. The two most common systems that use numbers assign them either in the *sequence* in which the references appear in the text or by numbering references in an *alphabetic* list.

The theme of these recommendations is uniformity. They are put forward as a compromise between the best of the innumerable systems that at present take up so much time when manuscripts are being prepared for publication. The suggestions are based on two premises first that it is much easier to convert the text from the name/date to one of the numeric systems than vice versa, and secondly that many authors use the name/date system in preparing the early drafts of their papers. (*Note:* Authors who are sure that their papers will be submitted to/accepted by a publication using a numeric system do not need to use the master typescript method but they should follow the recommendations made here in sections 3 and 4.)

The principal features of these suggestions are:

(1) A master typescript system is described that allows easy conversion of name/date citations to numeric citations;

(2) Reference lists should be arranged according to the system (numeric or alphabetic) required by the author's first target journal;

(3) References should always be made up of the same elements, arranged in the same order;

(4) References should be styled with a maximum of information and should conform as closely as possible with the conventions of spelling and punctuation of ordinary text in the language concerned. However, for easier transformation between reference systems, punctuation is here kept to a minimum;

(5) Other than as shown in the examples, capitalization and italicization of elements in references need not be indicated by authors but may be left for editors or copy editors to specify according to each publication's typographical style;

(6) An optional dual-access system is described that not only provides readers with a bibliography but also allows them to refer back from the reference list to find where the reference is cited in the text, using the reference list as an author index.

If a master typescript is prepared according to these recommendations, the author can - with a minimum of effort - alter copies of the master so that they conform to the requirements of individual serials or books, as follows:

(a) If the information contained in certain elements (e.g. titles of articles) that are listed here as standard parts of a reference is not wanted in a particular publication, the unwanted elements are deleted before the final typescript is submitted. If punctuation additional to that recommended in this document is required by the publication, punctuation marks are added as necessary.

(b) If the numeric-alphabetic system is required in the target publication, the list is numbered and numbers are substituted for the citations originally typed in parentheses in the text.

(c) If the sequential-numeric system is required in the target publication, the list is rearranged/retyped in order of citation (in practice, if the first journal to which an author submits a paper uses sequential numbering, the references in the list should be typed in the order of their appearance in the text). Name and date citations in the text are then changed to the relevant numbers.

RECOMMENDATIONS

1. *Citations in the text*

1.1 In the master typescript, citations should be in the name/date form.

Jack & Smith (1977) report ...

or

... as was recently reported (Jack & Smith 1977, Braun et al 1978)

If the target publication uses numeric system, several spaces should be left after each citation in the text.

After conversion to a numeric system (1.5), names of cited authors will appear in the text only if they were originally cited outside parentheses:

Jack & Smith (1977) report that ...

becomes

Jack & Smith (22) report that ...

But

as was recently reported (Jack & Smith 1977)

becomes

as was recently reported (22).

1.2 For citations in the text that refer to three or more authors, the form Brown et al. should be used every time such citations are made, including the first.

1.3 If citations refer to specific pages in books or in lengthy journal articles, those page numbers belong with the citations in the text, not in the reference list, and may be placed after the date. Reading is facilitated if the year or reference number is separated from the page number by adequate punctuation:

(Smith 1977 p 302), or (Smith 1977:p 302).

In a numeric system this would become:

(22:p 302) or (ref.22:p 302)

Alternatively (and perhaps preferable where superior numbers are used), a new number may be assigned each time a different page of the same book or article is cited in the text; each such reference in the list then gives the specific page number as well as the other elements of the reference.

1.4 Citations referring to communications which cannot be retrieved by readers do not belong in the reference list. If included at all, such communications should be described in the text as 'unpublished work' (not as 'in preparation') or 'personal communication' (not as a 'private communication') and the source should be included:

> ... as was found recently (J Smith, 1977, personal communication).

or (for a meeting without published proceedings):

> This project has produced significant results (M. Brown, unpublished paper presented at 3rd Congr Psychol Eng, 25 November 1977).

1.5 If a numeric system is required in the publication, appropriate numbers are substituted for the name/date citations in the text of the master typescript (see 2.1.5 and 2.2.1).

1.6 Where names/dates are used in the text, and if it is desirable for readers to be able to refer back from the reference list to the text, a dual-access system (Abbott 1974) may be used. In this system, numbers are placed *between* names and dates in the text, with the numbers running consecutively through the text (as in the sequential-numeric system, but with each citation being assigned a new number, however many times a reference may be cited in the text):

> Jack & Smith[1] (1977) report ...
>
> or
>
> ... as already described (Jack & Smith[5] 1977).

These numbers are then placed in parentheses at the end of each reference in the reference list in the way shown in 2.1.6.

2. *Reference lists*

2.1 Alphabetic lists

2.1.1 For publications requiring alphabetic reference lists, entries should be arranged in strict alphabetic order, *unless* there are several citations of articles/books by *three or more* different authors *with the same first author*. References of this latter kind, where 'Brown et al.' (for example) is followed by several dates in the text, should be arranged chronologically in the list, regardless of the number (*above two*) of authors and regardless of the alphabetical order of names of authors other than the first (since the names of the other authors are not known when the text is being read). References arranged in this alphabetic-chronological way should be placed after papers by Brown or by Brown & X, as shown in 2.1.2.

2.1.2 Arrangement of entries referring to the same first author (IUB-CEBJ 1973):

Citation in text	*Reference list*
Brown (1977)	Brown A 1977 ...
Brown & Green (1976)	Brown A, Green B 1976 ...
Brown & White (1975)	Brown A, White C 1975 ...
Brown et al (1974)	Brown A, White C, Green B 1974 ...

Citation in text	*Reference list*
Brown et al (1975)	Brown A, Black D, White C, Green B 1975 ...
Brown et al (1976)	Brown A, Black D, Green B 1976 ...
Brown et al (1977a)	Brown A, White C, Green B 1977a Why wheels turn. Gen. Eng. 7:59-65
Brown et al (1977b)	Brown A, Green B, White C 1977b How wheels turn. Gen. Eng. 8:10-15

2.1.3 Authors' surnames should be given and alphabetized exactly as spelt, with prefixes. 'Mc' and 'M' should not be alphabetized as 'Mac'. Names beginning with prepositions such as de, De, van, von, O', etc. should be alphabetized under the first letter of the prefix. The term 'Jr' or roman numerals may be added after initials. The important principle is conformity between citation and reference list entry.

2.1.4 Terms such as 'ibid.', 'idem'. 'op. cit.' and 'loc. cit.' should never be substituted for complete references in reference lists. In the text, if used at all, they should be used very sparingly and with great caution. Later changes in the text can make such references misleading.

2.1.5 If the numeric-alphabetic system is required, the alphabetically arranged references are first numbered:

1. Adam E 1976 ...
2. Adamson J 1975 ...

The numbers are then substituted for the name/date citations in the text.

2.1.6 If the dual-access system described in 1.6 is used, the numbers from the text are inserted in parentheses at the end of each reference in the reference list:

Jack V F, Smith J 1977 Structure of seeds of Stellaria media (L.) Vill. J Bot 35:290-295 (1,5,149)

2.2 Sequential-numeric lists

2.2.1 If the sequential-numeric system is required, the name/date text citations are numbered sequentially in the order of citation in the text; the references in the list are numbered accordingly and the list is arranged so that the entries are in numerical order, regardless of which letter of the alphabet the (first) author's name starts with. When a reference is cited more than once in the sequential system, either (1) the reference may retain the same number, thus breaking with the strict sequential principle, or (2) it may appear as many times in the list as it is cited in the text, each time numbered according to the sequence of citation. Before conversion or preparation of the typescript the author must consult the instructions to authors of the target publication to see how it handles multiple citation of a single reference.

3. *References to articles in serials*

3.1 References to articles in serials should take the following forms:

Jack V F 1977 The seeds of Stellaria media (L.) Vill. J Bot 35:257-261

Abel B C, Cain A D, Adam E, Eve A 1976 Bark thickness in apple trees. Tree J 12:90-99

3.2 This system uses the minimum punctuation necessary to present the information clearly. It also takes into consideration that punctuation marks can be added easily if the journal's style so requires, and that words and characters can be deleted more easily than they can be added (see 3.4). The references in 3.1 including the following features:

Surnames and intials of all authors are inverted.

The initials are spaced to permit punctuation if required.

A comma appears between the initials of one author and the surname of the next author.

No ampersand or 'and' is needed between authors' names but space should be left for its subsequent insertion if required.

A stop is placed only at the end of titles of articles.

No elements are underlined.

3.3 The date of publication is placed immediately after the authors' names. These two points of information are the most valuable in identifying or appraising a reference and are the key elements in locating a reference in a reference list when name/date citations are used.

3.4 Titles of articles in serial publications should be given, as many readers consider this to be essential information. However, if the first target publication will not print them, the article titles can easily be deleted or omitted.

3.5 Journal titles should *either* be given in full, *or* be abbreviated in accordance with ISO 833, ANSI Z39.5(1969), and BS 4148 (1970, 1975) (these standards are compatible with each other) or with lists (such as Index Medicus, BIOSIS List of Serials, etc.) based on the principles set out in those standards. One-word titles are never abbreviated.

3.6 First *and* last pages of articles in journals and chapters in books should be given. This information tells readers the length of the article, i.e. whether it is a short note or letter of a full-length paper, and indicates how much a photocopy is likely to cost.

3.7 The issue number of a journal is redundant information unless each issue is paged separately; if issues *are* paged separately, the issue numbers should be inserted (in parentheses) immediately after the volume number:

12. Jack V F 1967 Trees. Sci Am 17(3):38-47

For newspapers or popular weeklies the full date of the issue may be given instead of the volume number:

Smith A 1974 Creating wealth. Times(Lond.) 25 November:p 3

4. *References to books*

4.1 References to books should take the following basic forms:

4.1.1 Books by one or more authors, without an editor:

Jack V F 1977 Monograph of Stellaria media (L.) Vill. Pergamon, Oxford

Zorba A, Quinn A 1976 Bone structure of early Cretans, 3rd edn. Elsevier/Excerpta Medica/ North-Holland, Amsterdam

4.1.2 Chapters of sections of edited books:

Jack V F 1977 Seeds of Stellaria media (L.) Vill. In: Smith J (ed) Anatomy of Caryophyllaceae. Springer, Berlin, p 250-280

or (preferred if there are references to several contributions to the same book)

Jack V F 1977 Seeds of Stellaria media (L.) Vill. In: Smith (1977) p 250-280

...

...

Smith J (ed) 1977 Anatomy of Caryophyllaceae. Springer, Berlin

4.1.3 Chapters in edited books forming part of a series:

Bull F, Friend A 1975 Jumping over the moon. In: Jones M, Lloyd P (eds) Space travel. Galaxy Press, Houston (Soc Space Sci Symp 21) vol 3:1-24

4.2 The references in 4.1 include the following features:

Name(s) and initials of editor(s) are given as shown in 4.1.2 and 4.1.3.

'Editor', 'editors', and 'edition' may be abbreviated as 'ed', 'eds', and 'edn' respectively.

A stop is placed at the end of chapter titles.

The publisher's name, shortened if necessary, is placed before the place of publication. For publishers operating in several places, it is sufficient to give the name of the first place listed on the title page of the cited book as the publisher's address.

Page numbers, or volume and page numbers referred to, are placed after the place of publication.

No full stop is used at the end of the entries.

No elements are underlined.

REFERENCES

Abbott K M 1974 References - a new system used by CSIR. Scientiae 15:27-28

ANSI Z39.5 1969 Abbreviation of titles for periodicals. American National Standards Institute, New York

BIOSIS list of serials. BIOSIS/Biological Abstracts, Philadelphia

BS 4148 1970 The abbreviation of titles of periodicals. Part 1. Principles. British Standards Institution, London

BS 4148 1975 The abbreviation of titles of periodicals. Part 2. Word-abbreviation list. British Standards Institution, London

ELSE Working Group 1977 Recommendations on references. Earth Life Sci Ed 5:7-8

Index Medicus. US Dept of Health, Education and Welfare, Public Health Service, National Institutes of Health. US Government Printing Office, Washington, DC

ISO DIS 690 Bibliographic references to monographs and serials

ISO 833 Documentation - International list of periodical title word abbreviations

IUB-CEBJ 1973 The citation of bibliographic references in biochemical journals. Recommendations (1971). Biochem J 135:1-3 [and elsewhere]

This document evolved from discussions before, during and after an ELSE-Ciba Foundation Workshop on the Consolidation and Adoption of a Rational System for References, held at the Ciba Foundation, London, on 25 November 1977. The Workshop, funded largely by the Ciba Foundation, was held in order that the ELSE Working Group's recommendations on references (1977) and the ensuing discussions and attempts at formulating generally acceptable guidelines could be discussed by a wider group. The participants in the Workshop represented the earth sciences, life sciences, chemistry and biochemistry, physics and engineering.

ELSE-Ciba Foundation Workshop on Consolidation and Adoption of a Rational System for References held at the Ciba Foundation, London on 25 November 1977.

LIST OF PARTICIPANTS*

Miriam Balaban	Editor, *Desalination;* National Council for Research and Development, Jerusalem, Israel and Boston University
D H M Bowen	Director, Book and Journals Division, American Chemical Society, Washington, DC, USA
L W Carter	Chief copy editor, Springer-Verlag, Heidelberg, West Germany
Anita DeVivo	Executive Editor, Journals of the American Psychological Association, Washington, DC, USA
W G Evans	Royal Society, London, UK
K Faegri (Chairman)	Editor, *Naturen*, University of Bergen, Norway
H Götze	President, Springer-Verlag, Heidelberg, West Germany
N Hankins	Institute of Physics, Bristol, UK
A Hopkinson	UNISIST Centre for Bibliographic Descriptions, British Library Research and Development Dept., London, UK
E J Huth	Editor, *Annals of Internal Medicine*, American College of Physicians, Philadelphia, USA
C Liébecq	Chairman, IUB Committee of Editors of Biochemical Journals, Liège, Belgium
S Lock	Editor, *British Medical Journal*, London, UK
A Martinsson	Editor, *Lethaia*, Dept of Palaeobiology, University of Uppsala, Sweden
Maeve O'Connor*	Senior Editor, Ciba Foundation, London, UK
D Sharp	Deputy Editor, *The Lancet*, London, UK
J St Aubyn	Editor in Chief, Institution of Electrical Engineers, Stevenage, Herts, UK

*N.B. Not all the participants agree with all the details of the suggestions made in the document. Contact for further information.

ELSE WORKING GROUP ON REFERENCES

H Barnes	Dunstaffnage Marine Research Laboratory, Oban Argyll, UK
G N J Beck	Former Editor, *Endeavour*, London, UK
K Faegri (Convenor)	Editor, *Naturen*, University of Bergen, Norway
Olive M Harrison	Managing Editor, *British Medical Bulletin*, London, UK
J P Hudson	Former Director, Long Ashton Research Station, Bristol, UK
E J Huth	Editor, *Annals of Internal Medicine*, Philadelphia, USA
J C Rigg	Pudoc Centre for Agricultural Documentation, Wageningen, The Netherlands
B J Walby	Editor-in-chief, CSIRO Editorial and Publications Service, East Melbourne, Australia

LIST OF PARTICIPANTS

JEAN-PAUL ABADIE, Directeur Adjoint des Publications Scientifiques aux Editions Masson, 120 Boulevard Saint-Germain, 75280 Paris Cedex 06, France

PROF. JOSEPH AGASSI, Department of Philosophy, Boston University, Boston, Mass. 02215 and Tel Aviv University, Department of Philosophy, Tel Aviv, Israel

DR. PHILIP L. ALTMAN and Mrs. Altman, Director, FASEB Office of Biological Handbooks, Federation of American Societies for Experimental Biology, 9650 Rockville Pike, Bethesda, Maryland 20014

IRVING AMRON and Mrs. Amron, Editor of Information Services, American Society of Civil Engineers, 345 East 47th Street, New York, NY 10017

EDWARD ASLAKSEN, Universitetsforlaget (Oslo University Press Publisher), Stensberggt 25, Postboks 2670, St. Hanshaugen, Oslo 1, Norway

PAUL NIJHOFF ASSER and Mrs. Asser, Secretary General STM (International Group of Scientific Technical and Medical Publishers), 462 Kaizersgracht, Amsterdam, The Netherlands

MIRIAM BALABAN, Editor, Desalination, National Council for Research and Development Jerusalem, POB 4059, Jerusalem and Center for the Philosophy and History of Science, Boston University, Boston, MA 02215

S. PETER BAKKER, Junk Publishers, POB. 3713, 9 Lange Voorhout, The Hague, The Netherlands

JANINA BARD, International Irrigation Information Center, Volcani Center, POB 49, Bet Dagan, Israel

DR. HAROLD BARNES, The Dunstaffnage, Marine Research Lab., Oban, Argyll, Scotland

BRUCE BATTRICK, Scientific & Technical Publications Branch, European Space Agency, Domeinweg, Noordwijk, The Netherlands

W.J. BAULKWILL, Ministry of Overseas Development, Land Resources Div., Tolworth Tower, Surbiton, Surrey KT6 7DY, England

PROF. ROBERT N. BECK and Mrs. Beck, Editor, Idealistic Studies, Department of Philosophy, Clark University, Worcester, MA

PROF. CYRIL S. BELSHAW, Editor, Current Anthropology, University of British Columbia, Vancouver, B.C. Canada, V6T 1W5

DR. STEFAN BENGTSON, Managing Editor, Lethaia, Department of Palaeobiology, Box 564, S-751 22, Uppsala, Sweden

MRS. SHIRLEY M. BLATCHFORD, Tropical Products Institute, 56-62 Gray's Inn Road, London WC1X 8LU, England

URI BLOCH, Head of Information and Documentation, Ministry of Defence, Armament Development Authority, POB 962, Haifa, Israel

MARY L. BORYSEWICZ, Managing Editor, American Journal of Ophtalmology, 233 E. Ontario Street, Suite 1401, Chicago, Ill.

DR. D.H. MICHAEL BOWEN, Director, Books & Journals Div., American Chemical Society, 1155 16th Street, Washington, D.C.

DR. JOSEPH PALTI, Volcani Institute for Agricultural Research, Bet Dagan, Israel

DR. PAOLINO PAPALI, Ed. Nuovo Cimento, Via degli Andalo 2, Bologna, Italy

MARJO PELTENBURG-JAGER, International Union of Geological Sciences, POB 379, Haarlem, Netherlands

KURT PAULUS and Mrs. Paulus, Executive Editor, The Institute of Physics, Publ. Div., Netherton House, 23 Marsh St., Bristol BS1 4BT, England

ITAMAR PERATH, Science Editor, The Geological Survey of Israel, 30 Malkei Israel St., Jerusalem, Israel

GERARD PIEL and Mrs. Eleanor Piel, Publisher, Scientific American, 415 Madison Avenue, New York, NY 10017

INNA POMMERANTZ, Dept. of Antiquities, POB 586, Jerusalem, Israel

DR. FRANKLIN H. PORTUGAL, Assoc. Man. Ed., Proc. of the National Academy of Sciences, 2101 Constitution Ave., Washington DC 20418

JEANNE POYEN, Secretary General, ICSU AB, 17 rue Mirabeau, 75016 Paris, France

VIVIAN PRIEL, Volcani Institute for Agricultural Research, Rehovot, Israel

SIMA RANAA, Science Editor - Yediot Acharonot, 81 Margolin St., Yad Eliahoo, Tel Aviv

JACQUES G. RICHARDSON, Ed. Impact of Science on Society, UNESCO, 7 Place de Fontenoy, 75700 Paris, France

CHRISTOPHER J. RIGG, PUDOC Center for Agricultural Documentation, POB 4 Wageningen, The Netherlands

ZVI RON, Bank of Israel, Jerusalem, Israel

ARTHUR ROSENTHAL, Director, Harvard University Press, 79 Garden Street, Cambridge, Mass. 02138

MARGARET ANN ROSENTHAL, Harvard University Press, 79 Garden St., Cambridge, MA

DR. LIVIA ROTHKIRCHEN, Editor, Yad Vashem Studies, Yad Vashem, Jerusalem, Israel

N. RUBIN, Oil Institute of Israel, Tel-Aviv, Israel

MRS. JEAN SACKS, Director of Journals, Chicago University Press, Chicago, Ill.

PROF. MARTIN SACKS, The Hebrew University - Hadassah Medical School, Jerusalem, Israel

DR. SAMULOWITZ, Documentationswesen, 6000 Frankfrut Am M, Niederrad Herriotstrasse 5, Germany

NORMA SCHNEIDER, Israel Academy of Sciences & Humanities, 43 Jabotinsky St., Jerusalem, Israel

DR. HANS BRATTSTROM, Ed. Sarsia, Biologisk Stasjon, Espegrend, N-5065 Blomsterdalen, Norway

PROF. SHU-TING CHANG, Dean of Science Faculty, Department of Biology, The Chinese University of Hong Kong, Shatin, New Territories, Hong Kong

PROF. NORMAN CHIGIER, Editor, Progress in Energy and Combustion Science - An International Review Journal, Dept. of Chemical Engineering and Fuel Technology, The University of Sheffield, Mappin St., Sheffield S1 3JD, England

DR. LOUIS COHEN, Executive Secretary, The Institute of Physics, 47 Belgrave Sq., London SW1X 8QX, England

PROF. ELI COHN, Editor, Phytoparasitica, Division of Nematology, Agricultural Research Organization, The Volcani Center, POB 6 Bet-Dagan, Israel

DR. MARTIN CREMER and Mrs. Cremer, Director, Institut für Dokumentationswesen, 6 Frankfurt (Main), 71, Postfach 710350, Germany

DR. AMANDO M. DALISAY, Treasurer, EDITEAST, Association of Editors in Science in Southeast Asia, Australasia and Oceania, National Research Council, PTRI Bldg., Bicutan, Tagig, Rizal, Philipines

DR. P.A. DAVENPORT, Editor, Plasma Physics, Culham Laboratory, Abingdon, Berkshire OX14 3DB, England

PROF. A. MICHAEL DAVIES, Dept. of Medical Ecology, The Hebrew University - Hadassah Medical School, POB 1172, Jerusalem, Israel

J.W. DE BLOK, Netherlands Institute for Sea Research, POB 5g, Texel, The Netherlands

PROF. ARYEH DVORETZKY, President, Israel Academy of Sciences and Humanities, Dept. of Mathematics, The Hebrew University of Jerusalem, Israel

AGNITA EKMAN, Uppsala Universitet, Kvartargeologiska Avdelningen, Box 555, S-751 22, Uppsala, Sweden

DR. CECIL EWING, Editor, Can. J. of Ophtalmology, Eye Department, University Hospital, Saskatoon S7 N OW8, Canada

PROF. KNUT FAEGRI & Mrs. Faegri, Botanical Museum, Bergen Univ., Pres. IUBS, Bergen, N5014 Norway

PROF. ALFRED P. FISHMAN and Mrs. Fishman, Chmn. Publ. Board, Am. Physiol Soc., Director, Cardiovascular-Pulmonary Div., Dept. of Medicine, Hospital of Univ. of Pennsylvania, 3400 Spruce St., Philadelphia, PA 19104

DR. CHAIM FORGACS, Research & Devel. Authority, Ben Gurion University of the Negev, POB 1025, Beer Sheva 84110, Israel

MARGARET FOTI, Managing Editor, Cancer Research, Editorial Office, 314 Jones Hall, Temple Univ., School of Medicine, Philadelphia, PA 19140

DR. EUGENE GARFIELD, President, Institute for Scientific Information, 325 Chestnut St., Philadelphia, PA 19106

J. MICHEL GIBB, Commission of the European Communities, Directorate General for Scientific & Technical Information and Information Management, DG XIII, European Centre, Kirchberg, Luxembourg

PROF. JOSEPH GILLIS, Editor, Math. Teaching J., The Weizmann Institute of Science, Dept. of Applied Mathematics and of Science Teaching, Rehovot, Israel

PROF. NATHAN GOLDBLUM, Vice President, The Hebrew University of Jerusalem, Dept. of Virology, Israel

DR. MICHAEL GORDON, University of Leicester, Primary Communications Research Center, Dept. of Astronomy, Univ. Road, Leicester LE1 7RH, England

PAUL GREENBERG, Weizmann Science Press, POB 801, Jerusalem, Israel

ANNIE GRILL, Atomic Energy Commission, Nuclear Research Center, POB 9007, Beer Sheva, Israel

DR. HAROLD HAENDLER, Director, Dokumentationssteller der Universität Hohenheim (Agric. High School), Paracelsusstrasse 2, D 7000, Suttgart 70 Germany

DR. EDWARD H. HARE, and Mrs. Hare, Editor, British J. of Psychiatry, The Bethlem Royal Hospital & the Maudsley Hospital, Denmark Hill, London, S.E.5., England

MICHAEL HARRIS, V.P. Wiley-Interscience, Wiley Publishers, 605 Third Ave., New York, NY 10016

DR. ULRICH HERZFELD, Springer-Verlag, Medical Ed., Postfach 105280, Neuenheimer Landstr. 28-30, 6900 Heidelberg, Germany

LORE HENLEIN, Advanced Book Program, Addison-Wesley Publishing Co., Reading MA 01867

DR. FREDERIC W. HILL, Ed. J. Nutrition, Dept. of Nutrition, University of California, Davis, Cal. 95616

MARY HILL, Information Office, U.S. Dept. of the Interior, Geological Survey, Western Region, Menle Park, Calif. 94025

PROF. DR. KLAUS HELBIG, Ed. Geophysical Prospecting, Vening Meinesz Laboratorium voor Geofysica en Geochemie, Rijksuniversiteitte Utrecht, Lucas Bolwerk 7, The Netherlands

SYNNØVE IRGENS-JENSEN, Nordic Publishing Board in Science, NAVG, Munthes gt. 29, Oslo 2, Norway

DR. ANDREAS JESSE, Avenue Wendt 22, CH=1203 Geneva, Switzerland

PROF. M. JEUKEN, Institute for Theoretical Biology, Stationsweg 25, Leiden, The Netherlands

FINN A. JORSTAD, Norwegian Geotechnical Institute, POB 40, Tassen, Oslo 8, Norway

DR. THOMAS KARLSSON, Ed. Opera Botanica, Swensk Botanisk Tidskriff, Dept. of Plant Taxonomy, Botaniske Notiser, O Vallgatan 18-20, S-223, Sweden

PROF. ASA KASHER, Ed. Philosophia, Dept. of Philosophy, Bar-Ilan Univ., Ramat-Gan, Israel

SHAUL KATZ, Dept. of Sociology, The Hebrew University of Jerusalem, Israel

DR. PAUL KATZENBERGER, Max-Planck Institut für Ausländisches und Internationales Patent-Urhebest und Wetterbewerbstrecht, 8 Munchen 80, Sieberstrasse 3, Germany

MYRA KAYE, Israeli Institute for Biological Research, POB 19, Ness Ziona, Israel

PROF. ORA KEDEM, The Weizmann Institute of Science, Rehovot, Israel

ARTUR ISENBERG, Editor, Kidma Is. J. Development, POB 13130, 3 Moshe Wallach St., Jerusalem, Israel

DR. MARK KELLER & Mrs. Sally Keller, Editor, J. Studies on Alcohol, Rutgers Center of Alcohol Studies, Smithers Hall, Allison Road, Rutgers, State University of New Jersey, Piscataway, J.A. 08854.

DR. H.E. KENNEDY & Mrs. Dorothy Kennedy, Director, Biological Abstracts, 2100 Arch St., Philadelphia, PA 19103

CARL KEREN, Director, National Center for Scientific and Technological, POB 20125, Tel-Aviv, Israel

PROF. STEFAN KERTES, Dept. of Chemistry, The Hebrew University of Jerusalem, Givat Ram, Jerusalem, Israel

PROF. MANFRED KOCHEN, Mental Health Research Institute, University of Michigan, Ann Arbor, Michigan 48104

PROF. ALEXANDER KOHN, Editor, J. Irreproducible Results, Israel Institute for Biological Research, POB 19, Ness Ziona, Israel

PROF. LARS-KONIG KONIGSSON, Uppsala Universitet, Kvartargeologiska avdelningen, Box 555, S-751 22, Uppsala, Sweden

PROF. ANDERS KYLIN and Mrs. Kylin, Institute of Plant Physiology Fack, S-220 07 Lund 7, Sweden

DR. C.A. LANG, Physical Sciences Editor, Cambridge University Press, The Pitt Building, Trumpington St., Cambridge CB2 1RP, England

DR. HANS DIETRICH LANG, Bundesanstalt für Geowissenschaften und Rohstoffe, POB 510153 D-3000 Hannover 51, Germany

PETER W. LEA, Manchester Polytechnic, Faculty of Community Studies, Bracken House, Charles Street, Manchester M1 7DF, England

DR. KLAUS-DIETER LEHMANN and Mrs. Lehmann, Vice Director, University Library, Bockenheimer Landstr., Frankfurt/M.

PETER LENGYEL, Editor, Int. Soc. Science J., UNESCO, 7 Place de Fontenoy, 75700 Paris, France

LARRY LESTER, Executive Editor, Weizmann Science Press, POB 801, Jerusalem, Israel

DR. STEPHEN LOCK and Mrs. Lock, British Medical Journal, BMA House, Tavistock Square, London WC1H 9JR, England

DR. WOLFGANG LÖHNER, UNESCO, Division of the General Information Programme, 7 Place de Fontenoy, 75700, Paris, France

PROF. SHNEIOR LIFSON, The Weizmann Institute of Science, Dept. of Chemical Physics, Rehovot, Israel

DR. PAUL A.M. MAAS, Executive Editor, Delft University Press, Mijnbouwplein 11, NL-2208 Delft, The Netherlands

L. REGINALD MACINTYRE, Publications Coordinator, International Development Research Centre, POB 8500, Ottawa, Canada

ERNEST J. MANN, Director, Commonwealth Bureau of Dairy Science and Technology, Ed. Food Abst., Shinfield, Reading RG2 9AT, England

DR. ARIE A. MANTEN, Elsevier Scientific Publishing Co., POB 330, Amsterdam

DR. DAVID MARGALITH, Ed. Koroth, 37 Gedalyahu St., Neve Shaanan, Maon Harofe, Haifa, Israel

TOVA MARKOWITZ, National Center for Scientific Technology, 84 Hahashmonaim St., Tel Aviv, Israel

PROF. ANDERS MARTINSSON, Ed. Lethaia, Dept. of Palaeobiology, University of Uppsala, POB 564, S-751, 22 Uppsala, Sweden

JOANNA M. MARX, Bureau of Publications, Foundation for Education Science and Technology, Didacta Bldg., 211 Skinner St., POB 1758, Pretoria 0001, South Africa

PROF. JAMES L. McCARTNEY, Dept. of Sociology, University of Missouri, Columbia, MO 65201

MRS. HILDE McGRATH and Mr. McGrath, Ed. Plant Disease Reporter, Applied Plant Pathology Laboratory, Room 24, Bldg. 001, Beltsville Agricultural Research Center, BARC West, Beltsville, MD 20705

PROF. WILLIAM J. MELLMAN, Ed. in Chief, Am. J. Human Genetics, The School of Medicine, Dept. of Human Genetics, Richards Bldg. G4, University of Pennsylvania, Philadelphia, PA 19174.

STEPHEN LAWRENCE MERCER, Pans Office Center for Overseas Pest Research, College Home, Wrights, London W8 5SJ, England

JACQUES MICHEL, Bureau National de l'Information Scientifique et Technique, 8-10 rue Crillon, 75194 Paris Cedex 04, France

DR. DOROTHY U. MIZOGUCHI, Tech. Ed. Gann, Jap. J. Cancer REs., Cancer Therapy Information Center, Cancer Institute, Kami-Ikebukuro 1-37-1, Toshima-ku, Tokyo 170, Japan

NANCY MORRIS, Secretary, ELSE-Editerra, Bruins, 30 Longdown Road, Lower Bourne, Farnham, Surrey GU10 3JL, England

MAEVE O'CONNOR, The Ciba Foundation, 41 Portland Pl., London, W1N 4BN, England

DR. RAGNAR OLSSON, Exec. Ed. Acta Zoologica, University of Stockholm, Dept. of Zoology, Box 6801 S-113 86, Stockholm, Sweden

GILLIAN PAGE, Director of Journals, Cambridge University Press, POB 92, London NW1 2DB, England

DR. UDO SCHÜTZSACK, Institut fur Documentationswesen, 600 Frankfurt Am M, Niederrad Herriotstrasse 5, Germany

PROF. ELEAZAR SHAFRIR, Ass. Ed. Israel J. Med. Sci., Department of Biochemistry, Hadassah University Hospital, Jerusalem, Israel

DR. DAVID A.E. SHEPHARD, Ex. Ed. Clin. Invest. Med., Kellogg Centre for Advanced Studies in Primary Care, Livingston Hall, The Montreal General Hospital, 1650 Cedar Avenue, Montreal, Quebec H3G 1A4, Canada

DR. ANDREW SHERRINGTON and Mrs. Sherrington, Ed. Board Can. J. of Pub. Health, 211 Wurtemburg, Apt. 2203, Ottawa, Ontario, Canada K1N 8R4

SCHLOMO SIRKIS, Bank of Israel, Jerusalem, Israel

DR. M. SKULSI, Inland Waters Directorate, Adm. Branch, Ottawa, K1A OE7, Canada

DR. M. THERESE SOUTHGATE, Dep. Ed. J. Am. Med. Ass., 535 N. Dearborn St., Chicago, Ill. 60610.

J.D. St. AUBYN, Institution of Electrical Engineers, Herts SG1 1HQ, England

DR. RONALD SWARTZ, School of Education, Oakland University, Rochester, Michigan 48063

PROF. EART J. TAFT, Ed. Communications in Algebra, Rutgers, The State University of N.J., Dept. of Mathematics, Hill Center for the Mathematical Sciences, Busch Campus, New Brunswick, NJ 08903

DR. ELIEZER TAL, Director, National Council for Research and Development, Jerusalem, Israel

YERAH TAL, Nordau St. 20, HAARETZ, Haifa, Israel

Y.J. TAUB, The Economic Quarterly, Neve Shaanan 7, Jerusalem

DR. R.E.M. VAN DEN BRINK, Chairman Board of Directors, Elsevier Publishing Co., 335 Jan van Galenstraat, Amsterdam, The Netherlands

DR. A.J. VAN LOON, Inst. Aardwetenschepen vrye Universiteit de Boelelaan 1085, Amsterdam, The Netherlands

LYDIA VILENTCHUK, Head Dept. of Publications, National Center of Scientific and Technological Information, POB 20125, Tel Aviv

DR. DIMITRI VIZA, Ed. Differentiation, Laboratory of Immunobiology, School of Medicine, 15 rue Ecole de Medecine, B.P. 254, 75264 Paris Cedex 06 France

ALBRECHT VON HAGEN, Publisher, Center for Advanced Publishing, McGraw-Hill International Book Co., 100 Graf Adolf Strassse, 4000 Dusseldorf, Germany

BASIL J. WALBY, Director, CSIRO Editorial and Publications Service, 372 Albert St., East Melbourne, Victoria 3002, Australia

DR. JEFFREY WATSON, Ed. J. Fisheries Res. Board of Can., 116 Lisgar, Ottawa K1A OH3, Canada

DR. YEHEZKEL WEILER, Ed, Israel J. Earth Sci., Weizmann Science Press, Dept. of Geology, The Hebrew University of Jerusalem, Israel

DR. ROBERT WEINTRAUB, Research and Development Authority, Ben Gurion University of the Negev, POB 1025, Beersheva 84110, Israel

R.B. WEISS, Centre for Agricultural Strategy, University of Reading, Reading RG8 2AT, England

PROF. YEHUDAH L. WERNER, Ed. Israel J. Zoology, Dept. of Zoology, The Hebrew University of Jerusalem, Israel

PROF. EDGAR F. WESTRUM JR., & Mrs. Westrum, Sec. Gen., Committee on Data for Science and Technology (CODATA), Dept. of Chemistry, University of Michigan, Ann Arbor, Mich. 48109

JOHN R. WETTERSTEN, University of South Carolina, Dept. of Philosophy, Columbia, S.C. 29208

MOONYEN WHIFFIN, Ed., Royal Meteorological Society, James Glaisher House, Grenville Place, Bracknell, Berkshire, RK12 1BX, England

PROF. ARTHUR J.C. WILSON, Ed. Acta Crystallographica, Dept. of Physics, Chancellor's Court, University of Birmingham, POB 363, Birmongham B15 2TT, England

ANTHONY M. WOODWARD, Aslib, 3 Belgrave Square, London SW1X 8PL, England Present address: Agricultural Documentation Centre, POB 30028, Nairobi, Kenya

PROF. JOHN M. ZIMAN, F.R.S., University of Bristol, H.H. Wills Physics Laboratory, Tyndall Avenue, Bristol BS8 1TL, England

RAMA ZOTA, Bank of Israel, Jerusalem, Israel

E D I T O R S' A S S O C I A T I O N S

LIFE SCIENCES

CBE - Council of Biology Editors (North America)

President: Dr. H. Edward Kennedy, Executive Director, Biosciences Information Service, 2100 Arch Street, Philadelphia, Pa. 19103
Secretary: Prof. A.J. Ladman, Department of Anatomy, University of New Mexico Medical School.
Treasurer: Gisella Pollack

ELSE - European Association of Editors of Biological Periodicals

President: Herbert J. Kramer, Medizinische Univ. Poliklinik, 5300 Bonn, Wilhelmstr. 35-37, FRG
Vice President: Gillian Page, Director of Journals, Cambridge University Press, 200 Euston Road, London NW1 2DB, UK
Secretary: Nancy Morris, Bruins, 30 Longdown Road, Lower Bourne, Farnham, Surrey, GU10 3JL, UK
Treasurer, Miriam Balaban, National Council for Research and Development, POB 4059, Jerusalem, Israel

FOOD AND AGRICULTURE

Association of Food and Agriculture Journals (in formation)

Ernest Mann, Commonwealth Bureau of Dairy Science & Technology, Shinfield, Reading RG2, UK

EARTH SCIENCES

Editerra (Europe and adjoining areas)

President: Prof. Svend Saxov, Laboratory of Geophysics, 6-8 Finlandsgade, DK-8200 Aarhus-N, Denmark
Secretary: Nancy Morris, Bruins, 30 Longdown Road, Lower Bourne, Farnham, Surrey GU10 3JL, UK
Treasurer: Anthony P. Harvey, Ragstones, Broadoak, Heathfield, Sussex TN21 8UD, UK

AESE - Association of Earth Science Editors (North America)

President: Dr. Stuart E. Jenness, POB 7376, Ottawa, Ontario KiL 8E4, Canada
Secretary-Treasurer: Mr. John Heller, U.S. Geological Survey, Denver Federal Center, Stop 303, Box 25046, Denver, Colo. 80225

IUGS - Advisory Board for Publication

Chairman: Michael G. Bassett, Department of Geology, National Museum of Wales, Cardiff CF1 3NP, UK

Alegeo (Latin America)

Dr. Cecily Petzall, Dirección de Geología MMH, Centro Simón Bolívar, Torre Norte, piso 19, Caracas, Venezuela

BIOCHEMISTRY

CEBJ - IUB Committee of Editors of Biochemical Journals (International)

Prof. Claude Liébecq, 69/054 Bd. de la Constitution, B-4020 Liège, Belgium

CHEMISTRY

ACS-CAS - American Chemical Society and Chemical Abstracts (US)

There is no association, but Dr. D.H. Michael Bowen is on IFSEA organizing committee, and ACS-CAS is actively involved in editors' affairs.

IUPAC - International Union of Pure and Applied Chemistry

Considering formation of editors' committee
Executive Secretary: Dr. Maurice Williams, IUPAC Secretariat, 2-3 Pound Way, Cowley Center, Oxford

EdEuChem - Editors of European Chemistry

President: Prof. E. Cherbuliez, Fossard 48, CH-1231 Conches/Geneva
Secretary: Dr. Edwin Kisman, Sigma Chemie, Publishing Foundation, Riouwstraat 153, POB 1767, The Hague

IFCC - International Federation of Clinical Chemistry

CEJCCC - Commission of Editors of Journals Concerned with Clinical Chemistry

Chairman: Dr. R. Dybkaer, Dept. of Clinical Chemistry, de Gamles By, Nørre Allee 41, DK-2200, Copenhagen N, Denmark
Secretary: Mr. P.M.G. Broughton, Wolfson Research Laboratories, Department of Clinical Chemistry, Queen Elizabeth Medical Centre, Birmingham B15 2TH, UK

PHYSICS

There is no editors' association but the European Physical Society has a scheme for standards for the Europhysics journals and the American Institute of Physics is an active publications center.

IUPAP - Prof. B.R. Coles, Department of Solid State Physics, Imperial College, Prince Consort Road, London SW7, UK

EPS - Publications Advisory Committee

Chairman: Prof. E.R. Dobbs, Bedford College, London
EPS Secretariat, POB 39, CH-1213 Petit Lancy 2, Siwtzerland

UK IP - Executive Secretary: Dr. Louis Cohen, The Institute of Physics, 47 Belgrave Sq., London SW1X 8QX
Prof. J. Ziman, University of Bristol, H.H. Wills Physics Laboratory, BS8 1TL
Vice-President: Prof. R.G. Chambers, University of Bristol
Director of Publications: C.I. Pedersen, Techno House, Redcliffe Way, BS1 6NX, Bristol

USA AIP - Dr. A.W. Kenneth Metzner, Director, Publications Division, American Institute of Physics, 335 E 45th St., New York, NY 10017
Dr. William Koch, Director of American Institute of Physics

CRYSTALLOGRAPHY

International Union of Crystallography

Executive Secretary: Dr. J.N. King, 13 White Friars, Chester CH1 1NZ
Former Editor: Prof. A.J.C. Wilson, Dept. of Physics, University of Birmingham, POB 363, Birmingham, UK
Editor, Acta Crystallographica, Prof. S.C. Abrahams, Bell Labs., 600 Mountain Ave., Murray Hill, NJ 07974
Prof. Andre Guinier, Laboratoire de Physique des Solides, Université Paris Sud Orsay, France

MATHEMATICS

No editors' association but Mathematical Association of America (MAA) and American Mathematical Society (AMS) active and interested.

MAA - Mathematical Association of America

Executive Director: Dr. Alfred B. Willcox, 1225 Connecticut Ave., NW, Washington, DC 20036
Prof. Earl J. Taft, Rutgers, The State University of New Jersey, Department of Mathematics, Hill Center for Mathematical Sciences, Busch Campus, New Brunswick, N.J. 08903

AMS - American Mathematical Society

Dr. William J. Leveque, POB 6248, Providence, RI 02940

ANTHROPOLOGY

CIPA - Committee on International Publication in Anthropology (International)

Prof. Cyril S. Belshaw, Editor, Current Anthropology, Dept. of Anthropology and Sociology, University of British Columbia, Vancouver, B.C. V6T 1W5
Dr. Isaac Chivas, Laboratoire d'Anthropologie Sociale, Collège de France, 11 Place Marcelin Berthelot, Paris 5w

SOCIOLOGY

Ad hoc Council of Social Science Editors (International)

Prof. James L. McCartney, Department of Sociology, University of Missouri, Columbia, MO 65201

PHILOSOPHY

Association of Philosophy Journal Editors

President: Prof. Robert N. Beck, Department of Philosophy, Clark University, Worcester, MA 01610

HUMANITIES

CELJ - Conference of Editors of Learned Journals

President: Marilyn Gaull, Department of English, Temple University, College of Liberal Arts, Philadelphia, Pa 19122

PSYCHOLOGY

Anite DeVivo, American Psychological Association, 1200 Seventeenth St., NW, Washington, DC 20036

WATER RESOURCES KNOWLEDGE TRANSFER

Dr. Neil Grigg, Water Resources Research Institute,
124 Riddick Bldg., North Carolina State University
Raleigh, NC 27607

ENGINEERING

IEEE - Institute of Electrical and Electronic Engineers (North America)

Elwood Gannett, Institute of Electrical and Electronics Engineers, 354 E. 47th St., New York, NY 10017

Institution of Electrical Engineers (UK)

Editor in Chief: J.D. St. Aubyn, Institution of Electrical Engineers, Stevenage, Herts., SG1 1HQ, UK

GENERAL

ASJ - Association of Scientific Journals (North America)
(US Association of Scholarly Publishing contemplated -
John Strawhorn, Capital Systems Group,
6110 Executive Blvd., Rockville, Maryland 20852)

James Lufkin, G2118 Honeywell Plaza, Minneapolis MN 55408

Editeast - Association of Editors in Science in Southeast Asia, Australia and Oceania

President: Dr. Santokh Singh, Geological Survey of Malaysia, Jalan Guerney, Kuala Lumpur, Malaysia
Treasurer: Dr. Amando M. Dalisay, Exec. Dir., National Research Council of the Philippines, PTRI Building, Bicutan, Tagig, Rizal, Philippines
Secretary: Dr. Bakri Abbas, LIPI, Medan Merdeka Selatan II, Jakarta, Malaysia

IFSEA - INTERNATIONAL FEDERATION OF SCIENTIFIC EDITORS' ASSOCIATIONS

Organizing Committee

Miriam Balaban, National Council for Research & Development, POB 4059, Jerusalem (for correspondence)

D.H. Michael Bowen, American Chemical Society,
1155 Sixteenth St., Washington, D.C. 20036

Martin Cremer, Gesellschaft für Information und
Dokumentation MBH, 6000 Frankfurt/Main 71 (Niederrad),
Herriotstrasse 5, W. Germany

Louis Cohen, Institute of Physics,
47 Belgrave Sq., London SW1X 8QX

Knut Faegri, Botanical Museum
Bergen University, N 5014 Bergen, Norway

H. Edward Kennedy, Biological Abstracts,
2100 Arch St., Phila, Pa. 19103

Wolfgang Löhner, UNESCO, 7, place de Fontenoy, 75700 Paris

Ernest Mann, Commonwealth Bureau of Dairy Science and
Technology, Shinfield, Reading RG2

Anders Martinsson, Department of Palaeobiology,
University of Uppsala, POB 564, S-751, 22 Uppsala

Jacques Michel, Bureau National de l'Information Scientifique
et Technique, 8-10 rue Crillon, Paris Cedex 04

Gillian Page, Cambridge University Press,
POB 92, London NW1 2DB

Udo Schützsack, Gesellschaft für Information und
Dokumentation MBH, 6000 Frankfurt/Main 71 (Niederrad)
Herriotstrasse 5

ORGANIZATIONS INTERESTED IN EDITORS' AFFAIRS - EITHER MEMBERS OR OBSERVERS

UNESCO, Division of the General Information Programme
UNESCO, 7, place de Fontenoy, 75700 Paris

Dr. Adam Wysocki
Dr. Wolfgang Löhner

Commission of the European Communities, Directorate General for Scientific and Technical Information and Information Management

J. Michel Gibb, DG XIII, European Centre, Kirchberg, Luxembourg

ICSU - International Council of Scientific Unions

F.W.G. Baker, Executive Secretary, ICSU, 51 Blvd. de Montmorency, 75016, Paris

GID (formerly IDW) - Gesellschaft für Information and Dokumentation MBH

Dr. Martin Cremer, Bischofsweg 31, Frankfurt/Main

BNIST - Bureau National de l'Information Scientifique et Technique
Ministère de l'Industrie et de la Recherche,
8-10 rue Crillon, Paris, Cedex 04

Jacques Michel
Lucette Degail

ISSC - International Social Science Council

Dr. Samy Friedman, International Social Science Council,
1, rue Miollis, 75015 Paris

NSF - National Science Foundation, Division of Science Information Services, 18th & G Streets, Washington DC 20550

Sarah Rhodes
William Savin

STM - International Group of Scientific, Technical and Medical Publishers

Paul Nijhoff Asser, Secretary General, 462 Keizersgracht, Amsterdam

ISI - Institute for Scientific Information

Dr. Eugene Garfield, President, ISI, 325 Chestnut St.,
Philadelphia, Pa. 19106

Ciba Foundation

Maeve O'Connor, Senior Editor, The Ciba Foundation
41 Portland Place, London W1N 4BN

NOP - Nordic Publishing Board in Science (Denmark,Finland,Norway,Sweden)

Dr. Walter Schytt, Mrs. Synnøve Irgens-Jensen,
NOP, NAVG, Munthes gt. 29, Oslo 2, Norway

NOP(M) - NOP for Medicine, Karolinska Institutet
S-104 01 Stockholm 60, Sweden

Pacific Science Association - a regional body consisting of the National Research Councils and Academies of the Pacific region - primarily concerned with scientific communication

Dr. Roland Force, Secretary General,
The British Museum, Honolulu

STC - Society for Technical Communication

H. Kenneth Hansen, Chairman of STC International Liaison Committee
9500 N. Oliver Avenue, Minneapolis, MN 55444

NAS - National Academy of Sciences USA (International Information Policy)

Joel Lloyd, National Academy of Sciences,
2101 Constitution Ave., Washington, DC 20418

Royal Society UK

Dr. R. Keay, 6 Carlton House Terrace, London SW1

National Enquiry into Scholarly Communication
(Sponsored by the American Council of Learned Societies)

Edward E. Booher, Director, POB 2067, Princeton, NJ 08540

Codata

Prof. Edgar F. Westrum Jr., Department of Chemistry,
University of Michigan, Ann Arbor, Mich. 48109
Codata Secretariat, 51 Blvd. de Montmorency, 75016 Paris

ICSU AB POGSI

Jeanne Poyen, Secretary General, 17 rue Mirabeau, 75016 Paris

ISO - International Organization for Standardization
Central Secretariat, 1,rue de Varembe, C.P.56, CH-1211 Geneva
Mr. B. Norbrink, Director, Information and Public Relations

Mrs. Johanna Eggert, Secretary, ISO/TC-46, DIN Deutsches Institut für Normung, Burggrafenstrasse 4/7, 1 Berlin 30, W. Germany

IDRC - International Development Research Centre, Box 8500, Ottawa K1G 3H9
Reginald MacIntyre, Associate Director, Publication Division

CSG - Capital Systems Group
John Strawhorn, 6110 Executive Blvd., Rockville, Maryland 20852

INSPEC - D. Barlow, The Institution of Electrical Engineers, Savoy Place, London WC2R 0BL

ESA - European Space Agency, ESTEC (European Space Research and Technology Centre), Noordwijke, The Netherlands

William Burke

PUDOC - Centre for Agricultural Publishing and Documentation
POB 4, Wageningen, The Netherlands

Aart Rutgers, Director

STYLE MANUALS AND GUIDES EXHIBITED

The following is a list of style manuals and guides exhibited. We would appreciate receiving updated editions of these as they are issued and others which do not appear here, in preparation for the Second Conference, so that we can prepare a current bibliography from time to time and for reference in our efforts towards consolidation and cooperation.

A Manual of Style, University of Chicago Press, Chicago, Ill., 12th edition completely revised 1969, 546 pp.

CBE (Council of Biology Editiors) Style Manual, 4th edition 1977, American Institute of Biological Sciences, Washington D.C. 20016, $12.75 incl. postage.

Scientific Writing for Graduate Students, a CBE Manual, by F. Peter Woodford, The Rockefeller University Press, N.Y., 1st ed., 1968; 2nd ed., 1976, CBE (paperback) 190 pp., $7.00 incl. postage.

Writing Scientific Papers in English, An Else-Ciba Foundation Guide for Authors, Maeve O'Connor & F. Peter Woodford, Elsevier, Amsterdam, 1975, 108 pp.

Editing Scientific Books and Journals, (Prepublication notice), An Else-Ciba Foundation Guide for Editors, F. Peter Woodford and Maeve O'Connor, Pitman, London, in preparation.

Style book/Editorial Manual, American Medical Association, Publishing Sciences Group, Acton, Mass., 1976, 161 pp.

A Manual for Auth rs of Mathematical Papers, American Mathematical Society, 1962, 23pp.

Mathematics into Type, Ellen Swanson, American Mathematical Society, Providence, Providence, R.I., 1971, 98 pp.

New Alphabets and Symbols for Typesetting Mathematics, American Mathematical Society, Providence, R.I.

Notes for Authors, The Institute of Physics, Publishing Division, Bristol, 1976, 30 pp.

Style Manual, The Institute of Physics, Publishing Division, Bristol, 1976, 69 pp.

Style Manual for Guidance in the Preparation of Papers, American Institute of Physics, New York, N.Y.

International Union of Crystallography - Notes for Authors, reprinted from Acta Crystallographica Vol. A29 part 1, and Vol. B29, Part 1, 1973, and Journal of Applied Crystallography , Vol. 6 part 1, 1973, 11 pp.

Proposal for a Reference Manual for the Preparation of CERN Scientific Papers, A. Jesse, CERN-Data Handling Division, Geneva 1976, 63 pp.

Journal of Chemical Research, Instructions for Authors, The Chemical Society, London, 1977, 12 pp.

International Union of Pure and Applied Chemistry, Notes to Contributors, Butterworths, London, 1969, 6 pp.

General Guidelines to Authors for writing a review article for Progress in Energy and Combustion Science, Pergamon, Oxford.

Richtlinien fur die Verfasser geowissenschaftlicher Veroffentlichungen, Hans Dietrich Lang, Arnold Rabien, Wolfgang Struve and Egon Wiegel, E. Schweizerbart'sch Verlagsbuchhandlung, Stuttgart, 1976, 36 pp.

How to write for Geophysical Prospecting, K. Helbig, reprinted from Geophysical Prospecting, 22 (2) (1974), 6 pp.

Authors Guide to the Publications of ASCE, American Society of Civil Engineering, New York, N.Y., revised 1975, 37 pp.

Contributions to the Journal of the Association for Computing Machinery, 1976

Publication Manual of the American Psychological Association, second edition, 1974, Washington, D.C., 136 pp.

American Anthropological Association - Guidelines for Articles and for Discussion and Debate, American Anthropologist, editor-in-chief Richard Woodbury, Massachusetts 1976, 4 pp.

Information for Authors - Current Anthropology.

Handbook and Style Manual for ASA, CSSA, and SSSA Publications; American Society of Agronomy, Crop Science Society of America, Soil Science Society of America, Madison, Wisc., 1976, 97 pp., $ 3.50. incl. postage.

Australian Journal of Experimental Agriculture and Animal Husbandry - Notes for contributors, 1976, 15 pp.

Guide for Preparing Fisheries and Marine Service Scientific and Technical Reports, D.G. Cook, T.M. Reinhart, and J. Watson, Fisheries and Marine Service, Ottawa, 1977.

Journal of the Fisheries Research Board of Canada - Guide for Authors, Guide for Typists Preparing Copy, Fisheries and Marine Service, Ottawa

ISO Standards, International Organization for Standardization, 1 rue de Varembe, Geneva.

Bibliographic Guide for Editors and Authors, The American Chemical Society, Washington, D.C., 1974, 362 pp.

Units, Symbols and Abbreviations, The Royal Society of Medicine.

Guide pour L'Utilisation due systeme International D/Unites (SI), Organisation Mondiale de la Sante Geneve.

Guide to International Recommendations on Names and Symbols for Quantities and on Units of Measurement, World Health Organisation, Geneva 1975.

Proposed American National Standard for Scientific and Technical Reports - Format and Production, Secretariat, Council of National Library Associations, American National Standards Institute Inc., 1971, 26 pp.

Style and Information Sheet for Editors of International Social Science Journals and Promotion, Lengyel, UNESCO, Paris.

Draft Guidelines for Editors of Scientific and Technical Journals, Helmut Grünewald, UNESCO, Paris, 1977, 37 pp.

Steps Toward Better Scientific Illustrations, Arly Allen, Allen Press Inc., revised second edition 1977, Kansas, 33 pp.

Information flow between primary journals and secondary services in the biological field, N. Landau and R.B. Weiss, The Library, University of Reading, 1976, 89 pp.

Report Construction, Mary Fran Buehler, Foothill Publications, California, 1970, 57 pp.

The Levels of Edit, Robert Van Buren, Mary Fran Buehler, Jet Propulsion Laboratory, California Institute of Technology, Pasadena, Calif., 1976, 27 pp.

A Guide to Writing Research Papers, Dorothea M. Berry, Gordon P. Martin, McGraw-Hill Book Company, U.S.A., 1971, 161 pp.

JOURNALS

Scholarly Publishing - a Journal for Authors & Publishers, University of Toronto Press, Toronto.

Journal of Research Communication Studies, Elsevier, Amsterdam.

Journal of Technical Writing and Communication, Jay R. Gould, Ed., Baywood, New York, N.Y.

Information Reports and Bibliographies, Science Associates/International.

IEEE Transactions on Professional Communication, Inst. of Electrical and Electronics Engineers, New York, N.Y.

NEWSLETTERS

CBE Newsletter, bimonthly, Bernard K. Forscher, Editor, Proc. Natl., Acad. Sci., USA, 2101 Constitution Ave., Washington DC., 20418

Earth and Life Science Editing, Newsletter of the European Association of Earth Science Editors & the European Life Science Editors' Association, Nancy Morris, Secretary, Bruins, 30 Longdown Road, Lower Bourne, Farnham, Surrey GU10 3JL, UK

Blueline, The Newsletter of the Association of Earth Science Editors, c/o Society of Exploration Geophysicists, P.O. Box 3098, Tulsa, OK 74101, USA

DATA BOOK AND JOURNAL

Biology Data Book, compiled and edited by P.L. Altman and D.S. Dittmer, Federation of American Societies for Experimental Biology, 9650 Rockville Pike, Bethesda, Maryland 20014, 1600 pp.

Codata Bulletin, Committee on Data for Science and Technology, ICSU, Bertrand Dreyfus, Editor, Codata Secretariat, 51 Boulevard de Montmorency, 75016, Paris, France

AUTHOR INDEX

* Numbers indicated by an asterisk refer to papers; others refer to discussants.